Computational Science and Its Applications

Computational Science and Its Applications

Edited by
A. H. Siddiqi
R. C. Singh
G. D. Veerappa Gowda

CRC Press
Taylor & Francis Group
Boca Raton London New York

CRC Press is an imprint of the
Taylor & Francis Group, an **informa** business

A CHAPMAN & HALL BOOK

MATLAB® is a trademark of The MathWorks, Inc. and is used with permission. The MathWorks does not warrant the accuracy of the text or exercises in this book. This book's use or discussion of MATLAB® software or related products does not constitute endorsement or sponsorship by The MathWorks of a particular pedagogical approach or particular use of the MATLAB® software.

First edition published 2021
by CRC Press
6000 Broken Sound Parkway NW, Suite 300, Boca Raton, FL 33487-2742

and by CRC Press
2 Park Square, Milton Park, Abingdon, Oxon, OX14 4RN

© 2021 Taylor & Francis Group, LLC

CRC Press is an imprint of Taylor & Francis Group, LLC

ISBN: 978-0-367-25623-4 (hbk)
ISBN: 978-0-429-28873-9 (ebk)

Typeset in Palatino
by Deanta Global Publishing Services, Chennai, India

Dedication

The book is dedicated to Professor K. R. Sreenivasan

Contents

Preface..xi

Professor K. R. Sreenivasan..xv

Editors..xvii

List of Contributors.. xix

1. Ingredients for Applied Fourier Analysis..1
   Hans G. Feichtinger

2. Industrial Applications of Optimal Control for Partial Differential
   Equations on Networks: Reduction and Decomposition Methods Applied to
   the Discrete–Continuous Control of Gas Flow in Complex Pipe Systems 25
   Günter Leugering

3. A Model to Assess the Role of Spatial Urban Configurations on Crowd
   Evacuation Dynamics During Terrorist Attacks 41
   D. Provitolo, R. Lozi, and E. Tric

4. Free Radical Processes in Medical Grade UHMWPE....................................61
   M. Shah Jahan, Benjamin Walters, Afsana Sharmin, and Saghar Gomrok

5. Inverse Problems Involving PDEs with Applications to Imaging...........................77
   Taufiquar Khan

6. On an Efficient Family of Steffensen-Like Methods for Finding
   Nonlinear Equations...91
   Vali Torkashvand and Manochehr Kazemi

7. Computational Methods for Conformable Fractional Differential Equations...........103
   A. H. Siddiqi, R. C. Singh, and Santosh Kumar

8. Pan-Sharpening Using Modified Nonlocal Means-Based Guided Image
   Filter in NSCT Domain 117
   Tarlok Singh and Pammy Manchanda

9. Moser-Trudinger and Adams Inequalities...135
   Kunnath Sandeep

10. Computational Ship Hydrodynamics: Modeling and Simulation...........................149
    Sashikumaar Ganesan, Bhanu Teja, and Thivin Anandh

11. DNA—Drug Interaction: Insight from Computer Simulations..............................171
    Anurag Upadhyaya, Shesh Nath, and Sanjay Kumar

12. A Graphical User Interface for Palmprint Recognition...195
Rohit Khokher and R. C. Singh

13. Wavelet Approach for the Classification of Autism Spectrum Disorder...............209
Noore Zahra, Hessa N. Al Eisa, Kahkashan Tabassum, Sahar A. EI-Rahman, and Mona Jamjoom

14. De-Noising Raman Spectra Using Total Variation De-Noising with Iterative Clipping Algorithm..225
Padmesh Tripathi, Nitendra Kumar, and A. H. Siddiqi

15. An Efficient Identity-Based Mutual Authentication Protocol for Cloud Computing..233
Vinod Kumar and Musheer Ahmad

16. Development of a Statistical Yield Forecast Model for Rice Using Weather Variables Over Gorakhpur District of Eastern Uttar Pradesh.................................243
R. Bhatla, Rachita Tulshyan, Babita Dani, and A. Tripathi

17. Cognitive Computing Agent Systems: An Approach to Building Future Real-World Intelligent Applications...257
A. Chandiok, A. Prakash, A. H. Siddiqi, and D. K. Chaturvedi

18. Controllability of Quasilinear Stochastic Fractional Dynamical Systems in Hilbert Spaces...275
R. Mabel Lizzy and K. Balachandran

19. The Advection-Dispersion Equation for Various Seepage Velocity Patterns in a Heterogeneous Medium..291
Amit Kumar Pandey and Mritunjay Kumar Singh

20. On the Application of Genetic Algorithm and Support Vector Machine for Classification of MRI Images for Brain Tumors..311
D. Gupta and M. Ahmad

21. Proximinality and Remotality in Abstract Spaces..325
Sangeeta and T. D. Narang

22. Conformable Fractional Laguerre and Chebyshev Differential Equations with Corresponding Fractional Polynomials...335
Ajay Dixit and Amit Ujlayan

23. Computational Linguistics: Inverse Problems and Emerging Mathematical Concepts...345
Sana A. Ansari and Masood Alam

24. Legendre Wavelets Method for Solution of First-Kind Volterra Problems...........357
Pooja, J. Kumar, and P. Manchanda

25. **Computational Linguistic Analysis of Retail E-Commerce** 371
    Rashmi Bhardwaj and Aashima Bangia

26. **Impact of Data Assimilation on Simulation of a Monsoonal Heavy Rainfall
    Event over India Using ARW Modeling System** 381
    Shilpi Kalra, Sushil Kumar, and A. Routray

Index 393

Preface

Computational science deals with the application of computers, networks, storage devices, software, and algorithms to solve real-world problems, perform simulations, build things, or create new knowledge. Computational science presents a broader view, implying science and engineering that is computational as opposed to experimental or theoretical in nature. Computational science is a relatively new discipline, and there is currently no consensus on a precise definition of what computational science actually is. In broad terms, computational science involves using numerical methods and computers to study scientific problems and complements the areas of theory and experimentation in traditional scientific investigation. Computational science seeks to gain understanding of science principally through the use and analysis of mathematical models on high-performance computers.

This edited volume is mainly based on talks delivered during an international workshop on Computational Science and Applications at Sharda University, Greater Noida, Knowledge Park-3, India during February 5–6, 2018, organized under the auspices of the Centre for Advanced Research in Applied Mathematics and Physics. The Chancellor of Sharda University, Mr. P. K. Gupta, and the Vice Chancellor of Sharda University, Prof. B. S. Panwar graced the inaugural function of the workshop as well as Prof. K. L. Chopra, former Director IIT Kharagpur (Padama Shri and recipient of Bhatnagar award), Prof. N. K. Gupta (Padama Shri and former Vice-President, INSA), and several senior Professors of IIT Delhi and other national institutions.

The topics covered in the edited volume are: Ingredients for applied Fourier analysis; Industrial applications of optimal control for partial differential equations on networks reduction and decomposition methods; models to access the role of spatial urban configurations on crowd evacuation dynamics during terrorist attacks; the application of a polymeric biomaterial to medical science; the application of computational algorithm inverse problem related to MRI, fractional derivatives and their real-life applications; Computational ship hydrodynamics—modeling and simulation; and applications of wavelets and fractals to diverse fields, especially in medical science and computational linguistics. Alongside several speakers from Sharda University and other universities of Northern India, the list of speakers included distinguished researchers in the field, namely: Prof. Pavel Exner, President, European Mathematical Society; Prof. G. Leugering, Vice-President, Friedrich-Alexander University, Germany; Prof. Stephane Jaffard, University of Paris, East, France; Prof. M. Shahjahan, Memphis University, USA; Prof. H. Feichtinger, Vienna University, Austria; Prof. R. Lozi, Nice University, France; Dr. Damienne Provitolo, Permanent Researcher CNRS, Nice, France; Prof. K. Sandeep, (TIFR-CAM, Bangalore); Prof. S. Ganesan (IISc, Bangalore); Prof. K. Balachandran, Bharathiyar University, Coimbatore.

The last three speakers were part of a symposium during the workshop organized by Prof. G. D. Veerappa Gowda, Dean, TIFR CAM, Bengaluru in honor of Prof. K. R. Sreenivasan, former Director, ICTP, Trieste, Italy on his seventieth birthday. Prof. K. R. Sreenivasan is a renowned expert of interdisciplinary studies who has the rare honor of being professor of three disciplines—Mechanical Engineering, Applied Mathematics, and Physics. He is well known throughout the world for his research contribution to the field of digital fluid mechanics and computational fluid dynamics.

There are 26 chapters in this edited volume. Chapter 1 is contributed by Prof. Hans G. Feichtinger of Vienna University, who is well-known throughout the world for his work in the field of numerical harmonic analysis. His research collaborators are spread all over Europe and even in India and North America. He is the Editor-in-Chief of Springer's *Journal of Fourier Analysis and Applications*. He has written an interesting chapter on "Ingredients for Applied Fourier Analysis."

In Chapter 2, Prof. Guenter Leugering of Friedrich-Alexander University, Erlangen, Germany, has presented his study on industrial applications of optimal control for partial differential equations on networks: reduction and decomposition methods applied to the discrete continuous control of gas flow in complex pipe systems.

Chapter 3, by Professors D. Provitolo and R. Lozi, is devoted to a comparatively new type of problem, namely, a model to access the role of spatial urban configurations on crowd evacuation dynamics during terrorist attacks.

Prof. Lozi is a well-known researcher who has collaborated with Prof. Leon Chua (inventor of Chua Circuit and Memristor and Alexander Sharkovsky, who introduced "Sharkovsky's order"). Lozi has discovered a particular mapping of a simple strange attractor now known as the "Lozi Map."

Prof. M. Shah Jahan of Memphis University, who has established a well-reputed department of Physics and Material Science, has presented his work on free radical processes in medical grade UHMWPE in Chapter 4.

Chapter 5 deals with the work of Prof. Taufiquar Khan of Clemson University, USA on computational algorithms for ill-posed inverse problems.

Kazemi and Gulshan of Iran have presented their work on an efficient family of Steffensen-like methods for finding nonlinear equations in Chapter 6.

In Chapter 7, Siddiqi, Singh, and Kumar have discussed their work on computational methods for conformable fractional differential equations.

Tarlok Singh and Manchanda present their work on pan-sharpening using modified nonlocal means based guided image filter in NSCT domain in Chapter 8.

Bhatnagar awardee Prof. Kunnath Sandeep of TIFR, CAM, Bengaluru, has presented his work on Moser–Trudinger and Adams inequalities in Chapter 9.

Prof. Ganesan et al·, IISc Bengaluru, a leading expert in computational science and data analysis, introduce their work on computational ship hydrodynamics: modeling and simulation, in Chapter 10.

Upadhyaya, Nath, and Sanjay Kumar (of BHU, a well-known expert in his field) have presented their work on computer simulations of DNA-drug interaction in Chapter 11.

In Chapter 12, Khokhar and Singh introduce a graphical user interface for palmprint recognition.

In Chapter 13 Noor Zahra (currently at Prince Nura University Riyadh, Saudi Arabia) et al· have studied wavelet approach for the classification of autism spectrum disorder.

Chapter 14, by Tripathi et al·, presents de-noising Raman spectra using total variation, de-noising the iterative clipping algorithm.

Chapter 15, by Kumar and Ahmad, is devoted to cloud computing.

Bhatla et al· have studied the development of a statistical yield forecast model for rice using weather variables in Gorakhpur district of eastern Uttar Pradesh in Chapter 16.

In Chapter 17, Chandiok et al· have studied cognitive computing agent systems.

Lizzy and Balachandran (expert in fractional calculus and analysis) have studied the controllability of quasi-linear stochastic fractional dynamical systems in Hilbert spaces in Chapter 18.

In Chapter 19 the advection-dispersion equation with zero-order production term and various seepage velocity patterns in a heterogeneous medium has been studied by Pandey and Singh.

Chapter 20, devoted to the application of genetic algorithm and support vector machine for brain-tumor classification of MRI images, has been written by Gupta and Ahmad.

Sangeeta and T. D. Narang have presented their work on proximity and remotality in abstract spaces in Chapter 21.

Chapter 22, by Dixit and Ujlayan, is devoted to conformable fractional Lauguerre and Chebyshev differential equations.

Chapter 23, by Ansari and Alam, is essentially related to the role of mathematics in Computational Linguistics.

Pooja et al· have studied Legendre Wavelets methods for solution of first-kind Volterra problems in Chapter 24.

In Chapter 25, Bhardwaj and Bangia carry out computational linguistic analysis of retail e-commerce using word clouds.

Chapter 26 deals with the impact of data assimilation on the simulation of a monsoonal heavy rainfall event over India using the ARW modelling system.

We take this opportunity to thank Ms. Aastha Sharma of Taylor & Francis for providing continuous guidance and help. We will be failing in our duty if we don't express our gratitude to Mr. P. K. Gupta, Chancellor, Sharda University, Greater Noida and Dr. Nadeem Tarin, alumni of AMU based in Saudi Arabia for providing valuable financial help in organizing academic activity yielding a valuable volume.

The Late Prof. A. H. Siddiqi
Sharda University Greater Noida, India

Prof. R. C. Singh
Sharda University Greater Noida, India

Prof. G. D. Veerappa Gowda
TIFR, CAM Centre of Applicable Mathematics Bengaluru, India

MATLAB® is a registered trademark of The MathWorks, Inc. For product information, please contact:

The MathWorks, Inc.
3 Apple Hill Drive
Natick, MA 01760-2098 USA
Tel: 508 647 7000
Fax: 508-647-7001
E-mail: info@mathworks.com
Web: www.mathworks.com

Professor K. R. Sreenivasan

This book, "Computational Science and Its Applications" is dedicated to Professor Katepalli R. Sreenivasan to mark his 70th birthday. He is one of the most well-known scientists in his field and has made outstanding interdisciplinary contributions in turbulent flow, intermittency, cryogenic helium, and nonlinear dynamics. Professor Sreenivasan is Dean of the NYU Tandon School of Engineering and Eugene Kleiner Professor for Innovation. He is a professor at the Department of Physics and at the Courant Institute for Mathematical Sciences. He was director of the Abdus Salam International Center for Theoretical Physics (ICTP) in Trieste, Italy, between 2003 and 2009.

In recognition of his scientific contributions, he has received numerous awards, including the Guggenheim Fellowship, the Otto Laporte Memorial Award, the Dwight Nicholson Medal from the American Physical Society, the UNESCO Medal for the Promotion of International Scientific Cooperation and World Peace from the World Heritage Center, and the Award for Scientific Cooperation of the American Association for the Advancement of Science. He has been elected to many prestigious academies including the US National Academy of Sciences, US National Academy of Engineering, American Academy of Arts and Sciences, Indian Academy of Sciences, Indian National Science Academy, Indian National Academy of Engineering, the Academy of Sciences for the Developing World (TWAS), the Accademia dei Lincei in Rome, and the African Academy of Sciences.

G. D. Veerappa Gowda
TIFR-CAM, Bangalore

Editors

The late **Prof. A. H. Siddiqi** was an Indian mathematician and Professor of Applied Mathematics. He serred as Professor Emeritus in the School of Basic Sciences and Research at Sharda University, Greater Noida, India. He also served as a former Pro Vice-Chancellor, Aligarh Muslim University. Siddiqi was also the President of the Indian Society of Industrial and Applied Mathematics (ISIAM). He authored 12 books, edited 12 books, and published more than 100 research papers. He has also held important administrative positions in AMU and other institutions of repute.

Dr. R. C. Singh has 30 years of experience as a researcher, of which more than 10 years were as Professor of Physics. Dr. Singh is a Professor of Physics at and the Controller of Examinations of Sharda University. Dr. Singh obtained his doctorate from Banaras Hindu University (BHU), Varanasi in theoretical Condensed Matter Physics. He obtained his B.Sc. (Hons.) and M.Sc. degrees also in Physics from Banaras Hindu University. Dr. Singh has published more than 30 research papers in peer-reviewed international journals and conference proceedings. He has authored one book and co-edited seven conference proceedings. His area of research interest includes study of phase transitions in molecular liquids using density-functional theory; time-series analysis using wavelets; biometrics and solar cells.

Dr. G. D. Veerappa Gowda is a Professor at the TIFR Centre for Applicable Mathematics, Bangalore. He has done fundamental work in the areas of Hyperbolic conservation laws, control theory, and numerical approximation of conservation laws. Dr. Gowda obtained their M.Sc in Physics from Mysore India, in 1978. They obtained their P.hD. from the University of Paris-Dauphine, France in 1988. Dr. Gowda obtained an explicit formula for the entropy weak solution of convex conservation laws with boundary conditions, extending the earlier work of Hopf and Lax for the initial value problem. Dr. Gowda has developed numerical schemes to solve problems arising in oil recovery and sand pile problems on a vertical source.

List of Contributors

Musheer Ahmad
Department of Applied Sciences and
   Humanities
Jamia Millia Islamia
New Delhi, India

Masood Alam
Sultan Qaboos University
Muscat, Oman

Hussah N. Al-Eisa
Princess Nourah Bint Abdulrahman
   University
Riyadh, Saudi Arabia

Thivin Anandh
Department of Computational and Data
   Sciences
Institute of Science
Bangalore, India

Sana A. Ansari
Aligarh Muslim University
Aligarh, India

K. Balachandran
Department of Mathematics
Bharathiar University
Coimbatore, India

Aashima Bangia
Guru Gobind Singh Indraprastha
   University
Delhi, India

Rashmi Bhardwaj
Guru Gobind Singh Indraprastha
   University
Delhi, India

R. Bhatla
Department of Geophysics
Banaras Hindu University
Varanasi, India

A. Chandiok
Banasthali University
Vanasthali, India

D. K. Chaturvedi
Dayalbagh Educational Institute
Agra, India

Babita Dani
Department of Geology
Banaras Hindu University
Varanasi, India

Ajay Dixit
KIET Group of institutions
Ghaziabad, India

Sahar A. EI-Rahman
Princess Nourah Bint Abdulrahman
   University
Riyadh, Saudi Arabia
and
Faculty of Engineering
Shoubra Benha University
Cairo, Egypt

Hans G. Feichtinger
Faculty of Mathematics
University Vienna
Vienna, Austria

Sashikumaar Ganesan
Department of Computational and Data
   Sciences
Institute of Science
Bangalore, India

Saghar Gomrok
Department of Physics and Materials Science
University of Memphis
Memphis, Tennessee

D. Gupta
Department of Mathematics
Jamia Millia Islamia
New Delhi, India

M. Shah Jahan
Department of Physics and Materials
    Science
University of Memphis
Memphis, Tennessee

Mona Jamjoom
Princess Nourah Bint Abdulrahman
    University
Riyadh, Saudi Arabia

Shilpi Kalra
Gautam Buddha University
Greater Noida, India

Manochehr Kazemi
Department of Mathematics and Statistics
University of North Carolina
Charlotte, North Carolina

Taufiquar Khan
Department of Mathematics and Statistics
University of North Carolina
Charlotte, North Carolina

Rohit Khokher
Vidya Prakashan Mandir (P) Ltd.
Meerut, India

J. Kumar
Guru Nanak Dev University
Amritsar, India

Nitendra Kumar
Sharda University
Greater Noida, India

Sanjay Kumar
Banaras Hindu University
Varanasi, India

Santosh Kumar
Center for Advanced Research in Applied
    Mathematics and Physics
Sharda University
Greater Noida, India

Sushil Kumar
Gautam Buddha University
Greater Noida, India

Vinod Kumar
Department of Applied Sciences and
    Humanities
Jamia Millia Islamia
New Delhi, India

Günter Leugering
Lehrstuhl Angewandte Mathematik
Universität Erlangen-Nürnberg
Erlangen, Germany

R. Mabel Lizzy
Department of Mathematics
Bharathiar University
Coimbatore, India

R. Lozi
Laboratoire J.A. Dieudonné
Université Côte d'Azur
Nice, France

Pammy Manchanda
Department of Mathematics
Guru Nanak Dev University
Amritsar, India

T. D. Narang
Guru Nanak Dev University
Amritsar, India

Shesh Nath
Banaras Hindu University
Varanasi, India

Amit Kumar Pandey
Department of Applied Sciences
Krishna Group of Institutions
Ghaziabad, India

Pooja
Guru Nanak Dev University
Amritsar, India

A. Prakash
Dayalbagh Educational Institute
Agra, India

D. Provitolo
Laboratoire Géoazur
Université Côte d'Azur
Nice, France

A. Routray
National Centre for Medium Range
 Weather Forecasting
Gautam Buddha University
Noida, India

Kunnath Sandeep
TIFR Centre for Applicable Mathematics
Bangalore, India

Sangeeta
Amardeep Singh Shergill Memorial College
Mukandpur, India

Afsana Sharmin
Department of Physics and Materials
 Science
University of Memphis
Memphis, Tennessee

A. H. Siddiqi
Department of Mathematics
Sharda University
Greater Noida, India

Mritunjay Kumar Singh
Department of Applied Mathematics
Indian Institute of Technology (Indian
 School of Mines)
Dhanbad, India

R. C. Singh
Department of Physics
Sharda University
Greater Noida, India

Tarlok Singh
Department of Mathematics
Guru Nanak Dev University
Amritsar, India

Kahkashan Tabassum
Princess Nourah Bint Abdulrahman
 University
Riyadh, Saudi Arabia

Bhanu Teja
Department of Computational and Data
 Sciences
Institute of Science
Bangalore, India

Vali Torkashvand
Faculty of Basic Science
Islamic Azad University
Hamedan, Iran

E. Tric
Laboratoire Géoazur
Université Côte d'Azur
Nice, France

A. Tripathi
Amity Centre for Ocean-Atmospheric
 Science and Technology (ACOAST)
Amity University
Jaipur, India

Padmesh Tripathi
IIMT College of Engineering
Greater Noida, India

Rachita Tulshyan
Department of Environmental Sciences
Purvanchal University
Jaunpur, India

Amit Ujlayan
Gautam Buddha University
Greater Noida, India

Anurag Upadhyaya
Banaras Hindu University
Varanasi, India

Benjamin Walters
Department of Physics and Materials
 Science
University of Memphis
Memphis, Tennessee

Noore Zahra
Princess Nourah Bint Abdulrahman
 University
Riyadh, Saudi Arabia

1

Ingredients for Applied Fourier Analysis

Hans G. Feichtinger

CONTENTS

1.1 Introduction...1
1.2 The Status Quo..2
1.3 A Short Historical Glimpse 2
1.4 Ingredients for Applied Fourier Analysis...3
    1.4.1 The Pre-Assumptions...4
1.5 System Theory and Fourier Transforms...5
1.6 The Computational Side 6
1.7 The Goals of Fourier Analysis 6
    1.7.1 Spectral Analysis and Synthesis...8
1.8 The Shortcomings of Classical Fourier Analysis 9
1.9 A Guide to the Literature..10
    1.9.1 Renewal of our Approach to Fourier Analysis......................10
    1.9.2 A Reading List on the Subject...11
1.10 Levels of Explanation 14
    1.10.1 Limitations of the Approach 14
1.11 Arguments for a Limited Distribution Theory 15
1.12 Things That *Can* Be Done Using $S_0'(\mathbb{R}^d)$ 16
1.13 Summary...18
Acknowledgments..19
References...20

1.1 Introduction

This chapter outlines some ideas of material that should be covered by an application-oriented course on Fourier Analysis. It also comments on the topics that are seen as crucial for an understanding of Fourier Analysis (beyond the technical treatment of the subject). In some sense the article tries to further develop the ideas that have been already described in [1].

The difficulty is to find a balance between pure theory (which may require a huge amount of mathematical background and preparation) and the rather formal use of mathematical symbols in engineering or physics books. Evidently, this situation calls for a reconciliation of the two worlds.

We consider it as a wrong attitude that good theory does not need to show its usefulness for applications, but it is equally misleading to think that applied scientists need not worry about a conceptual background when they build their devices. Of course, one can proceed for a while using established recipes, but if there is a change of technology, or if new possibilities or combinations of existing possibilities arise, then a simple experimental exploitation without an underlying concept often proves to be cumbersome and inefficient. It is true that engineers or physicists sometimes find—based on intuition or experiments—interesting new ways of overcoming existing problems, but in the long run those empirical facts have to be complemented by a mathematical model in order to capture fine details and to find optimal regimes of operation.

1.2 The Status Quo

The current situation is really deplorable in the sense that pure mathematics provides a full-blown theoretical structure far from applications, while applied scientists make use of formulas and algorithms based on Fourier analytic methods, which are often viewed as a mystery or make use of "non-existent" objects (more correctly: not well-defined symbols) like the Dirac delta, described as $\delta(t)$ in a schematic way to "derive" results of practical relevance.

Taking this viewpoint further, one can say that distribution theory, especially in the form of tempered distributions as introduced by Laurent Schwartz some 70 years ago ([2]) has become a cornerstone of the mathematical treatment of partial differential equations (abbreviated as PDEs), especially through the work of Lars Hörmander (starting with [3]), but has found little attention in the engineering community, although it is the ideal framework to treat continuous and discrete, periodic and non-periodic signals on an equal footing. Abstract Harmonic Analysis, as proposed by Andre Weil ([4]) allows an understanding—based on group theoretical considerations—of the analogy between the different settings (see also [5] for an applied variant of this perspective), but it does not give any hint about the connection between the Fourier transform defined for integrable functions on the real line (i.e. $f \in L^1(\mathbb{R})$) and its numerical realization using the DFT (discrete Fourier transform), usually nowadays in the form of the FFT (as introduced by Cooley/Tuckey [6]) of the FFTW ([7]). The connection is sometimes described in a rather vague way, starting from a Riemanian sum for the Fourier transform defined through the usual integral:

$$\hat{f}(s) = \int_{\mathbb{R}} f(t)e^{-2\pi ist}dt, \quad t,s \in \mathbb{R}. \tag{1.1}$$

1.3 A Short Historical Glimpse

Fourier Analysis will soon celebrate 200 years of the original publication by J. B. Fourier (1822), claiming that every periodic function f allows a representation as a superposition of pure frequencies, i.e. of cosine and sine-functions with $k \in \mathbb{N}$ oscillations per period, or using Euler's rule and complex-valued exponential functions as a series of function of the form $\exp(2\pi in/\alpha), n \in \mathbb{Z}$, where α is the period of f.

It took a while to understand how this "series representation" is valid (e.g. in the pointwise sense, or in the quadratic mean-value sense, for $f \in L^2(\mathbb{T})$). When it comes to numerical computations it is plausible that one should use the FFT (Fast Fourier transform) as an implementation of the DFT (Discrete Fourier transform). But clear prescriptions of how to sample a continuous function correctly in order to get approximately the continuous Fourier transform are rare. In Abstract Harmonic Analysis this problem does not even show up, because from a mathematical viewpoint it does not make sense to compare objects living in different groups, namely the continuous and hopefully integrable function f whose Fourier transform is defined by the above integral, and a complex vector of finite length, to which the FFT can be applied giving another vector of equal length. But things are a bit more complicated than just the naive hope that one can get the samples of $\hat{f}(s)$ (a function defined over the frequency domain $s \in \mathbb{R}$) by applying the FFT to a sequence of samples of the original function f. And in which sense can one guarantee that taking more and more samples and applying to them longer and longer FFTs allows to derive more precise information concerning the sampling values of $\hat{f}$ over a suitable lattice in the Fourier domain? A good qualitative answer making use of $S_0(\mathbb{R}^d)$ has been presented by Kaiblinger [8], based on the results of Feichtinger [9].

1.4 Ingredients for Applied Fourier Analysis

It is the purpose of this section to outline a sequence of courses (resp. a reading list concerning existing material or material in preparation) that would cover most of the topics within Fourier Analysis that are relevant for applications, while still being mathematically well justified. Corresponding material is and will become available step by step at the NuHAG web-page (www.nuhag.eu).

We are convinced that also the applied scientist should get an idea of the concept of generalized functions. We do not think that is more abstract than the idea of point charges or the concentration of mass in a given point (center of gravity of a body), or even the idea of a force acting on a given piece of matter. We are comfortable with models of the gravitational laws (attracting objects on earth toward the center of the mass of the earth), but sometimes Dirac measures are treated as magic objects or inventions of mathematicians (inspired by the "sloppy" but ingenious ideas of Paul Dirac!).

In contrast, aside from technicalities, distributions (in the way described in this section) are the natural objects, which can only be described indirectly by a large number of measurements, which hopefully satisfy the superposition principle, or in other words through linear measurements.

As mathematicians we are trained to identify the concept of signals with the concept of "functions," which means, according to established mathematical tradition, the unique assignment of the function value $f(x)$ to any given point in x in the domain of f. When we want to describe the Fourier transform we write an integral and require integrability of the function and so on. But how realistic (aside from its mathematical elegance) is such a modelling.

Let me give a few examples. In fact, a few similar ideas appear in the literature on distribution theory.

Is the room temperature in a given room a function, well-defined in any (tiny) point in a three-dimensional coordinate system, and at any moment in time, at a millisecond scale? It should have this property if we think of the solution to the heat equation as an infinitely differentiable function, but the identification of this "function," even by very fine thermometers placed at a fine grid in the room, would not really allow us to describe (in a deterministic way) this function of four variables. Nevertheless it is of course meaningful to study the corresponding differential equation.

Or let us consider an acoustic signal. Listen to the noise around you. Can you describe the underlying function in a pointwise sense? But you can record it, and you can even produce a spectrogram from the record and thus identify a piece of music by listening to the output of your loud-speaker (even if there was data compression based on MP3 involved). But all you have recorded is a finite collection of (approximately) linear measurements of the given physical signal. These can be interpreted as samples of the STFT.

The same is true for a nice painting in the museum, but *is* it really just a pointwise function on some rectangular domain, even at the submicroscopic level? Is the recorded digital version of the image, taken with a high-quality digital camera not the image anymore, just a collection of measurements, or can we reproduce it even as a poster at high resolution, providing to the human observer all the relevant information down to the scale where we cannot distinguish details anymore with our senses.

This chapter tries to outline the relevant building blocks for a possible course preparing the ground for a better understanding of the mathematical foundations of signal processing, with an emphasis on the functional analytic side. It is mostly based on courses given in the spring of 2017 and again in the spring of 2018 at TUM, the Technical University of Munich, with an engineering audience, and at the time of writing at the Mathematical Institute of Charles University Prague, to a mathematical audience.

1.4.1 The Pre-Assumptions

As for any scientific text or specialized course there are assumptions to be made, i.e. the readers have to be familiar with certain mathematical concepts.

1. First of all one has to know real ($\mathbb{R}$) and complex numbers ($\mathbb{C}$);
2. Linear algebra: we have to assume that the reader is familiar with basic concepts from linear algebra, such as linear combinations, linear dependency of vectors, or spanning systems of vectors (all vectors are linear combinations of such a system), and the idea of a basis or coordinate system, which means: every vector has a unique representation as a linear combination of the elements of such a basis. Of course not only concrete vector spaces of column vectors are of interest, but any finite-dimensional vector space;
3. Concepts from analysis, such as continuity, Riemann integrals, and perhaps some ideas about the Lebesgue integral (but this is *not* needed!);
4. Concepts from (basic linear) functional analysis, normed vector spaces, completeness; bounded and invertible linear operators, Hilbert spaces, orthonormal bases; the space $\left(L^p(\mathbb{R}^d), \| \cdot \|_p \right)$, at least for $p = 1, 2, \infty$; of course one has to accept the use of the dual of a Banach space, occasionally endowed with the w^*-convergence, but this can be explained easily ad hoc and in the application context; just recall that a sequence of operators $T_n, n \geq 1$ is strongly convergent to a limit operator T_0 if one has $\lim_{n \to \infty} T_n f = T_0 f$ for any possible input. Similarly a sequence of linear functionals $(\sigma_n)_{n \geq 1}$ in a dual space is w^*-convergent to σ_0 if $\lim_{n \to \infty} \sigma_n(f) = \sigma_0(f)$ for all elements f of the underlying Banach space;
5. It is of course good to know that for any bounded (regular Borel) measure the integral $\mu(h) := \int_{\mathbb{R}^d} h(x) d\mu(x)$ is well defined, for any bounded and continuous, complex-valued function $h \in C_b(\mathbb{R}^d)$, with an estimate $|\mu(h)| \leq \| \mu \| \, \| h \|_\infty$, where the

total variation norm of μ is given by $\int_{\mathbb{R}^d} d\,|\mu|$, but we will work with the space of bounded measures simply by introducing it as the dual space of $\left(C_0(\mathbb{R}^d), \|\cdot\|_\infty\right)$.

1.5 System Theory and Fourier Transforms

One of the central topics of Fourier Analysis is clearly the concept of "convolution" and the so-called "convolution theorem," essentially stating that the Fourier transform is turning convolution into pointwise multiplication (and vice versa) which are used to characterize translation-invariant linear systems (TILS) as a kind of moving average or convolution. So we have essentially to take care of the following questions:

1. For which pairs of objects can one define a convolution?
2. For which objects can one define a Fourier transform?
3. Under which conditions can one prove that convolution is turned into pointwise multiplication and vice versa?

We avoid on purpose the question of whether the "convolution exists" or "whether a given function" (or distribution) "has a Fourier transform," because such statements become easily ambiguous and hence unclear. Also, the answer may depend on the technical machinery in the background which would have to be specified. In some cases a function may have a Fourier transform in the distributional sense which is even a function, without allowing the transform being realized as a Lebesgue integral pointwise (not even pointwise almost everywhere). We will try to avoid such questions here.

As we will see, these answers have easy and clear answers for "nice functions" but may be more delicate in a more general setting, even if the involved objects are quite reasonable functions.

We also did not call the "objects" functions or generalized functions in the above formulation, because to make these terms precise is one of the tasks of such a course and needs some preparation.

At first sight the answer to all these questions is provided by the use of the space Banach space $\left(L^1(\mathbb{R}^d), \|\cdot\|_1\right)$ of Lebesgue integrable functions, with the norm

$$\|f\|_1 = \int_{\mathbb{R}} f(x)\,|dx|, \quad f \in L^1(\mathbb{R}). \tag{1.2}$$

It appears as the right domain for convolution, because first of all one can show that the pointwise defined convolution integral

$$[g*f](x) = \int_{\mathbb{R}} f(x-y)g(y)dy, \quad x,y \in \mathbb{R}, \tag{1.3}$$

is defined almost everywhere whenever $f,g \in L^1(\mathbb{R})$, and furthermore the class (up to null functions) defined in this way (we call it $g*f$) is again an integrable function, with the important estimate

$$\|g*f\|_1 \le \|g\|_1 \cdot \|f\|_1, \quad f,g \in L^1(\mathbb{R}). \tag{1.4}$$

It appears also as natural to define the Fourier transform (as above) and then (using Fubini's Theorem) to verify the convolution theorem for this case

$$\widehat{f * g} = \hat{f} \cdot \hat{g}, \quad f, g \in L^1(\mathbb{R}).$$

Starting from this situation it can be viewed as a non-trivial extension to verify that it is possible to even extend the Fourier transform (then sometimes called Fourier-Stieltjes transform) to bounded (regular Borel) measures and then verify the convolution theorem for this extended (still commutative) operation, called convolution.

It is then even possible to define point-measures δ_x by their distribution function (a shifted Heavyside function with unit jump at $x \in \mathbb{R}$) and show that convolution of f by such a point measure is the same as shifting the function, or in a more engineering context: Convolution with a Dirac at x realizes a translation by x:

$$\int_{\mathbb{R}} \delta(y - x) f(y) dy = f(y - x) = [T_x f](y). \tag{1.5}$$

1.6 The Computational Side

An important part of modern digital signal processing is based on the use of computers. It is a plain statement that a computer—at least if viewed as a number-crunching machine—can only digest a finite set of vectors of finite length (and deal with time at the level of finite precision). Therefore one has to replace the continuous function by a sequence of samples each time a Fourier transform has to be computed approximately using the FFT. Recall that the mathematical description of the DFT (the Discrete Fourier Transform) is based on the rule $\mathbf{x} \mapsto \mathbf{y}$, with $\mathbf{x}, \mathbf{y} \in \mathbb{C}^n$:

$$y_k = \sum_{j=1}^{n} x_j \omega_N^{(k-1)(j-1)}, \quad k = 1, 2, .., n. \tag{1.6}$$

Sometimes this formula will be justified by some form of Riemann-sum argument, But most of the time it is (implicitly) suggested that the FFT applied to a sequence of regular samples corresponds to a sequence of samples of the Fourier transform (as defined by the integral transform). But this is true only approximately, and one has a quantitative control on the deviation, which of course depends mostly on the signal length n and certain properties of the involved function f (like smoothness or decay). But even if n and f are given the sampling strategy, this may strongly influence the result. A qualitative control in a rather general setting is provided by the paper [10], but further quantitative investigation should be carried out.

1.7 The Goals of Fourier Analysis

The key properties for the Fourier transform are certainly to get a handle on convolution, which in turn allows an understanding of the behavior of translation invariant operators via the Fourier transform.

Recalling the basic idea of matrix analysis resp. linear operators (e.g. on Hilbert spaces), one can fairly say, that for any given operator one would like to know its range, its null-space, to find out whether it is invertible, and how to compute the inverse linear mapping in order to solve the standard equation $A * x = b$.

This task is easily solved if A is a diagonal matrix, with diagonal entries of the form $[d_1, ..., d_n]$, because then $A * x = (d_k x_k)_{n_{k=1}}$ and all the questions are easily answered. Invertibility is equivalent to the condition $d_k \neq 0$ for all k. For the case of some zeros the (pseudo)-inverse is obtained by inverting exactly the non-zero coordinates.

We all know that it is hence a great opportunity if we can diagonalize a given matrix, because then our main questions are reduced to the diagonal case, making use of a basis of eigenvectors, and the elements in the diagonal matrix are just the eigenvalues of the given matrix A.

Of course it may be cumbersome to compute a set of eigenvectors, perhaps more complicated than solving the equation right away. But in some (lucky) cases, and for us most importantly, the structure of a matrix (in fact a commutative algebra of matrices) can already guarantee that there is a common set of eigenvectors which does not depend on the fine details of the given matrix. If then, in addition, there are fast algorithms to compute the eigenvalues of each such matrix A we have enormously reduced our computational workload.

This is exactly the case for so-called "circulant matrices," which have the property that their coordinates are constant along side-diagonals (understood in the cyclic sense). So if $j - k = \max(s, s + n)$ (the number of the side-diagonal, with $s \in [0, n - 1]$ (for an $n \times n$-matrix A) implies $a_{j,k} = b_s$. Such matrices are called "circulant matrices" and it is not difficult to verify that these are exactly the matrices that commute with the cyclic shift matrix S, mapping e_k to e_{k+1} for $k < n$ and e_n to e_1. Such a matrix can obviously also be written as $S = \sum_{k=0}^{n-1} b_k S^k$ (and, vice versa, any such matrix, being a polynomial in S, commutes with S).

This shows that the columns of the DFT-matrix are the (discrete) pure frequencies, i.e. the monomials $z^k, k \in [0, ..., n - 1]$, evaluated at the unit roots of order n (taken in the clockwise sense, in order to follow the established conventions).

The interesting thing is that in this way we can argue why the DFT is playing such an important role; we can even explain (not here) why the FFT is so much more efficient if $n = 2^k$ (or any other natural number having a relatively large number of divisors, because this implies that the cyclic group of order n has a rich subgroup structure, which in turn allows a deep recursion scheme, providing the efficiency of the FFT algorithm). There is no other choice for these (obviously very important) class of matrices. On the other hand, the concrete form of the DFT matrix implies that we can interpret the DFT, which is usually seen just as a unitary (up to normalization by the factor $\sqrt{n}$) mapping from $\mathbb{C}^n$ to $\mathbb{C}^n$ also as the evaluation mapping, which maps a given vector $b = (b_k)_{k=0}^{n-1}$ to the values of the associated polynomial $q_b(z) = \sum_{k=0}^{n-1} b_k z^k$ at the unit roots (ω_n^k) of order n, starting with $\omega^0 = 1$, in the clockwise sense.

Thus the inverse DFT computes the coefficients of a polynomial from its values at exactly those unit roots, and consequently identifies the polynomial uniquely if its degree is at most $n - 1$ (resp. *order*, i.e. number of coefficients at most n).

In this interpretation it is clear that the pointwise multiplication of the values of two such polynomials $p(x)$ and $q(x)$ gives us the values of product polynomial $r(x) = p(x)q(x)$, which, by the usual multiplication procedure of polynomials (also known as the Cauchy product of the corresponding coefficients), with

$$c_m = \sum_{j+k=m} a_j b_k, \quad m \geq 0$$

is just the discrete version of the convolution theorem.

1.7.1 Spectral Analysis and Synthesis

When looking at papers or books on Fourier analysis it is not always clear right away why one should consider things like convolution or the Fourier transform. All too often the nice properties of the (classical) Fourier transform on the classical spaces, in particular on $\left(L^2(\mathbb{R}^d), \|\cdot\|_2\right)$ (in the form of Plancherel's Theorem) or its the behaviour on other L^p-spaces (like the Hausdorff–Young Theorem, which only applies for $1 \leq p \leq 2$) are in the center of the presentation, but very unfortunately not so much the goals that would be mentioned by more applied scientists.

Let us try to formulate a few of these goals, in particular those which are related to the questions of spectral analysis and spectral synthesis. We formulate some of these question first in an informal style:

1. What are the frequencies in a given signal, respectively can be filtered from the signal in a suitable sense, and, as a counterpart, which ones do not play a role for the given signal?

2. How can one recover the signal from the information about the frequencies contained in the signal? In other words, what are the most important or most useful methods to recover a function or distribution from its Fourier transform?

3. Can we separate signals from each other in a translation invariant way (i.e. by a so-called filtering process, or moving average)? The answer to this question is of course a clear "yes" if the different signals are concentrated in disjoint frequency bands: such considerations are the basis for modern digital radio techniques which rely on the idea that particular radio stations get certain frequency bands assigned.

4. Spectral Analysis should also help us to explain Shannon's Sampling Theorem for band-limited functions and describe a well-localized series representation of band-limited signals, if they are sampled at least at the Nyquist rate, combined with a possibility to estimate the aliasing errors for the case that the signal is not strictly band-limited to the assumed frequency interval.

5. The notion of convolution was not much present in the classical work on Fourier series expansions, so that it does not yet play a prominent role in the classical book of A. Zygmund ([11]). It became however more relevant in Norbert Wiener's influential work on *Generalized Harmonic Analysis* published in 1930 [12] and his book [13] (only 7 years before A. Weil's book [4]). Here, and in particular for the proof of Tauberian Theorems the pointwise inversion of a Fourier transform plays an important role, and thus intensive use of convolution is made. He also clearly indicates that convolutions can be approximated well by linear combinations of shift operators (in the strong operator topology).

6. The Sampling Theorem is also important in order to understand the mathematical principles governing Digital Signal Processing and also provides a sound basis for Digital Image Processing.

7. Fourier-based methods are also relevant for the design of (modern mobile) communication systems (this is why Fourier Analysis is such an important part of modern digital signal processing!).

8. The Fourier Slice Theorem is highly relevant for Medical Imaging, in particular for tomography and MRI (magnet resonance imaging).

1.8 The Shortcomings of Classical Fourier Analysis

The original question treated by J. B. Fourier, namely to find a solution to the heat equation, poses itself not naturally on a closed circle but rather for a long and extended rod. But it is a physically realistic approximation (allowing for a good mathematical treatment) to work with a periodic signal. In fact, if we assume a deviation from a constant (say zero) only on some finite interval, given the (practically) finite speed of propagation of heat in such a homogeneous rod the Fourier series description gives a very good approximation. Only in the long run does the finite length comes into play and clearly then the total heat imposed at the beginning is just equally spread out over the rod, while in the truly infinite case heat is diffusing toward infinity. We have, so to say, an infinitesimal increase of temperature everywhere.

Only ca. 100 years after Fourier the limit from periodic to non-periodic signals was taken (even nowadays this is often done in a hand-waving way, in order to motivate the formulas for the forward and inverse Fourier transform). After Riemann, the introduction of the Lebesgue integral allowed the definition of a Fourier transform for rather general functions in $L^1(\mathbb{R}^d)$ with a continuous frequency distribution $\hat{f}$ via the well-known integral transform.

Formulas found in books on the subject may read as follows:

$$\int_{-\infty}^{\infty}\int_{-\infty}^{\infty} e^{i2\pi\nu(t-\tau)} f(\tau)d\tau\,d\nu = f(t) \tag{1.7}$$

which is sometimes rewritten as

$$\int_{-\infty}^{\infty} e^{i2\pi\nu(t-\tau)}d\nu = \delta(\tau - t) \tag{1.8}$$

because this formula, combined with the so-called *sifting property* of the Dirac delta, namely the formula

$$\int_{-\infty}^{\infty} f(s)\,\delta(s-t)ds = \int_{-\infty}^{\infty} f(s)\,\delta(t-s)ds = f(t) \tag{1.9}$$

gives a "formal proof" of the Fourier inversion

$$\int_{-\infty}^{\infty}\int_{-\infty}^{\infty} e^{i2\pi v(t-\tau)} f(\tau)d\tau\,dv = \int_{-\infty}^{\infty}\int_{-\infty}^{\infty} e^{i2\pi v(t-\tau)}dv\, f(\tau)d\tau$$

$$= \int_{-\infty}^{\infty} \delta(\tau - t)f(\tau) = f(t).$$

In reality the converse is true. One first has to show that at least for a suitable space of test functions one can establish the Fourier inversion formula, i.e. the possibility of recovering f from its Fourier transform, ideally in the pointwise sense, by the formula

$$f(t) = \int_{\mathbb{R}} \hat{f}(s)e^{2\pi i s t}ds, \quad t \in \mathbb{R}. \tag{1.10}$$

We do not have enough space in this survey to explain in which sense (namely by considering distributional limits in the context of Banach Gelfand triples) these "obscure formulas" can be given a meaning. Instead I would like to remind the reader that already in the context of real and complex numbers we use "mystic identites" such as

$$e^{2\pi i} = 1, \tag{1.11}$$

where we take a complex exponent, in fact a purely imaginary one involving the irrational factor π of the irrational (real) Euler's number e and claim that this is exactly 1. Here everybody understands that this use of the exponential rule is expressing certain limiting relations for the partial sums describing the exponential function as a limit of polynomials, and not as really "multiplying $e\,2\pi i$ -times by itself as one is used to from the symbol $x^k, k \in \mathbb{N}$.

1.9 A Guide to the Literature

1.9.1 Renewal of our Approach to Fourier Analysis

The idea of taking a fresh look at Fourier Analysis has been promoted by the author in several publications in recent years. Partially the writing is and was educational or quite informal (like the current chapter), partially more in the usual style of presentation, providing mathematical results (or at least describing such results) rather than providing perspectives and making suggestions.

One has to be aware that such a renewal process, as we have it in mind, will take quite a long time, especially due to the fact that Fourier Analysis is a well-established field both within mathematics and in the engineering literature, and generations of scientists and practitioners have been educated in the traditional way.

It needs strong arguments, but also willingness to listen to the needs of those in the applied science (and perhaps sometimes a bit of phantasy and willingness to explore possibly new questions on the basis of your own intuition) in order to move things in that direction and provide (as we expect) more and more insight that the proposed approach is not only a hobby of this author, but—up to details—the best way to reunify the different directions of Fourier Analysis or its applications in the various sciences using it.

1.9.2 A Reading List on the Subject

The paper [14] makes use of the term "Postmodern." Clearly the next step after "Modern Analysis" (which was fashionable in the second half of the last century) is "Postmodern."

What we mean by this is the following: In the period of "Modern Analysis," after the Bourbaki movement, it was *clear* how (Fourier) Analysis should be taught, based on the theory of Lebesgue integration and using perhaps tempered distribution. Details had to be filled in, the theory had to be taken from the Euclidean case to the setting of LCA groups and so on. There was no question of how and where to apply it (at least not for most mathematicians working in the field), but rather to clarify the foundations. Lebesgue, Wiener, Weil, and many others had paved the way toward a general approach to Fourier analysis in the context of LCA groups and now the generation of Reiter ([15, 16]), Loomis ([17]), and Hewitt ([18, 19]) had to fill in the missing pieces to get to the complete picture, with all the necessary (technical) details and results.

In the postmodern area the questions comes up: What can we do with all the accumulated knowledge on the field?, and also: Now that the tools are clear, can we use them for real-world applications? Even if the possibility is often claimed and exists in a vague sense, one would like to be able to argue how it is done concretely, and of course all too often engineers find mathematical courses too abstract, even if they provide examples and establish (in a wider sense) connections to the applications. There is no doubt that the mathematical theory of PDE has made gigantic progress due to the tools that became available through the theory of L. Schwartz ([2]) and the efforts following the work of Hörmander ([3]), but there Fourier analysis is showing up as a servant, a method to transform problems into a more handy form and solve them there. The whole calculus of pseudo-differential operators is based on such refined mathematical tools, but obviously these aspects of Fourier analysis did not make it into the reality of engineering education, even if it is quite appropriate to describe a slowly varying channel as it appears in the description of mobile communication. By this we mean a linear operator that behaves locally like a (variable over time or position) moving average, and describe it as an underspread operator, i.e. an operator whose spreading distribution is highly concentrated and thus allows (for example) the system to be well approximated by a Gabor multiplier (see [20], or [21, 22]).

Given the variety of tools, we need some criteria for choosing the right tool, for example, a suitable family of function spaces. This question has been addressed in [23], more or less with the suggestion to think of what one might call "consumer reports" in real life. Which function space or which algorithm should be used to handle a particular type of question? What is the best practice to make use of those existing tools?

It is clear that a comparison of the performance of algorithms to solve concrete mathematical problems are quite popular among engineers or numerical analysts, but we rarely have discussions about the usefulness of particular function spaces for a given context. We use what has been used traditionally, and sometimes mathematicians think that engineers just do not want to use them, but sometimes it is also true that traditional spaces (like L^p-spaces) might not provide any kind of insight into a problem handled by an engineer.

The shortcoming of the, by now, also classical approach that is widely called "Abstract Harmonic Analysis" is addressed in the paper [24], proposing the term "Conceptual Harmonic Analysis." It mainly points to the fact that the pure analogy between Fourier Analysis on different groups does not really allow an understanding of how one should use the FFT in order to compute a good approximation of some continuous Fourier transform. After all, the FFT handles functions on a finite Abelian group which a priori has nothing to do with a continuous and integrable function on the real line. This does not say that the viewpoint is not useful for engineers, and the book [5] by Ciarolo on "Unified Signal Processing" is a good example.

The proposed way out of this problem is to view periodic function as periodic function on the real line and not identify them automatically with function on the torus. We know such a viewpoint from the theory of almost periodic functions which have a kind of non-harmonic Fourier series expansion, but better to view them as (tempered) distributions, which have a Fourier transform and where it is possible to view dilated version of a given periodic function simultaneously and observe convergence, if the dilation parameters are convergent. In this setting the different modes of Fourier Analysis (the periodic continuous case, the non-periodic continuous case, the periodic-discrete case etc.) can be compared against each other and can be mutually approximated (in the distributional sense).

This of course suggests the introduction of test functions, dual spaces, work with nuclear Frechet spaces, and so on, in order to understand (for example) the theory of tempered distributions as introduced by L. Schwartz. The papers [25, 26, 27] by J. Fischer try to reconnect the work on tempered distributions as it is found in [2, 28, 29], using heavy machinery from the theory of topological vector spaces, to the focus of engineers who worry about periodic or (regularly) sampled signals and who often have to find out which kind of Fourier transform should be used. Instead of answering (in the spirit of abstract harmonic analysis, so perhaps also [5]): "You have to choose the group which fits your data!," he suggests thinking of just one Fourier transform for all these objects, and that they all have their place in the universe of tempered distributions, i.e. in the Schwartz space $\mathscr{S}'(\mathbb{R}^d)$. This is in some sense also the spirit of Conceptual Harmonic Analysis ([24]), but there we want to take it further, by replacing the complicated topological structure of the space of test functions by the more handy properties of a (dual) Banach space; namely $\left(S_0'(\mathbb{R}^d), \|\cdot\|_{S_0'}\right)$ (also endowed naturally with the w^*-topology, which we can describe in an elementary way), but also by adding the idea that due to the w^*-density of either discrete, or periodic signals, but even of test-functions (smooth and decaying) resp. purely discrete and periodic signals (which can be transformed using the well-known FFT algorithm) one can always approximate one by the other and could build the whole universe of signals (in $S_0'(\mathbb{R}^d)$) by just taking appropriate limits of objects within one of these classes. We hope to come back to this (currently still quite a bit vague) idea in future publications and talks, so it might be worthwhile to contact the author (or his web page at NuHAG) if you are interested in these aspects of the theory (maybe years after this chapter has been published).

We can report that, on the basis of methods from time-frequency analysis,* it is possible to make use of a concrete setting, called the Banach Gelfand Triple, a specific Banach space of test functions denoted by $\left(S_0(\mathbb{R}^d), \|\cdot\|_{S_0}\right)$, its dual space $\left(S_0'(\mathbb{R}^d), \|\cdot\|_{S_0'}\right)$ and in-between them (even in the sense of complex interpolation in the middle!) the Hilbert space $\left(L^2(\mathbb{R}^d), \|\cdot\|_2\right)$, thus a "rigged Hilbert space" setting avoiding the world of topological vector spaces and making it hopefully easier for applied scientist to use this somewhat simplified form of distribution theory. We have started to call the elements of $S_0'(\mathbb{R}^d)$ "mild distributions," because they are more general than ordinary measures or even pseudo-measures

---

* For historical correctness one has to say that the corresponding ingredients predate the vivid development of TF-analysis in recent decades and this has rather enabled that development!

(these are tempered distributions arising as Fourier transforms of elements of $L^\infty(\mathbb{R}^d)$), but less general than the usual distributions.*

Details on this approach are currently found in [31], but as time passes we will get more and more material (such as course notes) making use of this setting.

Many of the questions arising in Gabor analysis or even in time-frequency analysis in general (e.g. the existence of a spreading representation of operators, the use of time-frequency shifts) can be also described in a linear algebra context. For a redundant system like an "overcomplete Gabor family" one can ask how to represent a given vector using the minimal norm representation as realized using the (Moore–Penrose) pseudo-inverse of a rectangular matrix, alias the dual frame. This background, which helps to understand how one should deal with overcomplete systems (later on, "frames") is described in [32]. Clearly, at least in principle, the same argument can be used for Gabor Analysis over any finite Abelian group G. This requires some group theoretical concepts. The basic results about this context (still avoiding the use of infinite sums which could be divergent or the use of continuous variables, so one can use "unit vectors" instead of Dirac measures even at this generality) are provided by the paper [33].

The paper [34] (back to the continuous domain) provides "A novel mathematical approach to the theory of translation invariant linear systems." It demonstrates how one can start from a given locally compact Abelian group such as $\mathbb{R}^d$ with vector addition and start describing the translation invariant linear systems on $\left(C_0(\mathbb{R}^d), \|\cdot\|_\infty\right)$, i.e. the linear operators from $C_0(\mathbb{R}^d)$, the space of continuous complex-valued functions vanishing at infinity, endowed with the maximum norm $\|f\|_\infty := \max_{x \in \mathbb{R}^d} |f(x)|$, which commute with all the translation operators, as "moving averages," or equivalently as convolution operators, for a uniquely determined bounded linear functional, i.e. a bounded measure μ on $\mathbb{R}^d$. This approach has also been the starting point for recent courses on the subject.

Of course, the choice of function spaces plays, according to our understanding, an important role for the proper treatment of questions arising in mathematical signal processing. Aside from the usual spaces like $L^1(\mathbb{R}^d)$, $L^2(\mathbb{R}^d)$, $L^\infty(\mathbb{R}^d)$ one should consider the Wiener algebra $W(C_0, \ell^1)(\mathbb{R}^d)$ and of course $S_0(\mathbb{R}^d)$ as important examples. They are all Banach spaces with respect to their natural norm. For a better orientation within this "landscape of function spaces" one can recommend the article entitled "Choosing Function Spaces" [23].

Of course one can find relevant properties of the spaces $\left(S_0(\mathbb{R}^d), \|\cdot\|_{S_0}\right)$ and its dual in various papers on modulation spaces or just on Gabor Analysis, starting with [35], of course the original paper [36] and the recent papers by M. Jakobsen ([37, 38]), and of course the book by K. Gröchenig [39]. Gabor frames appear also as prominent examples in the book of O. Christensen [40].

The role of "Group theoretical methods and wavelet theory (coorbit theory and applications)" is highlighted in [41]. It helps to establish the connection between Gabor Analysis and abstract coorbit theory. Important references in this direction are certainly [42] and [43], with the last one pointing out the connection between the group theoretical view (as

---

* In this sense we have to warn the readers: some authors, like Zemanian [30] restrict the use of the word "distribution" to the elements of dual spaces of space of smooth, i.e. infinitely differentiable functions, and call the other cases just "generalized functions." So he might argue that we are only describing a Banach space of generalized functions and not a space of distributions. However, the adjective "mild" indicates clearly that we are just treating a special form of distribution and that the show a mild behaviour, i.e. the cannot be too rough. In particular derivatives of the Dirac distribution are excluded from our consideration.

employed in the context of coorbit spaces) and the time-frequency interpretation (which leads to projective representation of phase space in contrast to the unitary, square integrable group representaton (sometimes named the Schrödinger representation) of the reduced Heisenberg group.

1.10 Levels of Explanation

When proposing an alternative approach with simplified methods one has to address a number of questions.

First of all, in some cases it is necessary but fortunately also possible to use simplified definitions. For example, it should not be required to make use of Lebesgue integration for the definition of the space of test function (so it will *not* be defined a priori as a subspace of $L^2(\mathbb{R}^d)$ or $L^1(\mathbb{R}^d)$, but rather using spaces of continuous and Riemann integrable functions. Based on this one can introduce a suitable space of distributions, which are described either as the weak limits of ordinary functions or the elements of the dual space of the (normed, in fact Banach) space of test functions, which can be embedded into this dual space in a natural way (using the Haar measure on $\mathbb{R}^d$, i.e. the Riemann integral).

In a further step one has to relate the new approach to the well-established approach. For example, the set of distributions which can be approximated by test functions through bounded sequences of integrable functions (e.g. in the distributional sense) can be identified with the Lebesgue space $\left(L^1(\mathbb{R}^d), \|\cdot\|_1\right)$, due to the uniqueness of the completion of a normed space, in this case the space of test functions (or just $C_c(\mathbb{R}^d)$), endowed with the $k \mapsto \|k\|_1$, defined using the Riemann integral.

It is clear that in some cases the justification of certain simplifications itself will not be possible using elementary methods. Instead one will have to relate abstract results (involving arguments concerning topological vector spaces, e.g. dual spaces with weak or w*-convergence) to equivalent ways to describe convergence (e.g. using sequences only). Of course, this discussion has to be carried out in the background, and only once, until it is clear that a teacher using the simplified method is not defining simplified objects but rather provides a simplified description of the (mathematically) correct objects.

1.10.1 Limitations of the Approach

First of all, let us mention that it has been already recognized by e.g. A. Papoulis (see the preface to his book [44] published in 1962) that distribution theory is potentially important for engineers interested in signal processing or communication theory, but that the theory as proposed by Laurent Schwartz is considered too cumbersome, so that he gave a short summary as an Appendix, somehow better accessible to a mathematically educated engineer. There are also, of course, other attempts to simplify the approach to distribution theory by using e.g. the sequential approach (as promoted by Lighthill, see [45]), but in such a setting some of the important results are again cumbersome to prove and the independence of the representative sequence defining a generalized function is a boring and repetitive task that one should avoid. Sometimes, of course, the topological setup combined with operational realization using regularized sequences is the best combination of the two different descriptions of the same (from an abstract, mathematical viewpoint) objects.*

Of course one should mention some of the limitations of the proposed approach (and how to overcome them). It is clear that choosing one fixed Banach space of test functions and its dual (with the Hilbert space in the middle) one cannot expect to be able to get boundedness of effectively unbounded (in whatever sense) operators.

If one needs operators such as differentiation combined with the Fourier transform one has to resort more or less to constructions very similar to that of the Schwartz space $\mathscr{S}(\mathbb{R}^d)$, with a countable family of seminorms (hoping to get at least a complete metric space, i.e. a Frechet topology, by this approach), even if that metric is only used in order to provide analytic claims and not in order to do concrete estimates of operators. In short, we cannot describe the properties of $\mathscr{S}(\mathbb{R}^d)$ or satisfy the needs of PDE people using the single Banach Gelfand triple $(S_0, L^2, S_0')(\mathbb{R}^d)$.

1.11 Arguments for a Limited Distribution Theory

In discussions with colleagues sometimes the following question arises: "What is a distribution theory good for if it does not allow being able to deal with differential operators?" This question is, of course, well justified, because most of the time the idea of distributions is connected with the space of test functions $\mathscr{D}(\mathbb{R}^d)$ of infinitely differentiable, compactly supported functions. This space is obviously closed under differentiation, and differentiation is even a continuous linear procedure on such a space. The same is true for the Schwartz space $\mathscr{S}(\mathbb{R}^d)$.

Via transposition, differentiation can be applied to the dual space $\mathscr{D}'(\mathbb{R}^d)$ resp. $\mathscr{S}'(\mathbb{R}^d)$, turning such a dual space into a collection of objects which allow differentiation without leaving the vector space. In fact, in some sense the new (extended) operation is in some sense the only reasonable way of extending differentiation as it is known from smooth functions to more general objects. Since this is a feature of most variants of the usual distribution spaces (classical distributions, or tempered distributions, etc.) which are important for the distributional treatment of PDEs, it is plausible that most mathematicians familiar with distribution theory think of this as a key property of distribution theory.

In contrast, we take the liberty of calling a vector space a space of "distributions" if it is just a natural extension of a space of test functions, i.e. if the space of test functions is rich enough and can be naturally embedded into its dual space with the help of the (Riemann) integral:

$$k \mapsto \sigma_k : \quad \sigma_k(f) = \int_{\mathbb{R}^d} f(x)k(x)dx.$$

Since we restrict our attention to spaces of continuous, bounded, and Riemann integrable functions we can be sure that any integrable function and, in fact, any bounded function defines a generalized function, and that all the test functions have a Fourier transform in

---

* Very much like rational numbers and periodic decimal expressions, which are the same, although they look different.

the usual sense. Practically speaking we restrict our attention to spaces of test functions inside the Wiener algebra $W(C_0, \ell^1)(\mathbb{R}^d)$, but we do not discuss this aspect here.

To come back to our argument: we prefer to work with $S_0(\mathbb{R}^d)$, which is a space of continuous and Riemann integrable functions with the additional property of being Fourier invariant, which in turn implies that also the dual space is invariant under the (extended) Fourier transform, defined simply by the relation

$$\widehat{\sigma}(f) = \widehat{\sigma}(f), \quad f \in S_0(\mathbb{R}^d), \quad \sigma \in S_0'(\mathbb{R}^d).$$

The elements of $S_0(\mathbb{R}^d)$ are not smooth, in the sense that they may not have a derivative. The most simple example is the "triangular function," i.e. the convolution product of a rectangular function (the indicator function of some interval $[a, b] \subset \mathbb{R}$) by itself, because this function is obviously in the Fourier algebra (its Fourier transform is of the form $\sin(\pi t)^2 / (\pi t)^2 \in L^1(\mathbb{R})$) and it has compact support.

This lack of smoothness at the level of test functions should not only be seen as a drawback. It also has advantages at the level of the dual spaces. Among others it is the reason why one can prove that any distribution with finite support is just a finite discrete measure, i.e. it is of the form $\mu(f) = \sum_{k=1}^{n} c_k \delta_{x_k}$, a finite linear combination of Dirac measure.

Distributions (in fact bounded, continuous functions) whose Fourier transform (in the $S_0'(\mathbb{R}^d)$-sense) is such a finite discrete measure are called a "trigonometric polynomial." In fact the uniform limits of such trigonometric polynomials are known as "almost periodic functions" in the classical literature. The formulas to determine the frequencies contained in such an almost periodic function can also be well explained by means of this generalized Fourier transform. Individual discrete contributions on the Fourier side are filtered out by functions which are concentrated near such a Dirac measure and mostly ignoring the nearby contributions. On the time-side these filters can be characterized as convolutions with widely stretched envelope windows modulated by the corresponding frequency which has to be filtered out in a kind of "tuning process" comparable to the selection of a radio channel in a receiver.

A similar statement is valid for distributions $\sigma \in S_0'(\mathbb{R}^d)$ supported on a discrete lattice Λ. This is the basis for the proof of the so-called "Janssen representation" for the Gabor frame operator as given in [46] (an argument which does not work in the setting of $\mathscr{S}'(\mathbb{R}^d)$, because in this case also partial derivatives of Dirac measures are among the distributions concentrated in a single point).

1.12 Things That *Can* Be Done Using $S_0'(\mathbb{R}^d)$

Let us try to provide a list of things that one can realize with the help of the spaces $\left(S_0(\mathbb{R}^d), \|\cdot\|_{S_0}\right)$ and $\left(S_0'(\mathbb{R}^d), \|\cdot\|_{S_0'}\right)$ (we call $\sigma \in S_0'(\mathbb{R}^d)$ still a distribution, despite the above comment), sometimes not at all using the classical function spaces, or sometimes only using specific constructions for the particular purpose.

1. First of all the spaces $L^p(\mathbb{R}^d)$, with $1 \le p \le \infty$ are continuously embedded into $S_0'(\mathbb{R}^d)$; consequently for any of these function space (in fact for any Banach space of

distributions which is isometrically invariant under translations and modulations) the Fourier transform is well defined, because $S_0'(\mathbb{R}^d)$ is the maximal space in this family. Very unfortunately the well-known L_p-spaces do not behave well with respect to the Fourier transform. This is, among others, a reason why Fourier Analysis is perceived as a very sophisticated branch of analysis;

2. The space $S_0(\mathbb{R})$ of test functions is also relevant for a number of classical problems. It provides an ideal reservoir of summability kernels, and all the known (good) summability kernels (except for the Dirichlet kernel) belong to the space $S_0(\mathbb{R})$ (see the work of F. Weisz, e.g. [47, 48]);

3. The space $S_0(\mathbb{R}^d)$ is also a suitable domain for the validity of Poisson's formula and has been shown in [49]; the only counter-examples described in the book of Y. Katznelson ([50]), and later in a more comprehensive way in [51], arise "outside of $S_0(\mathbb{R}^d)$," i.e. in situations where the combined (weak) decay conditions on f and $\hat{f}$ are not sufficient to guarantee $f \in S_0(\mathbb{R}^d)$, see also [39];

4. From a distributional point of view the Poisson formula is equivalent to the fact that as a member of $S_0'(\mathbb{R}^d)$ the Dirac comb Ш_Λ is mapped under the (generalized) Fourier transform to $\text{Ш}_{\Lambda^\perp}$*. Consequently pointwise multiplication by Ш_Λ is turned into convolution by $\text{Ш}_{\Lambda^\perp}$, which is the same as periodization. This is the basis for a mathematically correct proof of Shannon's Sampling Theorem in the setting of L^p-spaces, which in turn is one of the cornerstones of digital signal processing. It can be used to explain why CDs store 44100 samples per second, and is also useful for subsampling of (smooth) images;

5. For any pair of L^p-spaces the space of translation invariant operators can be characterized by convolution using certain elements from $S_0'(\mathbb{R}^d)$, at least in their action on test functions $k \in S_0(\mathbb{R}^d)$;

6. These convolution kernels or impulse response distributions clearly have a Fourier transform, hence any such operator is uniquely related to a transfer function;

7. The kernel theorem allows to describe any operator from $S_0(\mathbb{R}^d)$ to $S_0'(\mathbb{R}^d)$ via some distributional kernel $\sigma \in S_0'(\mathbb{R}^{2d})$ (see e.g. [52, 38]);

8. There are further representations, e.g. the spreading functions or Kohn–Nirenberg symbols of a given linear operator, which are very useful ways to analyze the properties of a given e.g. pseudo-differential operator. The theory of slowly varying operators can be well described by the concept of underspread operators. These operators could be vaguely described as linear operators which behave locally very much like a convolution operator, but now with a convolution kernel (or moving average profile) which is slowly changing while it is moving over the signal to be smoothed;

9. The distributional framework for the Banach Gelfand Triple $(S_0, L^2, S_0')(\mathbb{R}^d)$ is of great use for Gabor analysis and time-frequency analysis, the area where these spaces showed their usefulness first. Even for the description of the Gabor frame operator for general $L^2(\mathbb{R}^d)$-functions only provides an operator which is bounded from $S_0(\mathbb{R}^d)$ to $S_0'(\mathbb{R}^d)$, while one requires additional properties on the window function to be used in order to have so-called "Bessel families," see [53, 46];

---

* Up to some constant. Here $\Lambda^\perp$ denotes the orthogonal lattice to the given lattice Λ.

10. There are many good arguments why one should use Gabor windows (or Gabor atoms) from $S_0(\mathbb{R}^d)$ only (described as modulation space $M^1(\mathbb{R}^d)$ in [39]), among others, because this guarantees the Bessel property for Gabor systems with such windows for general lattices. This is also the basis for the discussion with respect to jitter errors and in particular a variation of lattices ([54, 55]);

11. Larsen's book on the multiplier [56] provides a description of convolution operators (and also) of the corresponding transfer functions from $L^p(\mathbb{R}^d)$ to $L^q(\mathbb{R}^d)$ ($1 \le p, q < \infty$) using quasi-measures, as introduced by Gaudry [57, 58, 59]. But the (unlimited) global structure of quasi-measures (one could view this space equivalently as $\mathcal{F}L^\infty_{\text{loc}}(\mathbb{R}^d)$). The introduction of the space $S_0'(\mathbb{R}^d)$ helps us to come back to the expected statement: The transfer function is the Fourier transform of the impulse response (and vice versa);

12. There was a theory of transformable measures put forward by L. Argabright and J. Gil de Lamadrid, e.g. in their papers [60, 61, 62]. This can be also treated well in our context.

13. Another generalization of measures on multi-dimensional spaces are the so-called *bimeasures*, which appear in the following papers [63, 64, 65, 66, 67, 68, 69, 70, 71]. They also find a natural interpretation in our context;

14. In addition to the deterministic world of functions and signals questions arising in stochastic signal processing can be well treated in the context of the Banach Gelfand triple. While generalized functions allow average values over time to be described, stochastic processes define the change of statistical properties over time. Thus generalized stochastic processes combine these two features in a unified way. The following table captures the essence of this view-point.

Toward Generalized Stochastic Processes	
Ordinary Functions on $\mathbb{R}^d$	Stochastic Processes on $\mathbb{R}^d$
$\mathbb{R}^d \mapsto \mathbb{C}$	$\mathbb{R}^d \mapsto \mathscr{H}$
$t \mapsto f(t)$	$t \mapsto X(t, \omega) \in L^2(\Omega, \Sigma, P)$
Generalized Functions	*Generalized Stochastic Processes*
$C_c(\mathbb{R}^d) \mapsto \mathbb{C}$	$\rho : S_0(\mathbb{R}^d) \mapsto \mathscr{H} : f \mapsto \rho(f)$
$k \mapsto \sigma(f) := \int_{\mathbb{R}^d} k(x) f(x) dx$	e.g. $f \mapsto \int_{\mathbb{R}^d} X(t, \omega) k(t) dt$

Let us give a few references in that direction: [72], [73], with a recent summary in [74], or [75], [76], and in particular the papers by P. Wahlberg: [77], [78, 79, 80] and [81].

1.13 Summary

We are convinced that the theory of distributions is a central part of any mathematical foundation for signal processing. After all, for almost all the concrete situations signals are things that can be measured. Physical processes are not so much pointwise (almost everywhere) defined functions in the mathematical sense, even if the modelling using function spaces like $\left(L^2(\mathbb{R}^d), \|\cdot\|_2 \right)$ provides mathematically pleasing answers. But does it really describe well the e.g. physical reality of a long piece of music recorded with a high-quality microphone?

At a technical level it is clear that vector spaces of signals are not just finite dimensional, and consequently it is not possible to describe e.g. linear systems via ordinary linear algebra, using (finite) matrices or bases for the involved function spaces. Again this points to the use of methods from (linear) functional analysis, as we have seen. Let us repeat the hint that concrete material can be found at the NuHAG web-page (www.nuhag.eu) and that further material may be available directly from the author by writing an email to hans.feichtinger@univie.ac.at.

Based on this insight we have tried to outline in a quasi-practical way the ingredients required to design a course providing a good basis for further mathematical studies of problems arising in signal processing.

In the spirit of Conceptual Harmonic Analysis such an approach does not limit the considerations to a catalog of settings, usually described in using the categories "discrete or continuous," "periodic or non-periodic,"* but rather asks for a mathematically correct description of the transition between the different settings, making use of distributional convergence. The most interesting aspect is related to the question of approximately computing the ordinary (integral transform called) Fourier transform of a smooth function in a numerical way, based on the FFT.

We expect that it will be important in the coming years to take care of a quantitative estimates for constructive realizations of many different operators or procedures, using a variety of function space norms (cf. [82, 23]). The present chapter should encourage readers to think about the general setting and find some suggestions that hopefully will help them to make use of a suitable set of tools, and to prepare their students for such tasks ahead.

In this sense we would like to communicate to the readers that Harmonic Analysis should not be seen as an abstract, bloodless, and perhaps esoteric subject of pure mathematics, where it is hard to find interesting mathematical problems, or only those which have been left over by earlier generations of mathematicians, or which require a heavy machinery. In contrast, a systematic recombination of methods from linear algebra, supported through numerical software such as MATLAB®, with good functional analytic foundations, including a certain collection of function spaces, mostly Banach spaces (according to our vision), quickly leads to a number of interesting questions which may not only be mathematically challenging but also relevant for applied scientists.

Acknowledgments

During the time of the preparation of this chapter, the author held positions at the Institute for Theoretical Information Sciences, TU Munich (spring 2018) and then (autumn 2018) at the Institute of Mathematics of Charles University, Prague, Czech Republic.

---

* As in [55], inspired by Abstract Harmonic Analysis.

References

1. H. G. Feichtinger. Banach Gelfand triples for applications in physics and engineering. Volume 1146 of *AIP Conf. Proc.*, pages 189–228. *Am. Inst. Phys.*, 2009.
2. L. Schwartz. *Théorie des Distributions. (Distribution Theory). Nouveau Tirage. 1.* Hermann, Paris, xii, 420p., 1957.
3. L. Hörmander. *Linear Partial Differential Operators. 4th Printing.* Springer, Berlin, Heidelberg, New York, 1976.
4. A. Weil. *L'integration dans les Groupes Topologiques et ses Applications.* Hermann and Cie, Paris, 1940.
5. G. Cariolaro. *Unified Signal Theory.* Springer, London, 2011.
6. J. Cooley and J. W. Tukey. An algorithm for the machine calculation of complex Fourier series. *Math. Comp.*, 19:297–301, 1965.
7. M. Frigo and S. Johnson. *FFTW User's Manual.* Massachusetts Institute of Technology, 1999.
8. N. Kaiblinger. Approximation of the Fourier transform and the dual Gabor window. *J. Fourier Anal. Appl.*, 11(1):25–42, 2005.
9. H. G. Feichtinger and N. Kaiblinger. Quasi-interpolation in the Fourier algebra. *J. Approx. Theory*, 144(1):103–118, 2007.
10. R. P. Kanwal. *Generalized Functions. Theory and Applications*, 3rd Revised ed. Birkhäuser, MA, 2004.
11. A. Zygmund. *Trigonometrical Series*, 2nd ed. Dover Publications Inc., New York. 1955.
12. N. Wiener. Generalized harmonic analysis. *Acta Math.*, 55(1):117–258, 1930.
13. N. Wiener. *The Fourier Integral and Certain of Its Applications.* Cambridge University Press, Cambridge, 1933.
14. Gröchenig, K., Y. Lyubarskii, and K. Seip, eds. Operator-Related Function Theory and Time-Frequency Analysis: The Abel Symposium 2012. Vol. 9. Springer, 2014.
15. H. Reiter. *Classical Harmonic Analysis and Locally Compact Groups.* Clarendon Press, Oxford, 1968.
16. H. Reiter and J. D. Stegeman. *Classical Harmonic Analysis and Locally Compact Groups*, 2nd ed. Clarendon Press, Oxford, 2000.
17. L. Loomis. The spectral characterization of a class of almost periodic functions. *Ann. of Math.* 72:362–368, 1960.
18. E. Hewitt and K. A. Ross. *Abstract Harmonic Analysis I.* Number 115 in Grundlehren Math. Wiss. Springer, Berlin, 1963.
19. E. Hewitt and K. A. Ross. *Abstract Harmonic Analysis. Vol. II: Structure and Analysis for Compact Groups. Analysis on Locally Compact Abelian Groups.* Springer, Berlin, Heidelberg, New York, 1970.
20. W. Kozek. On the transfer function calculus for underspread LTV channels. *IEEE Trans. Signal Process.*, 45(1):219–223, January 1997.
21. M. Dörfler and B. Torresani. Representation of operators in the time-frequency domain and generalized Gabor multipliers. *J. Fourier Anal. Appl.*, 16(2):261–293, 2010.
22. M. Dörfler and B. Torrésani. Representation of operators by sampling in the time-frequency domain. *Sampl. Theory Signal Image Process.*, 10(1–2):171–190, 2011.
23. H. G. Feichtinger. Choosing Function Spaces in Harmonic Analysis. Volume 4 of *The February Fourier Talks at the Norbert Wiener Center, Appl. Numer. Harmon. Anal.*, pages 65–101. Birkhäuser/ Springer, Cham, 2015.
24. H. G. Feichtinger. Thoughts on numerical and conceptual harmonic analysis. In A. Aldroubi, C. Cabrelli, S. Jaffard, and U. Molter, editors, *New Trends in Applied Harmonic Analysis. Sparse Representations, Compressed Sensing, and Multifractal Analysis*, Applied and Numerical Harmonic Analysis, pages 301–329. Birkhäuser, Cham, 301–329, 2016.
25. J. V. Fischer. On the duality of discrete and periodic functions. *Mathematics*, 3(2):299–318, 2015.
26. J. V. Fischer. On the duality of regular and local functions. *Mathematics*, 5(41), 2017.

27. J. Fischer. Four particular cases of the Fourier transform. *Mathematics*, 12(6):335, 2018.
28. J. Horvath. *Topological Vector Spaces and Distributions. Vol. I.* Addison-Wesley, Reading, 1966.
29. F. Treves. *Topological Vector Spaces, Distributions and Kernels.* Number 25 in Pure Appl. Math. Academic Press, New York, 1967.
30. A. H. Zemanian. *Distribution Theory and Transform Analysis. An Introduction to Generalized Functions, with Applications. Reprint, Slightly Corrected.* Dover Publications, Inc., New York, 1987.
31. E. Cordero, H. G. Feichtinger, and F. Luef. Banach Gelfand triples for Gabor analysis. In *Pseudo-Differential Operators*, volume 1949 of *Lecture Notes in Mathematics*, pages 1–33. Springer, Berlin, 2008.
32. H. G. Feichtinger, F. Luef, and T. Werther. A guided tour from linear algebra to the foundations of Gabor analysis. In *Gabor and Wavelet Frames*, volume 10 of *Lect. Notes Ser. Inst. Math. Sci. Natl. Univ. Singap.*, pages 1–49. World Sci. Publ., Hackensack, NJ, 2007.
33. H. G. Feichtinger, W. Kozek, and F. Luef. Gabor analysis over finite Abelian groups. *Appl. Comput. Harmon. Anal.*, 26(2):230–248, 2009.
34. H. G. Feichtinger. A novel mathematical approach to the theory of translation invariant linear systems. In Peter J. Bentley and I. Pesenson, editors, *Novel Methods in Harmonic Analysis with Applications to Numerical Analysis and Data Processing*, Birkhäuser, Cham, 2017. 483–516.
35. H. G. Feichtinger. Modulation spaces of locally compact Abelian groups. In R. Radha, M. Krishna, and S. Thangavelu, editors, *Proc. Internat. Conf. on Wavelets and Applications*, pages 1–56, Chennai, January 2002. Allied Publishers, New Delhi, 2003.
36. H. G. Feichtinger. On a new Segal algebra. *Monatsh. Math.*, 92:269–289, 1981.
37. S. Jahan. *Approximative K-Atomic Decompositions and Frames in Banach Spaces.* 2018.
38. H. G. Feichtinger and M. S. Jakobsen. Distribution theory by Riemann integrals. In Pammy Machanda et al., editors, *ISIAM Proceedings*, pages 1–42. 2019.
39. K. Gröchenig. *Foundations of Time-Frequency Analysis.* Appl. Numer. Harmon. Anal. Birkhäuser, Boston, MA, 2001.
40. O. Christensen. *An Introduction to Frames and Riesz Bases.* Applied and Numerical Harmonic Analysis. Birkhäuser, Basel, Second edition, 2016.
41. H. G. Feichtinger. Group theoretical methods and wavelet theory (coorbit theory and applications). In *SPIE Defense, Security, and Sensing. International Society for Optics and Photonics (Baltimore)*, pages 1–10. SPIE, 2013.
42. H. G. Feichtinger and K. Gröchenig. Banach spaces related to integrable group representations and their atomic decompositions, I. *J. Funct. Anal.*, 86(2):307–340, 1989.
43. H. G. Feichtinger and K. Gröchenig. Gabor wavelets and the Heisenberg group: Gabor expansions and short time Fourier transform from the group theoretical point of view. In C. K. Chui, editor, *Wavelets: A Tutorial in Theory and Applications*, volume 2 of *Wavelet Anal. Appl.*, pages 359–397. Academic Press, Boston, MA, 1992.
44. A. Papoulis. *The Fourier Integral and Its Applications.* McGraw-Hill Book Co., Inc., New York, pp. ix, 318, 1962.
45. M. J. Lighthill. *Introduction to Fourier Analysis and Generalised Functions.* Cambridge Monographs on Mechanics and Applied Mathematics. Cambridge University Press, New York, 1958.
46. H. G. Feichtinger and W. Kozek. Quantization of TF lattice-invariant operators on elementary LCA groups. In H. G. Feichtinger and T. Strohmer, editors, *Gabor Analysis and Algorithms*, Appl. Numer. Harmon. Anal., pages 233–266. Birkhäuser, Boston, MA, 1998.
47. H. G. Feichtinger and F. Weisz. Wiener amalgams and pointwise summability of Fourier transforms and Fourier series. *Math. Proc. Cambridge Philos. Soc.*, 140(3):509–536, 2006.
48. H. G. Feichtinger and F. Weisz. The Segal algebra $S_0(R^d)$ and norm summability of Fourier series and Fourier transforms. *Monatsh. Math.*, 148:333–349, 2006.
49. K. Gröchenig. An uncertainty principle related to the Poisson summation formula. *Studia Math.*, 121(1):87–104, 1996.
50. Y. Katznelson. *An Introduction to Harmonic Analysis.* New York/London/Sydney/Toronto: John Wiley and Sons, Inc, 1968.
51. J.-P. Kahane and P. G. Lemarié Rieusset. Remarks on the Poisson summation formula. (Remarques sur la formule sommatoire de Poisson.). *Studia Math.*, 109:303–316, 1994.

52. H. G. Feichtinger and M. S. Jakobsen. Distribution theory by Riemann integrals. Mathematical Modelling, Optimization, Analytic and Numerical Solutions. Springer, Singapore, 2020. 33–76.

53. H. G. Feichtinger and G. Zimmermann. A Banach space of test functions for Gabor analysis. In H. G. Feichtinger and T. Strohmer, editors, *Gabor Analysis and Algorithms: Theory and Applications*, Applied and Numerical Harmonic Analysis, pages 123–170. Birkhäuser, Boston, MA, 1998.

54. H. G. Feichtinger and N. Kaiblinger. Varying the time-frequency lattice of Gabor frames. *Trans. Am. Math. Soc.*, 356(5):2001–2023, 2004.

55. G. Ascensi, H. G. Feichtinger, and N. Kaiblinger. Dilation of the Weyl symbol and Balian-Low theorem. *Trans. Am. Math. Soc.*, 366(7):3865–3880, 2014.

56. R. Larsen. *An Introduction to the Theory of Multipliers*. Springer-Verlag, New York/Heidelberg, 1971.

57. G. I. Gaudry. Quasimeasures and operators commuting with convolution. *Pacific J. Math.*, 18:461–476, 1966.

58. G. I. Gaudry. Multipliers of type (p, q). *Pacific J. Math.*, 18:477–488, 1966.

59. G. I. Gaudry. Bad behavior and inclusion results for multipliers of type (p,q). *Pacific J. Math.*, 35:83–94, 1970.

60. L. N. Argabright and J. Gil de Lamadrid. Fourier transforms of unbounded measures. *Bull. Am. Math. Soc.*, 77:355–359, 1971.

61. L. N. Argabright and J. Gil de Lamadrid. Analyse harmonique des mesures non bornés sur les groupes abeliens localement compacts. In *Conf. Harmonic Analysis, College Park, Maryland 1971*, volume 266 of *Lect. Notes Math.*, pages 1–16. Springer, New York, 1972.

62. L. N. Argabright, P. Eymard, H. G. Feichtinger, and J. Gil de Lamadrid. Winterschule 1979 Internationale Arbeitstagung über Topologische Gruppen und Gruppenalgebren Tagungsbericht, 1979.

63. D. Dehay and R. Moché. Trace measures of a positive definite bimeasure. *J. Multi. Anal.*, 40(1):115–131, 1992.

64. C. C. Graham and B. M. Schreiber. Projections in spaces of bimeasures. *Canad. Math. Bull.*, 31:19–25, 1988.

65. C. C. Graham and B. M. Schreiber. Sets of interpolation for Fourier transforms of bimeasures. *Colloq. Math.*, 51:149–154, 1987.

66. J. E. Gilbert, T. Ito, and B. M. Schreiber. Bimeasure algebras on locally compact groups. *J. Funct. Anal.*, 64:40, 1985.

67. C. C. Graham and B. M. Schreiber. Bimeasure algebras on LCA groups. *Pacific J. Math.*, 115:91–127, 1984.

68. H. Niemi. Diagonal measure of a positive definite bimeasure. In Kölzow, D., and Maharam-Stone, D. (eds.), *Measure Theory Oberwolfach 1981. Lecture Notes in Mathematics, vol 945.* Springer, Berlin, Heidelberg, 1982.

69. I. Kluvanek. Remarks on bimeasures. *Proc. Am. Math. Soc.*, 81:233–239, 1981.

70. H. Niemi. On the support of a bimeasure and orthogonally scattered vector measures. *Ann. Acad. Sci. Fenn. Ser. A I Math.*, 1:249–275, 1975.

71. M. Morse and W. Transue. The representation of a c-bimeasure on a general rectangle. *Proc. Natl. Acad. Sci. USA*, 42:89–95, 1956.

72. W. Hörmann. *Stochastic Processes and Vector Quasi-Measures*. Master's thesis, University of Vienna, July 1987.

73. W. Hörmann. *Generalized Stochastic Processes and Wigner Distribution*. PhD thesis, University of Vienna, (AUSTRIA), 1989.

74. H. G. Feichtinger and W. Hörmann. A distributional approach to generalized stochastic processes on locally compact abelian groups. In G. Schmeisser and R. Stens, editors, *New Perspectives on Approximation and Sampling Theory. Festschrift in Honor of Paul Butzer's 85th Birthday*, pages 423–446. Birkhäuser/Springer, Cham, 2014.

75. T. Steger. *Approximation of Generalized Stochastic Processes*. Master's thesis, University of Vienna, 2009.

76. B. Keville. *Multidimensional Second Order Generalised Stochastic Processes on Locally Compact Abelian Groups*. PhD thesis, Trinity College Dublin, 2003.

77. P. Wahlberg. The random Wigner distribution of Gaussian stochastic processes with covariance in $S_0(R^{2d})$. *J. Funct. Spaces Appl.*, 3(2):163–181, 2005.

78. P. Wahlberg and P. Schreier. Spectral relations for multidimensional complex improper stationary and (almost) cyclostationary processes. *IEEE Trans. Inform. Theory*, 54(4):1670–1682, 2008.

79. P. Wahlberg and P. Schreier. Gabor discretization of the Weyl product for modulation spaces and filtering of nonstationary stochastic processes. *Appl. Comput. Harmon. Anal.*, 26:97–120, January 2009.

80. P. Wahlberg and P. Schreier. On Wiener filtering of certain locally stationary stochastic processes. *Signal Process.*, 90(3):885–890, 2010.

81. P. Wahlberg. Regularization of kernels for estimation of the Wigner spectrum of Gaussian stochastic processes. *Probab. Math. Statist.*, 30(2):369–381, 2010.

82. H. G. Feichtinger and D. Onchis. Constructive realization of dual systems for generators of multi-window spline-type spaces. *J. Comput. Appl. Math.*, 234(12):3467–3479, 2010.

2

Industrial Applications of Optimal Control for Partial Differential Equations on Networks: Reduction and Decomposition Methods Applied to the Discrete–Continuous Control of Gas Flow in Complex Pipe Systems

Günter Leugering

CONTENTS

2.1 Introduction..25
    2.1.1 Modeling of Gas Flow in a Single Pipe..25
    2.1.2 Network Modeling 26
2.2 Optimal Control Problems and Outline..28
    2.2.1 Time Discretization 29
2.3 Domain Decomposition..32
2.4 Domain Decomposition for Optimal Control Problems........................34
    2.4.1 Optimal Control on the Decomposed System: A Jacobi-Type Procedure 34
    2.4.2 Decomposition of the Optimality System for Continuous Controls with Fixed Switching Structure *s* 35
    2.4.3 Virtual Controls 36
References..38

2.1 Introduction

2.1.1 Modeling of Gas Flow in a Single Pipe

The aim of this contribution is to provide an exemplary road map—in a nutshell—from a given industrial application, here the control of gas networks, which is far too complex for a direct approach, to a problem that can be actually handled using well-known methods in control theory. This road map involves hierarchies of models, time discretization, reduction of space–time optimal control problems to static ones by way of instantaneous controls. A second objective is to introduce into network modeling what appears to be a tool that is becoming more and more important. Finally, as networks in industrial or civil applications are typically very complex, decomposition techniques are critical for the numerical handling. This reduction is done here by an iterative non-overlapping domain decomposition of the network and the corresponding optimal control problem. Similar techniques apply to problems of sewer systems, freshwater networks, irrigation systems, as well as traffic networks.

We depart from a full space–time formulation of the problem with both discrete and continuous controls. After that we discretize the problem with respect to time using a simple implicit Euler step, for the sake of brevity. The corresponding optimal control problem, which still involves all time-steps is then reduced to what has come to be known as an instantaneous control problem that is defined on a single time-step, thereby decoupling the time history. The resulting optimal control problem can be viewed as a nonlinear (semi-linear or even linear) elliptic optimal control problem on the network. The network, in turn, is then decomposed into small subnetworks even to single links. As a result, at the core of the approach, elementary problems can be solved very effectively.

Clearly, in this chapter, we can only indicate the procedure and refer to mathematically rigorous results from the literature whenever possible and feasible.

To this end, we consider the Euler equations consisting of the continuity equation, the balance of moments, and the energy equation. (cf. [1, 2, 3, 4]) We denote by ρ the density, by v the velocity of the gas, and by p the pressure. We define λ as the friction coefficient of the pipe, D the diameter, a the cross section area. The unknowns of the system are given by ρ and the flux $q = a\rho v$. We also denote by c with $c^2 = \dfrac{\partial p}{\partial \rho}$ (for constant entropy) the speed of sound. For horizontal pipes

$$\frac{\partial \rho}{\partial t} + \frac{\partial}{\partial x}(\rho v) = 0$$

$$\frac{\partial}{\partial t}(\rho v) + \frac{\partial}{\partial x}(p + \rho v^2) = -\frac{\lambda}{2D}\rho v |v|. \tag{2.1}$$

In the particular case, where we have a constant speed of sound $c = \sqrt{\dfrac{p}{\rho}}$, for small velocities $|v| \ll c$, we arrive at the semi-linear model

$$\frac{\partial \rho}{\partial t} + \frac{\partial}{\partial x}(\rho v) = 0$$

$$\frac{\partial}{\partial t}(\rho v) + \frac{\partial p}{\partial x} = -\frac{\lambda}{2D}\rho v |v|. \tag{2.2}$$

2.1.2 Network Modeling

Let $G = (V, E)$ denote the graph of the gas network with vertices (nodes) $V = \{n_j : j \in \mathcal{J}, |\mathcal{J}| = |V|\}$ and edges $E = \{e_i : i \in \mathcal{I}, |\mathcal{I}| = |E|\}$. We associate to each edge a direction. To this end, we introduce the edge-node incidence matrix

$$d_{ij} = \begin{cases} -1, & \text{if node } n_j \text{ is the left node of the edge } e_i, \\ +1, & \text{if node } n_j \text{ is the right node of the edge } e_i, \\ 0, & \text{else.} \end{cases}$$

In contrast to the classical notion of discrete graphs, the graphs considered here are known as metric graphs, in the sense that the edges are continuous curves. In fact, we consider here straight edges, along which differential equations hold. The index set of all edges incident at node n_j is $i \in \mathcal{I}_j := \{i \in 1, \dots E \,|\, d_{ij} \neq 0\}$. We introduce the edge degree $d_j := |\,\mathcal{I}_j\,|$ and distinguish between multiple nodes $\{n_j : d_j > 1\} = \{j \in \mathcal{J}^M\}$ and simple nodes $\{n_j : d_j = 1\} = \{i \in \mathcal{J}^S\}$. We decompose the set of simple nodes into Dirichlet nodes $\mathcal{J}_D^S$ and Neumann nodes $\mathcal{J}_N^S$. At multiple nodes, we assume continuity conditions

$$p_i(n_j,t) = p_k(n_j,t), \forall i,k \in \mathcal{I}_j, j \in \mathcal{J}^M, t \in (0,T). \tag{2.3}$$

The nodal balance for the fluxes can be written as

$$\sum_{i\in\mathcal{I}_j} d_{ij} q_i(n_j,t) = 0, \; j \in \mathcal{J}^M, t \in (0,T). \tag{2.4}$$

We assume that valves and compressors are serial nodes n_j, i.e. $j \in \mathcal{J}^M$ with $d_j=2$. At such a node we have an incoming edge with unique index $i \in \mathcal{I}_j^+$, where $\mathcal{I}_j^+ =: \{i \in \mathcal{I}_j : d_{ij} = 1\}$, and an outgoing edge with unique index $k \in \mathcal{I}_j^- =: \{i \in \mathcal{I}_j : d_{ij} = -1\}$. We label compressor and valve nodes by $\mathcal{J}_c, \mathcal{J}_v$, respectively.

With this, we now provide the network model of Equation (2.2).

$$\partial_t p_i(x,t) + \frac{c_i^2}{a_i} \partial_x q_i(x,t) = 0, \quad i \in \mathcal{I}$$

$$\partial_t q_i(x,t) + \partial_x p_i(x,t) = -\frac{\lambda c_i^2}{2D_i a_i^2} \frac{q_i(x,t)|q_i(x,t)|}{p_i(x,t)} \quad i \in \mathcal{I}$$

$$p_i(n_j,t) = p_k(n_j,t), \quad j \in \mathcal{J}^M \setminus (\mathcal{J}_c \cup \mathcal{J}_v), i,k \in \mathcal{I}_j$$

$$g_j(p_i(n_j,t),q_i(n_j,t)) = u_j(t), \quad j \in \mathcal{J}^S, i \in \mathcal{I}_j$$

$$\sum_{i\in\mathcal{I}_j} d_{ij} q_i(n_j,t) = 0, \quad j \in \mathcal{J}^M \tag{2.5}$$

$$s_j^v(t)\big(p_i(n_j,t) - p_k(n_j,t)\big) + (1-s_j^v(t))q_i(n_j,t) = 0, \quad j \in \mathcal{J}_v, i \in \mathcal{I}_j^-, k \in \mathcal{I}_j^+$$

$$s_j^c(t)\left(u_j - C\left(\left(\frac{p_k(n_j,t)}{p_i(n_j,t)}\right)^{sign(q_k(n_j,t))\kappa} - 1\right)\right)$$

$$+(1-s_j^c(t))\big(p_i(n_j,t)-p_k(n_j,t)\big) = 0, \quad j \in \mathcal{J}_c, i \in \mathcal{I}_j^-, k \in \mathcal{I}_j^+$$

$$p_i(x,0) = p_{i,0}(x), q_i(x,0) = q_{ij}(x), \quad i \in \mathcal{I}$$

It is obvious from Equation (2.5) that for $s_j^v(t) = 1$, i.e. the case in which the valve at node n_j is open, the classical transmission conditions hold, while for $s_j^v(t) = 0$, the outgoing flow and—according to the Kirchhoff condition, which still holds—the incoming flow is zero. Similarly, for $s_j^c(t) = 1$, the compressor is active, resulting in pressure control such that the pressure in the outgoing pipe is increased with respect to the pressures of the incoming pipes. Equation (5) is an example of a dynamical system evolving on a metric graph according to discrete and continuous controls. To the best knowledge of the author, Equation (2.5) with switching functions $s_j^v(t), s_j^c(t) \in \{0,1\}$, even for simple but non-trivial networks has not been considered for the semilinear problem so far. Even smooth relaxations of the discrete control variables $s_j^v(\cdot)$ and $s_j^c(\cdot)$ do not seem to have been investigated for that system. See [5] for a more detailed discussion.

Note that we can replace the transmission conditions at the compressor node by the bilinear transmission conditions as follows:

$$u_j - C\left(\left(\frac{p_k(n_j,t)}{p_i(n_j,t)}\right)^{sign(q_k(n_j,t))\kappa} - 1\right) = 0$$

$$\Leftrightarrow \left(\frac{u_j + C}{C}\right)^{\frac{1}{\kappa}sign(q_k(n_j,t))} = \frac{p_k(n_j,t)}{p_i(n_j,t)}.$$

If we replace u_j by

$$u_j =: \left(\frac{u_j + C}{C}\right)^{\frac{1}{\kappa}sign(q_k(n_j,t))}$$

and ensure $u_j \geq 1$, the original transmission condition at the compressor node can be replaced with

$$p_i(n_j,t)u_j(t) - p_k(n_j,t) = 0$$

if the compressor is active. Otherwise, the classical continuity condition for the pressure holds. This is then a bilinear boundary control. Even in this simplified version, such problems have not been considered rigorously so far. This already indicates that a lot of interesting mathematical problems can be derived from the context of applications.

2.2 Optimal Control Problems and Outline

We now formulate optimal control problems. In real industrial gas-network applications, there are two types of controls to be considered: optimizing decision variables such as on–off states for valves and compressors, zero full-supply, and demand variables for input and exit nodes and continuous profile nodal controls. Valves and compressors, in turn, can be modeled as transmission conditions at a serial node, as above (see [5, 6, 7]). We now describe the general format for an optimal control problem associated with the semi-linear model equations.

$$\min_{(p,q,u,s)\in\Xi} I(p,q,u,s) := \sum_{i\in\mathcal{I}}\int_0^T\int_0^{\ell_i} I_i(p_i,q_i,u_i,s_i)dxdt + \frac{\nu}{2}\sum_{j\in\mathcal{J}^S}\int_0^T |u_j(t)|^2\,dt + \frac{\mu}{2}\sum_{j\in\mathcal{J}_v\cup\mathcal{J}_c}\int_0^T s_j(t)^2\,dt$$

$$s.t. \tag{2.6}$$

$$(p,q,u,s) \text{ satisfies } (2.5),$$

$$\Xi := \{(p,q,u,s): \underline{p}_i \leq p_i \leq \overline{p}_i, \underline{q}_i \leq q_i \leq \overline{q}_i, \underline{u}_i \leq u_i \leq \overline{u}_i, i\in\mathcal{I}, s_j(t)\in\{0,1\}, j\in\mathcal{J}_v\cup\mathcal{J}_c\}. \tag{2.7}$$

Here, $v, \mu > 0$ are penalty parameters and $I_i(\cdot, \cdot, \cdot, \cdot)$ a continuous function on (p, q, u, s). The set Ξ of admissible states and controls involves box constraints: $\underline{p}_i, \underline{q}_i, \overline{p}_i, \overline{q}_i$ are given constants that determine the feasible pressures and flows in the pipe i, while $\underline{u}_i, \overline{u}_i$ describe control constraints. In the continuous-time case the inequalities are considered as being satisfied for all times and everywhere along the pipes. Optimal control problem Equation (2.6) is a fully nonlinear (in fact semi-linear) discrete–continuous problem. Again, there does not seem to be any result in the literature. The problem is the intrinsic coupling of integer controls, continuous controls, and nonlinear dynamics on a metric graph. The idea behind handling such problems consists in applying a decomposition into an integer optimal control problem, where the continuous control variables are kept fixed in the admissible set $\Xi(p, q, u, s) = \Xi(p, q, u^*, s)$ and the optimization is performed with respect to the discrete (integer) control variables. The result is then an optimal switching structure s^*, which is then fixed in the next step, where the optimization is done with respect to the continuous controls. This results in a pair (s^*, u^*) and the procedure is repeated until a sufficient descent can be observed. Clearly, even though the discrete part of the procedure may result in a global optimum s^*, depending of course on the nonlinear structure of the underlying control problem, the continuous control part typically results in a local minimum if, by some reasonable assumptions, the problem is not convex. Therefore, one cannot expect the overall procedure to converge to a global optimum. In fact, for the proposed alternating direction approach, not even convergence per se is guaranteed. On the extreme end, if the problem can be reduced to one on a single edge with linear dynamics, we clearly have a unique global optimum. This discussion underlines the importance of further research in the direction of mixed-integer nonlinear programming for PDE-problems on networks. We refer to [5] and to the entire collaborative research cluster funded by the German Science Foundation DFG: TR154.

2.2.1 Time Discretization

According to the outline, we now consider the time discretization of Equation (2.5) such that $[0, T]$ is decomposed into break points $t_0 = 0 < t_1 < \ldots < t_N = T$ with widths $\Delta t_n := t_{n+1} - t_n, n = 0, \ldots, N - 1$. Accordingly, we denote $p_i(x, t_n) =: p_{i,n}(x), q_i(x, t_n) =: q_{i,n}(x)$, $n = 0, \ldots, N - 1$. We consider a mixed implicit–explicit Euler scheme which takes p_i in the friction term in an explicit manner.

$$\frac{1}{\Delta t} p_{i,n+1}(x) + \frac{c_i^2}{a_i} \partial_x q_{i,n+1}(x) = \frac{1}{\Delta t} p_{i,n}(x), \, x \in (0, \ell_i), i \in \mathcal{I}$$

$$\frac{1}{\Delta t} q_{i,n+1}(x) + \partial_x p_{i,n+1}(x) = -\frac{\lambda c_i^2}{2 D_i a_i^2} \frac{q_{i,n+1}(x) |q_{i,n+1}(x)|}{p_{i,n}(x)} + \frac{1}{\Delta t} q_{i,n}(x), \, x \in (0, \ell_i), i \in \mathcal{I}$$

$$g_j(p_{i,n+1}(n_j), q_{i,n+1}(n_j)) = u_{j,n+1}, \, i \in \mathcal{I}_j, j \in \mathcal{J}^s$$

$$p_{i,n+1}(n_j) = p_{k,n+1}(n_j), \forall i, k \in \mathcal{I}_j, , j \in \mathcal{J}^M \setminus (\mathcal{J}_c \cup \mathcal{J}_v)$$

$$\sum_{i \in \mathcal{I}_j} d_{ij} q_{i,n+1}(n_j) = 0, j \in \mathcal{J}^M, \tag{2.8}$$

$$s_{j,n+1}^v \left(p_{i,n+1}(n_j) - p_{k,n+1}(n_j) \right) + (1 - s_{j,n+1}^v) q_{i,n+1}(n_j) = 0, j \in \mathcal{J}_v, i \in \mathcal{I}_j^-, k \in \mathcal{I}_j^+$$

$$s_{j,n+1}^c \left(u_j - C \left(\left(\frac{p_{k,n+1}(n_j)}{p_{i,n+1}(n_j)} \right)^{sign(q_{k,n+1}(n_j))\kappa} - 1 \right) \right)$$

$$+ (1 - s_{jn+1}^c) \left(p_{i,n+1}(n_j) - p_{k,n+1}(n_j) \right) = 0 \quad j \in \mathcal{J}_c, i \in \mathcal{I}_j^-, k \in \mathcal{I}_j^+$$

$$p_i(x, 0) = p_{i,0}(x), q_i(x, 0) = q_{ij}(x), \quad i \in \mathcal{I} \, x \in (0, \ell_i), i \in \mathcal{I}.$$

We then obtain the optimal control problem on the time-discrete level:

$$\min_{(p,q,u,s)} I(p,q,u,s) := \sum_{i\in\mathcal{I}}\sum_{n=1}^{N}\int_0^{\ell_i} I_i(p_{i,n},q_{i,n})dx + \frac{v}{2}\sum_{j\in\mathcal{J}^S}\sum_{n=1}^{N}|u_j(n)|^2 + \frac{\mu}{2}\sum_{j\in\mathcal{J}_v\cup\mathcal{J}_c}\sum_{n=1}^{N}s_{j,n}^2$$

$$s.t. \tag{2.9}$$

$$(p,q,u,s) \text{ satisfies } (2.8).$$

In Equation (2.4), we consider edgewise given cost functions e.g.

$$I_i(p_{i,n},q_{i,n})(x) := \frac{\kappa_i}{2}\left\{|p_{i,n}(x)-p_{i,n}^d(x)|^2 + |q_{i,n}(x)-q_{i,n}^d(x)|^2\right\}, x\in(0,\ell_i), i\in\mathcal{I}.$$

It is clear that (2.4) involves all time-steps in the cost functional. We would like to reduce the complexity of the problem even further. To this aim we consider what has come to be known as "instantaneous control." This amounts to reducing the sums in the cost function of Equation (2.9) to the time-level t_{n+1}. This strategy has is known as a "rolling horizon approach," the simplest case of the "moving horizon" paradigm, see e.g. [8, 9]. Thus, for each $n = 1,\dots,N-1$ and given $p_{i,n},q_{i,n}$, we consider the problems

$$\min_{(p,q,u,s)} I(p,q,u,s) := \sum_{i\in\mathcal{I}}\int_0^{\ell_i} I_i(p_i,q_i)dx + \frac{v}{2}\sum_{j\in\mathcal{J}^S}|u_j|^2 + \frac{\mu}{2}\sum_{j\in\mathcal{J}_v\cup\mathcal{J}_c}s_j^2$$

$$s.t. \tag{2.10}$$

$$(p,q,u,s) \text{ satisfies } (8) \text{ at time level } n+1.$$

It is now convenient to discard the actual time level $n+1$ and redefine the states at the former time as input data. To this end, we introduce $\alpha_i := \frac{1}{\Delta t}$, $f_i^1 := \frac{1}{\Delta t}p_{i,n}(x)$, $f_i^2 := \frac{1}{\Delta t}q_{i,n}(x)$, $\gamma_i(x) := \frac{\lambda c_i^2}{2D_i a_i^2}\frac{1}{p_{i,n}(x)}$ and rewrite Equation (2.8) as

$$\alpha_i p_i(x) + \frac{c_i^2}{a_i}\partial_x q_i(x) = f_i^1, x\in(0,\ell_i), i\in\mathcal{I}$$

$$\alpha_i q_i(x) + \partial_x p_i(x) + \gamma_i(x)q_i(x)|q_i(x)| = f_i^2, x\in(0,\ell_i), i\in\mathcal{I}$$

$$g_j(p_i(n_j),q_i(n_j)) = u_j, i\in\mathcal{I}_j, j\in\mathcal{J}^S$$

$$p_i(n_j) = p_k(n_j), \forall i,k\in\mathcal{I}_j, j\in\mathcal{J}^M\setminus(\mathcal{J}_c\cup\mathcal{J}_v)$$

$$\sum_{i\in\mathcal{I}_j} d_{ij}q_i(n_j) = 0, j\in\mathcal{J}^M \tag{2.11}$$

$$s_j^v\left(p_i(n_j)-p_k(n_j)\right)+(1-s_j^v)q_i(n_j) = 0, j\in\mathcal{J}_v, i\in\mathcal{I}_j^-, k\in\mathcal{I}_j^+$$

$$s_j^c\left(u_j - C\left(\left(\frac{p_k(n_j)}{p_i(n_j)}\right)^{sign(q_k(n_j))\kappa}-1\right)\right)$$

$$+(1-s_j^c)\left(p_i(n_j)-p_k(n_j)\right) = 0 \quad j\in\mathcal{J}_c, i\in\mathcal{I}_j^-, k\in\mathcal{I}_j^+.$$

We now differentiate the first equation in (11) with respect to x and insert the result into the second equation. After renaming $f_i := f_i^2 - \dfrac{1}{\alpha_i}\partial_x f_i^1$, $\mathcal{I} = \{i = 1, \ldots, m\}$ and introducing $\beta_i = \dfrac{c_i^2}{a_i \alpha_i}$ we have $p_i = -\beta_i \partial_x q_i + p_i$. We consider the semi-linear elliptic problem on the graph G with Neumann controls at simple nodes.

$$
\begin{aligned}
&\alpha_i q_i(x) - \beta_i \partial_{xx} q_i(x) + g_i(x; q_i(x)) = f_i(x),\, x \in (0, \ell_i), i \in \mathcal{I}\\
&\beta_i \partial_x q_i(n_k) = \beta_j \partial_x q_j(n_k),\, i \neq j \in \mathcal{I}_k,\, k \in \mathcal{J}^M \setminus (\mathcal{J}_c \cup \mathcal{J}_v)\\
&q_i(n_k) = 0,\, i \in \mathcal{I}_k,\, n_k \in \mathcal{J}_D^S\\
&\partial_x q_i(n_k) = u_k,\, i \in \mathcal{I}_k,\, n_k \in \mathcal{J}_N^S\\
&\sum_{i \in \mathcal{I}_k} d_{ik} q_i(n_k) = 0,\, k \in \mathcal{J}^M,
\end{aligned}
$$

$$\tag{2.12}$$

$$
s_j^v\left(-\beta_i\partial_x q_i(n_j) + \beta_k\partial_x q_k(n_j)\right) + (1 - s_j^v)q_i(n_j) = -s_j^v\left(\bar{p}_i(n_j) - \bar{p}_k(n_j)\right),\, j \in \mathcal{J}_v,\, i \in \mathcal{I}_j^-,\, k \in \mathcal{I}_j^+
$$

$$
s_j^c\left(u_j - C\left(\left(\frac{-\beta_k\partial_x q_k(n_j) + \bar{p}_k(n_j)}{-\beta_i\partial_x q_i(n_j) + \bar{p}_i(n_j)}\right)^{\text{sign}(q_k(n_j))\kappa} - 1\right)\right)
$$

$$
+ (1 - s_j^c)\left(-\beta_i\partial_x q_i(n_j) + \beta_k\partial_x q_k(n_j)\right) = (1 - s_j^c)(\bar{p}_i(n_j) - \bar{p}_k(n_j))\quad j \in \mathcal{J}_c,\, i \in \mathcal{I}_j^-,\, k \in \mathcal{I}_j^+
$$

where we set $g_i(x; s) := \gamma_i(x)s|s|$. We then consider the following optimal control problem:

$$
\min_{(p,q,u,s)} I(p,q,u,s) := \sum_{i\in\mathcal{I}} \int_0^{\ell_i} I_i(p_i, q_i)dx + \frac{\nu}{2}\sum_{j\in\mathcal{J}^S}|u_j|^2 + \frac{\mu}{2}\sum_{j\in\mathcal{J}_v\cup\mathcal{J}_c} s_j^2
$$

s.t.

$$\tag{2.13}$$

(p,q,u,s) satisfies (2.12).

In Equation (2.8), we have arrived at a sequence of mixed-integer optimal control problems for elliptic semi-linear systems on a metric graph. We have deliberately discretized with respect to time the nonlinear part in such a way that the denominator of the expression contains p_i at the previous time level. This choice leads to monotone nonlinearity. Therefore, for given controls (u, s), we can apply the celebrated Brezis theorem for the surjectivity of nonlinear operators [10]. In this spirit, we can extend a wellposedness result in [7, 11] to this situation.

Let $(u, s) \in \mathbb{R}^{|\mathcal{J}_N^S \cup \mathcal{J}_c|} \times \{0, 1\}^{|\mathcal{J}_v \cup \mathcal{J}_c|}$ be given controls and $f \in \Pi_{i=1}^N L^2(0, \ell_i)$ given right-hand sides. Then there is a unique solution $q \in \Pi_{i=1}^N H^2(0, \ell_i)$ of Equation (2.12). Moreover, the optimal control problem in Equation (2.8) admits a unique optimal solution (u, s).

We realize that even under the instantaneous control paradigm, the optimal control problem Equation (2.13) is far too complex for computations in the context of realistic networks. We therefore resort now to domain decomposition techniques in order to reduce the graph to suitable subgraphs. In particular, we choose the subgraphs G_{sub} such that the integer controls—which operate by construction at given serial nodes $\mathcal{J}_v \cup \mathcal{J}_c$ (those with edge-degree 2)—are executed at interior nodes of the subgraph. That is to say, we decompose at multiple nodes $\mathcal{J}^M \setminus (\mathcal{J}_c \cup \mathcal{J}_v)$. The method can be written down in such a way that the entire graph G is disjointly decomposed in subgraphs G_k, i.e. $G = \dot\cup_k G_k$. At the interface nodes between the subgraphs—those can have multiplicity two or more—no discrete control is assumed. To be very precise, the method requires even more notation, which we dispense with here because of space limitations. The full picture will be published elsewhere.

2.3 Domain Decomposition

We provide an iterative non-overlapping domain decomposition that can be interpreted as an Uzawa method (Alg3, in the sense of Glowinski). See the monograph [12] for details. The idea for this algorithm originates from a decoupling of the transmission conditions. To this end, we define the flux vector $q^k := (d_{ik}q_i(n_k), i \in \mathcal{I}_k)^T$ and the pressure vectors $\partial q^k := (\beta_i \partial_x q_i(n_k), i \in \mathcal{I}_k)^T$ at a given node $n_k, k \in \mathcal{J}^M \setminus (\mathcal{J}_c \cup \mathcal{J}_v)$. Given a vector $z := (z_i, i \in \mathcal{I}_k)$, we define

$$(\mathcal{S}^k(z))_i := \frac{2}{d_k} \sum_{j \in \mathcal{I}_k} z_j - z_i. \tag{2.14}$$

Then $(\mathcal{S}^k)^2 = I$ and $\mathcal{S}^k(1) = 1$ for $1 := (1, \ldots, 1) \in \mathbb{R}^{d_k}$. With this notation, the general concept is easily established. We set for any $\sigma > 0$:

$$q^k + \sigma \partial q^k = \sigma \mathcal{S}^k(\partial q^k) - \mathcal{S}^k(q^k). \tag{2.15}$$

Applying $\mathcal{S}^k$ to both sides of (2.15), we obtain

$$\sum_{i \in \mathcal{I}_k} d_{ik} q_i(n_k) = 0. \tag{2.16}$$

But then Equation (2.15) reduces to

$$\partial q_i(n_k) = \frac{1}{d_k} \sum_{k \in \mathcal{I}_k} \partial q_j(n_k), i \in \mathcal{I}_k,$$

which, in turn, implies

$$\beta_i \partial_x q_i(n_k) = \beta_j \partial_x q_j(n_k), i \neq j \in \mathcal{I}_k, k \in \mathcal{J}^M \setminus (\mathcal{J}_c \cup \mathcal{J}_v). \tag{2.17}$$

Clearly, if the transmission conditions of Equations (2.16) and (2.17) hold at the multiple node n_k, then Equation (2.15) is also fulfilled. Thus, Equation (2.15) is equivalent to the transmission conditions of Equations (2.16) and (2.17). These new conditions of Equation (2.15) are now relaxed in an iterative scheme as follows. We use l as iteration number.

$$(p^k)^{l+1} + \sigma \partial (q^k)^{l+1} = \sigma \mathcal{S}^k((\partial q^k)^l) - \mathcal{S}^k((q^k)^l) =: (g^k)^{l+1}. \tag{2.18}$$

We have the following relations:

$$(g^k)^{l+1} = \mathcal{S}^k(2\sigma\partial(q^k)^l - (g^k)^l). \tag{2.19}$$

This gives rise to the definition of a fixed point mapping. To this end, we need to look into the behavior of the interface in terms of g^k, $k \in \mathcal{J}^M \setminus (\mathcal{J}_c \cup \mathcal{J}_v)$, that is

$$g \in \mathcal{X} := \Pi_{k \in \mathcal{J}^M \setminus (\mathcal{J}_c \cup \mathcal{J}_v)} \Pi_{i \in \mathcal{I}_k}, \| g \|_{\mathcal{X}}^2 := \sum_{k \in \mathcal{J}^M \setminus (\mathcal{J}_c \cup \mathcal{J}_v)} \sum_{i \in \mathcal{I}_k} \frac{1}{\sigma} | g_{i,k} |^2, \tag{2.20}$$

$$\mathcal{T} : \mathcal{X} \to \mathcal{X}, \tag{2.21}$$

$$(\mathcal{T}g)_{i,k} := \mathcal{S}^k(2\sigma\partial(q^k) - g^k)_i, \ k \in \mathcal{J}^M \setminus (\mathcal{J}_c \cup \mathcal{J}_v), i \in \mathcal{I}_k,$$

$$(\mathcal{T}g)_k = \{(\mathcal{T})_{i,k}, i \in \mathcal{I}_k\},$$

$$\mathcal{T}g = \{(\mathcal{T}g)_k, k \in \mathcal{J}^M \setminus (\mathcal{J}_c \cup \mathcal{J}_v)\}.$$

The relaxed version of a fixed point iteration is as follows: for $\varepsilon \in [0,1)$

$$g^{l+1} = (1-\varepsilon)\mathcal{T}(g^l) + \varepsilon g^l. \tag{2.22}$$

For the analysis of the convergence of the iterates, we need to specify the equations. In order to prove convergence, one needs to address the errors between the solutions to the problem in Equation (2.12) and

$$\begin{aligned}
\alpha_i q_i^{l+1}(x) - \beta_i \partial_{xx} q_i^{l+1}(x) + g_i(q_i^{l+1}) &= f_i(x), \ x \in (0, \ell_i), i \in \mathcal{I} \\
q_i(n_k)^{l+1} &= 0, i \in \mathcal{I}_k, n_k \in \mathcal{J}_D^s \\
\partial_x q_i(n_k)^{l+1} &= u_k, i \in \mathcal{I}_k, n_k \in \mathcal{J}_N^s \\
d_{ik} q_i^{l+1}(n_k) + \sigma \partial_x q_i^{l+1}(n_k) & \\
&= \sigma\left(\frac{2}{d_k} \sum_{j \in \mathcal{I}_k} \beta_j \partial_x q_j^l(n_k) - \beta_i \partial_x q_i^l(n_k) \right) \\
- \left(\frac{2}{d_k} \sum_{j \in \mathcal{I}_k} d_{ik} q_j^l(n_k) - d_{ik} q_i^l(n_k) \right) &= g_{ik}^{l+1}, \ k \in \mathcal{J}^M \setminus (\mathcal{J}_c \cup \mathcal{J}_v) \\
\sum_{i \in \mathcal{I}_j} \beta_i q_i^{l+1}(n_j) &= 0, \quad j \in \mathcal{J}_c \cup \mathcal{J}_v
\end{aligned} \tag{2.23}$$

$$s_j^v\left(-\beta_i\partial_x q_i(n_j)^{l+1} + \beta_k\partial_x q_k(n_j)^{l+1}\right) + (1-s_j^v)q_i(n_j)^{l+1} = -s_j^v\left(\bar{p}_i(n_j) - \bar{p}_k(n_j)\right), j \in \mathcal{J}_v, i \in \mathcal{I}_j^-, k \in \mathcal{I}_j^+$$

$$s_j^c\left(u_j - C\left(\left(\frac{-\beta_k\partial_x q_k(n_j)^{l+1} + \bar{p}_h(n_j)}{-\beta_i\partial_x q_i(n_j)^{l+1} + \bar{p}_i(n_j)}\right)^{\text{sign}(q_k(n_j)^{l+1})\kappa} - 1\right) \right)$$

$$+(1-s_j^c)\left(-\beta_i\partial_x q_i(n_j)^{l+1} + \beta_k\partial_x q_k(n_j)^{l+1}\right) = (1-s_j^c)(\bar{p}_i(n_j) - \bar{p}_k(n_j)) \quad j \in \mathcal{J}_c, i \in \mathcal{I}_j^-, k \in \mathcal{I}_j^+.$$

It is important to understand that Equation (2.23) is an elliptic problem on a subgraph where the original transmission conditions are kept at controlled serial nodes. The only coupling between edges in Equation (2.23) is at nodes in $\mathcal{J}_c \cup \mathcal{J}_v$! Thus, we introduce $e^{l+1} := q^{l+1} - q$. Then e^{l+1} solves a nonlinear differential equation with nonlinearity

$g_i(e_i^{l+1} + q_i) - g_i(q_i)$, zero right-hand side and homogeneous boundary conditions at the simple nodes. We assume

$$(g_i(x;s) - g_i(x;t))(s-t) \geq 0, \forall x \in (0, \ell_i), i \in \mathcal{I}. \tag{2.24}$$

Then we can show (see Equation [13]), under assumption of Equation (2.24), for each $\varepsilon \in [0,1)$ the iteration of Equation (2.22) with Equation (2.23) and Equations (2.20) and (2.21) converges as $l \to \infty$.

2.4 Domain Decomposition for Optimal Control Problems

There are many ways to deal with the original optimal control problem in the context of domain decomposition. We outline two strategies. The first one can be seen as a Jacobi-type approach. Here we take the iteration with respect to the domain decomposition outside of the control problem and minimize just the part of the cost that is responsible for the edges which are included in the subgraphs G_k. In the second approach, we fix the integer controls s and decompose the corresponding optimality system for the entire graph into the subgraphs G_k by a another, but very similar, non-overlapping domain decomposition, as in [13]. This is the essence of a virtual control paradigm given in [12].

2.4.1 Optimal Control on the Decomposed System: A Jacobi-Type Procedure

Here, we introduce the following mixed-integer control problem on a subgraph $G_r = (V_r, E_r)$ with corresponding index sets $\mathcal{I}^r, \mathcal{J}^r$. We introduce the index-sets of multiple and simple nodes within G_r as $\mathcal{J}^{r,M}, \mathcal{J}^{r,S}$ and all other sets accordingly. Then we look at the following problem on the subgraph G_r:

$$\alpha_i q_i^{l+1}(x) - \beta_i \partial_{xx} q_i^{l+1}(x) + g_i(q_i^{l+1}) = f_i(x), x \in (0, \ell_i), i \in \mathcal{I}^r$$

$$q_i(n_k)^{l+1} = 0, i \in \mathcal{I}_k^r, n_k \in \mathcal{J}_D^{r,S}$$

$$\partial_x q_i(n_k)^{l+1} = u_k, i \in \mathcal{I}_k^r, n_k \in \mathcal{J}_N^{r,S}$$

$$d_{ik} q_i^{l+1}(n_k) + \sigma \partial_x q_i^{l+1}(n_k)$$

$$= \sigma \left(\frac{2}{d_k} \sum_{j \in \mathcal{I}_k} \beta_j \partial_x q_j^l(n_k) - \beta_i \partial_x q_i^l(n_k) \right)$$

$$- \left(\frac{2}{d_k} \sum_{j \in \mathcal{I}_k} d_{ik} q_j^l(n_k) - d_{ik} q_i^l(n_k) \right) = g_{ik}^{l+1}, k \in \mathcal{J}^M \setminus (\mathcal{J}_c^r \cup \mathcal{J}_v)^r \tag{2.25}$$

$$\sum_{i \in \mathcal{I}_j} \beta_i q_i^{l+1}(n_j) = 0, \quad j \in \mathcal{J}_c^r \cup \mathcal{J}_v^r$$

$$s_j^v \left(-\beta_i \partial_x q_i(n_j)^{l+1} + \beta_k \partial_x q_k(n_j)^{l+1} \right) + (1 - s_j^v) q_i(n_j)^{l+1} = -s_j^v \left(\bar{p}_i(n_j) - \bar{p}_k(n_j) \right), j \in \mathcal{J}_v^r, i \in \mathcal{I}_j^{r,-}, k \in \mathcal{I}_j^{r,+}$$

$$s_j^c \left(u_j - C \left(\left(\frac{-\beta_k \partial_x q_k(n_j)^{l+1} + \bar{p}_h(n_j)}{-\beta_i \partial_x q_i(n_j)^{l+1} + \bar{p}_i(n_j)} \right)^{sign(q_k(n_j)^{l+1})\kappa} - 1 \right) \right)$$

$$+ (1 - s_j^c) \left(-\beta_i \partial_x q_i(n_j)^{l+1} + \beta_k \partial_x q_k(n_j)^{l+1} \right) = (1 - s_j^c)(\bar{p}_i(n_j) - \bar{p}_k(n_j)) \quad j \in \mathcal{J}_c^r, i \in \mathcal{I}_j^{r,-}, k \in \mathcal{I}_j^{r,+}.$$

and pose the mixed-integer optimal control problem on the subgraph G_r

$$\min_{(p^r,q^r,u^r)} I(p^r,q^r,u^r) := \sum_{i\in\mathcal{I}^r}\int_0^{\ell_i} I_i(p_i^r,q_i^r)dx + \frac{\nu}{2}\sum_{j\in\mathcal{J}^{r,S}}|u_j|^2$$

$$s.t. \tag{2.26}$$

(p^r,q^r,u^r) satisfies (25) at iteration level $l+1$.

We refer to [7, 11] and [14] for partial results. It is clear that this procedure has to be applied to all subgraphs at the same iteration level. In the next step, after updating the nodal conditions, the iteration level is raised by 1. The full picture including convergence is subject to current research.

2.4.2 Decomposition of the Optimality System for Continuous Controls with Fixed Switching Structure s

In order not to obscure the procedure and the presentation, we neglect the discrete control entirely. It is clear that the subgraph decomposition, where discrete controls are present, can be handled by the method outlined below. One needs to keep in mind only that the switching structure provided by the discrete controls s is fixed in this approach. Therefore, in contrast to the mixed-integer problem above, we can focus on optimality systems. We follow the article [13] of the author. We pose the following optimal control problem with Neumann boundary controls:

$$\min_{(q,u)} I(q,u) := \frac{\kappa}{2}\sum_{i\in\mathcal{I}}\|q_i-q_i^0\|^2 + \frac{\nu}{2}\sum_{k\in\mathcal{J}_N^S}|u_k|^2$$

subject to

$$\alpha_i q_i(x) - \beta_i\partial_{xx}q_i(x) + g_i(x;q_i(x)) = f_i(x),\ x\in(0,\ell_i), i\in\mathcal{I}$$
$$q_i(n_k) = 0, i\in\mathcal{I}_k, n_k\in\mathcal{J}_D^S$$
$$\partial_x q_i(n_k) = u_k, i\in\mathcal{I}_k, n_k\in\mathcal{J}_N^S$$
$$\beta_i\partial_x q_i(n_k) = \beta_j\partial_x q_j(n_k), i\neq j\in\mathcal{I}_k, k\in\mathcal{J}^M,$$
$$\sum_{i\in\mathcal{I}_k}d_{ik}q_i(n_k) = 0, k\in\mathcal{J}^M.$$

The corresponding optimality system then reads as follows:

$$\alpha_i q_i - \beta_i\partial_{xx}q_i + g_i(q_i) = f_i \text{ in}(0,\ell_i), i\in\mathcal{I}$$
$$\alpha_i\rho_i - \beta_i\partial_{xx}\rho_i + g_i'(q_i)\rho_i = -\kappa(q_i-q_i^0) \text{ in}(0,\ell_i), i\in\mathcal{I}$$
$$q_i(n_k) = 0, \rho_i(n_k) = 0, i\in\mathcal{I}_k, n_k\in\mathcal{J}_D^S$$
$$\partial_x q_i(n_k) = \frac{1}{\nu}\rho_i(n_k), \partial_x\rho_i(n_k) = 0, i\in\mathcal{I}_k, n_k\in\mathcal{J}_N^S \tag{2.27}$$
$$\beta_i\partial_x q_i(n_k) = \beta_j\partial_x q_j(n_k), \beta_i\partial_x\rho_i(n_k) = \beta_j\partial_x\rho_j(n_k), i\neq j\in\mathcal{I}_k, k\in\mathcal{J}^M$$
$$\sum_{i\in\mathcal{I}_k}d_{ik}q_i(n_k) = 0 = \sum_{i\in\mathcal{I}_k}d_{ik}\rho_i(n_k), k\in\mathcal{J}^M.$$

The idea now is to use a domain decomposition similar to the original system on the network. The method allows the interpretation of the decomposed optimality system in Equation (28) as an edge-wise optimality system of an optimal control problem formulated on an individual edge. Therefore, it is then possible to parallelize the optimization problems rather than the forward and backward solves. To this end, we introduce the following local system:

$$\alpha_i q_i^{l+1} - \beta_i \partial_{xx} q_i^{l+1} = f_i \text{ in } (0, \ell_i), i \in \mathcal{I}$$

$$\alpha_i \rho_i^{l+1} - \beta_i \partial_{xx} \rho_i^{l+1} = -\kappa(q_i^{l+1} - q_i^0) \text{ in } (0, \ell_i), i \in \mathcal{I}$$

$$q_i^{l+1}(n_k) = 0, \ \rho_i^{l+1}(n_k) = 0, i \in \mathcal{I}_k, n_k \in \mathcal{J}_D^S$$

$$\partial_x q_i^{l+1}(n_k) = \frac{1}{\nu} \rho_i^{l+1}(n_k), \ \partial_x \rho_i^{l+1}(n_k) = 0, i \in \mathcal{I}_k, n_k \in \mathcal{J}_N^S$$

$$d_{ik} q_i^{l+1}(n_k) + \lambda_k \partial_x q_i^{l+1}(n_k) + \mu_k \partial_x \rho_i^{l+1}(n_k)$$

$$= \lambda \left(\frac{2}{d_k} \sum_{j \in \mathcal{I}_k} \beta_j \partial_x q_k^l(n_k) - \beta_i \partial_x q_i^l(n_k) \right) + \mu_k \left(\frac{2}{d_k} \sum_{j \in \mathcal{I}_k} \beta_j \partial_x \rho_k^l(n_k) - \beta_i \partial_x \rho_i^l(n_k) \right)$$

$$- \left(\frac{2}{d_k} \sum_{j \in \mathcal{I}_k} d_{ik} q_j^l(n_k) - d_{ik} q_i^l(n_k) \right) = g_{ik}^{l+1},$$

$$d_{ik} \rho_i^{l+1}(n_k) + \lambda_k \partial_x \rho_i^{l+1}(n_k) - \mu_k \partial_x q_i^{l+1}(n_k)$$

$$= \lambda \left(\frac{2}{d_k} \sum_{j \in \mathcal{I}_k} \beta_j \partial_x \rho_k^l(n_k) - \beta_i \partial_x \rho_i^l(n_k) \right) - \mu_k \left(\frac{2}{d_k} \sum_{j \in \mathcal{I}_k} \beta_j \partial_x q_j^l(n_k) - \beta_i \partial_x q_i^l(n_k) \right)$$

$$- \left(\frac{2}{d_k} \sum_{j \in \mathcal{I}_k} d_{ik} \rho_k^l(n_k) - d_{ik} \rho_i^l(n_k) \right) = h_{ik}^{l+1}, k \in \mathcal{J}^M.$$

(2.28)

It is important to remark that the decomposition involves the original state variables as well as adjoint variables. The same arguments that led from Equations (2.15) and (2.14) to (2.16) and (2.17) apply to show that, upon convergence as $l \to \infty$, Equation (2.28) tends to Equation (2.27). Now, Equation (2.28) decomposes the fully connected problem in Equation (2.27) to a problem on a single edge $i \in \mathcal{I}$ with inhomogeneous Robin-type boundary conditions.

2.4.3 Virtual Controls

Let us now consider the following optimization problems on a single edge. The idea is to introduce a virtual control that aims at controlling classical inhomogeneous Neumann condition including the iteration history at the interface as inhomogeneity to the Robin-type condition that appears in the decomposition. To this end, it is sufficient to consider three cases: (a) the edge i connects a controlled Neumann $j \in \mathcal{J}_N^S$ node with a multiple node $k \in \mathcal{J}^M$ at which the domain decomposition is active, (b) the edge i connects a controlled Neumann node $j \in \mathcal{J}_N^S$ with multiple node $k \in \mathcal{J}^M$ at which the domain decomposition is active, and (c) the edge i connects two multiple nodes $j, k \in \mathcal{J}^M$.

Case (a):

$$\min_{u_j, v_{ik}} I(q_i, u_j, v_{ik}) := \frac{\kappa}{2} \| q_i - q_i^0 \|^2 + \frac{\nu}{2} u_j^2 + \frac{1}{2\mu_k} v_{ik}^2 + \frac{1}{2\mu_k} (\mu_k \partial_x q_i(n_k) + h_{ik}^l)^2$$

$$\text{subject to} \tag{2.29}$$

$$\alpha_i q_i - \beta_i \partial_{xx} q_i + g_i(q_i) = f_i, \, x \in (0, \ell_i)$$

$$d_{ik} q_i(n_j) = u_j, \, d_{ik} q_i(n_k) = -\lambda_k \beta_i \partial_x q_i(n_k) + g_{ik}^l + v_{ik}.$$

Case (b):

$$\min_{u_j, v_{ik}} I(q_i, u_i, v_i) := \frac{\kappa}{2} \| q_i - q_i^0 \|^2 + \frac{\nu}{2} u_i^2 + \frac{1}{2\mu_k} v_{ik}^2 + \frac{1}{2\mu_k} (\mu_k \beta_i \partial_x q_i(n_k) + h_{ik}^l)^2$$

$$\text{subject to} \tag{2.30}$$

$$\alpha_i q_i - \beta_i \partial_{xx} q_i + g_i(q_i) = f_i, \, x \in (0, \ell_i)$$

$$d_{ij} \beta_i \partial_x q_i(n_j) = u_j, \, d_{ik} q_i(n_k) = -\lambda_k \partial_x q_i(n_k) + g_{ik}^l + v_{ik}.$$

Case (c):

$$\min_{v_{ij}, v_{ik}} I(q_i, v_{ij}, v_{ik}) := \frac{\kappa}{2} \| q_i - q_i^0 \|^2 + \frac{1}{2\mu_j} v_{ij}^2 + \frac{1}{2\mu_k} v_{ik}^2$$

$$+ \frac{1}{2\mu_k} (\mu_k \beta_i \partial_x q_i(n_k) + h_{ik}^l)^2 + \frac{1}{2\mu_j} (\mu_j \beta_i \partial_x q_i(n_j) + h_{ij}^l)^2$$

$$\text{subject to} \tag{2.31}$$

$$\alpha_i q_i - \beta_i \partial_{xx} q_i + g_i(q_i) = f_i, \, x \in (0, \ell_i)$$

$$d_{ij} q_i(n_j) = -\lambda_j \beta_i \partial_x q_i(n_j) + g_{ij}^l + v_{ij}, \, d_{ik} q_i(n_k) = -\lambda_k \beta_i \partial_x q_i(n_k) + g_{ik}^l + v_{ik}.$$

Remark 2.1

- If we write down the optimality systems for Equations (2.29), (2.30), and (2.31), respectively, and combine the results, we arrive at Equation (2.28).
- This shows that within the loop of iterations that restore the transmission conditions at the multiple nodes, we can reformulate Equation (2.28) as the optimality system of an optimal control problem formulated on a single edge, with input data coming from the iteration history that involves all nodes adjacent at the ends of the given edge.
- By the same procedure, we can consider the decomposition into subgraphs G_k, where the decomposition takes place at multiple nodes with uncontrolled transmission conditions. The virtual control problems then involve the subgraph rather than a single edge.
- This means that we can actually decompose the optimization problem given on the graph into a sequence of local optimization problems given on an individual edge.
- The resulting optimization problem on the individual edges are strictly convex, thus, admitting a unique global solution.

We end the section with a result that is taken again from [13]. The convergence problem differs from [12] in that the nonlinear term g_i has to be taken care of. The stiffness $g_{i'}(q_i)$ influences the choice of the parameters λ, μ. In particular, if $\alpha = \dfrac{1}{\Delta t}$ is large, which means small time-steps, κ_i is small and μ is large compared to λ, then in the case $\mu, \lambda > 0$, we have convergence. We refer to [13] for the precise statement.

Under the assumptions above, for $\lambda > 0, \mu > 0$, the iterations converge and the solutions $q^l = (q_i^l)_{i \in \mathcal{I}}$ of the iterative process (28), describing the local optimality systems on the individual edges, converge to the solution of the optimality system in Equation (2.27). q_i^l, ρ_i^l converge to q_i, ρ_i in the energy sense.

As for the general development of the project, we refer to the website of the collaborative research cluster DFG-TRR 154 *Modeling, Simulation and Optimization Using the Example of Gas Networks* (https://trr154.fau.de/index.php/en/).

References

1. J. Smoller. *Shock Waves and Reaction-Diffusion Equations*, volume 258 of *Grundlehren der mathematischen Wissenschaften*. Springer Verlag, 1983.
2. Jens Brouwer, Ingenuin Gasser, and Michael Herty. Gas pipeline models revisited: Model hierarchies, nonisothermal models, and simulations of networks. *Multiscale Modeling & Simulation*, 9(2):601–623, 2011.
3. R. J. Le Veque. *Numerical Methods for Conservation Laws*. Birkhäuser, 1992.
4. R. J. Le Veque. *Finite Volume Methods for Hyperbolic Problems*. Cambridge University Press, 2002.
5. Falk M. Hante, Günter Leugering, Alexander Martin, Lars Schewe, and Martin Schmidt. Challenges in optimal control problems for gas and fluid flow in networks of pipes and canals: From modeling to industrial applications. In Pammy Manchanda, René Lozi, and Abul Hasan Siddiqi, editors, *Industrial Mathematics and Complex Systems: Emerging Mathematical Models, Methods and Algorithms*, pages 77–122. Springer, Singapore, 2017.
6. Martin Gugat, Günter Leugering, Alexander Martin, Martin Schmidt, Mathias Sirvent, and David Wintergerst. Towards simulation based mixed-integer optimization with differential equations. *Technical Report*, 2016.
7. Martin Gugat, Günter Leugering, Alexander Martin, Martin Schmidt, Sirvent , and David Wintergerst. Mip-based instantaneous control of mixed-integer pde-constrained gas transport problems. *Computational Optimization and Applications*, 2017.
8. M. Hinze and S. Volkwein. Analysis of instantaneous control for the burgers equation. *Nonlinear Anal.*, 50(1):1–26, 2002.
9. Ralf Hundhammer and Günter Leugering. Instantaneous control of vibrating string networks. In Martin Grötschel, Sven O. Krumke, and Jörg Rambau, editors, *Online Optimization of Large Scale Systems*, pages 229–249. Springer Berlin Heidelberg, 2001.
10. T. Roubíček. *Nonlinear Partial Differential Equations with Applications*, volume 153 of *International Series of Numerical Mathematics*. Birkhäuser/Springer Basel AG, Basel, second edition, 2013.
11. Martin Gugat, Günter Leugering, Alexander Martin, Martin Schmidt, Mathias Sirvent, and David Wintergerst. Towards simulation based mixed-integer optimization with differential equations. *Networks*, 2018.
12. J. E. Lagnese and G. Leugering. *Domain Decomposition Methods in Optimal Control of Partial Differential Equations*, volume 148 of *International Series of Numerical Mathematics*. BirkhÃuser Verlag, 2004.

13. G. Leugering. Domain decomposition of an optimal control problem for semi-linear elliptic equations on metric graphs with application to gas networks. *Applied Mathematics*, 8:1074–1099, 2017.

14. Günter Leugering, Alexander Martin, Martin Schmidt, and Mathias Sirvent. Nonoverlapping domain decomposition for optimal control problems governed by semilinear models for gas flow in networks. *Control and Cybernetics*, 46(3):191–225, 2017.

3

A Model to Assess the Role of Spatial Urban Configurations on Crowd Evacuation Dynamics During Terrorist Attacks

D. Provitolo, R. Lozi, and E. Tric

CONTENTS

3.1 Introduction: Modern Terrorist Attacks 41
3.2 Weighted PCR System: A Model for Analyzing the Dynamics of Human
    Behavior During a Disaster 44
    3.2.1 Network of Squares and Streets Modelling Spatial Urban Configurations 44
    3.2.2 Equations of the WPCR System on Each Node 45
    3.2.3 Transitional Dynamics 47
3.3 Coupled WPCR System on a Network...48
    3.3.1 Flows and Bottleneck Coupling...48
    3.3.2 Fixed Points of the Three-Node System 51
3.4 Influence of the Spatial Configuration on the Pace of Evacuation 56
    3.4.1 Scaling the Parameters...56
    3.4.2 Influence of the Intermediate Place Capacity on the Evacuation Dynamics.......56
3.5 Conclusion 58
Acknowledgments..58
References..59

3.1 Introduction: Modern Terrorist Attacks

Terrorism has always existed in history and has taken different forms: anarchist attacks, political attacks, or attacks by colonial states, such as the terrible Jallianwala Bagh massacre, which took place on April 13, 1919 in Amritsar, Punjab, when a crowd of non-violent protesters were fired upon by troops of the British Indian Army.

In this article we focus specifically on the behavior of populations facing new forms of terrorism, such as the attacks perpetrated by sects, groups or "lone wolves" that are becoming more and more prevalent. Three types of acts can be distinguished:

- Shooting and bombing attacks, for example in bus and subway stations (Madrid, Spain, 2004; Paris, France, 1995, 2015; Nigeria, 2014; Belgium, 2016; Kashmir, 2019), at airports (Brussels airport, 2016; Atatürk airport, 2016), in urban areas with simultaneous terrorist acts (Mumbai, India, 2008, with 12 coordinated shooting and bombing attacks lasting four days).

- Attacks using CBRN (chemical, biological, radiological, and nuclear weapons) such as the sarin gas attack on the Tokyo subway committed by the Aum sect (March 20, 1995) (Dauphiné and Provitolo 2013).

- *Ad hoc* weapons are now being employed, such as suicide planes flown into the World Trade Center (2001), lorries or cars that are simply driven directly through a crowd massed for a cultural event or strolling peacefully. This is in response to tightened police controls aiming to reduce bombing attacks. These attacks are most often led by "lone wolves," who are difficult to detect beforehand and equally difficult to locate because they are "nested" among civilians. On July 14, 2016, 86 people who attended the Bastille Day fireworks on the "Promenade des Anglais" in Nice, France, were killed; 458 more were wounded.

These attacks, perpetrated in crowds, in frequented places, or on public transport at rush hour, give rise to diversified human reactions. One can observe behaviors of sideration (paralysing panic), of flight panic, of agitation in all directions, of reasoned flight, of sheltering, or mutually offered assistance (Crocq 2013), etc.

Except for sideration, sheltering and escape behaviors, whether taken under the influence of panic (flight panic) or of reason (reasoned escape), produce trajectories of displacement. When an attack happens, the topography of the area is also very important. The dynamics of human reactions and the associated displacements are indeed guided by the space and the alternatives that it offers, as displayed in Figure 3.1, especially in terms of evacuation, flight, and accessibility to refuge areas, such as safe places and squares.

There are few data available to identify the dynamics of the displacements associated with these behaviors. Of course, it is difficult to artificially reproduce a disaster, which would otherwise allow us to observe the diversity of human reactions that could occur, to follow the spatio-temporal dynamics, and to analyze the impact of spatial configurations on these dynamics. To overcome these limits, it is possible to develop mathematical models from which hypotheses are tested by means of a human behavior computer simulation model.

The Com2SiCa research team* proposed the Panic–Control–Reflex (PCR) model (Provitolo et al. 2015), as shown in Figure 3.2, which is a model that simulates the possible human behavior that can occur during sudden onset and unpredictable disasters, such as

FIGURE 3.1
Street and places in Mediterranean city of Italy. (© D. Provitolo, March, 2019. With permission.)

---

* https://geoazur.oca.eu/en/research-geoazur/2158-com2sica-how-to-comprehend-and-simulate-human-behaviors-in-areas-facing-natural-disasters

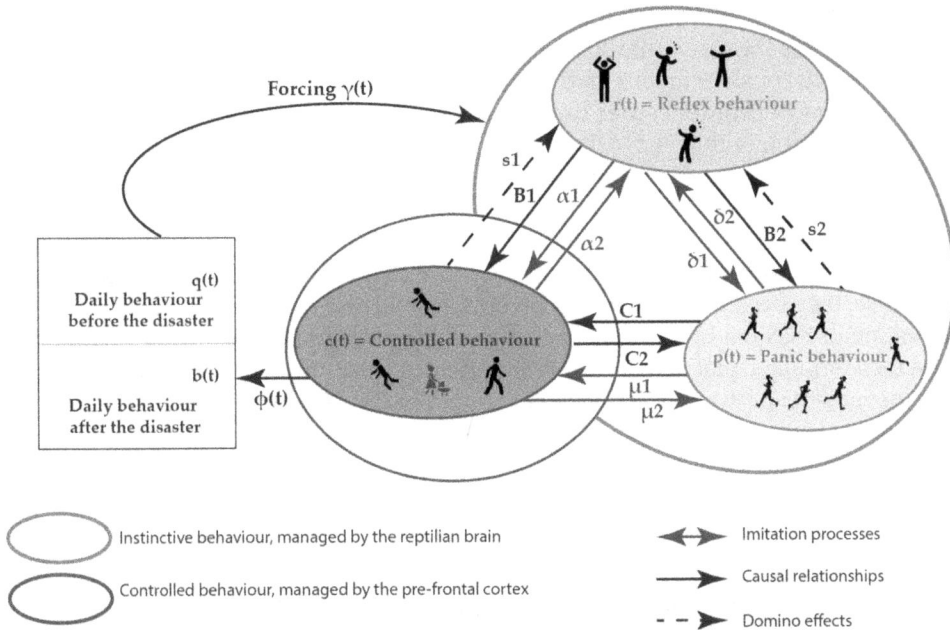

FIGURE 3.2
Graphic representation of the Panic-Control-Reflex behavior model (PCR) in the exceptional situation of disaster. (D. Provitolo et al. 2015. With permission.)

terrorist attacks. This model is formed by a system of ordinary differential equations to describe behavioral dynamics over time (Verdière et al. 2014, 2015, Cantin et al. 2016).

Neuroscience research explains that during a disaster, humans are rarely stuck in one type of behavior. The population switches between different behavioral states, some of which are the result of instinctive, others of reasoned, reaction.

As the brain switches from one behavioral state to another, in the context of terrorist attacks and therefore in a situation of sudden and unforeseen threat, the entire impacted population first adopts a behavioral reaction, Reflex $r(t)$, under the influence of surprise and the suddenness of the event, before transiting to the Panic reflex behavior $p(t)$, or Controlled behavior $c(t)$ (Provitolo et al. 2015). Reflex $r(t)$ and Panic $p(t)$ behavior are managed by the reptilian brain: they are instinctive, automatic reactions, allowing one to react extremely quickly to the threat, either by being stunned and paralyzed $r(t)$, or by fleeing as quickly as possible due to the panic fear $p(t)$. Controlled behavior $c(t)$ is managed by the pre-frontal cortex, which concerns reasoned and self-control reactions. These can take different forms during a catastrophe, for example, in the form of evacuation, escape, containment, sheltering, search for help, mutual aid, or, on the contrary, looting, etc. Despite their diversity, the PCR model aggregates all of these controlled behaviors.

In this article we propose an extension of the PCR model in order to take into account the influence of spatial configuration in the mathematical modeling of the dynamics of human reactions in the face of traumatic situations, such as terrorist attacks. We call this extended model the Coupled Weighted PCR (CWPCR).

In Section 3.2, we will present the Weighted Panic Control Reflex model in its graphic and mathematical form, as published by the Com2SiCa research team. This model is improved from the PCR model in order to take into account the role of spatial configurations on behavioral dynamics. In Section 3.3 we introduce the CWPCR model, which incorporates the pressure and counterpressure of the crowd in each place via a bottleneck effect, which takes into account the narrowness and the length of the streets and the size of places. We will consider an oriented network with three nodes representing three places or public squares of different sizes, linked by narrow streets. In this network, we compute the equilibrium points that show the motion of the crowd after a terrorist attack. Section 3.4 is therefore devoted to the analysis of the impact of the size of the intermediate place of this network on the evacuation of the population in the face of a terrorist attack, by means of numerical simulations. The numerical results highlight that, depending on their respective size, intermediate places modulate the dynamics and the speed of flow of the crowd. In this sense, they become strategic places, both for the planners who must think about the organization of the area to host public events and festivals, and also for the terrorists who can use these strategic places to multiply the effect of their harmful actions by trapping the escape movements between two areas of action. Finally, in Section 3.5, a brief conclusion will be drawn.

3.2 Weighted PCR System: A Model for Analyzing the Dynamics of Human Behavior During a Disaster

3.2.1 Network of Squares and Streets Modelling Spatial Urban Configurations

Just before a dramatic event like a terror attack in a city, the crowd is generally spread across several places, public squares and streets. In the aftermath of the initial shock, people rush through the streets to reach what they think will be more secure places, as one can see in Figure 3.3.

In this chapter, we define a city by a mathematical graph, where the places and public squares are called vertices or nodes and are denoted $N_1, N_2, \ldots, N_p$, and the streets, escalators, doors, and stairs are the oriented edges $(N_i \rightarrow N_j)$ linking these nodes. They are oriented, because the flow from one place toward another is not symmetric during terrorist attacks. Our aim is to model the motion of the crowd through such edges. To achieve this, we must introduce some "geographical" particularities of the city, like the size of places, the narrowness of streets, and the number of people initially present in each place. This is why we need to upgrade the standard PCR model, into the Weighted PCR (WPCR) model, by introducing new data, with Weight standing for the relative sizes of crowd, places, and streets.

First, on each node N_k, $k = 1, p$ we call $r_k(t), c_k(t), p_k(t), q_k(t), b_k(t)$ the *number* of people being in reflex, control, panic, daily usual behavior, and return to normal life and $V_k(t)$ the total number of people present at this node

$$V_k(t) = r_k(t) + c_k(t) + p_k(t) + q_k(t) + b_k(t) \tag{3.1}$$

Second, in order to more precisely model the characteristics of the city, we introduce W_k, the maximum capacity of the number of people who can be present in each node N_k (i.e. due to the size of the corresponding place). Each maximum capacity is a constant. Each time one must have

$$V_k(t) \leq W_k \tag{3.2}$$

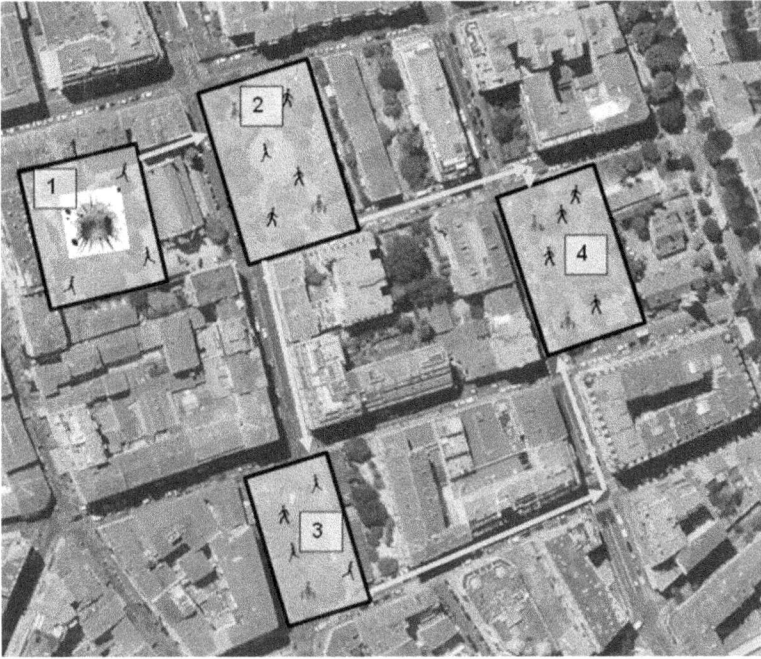

FIGURE 3.3
Possible paths of rushing people in a city network. (IGN—BD Orthophoto, 2017, 50 cm resolution.)

3.2.2 Equations of the WPCR System on Each Node

The WPCR simulates human behavior, in one specific place, during a catastrophic event. As explained in the introduction, it is based on neuroscience studies. During the event, the switches from one behavioral state to another are caused by transitional dynamics, as schematized in Figure 3.2 due to:

i. Causal relationships (B_1, B_2, C_1, C_2). Once the population is in the reflex behavior state, a part of it can evolve toward controlled behaviors at the rate B_1, while another part transitions towards panic behaviors at the rate B_2. Likewise, a part of the panicked population may switch to controlled behavior at the rate C_1. According to the evolution of the situation, individuals who have adopted a controlled behavior may switch back to panic behavior at the rate C_2. This process can be iterated many times.

ii. Processes of imitation and contagion, which are well known in crowd psychology and have been termed "emotional contagion" (Hatfield et al. 1994). The imitation processes are modelled identically to epidemiological propagation (Provitolo 2005). The imitation is valid in both directions and is modelled by the function $F(r,c)$ (see Equation (3.5)) for emotional contagion between reflex and controlled behavior (using the damping coefficients α_1 and α_2), by the function $G(r,p)$ (see Equation (3.6)) for emotional contagion between reflex and panic behavior (δ_1 and δ_2), and by the function $H(c,p)$ (see Equation (3.7)) for emotional contagion between controlled and panicked behavior (μ_1 and μ_2).

iii. Domino effects, which illustrate a succession of events (s_1 and s_2) correspond, for example, to a new attack in an urban area or to a closed door during an evacuation. In the PCR model, the parameters s_1 and s_2 are either constant or built in a periodic form.

The WPCR model is defined on each node k by

$$\dot{X} = \Phi(t, X) \tag{3.3}$$

with $\dot{X} = \dfrac{dX}{dt}$, $X = (r, c, p, q, b)^T \in \mathbb{R}^5$ and Φ given by $\Phi(t, X) = \left(\Phi_i(t, X)\right)^T$, $i = 1, \ldots, 5$, where the functions Φ_i are defined by

$$
\begin{cases}
\Phi_1(t, X_k) = \dot{r}_k(t) = \gamma(t)q_k(t)\left(W_k - r_k(t)\right) - (B_1 + B_2)r_k(t) + s_1(t)c_k(t) + s_2(t)p_k(t) \\
\qquad\qquad + F(r_k(t), c_k(t))r_k(t)c_k(t) + G(r_k(t), p_k(t))r_k(t)p_k(t) \\
\Phi_2(t, X_k) = \dot{c}_k(t) = -\varphi(t)c_k(t)(W_k - b_k(t)) + B_1 r_k(t) + C_1 p_k(t) - C_2 c_k(t) - s_1(t)c_k(t) \\
\qquad\qquad - F(r_k(t), c_k(t))r_k(t)c_k(t) + H(c_k(t), p_k(t))c_k(t)p_k(t) \\
\Phi_3(t, X_k) = \dot{p}_k(t) = B_2 r_k(t) - C_1 p_k(t) + C_2 c_k(t) - s_2(t)p_k(t) - G(r_k(t), p_k(t))r_k(t)p_k(t) \\
\qquad\qquad - H(c_k(t), p_k(t))c_k(t)p_k(t) \\
\Phi_4(t, X_k) = \dot{q}_k(t) = -\gamma(t)q_k(t)\left(W_k - r_k(t)\right) \\
\Phi_5(t, X_k) = \dot{b}_k(t) = \varphi(t)c_k(t)(W_k - b_k(t))
\end{cases} \tag{3.4}
$$

The parameters involved in Equation (3.4) are real positive coefficients:

$$B_i \geq 0, C_i \geq 0, i = 1, 2; \ \alpha_i \geq 0, \delta_i \geq 0, \mu_i \geq 0, i = 1, 2; \ s_i \geq 0, i = 1, 2.$$

The imitation functions F, G, and H are real valued functions defined on $\mathbb{R} \times \mathbb{R}$ by

$$F(r(t), c(t)) = -\alpha_1 f_1\left(\frac{r(t)}{c(t) + \varepsilon}\right) + \alpha_2 f_2\left(\frac{c(t)}{r(t) + \varepsilon}\right) \tag{3.5}$$

$$G(r(t), p(t)) = -\delta_1 g_1\left(\frac{r(t)}{p(t) + \varepsilon}\right) + \delta_2 g_2\left(\frac{p(t)}{r(t) + \varepsilon}\right) \tag{3.6}$$

$$H(c(t), p(t)) = \mu_1 h_1\left(\frac{c(t)}{p(t) + \varepsilon}\right) - \mu_2 h_2\left(\frac{p(t)}{c(t) + \varepsilon}\right), \tag{3.7}$$

where ε is a positive number and f_i, g_i, h_i for $i = 1$, 2 are real valued functions defined on $\mathbb{R}$. They have a decreasing shape indicating that the behavior imitation is symmetric. Moreover they are normalized,

$$0 \leq f_i(u) \leq 1, 0 \leq g_i(u) \leq 1, 0 \leq h_i(u) \leq 1, \quad \forall u \in \mathbb{R}, i = 1, 2 \tag{3.8}$$

Because this model does not take the mortality rate into account, the population remains constant. Therefore Equation (3.4) is considered when time is proceeding from an initial time $t_0 \geq 0$, with initial condition

$$r_k(t_0) + c_k(t_0) + p_k(t_0) + q_k(t_0) + b_k(t_0) = V_k(t_0)$$

$$= r_{k,0} + c_{k,0} + p_{k,0} + q_{k,0} + b_{k,0} = V_{k,0} \leq W_k \tag{3.9}$$

that satisfies the following properties

$$r(t_0) > 0, \; c(t_0) > 0, \; p(t_0) > 0, \; q(t_0) > 0, \; b(t_0) > 0 \tag{3.10}$$

We suppose that the characteristic parameters B_1, B_2, C_1, C_2, of each node have the same value, because they depend on cultural and psychological factors specific to each individual rather than to spatial configurations and crowd context. It is why we use function Φ_i instead of function $\Phi_{k,i}$.

3.2.3 Transitional Dynamics

Both forcing functions, γ and φ respectively, model the beginning of the disaster and the return to a quiescent daily behavior. Their shape can be adapted to various scenarios. When t is sufficiently large, they satisfy $\gamma(t) = \varphi(t) = 1$. In catastrophic situations it is considered that γ is a stiff function, ranging from 0 to 1 in a very brief interval of time (Verdière et al. 2014, 2015, Cantin et al. 2016) because if we consider a bomb attack, the entire crowd that is near the explosion passes from daily to reflex behavior in an instant, and it takes a very long time for people to return to their normal state.

Therefore, one can suppose that a terror attack is shaped by two characteristic times: t_s (for start) and t_e (for end) with $t_0 < t_s < t_e$ for which

$$\begin{cases} \gamma(t) = 1, & \forall t \geq t_s \\ \varphi(t) = 0, & \forall t < t_e \end{cases} \tag{3.11}$$

As an example for $I_{trans} = [3.2, 43.2]$, these functions can be defined by

$$\varphi(t) = \begin{cases} 0 & \text{if } 0 \leq x < 43.2 \\ \cos^2\left(2\pi \dfrac{x - 3.2}{160}\right), & \text{if } 43.2 \leq x \leq 83.2, \\ 1, & \text{if } x > 83.2 \end{cases} \tag{3.12}$$

$$\gamma(t) = \begin{cases} \cos^2\left(2\pi \dfrac{x - 3.2}{12.8}\right), & \text{if } 0 \leq x \leq 3.2, \\ 1, & \text{if } x > 3.2 \end{cases} \tag{3.13}$$

There are displayed in Figure 3.4.

Following Cantin et al. (2016) we keep the term *transitional dynamics* for the dynamics of the PCR model (likewise for both improved WPCR and the CWPCR model presented in Section 3.3) in the interval of time $I_{trans} = [t_s, t_e]$ (in terror attacks this interval of time can last from several minutes up to hours, as observed during the 2016 terrorist attack in Nice). Therefore in $\forall t \in I_{trans}$ functions φ and γ verify

$$\begin{cases} \gamma(t) = 1 \\ \varphi(t) = 0 \end{cases} \tag{3.14}$$

Hence, during the transitional dynamics, the population with daily behavior collapses and there is not yet a population that is back to daily behavior (i.e. $q(t) = b(t) = 0$).

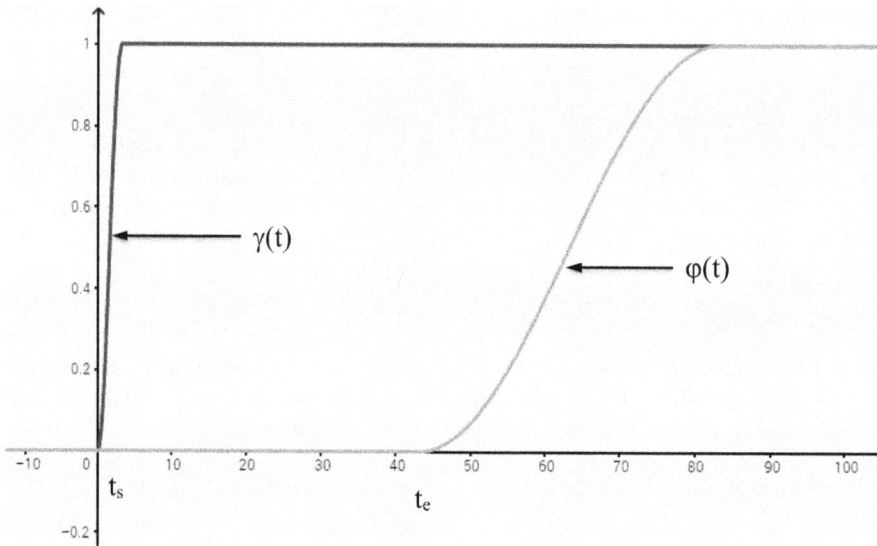

FIGURE 3.4

Forcing functions $\gamma(t)$, and $\varphi(t)$, $I_{trans} = [3.2, 43.2]$.

System (3.4) is reduced to

$$
\begin{cases}
\dot{r}(t) = -(B_1 + B_2)r(t) + s_1(t)c(t) + s_2(t)p(t) + \\
\qquad F(r(t), c(t))r(t)c(t) + G(r(t), p(t))r(t)p(t) \\
\dot{c}(t) = B_1 r(t) + C_1 p(t) - C_2 c(t) - s_1(t)c(t) - \\
\qquad F(r(t), c(t))r(t)c(t) + H(c(t), p(t))c(t)p(t) \\
\dot{p}(t) = B_2 r(t) - C_1 p(t) + C_2 c(t) - s_2(t)p(t) - \\
\qquad G(r(t), p(t))r(t)p(t) - H(c(t), p(t))c(t)p(t).
\end{cases}
\tag{3.15}
$$

3.3 Coupled WPCR System on a Network

3.3.1 Flows and Bottleneck Coupling

As previously defined, nodes are linked by edges. We now aim to model the motion of the crowd through such edges (i.e. streets, stairs, escalators, doors, etc.).

We also seek to identify the obstacles that slow the escape of the crowd in the aftermath of the initial shock by analysing the topology of the network of streets and places in a city. This information can potentially be used to improve the design of a city to facilitate the escape of a crowd towards more secure places.

For the sake of simplicity, as we show in Figure 3.5, we consider first a simplified oriented network with only three nodes $(N_1; W_1); (N_2; W_2); (N_3; W_3)$ and two edges $(N_1 \rightarrow N_2); (N_2 \rightarrow N_3)$. Such a simplified network can be straightforwardly complexified by adding as many nodes and edges as necessary, without any difficulty. However, it is better to first focus our attention on the nature of the obstacles in this simplified network.

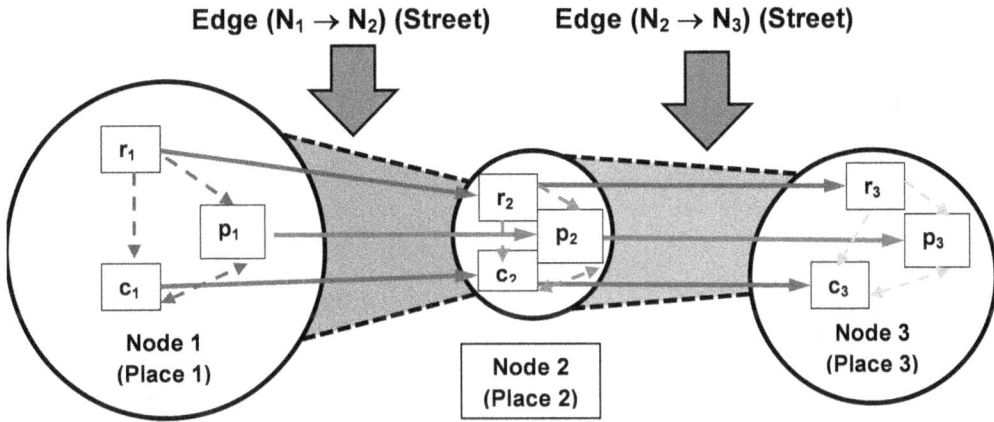

Edge (N₁ → N₂) (Street)   Edge (N₂ → N₃) (Street)

FIGURE 3.5
Three-node network: people in every state are rushing from node 1 towards node 2, and from node 2 towards node 3, keeping the same behavioral class.

We suppose that during the short interval of time when people are traveling inside one edge, they remain in the same behavioral class as displayed in Figure 3.5. In this figure, the "geographical" edges $(N_1 \rightarrow N_2); (N_2 \rightarrow N_3)$ (i.e. the streets linking node 1 to node 2, and node 2 to node 3) are split in three "behavioral edges" meaning that on the same street people in reflex, panic, or controlled behavior are escaping from node 1 to node 2 (and from node 2 to node 3), therefore people in each particular behavior in node 1 are meeting people in the same behavior in node 2 (and in the same manner from node 2 to node 3).

To continue focusing on the special coupling that we are introducing here, we suppose that imitation mechanisms are not activated (i.e. $F \equiv H \equiv G \equiv 0$, which is equivalent to $\alpha_i = \delta_i = \mu_i = 0$ for $i = 1, 2$), and, furthermore, there is no domino effect ($s_i = 0$, $i = 1, 2$) and we also suppose that we are in the interval $I_{trans} = [t_s, t_e]$ where only transitional dynamics are considered. Of course, it is easy to relax such limitations, which are not dependent upon the coupling, by not eliminating the corresponding terms in the equations.

In each node N_k, $k = 1, 2, 3$ such transitional dynamics are the solution of the system

$$\begin{cases} \dot{r}_k(t) = -(B_1 + B_2)r_k(t) \\ \dot{c}_k(t) = B_1 r_k(t) + C_1 p_k(t) - C_2 c_k(t) \\ \dot{p}_k(t) = B_2 r_k(t) - C_1 p_k(t) + C_2 c_k(t) \end{cases} \qquad (3.16)$$

which is the reduction of System (3.4) in the transition interval (and a particular case of Equation (3.15)).

3.1 Note: in the WPCR model, the terms $\dot{r}_k(t)$, $\dot{c}_k(t)$, $\dot{p}_k(t)$, $\dot{q}_k(t)$, $\dot{b}_k(t)$ can be considered as flows, because a flow is a quantity of something divided by a unit of time. There are two kinds of flow. In Equations (3.4), (3.15), and (3.16) flows are "behavioral," as they represent the quantities of people changing their behavior per unit of time. Now we consider also "motion" flows, which are the quantities of people in each behavior class, moving from one node to another node, per unit of time. Of course, both kinds of flow are combined to produce a global equivalent in the following equations.

Many studies on pedestrian flows have been published (Daamen et al. 2005, Dias et al. 2013, Kretz et al. 2006, Liao et al. 2014, Seyfried et al. 2005, Zhang et al. 2013, Zhou et al. 2014). We consider, in particular, the survey of Daamen et al. (2005), in which the graphs of six different experiments showing the relationship between the flow of pedestrians and their densities are displayed. All these graphs show clearly a nonlinear relationship of a logistic type between density and flow.

Moreover, Daamen and co-authors developed a first-order traffic flow theory to describe two-dimensional pedestrian flow operations in the case of an oversaturated bottleneck in front of which a large, high-density region has formed. Such a mathematical model also highlights the logistic relationship for any bottleneck width as one can see in Figure 3.6.

We now introduce our hypothesis for the coupled WPCR (CWPCR), based on this type of logistic relationship. Again, for the sake of simplicity, we consider only unidirectional motion of the crowd (i.e. motion on an oriented graph), as the one displayed in Figure 3.5, because it is supposed that a terrorist attack occurs in node 1 and that people try to escape from this node towards nodes 2 and 3.

When the crowd is moving from one node to another, its speed and the corresponding motion flow depends on three factors. The first factor reflects the narrowness and the length of the street. More people can go from one place to the next if the street is large, rather than in the case of a narrow street. This topological characteristic will be modelled by "roughness" coefficients $\eta_{1,2}$, $\eta_{2,3}$. In fact, in the considered coupling, we suppose that people cannot change their behavior when they move from one node to another (e.g. controlled people remain controlled, panicked people remain panicked, and so on), and thus we use three such kinds of roughness coefficients $\eta_{c,1,2}$, $\eta_{p,1,2}$, $\eta_{r,1,2}$, $\eta_{c,2,3}$, $\eta_{p,2,3}$, $\eta_{r,2,3}$ to refer to the population in reflex, panic, or controlled situations, respectively. Of course, they can have the same value. The second factor is proportional to the number of people present in the node from which the crowd is escaping. This is equivalent to the pressure of the crowd in these places. The third factor reflects the counterpressure due to the maximal capacity of the place toward which the people are fleeing, conjugated with the number of persons already present there. The combination of pressure and counterpressure gives a bottleneck effect.

We propose to model this bottleneck effect by a nonlinearity of a logistic type to keep the same philosophy as Daamen et al. (2005).

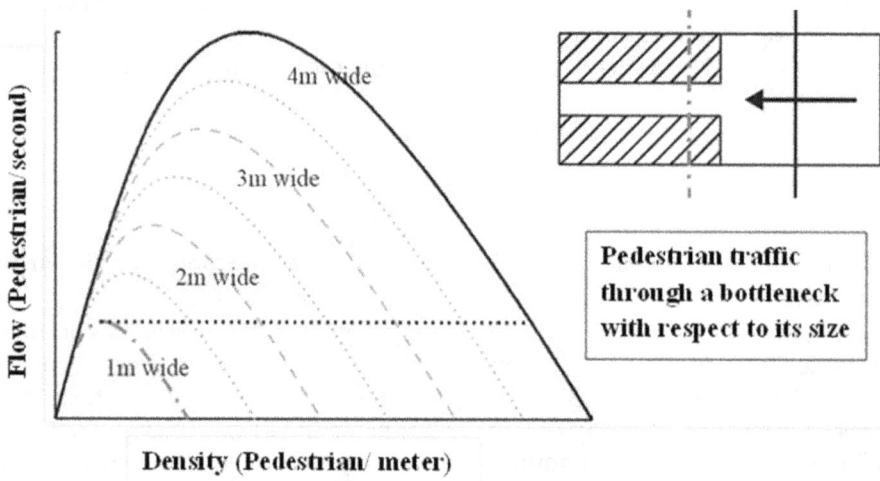

FIGURE 3.6
Relationship between flow and density of pedestrians going through an oversaturated bottleneck. (Adapted from Daamen et al. 2005.)

Thus, the bottleneck coupling corresponding to the one of Figure 3.5, is given by the system

$$
\begin{cases}
\dot{r}_1(t) = -(B_1 + B_2)r_1(t) - \eta_{r,1,2}\, r_1(t)(W_2 - c_2(t) - p_2(t) - r_2(t)) \\
\dot{c}_1(t) = B_1 r_1(t) + C_1 p_1(t) - C_2 c_1(t) - \eta_{c,1,2}\, c_1(t)(W_2 - c_2(t) - p_2(t) - r_2(t)) \\
\dot{p}_1(t) = B_2 r_1(t) - C_1 p_1(t) + C_2 c_1(t) - \eta_{p,1,2}\, p_1(t)(W_2 - c_2(t) - p_2(t) - r_2(t)) \\
\dot{r}_2(t) = -(B_1 + B_2)r_2(t) + \eta_{r,1,2}\, r_1(t)(W_2 - c_2(t) - p_2(t) - r_2(t)) \\
\qquad\quad - \eta_{r,2,3}\, r_2(t)(W_3 - c_3(t) - p_3(t) - r_3(t)) \\
\dot{c}_2(t) = B_1 r_2(t) + C_1 p_2(t) - C_2 c_2(t) + \eta_{c,1,2}\, c_1(t)(W_2 - c_2(t) - p_2(t) - r_2(t)) \\
\qquad\quad - \eta_{c,2,3}\, c_2(t)(W_3 - c_3(t) - p_3(t) - r_3(t)) \\
\dot{p}_2(t) = B_2 r_2(t) - C_1 p_2(t) + C_2 c_2(t) + \eta_{p,1,2}\, p_1(t)(W_2(t) - c_2(t) - p_2(t) - r_2(t)) \\
\qquad\quad - \eta_{p,2,3}\, p_2(W_3 - c_3(t) - p_3(t) - r_3(t)) \\
\dot{r}_3(t) = -(B_1 + B_2)r_3(t) + \eta_{r,2,3}\, r_2(t)(W_3 - c_3(t) - p_3(t) - r_3(t)) \\
\dot{c}_3(t) = B_1 r_3(t) + C_1 p_3(t) - C_2 c_3(t) + \eta_{c,2,3}\, c_2(t)(W_3 - c_3(t) - p_3(t) - r_3(t)) \\
\dot{p}_3(t) = B_2 r_3(t) - C_1 p_3(t) + C_2 c_3(t) + \eta_{p,2,3}\, p_2(t)(W_3 - c_3(t) - p_3(t) - r_3(t))
\end{cases}
\tag{3.17}
$$

with initial conditions satisfying

$$
r_{1,0} + c_{1,0} + p_{1,0} = V_{1,0} \leq W_1 \,;\, r_{2,0} + c_{2,0} + p_{2,0} = V_{2,0} \leq W_2 \,;\, r_{3,0} + c_{3,0} + p_{3,0} = V_{3,0} \leq W_3
\tag{3.18}
$$

In this system the bottleneck coupling, concerning, for example, the controlled population that is moving from node 1 to node 2 is given by the term:

$$
\eta_{c,1,2}\, c_1(t)(W_2 - c_2(t) - p_2(t) - r_2(t))
\tag{3.19}
$$

in the second equation of (3.17)

$$
\dot{c}_1(t) = B_1 r_1(t) + C_1 p_1(t) - C_2 c_1(t) - \eta_{c,1,2}\, c_1(t)(W_2 - c_2(t) - p_2(t) - r_2(t))
$$

In this bottleneck coupling, Equation (3.19), $\eta_{c,1,2}$ is the parameter that models the topological characteristic of the street linking node 1 to node 2. The second factor $c_1(t)$ of Equation (3.19) reflects the pressure of controlled people in node 1 willing to escape towards node 2 and also the proportionality of people escaping with respect to people staying in node 1. Finally, the factor $(W_2 - c_2(t) - p_2(t) - r_2(t))$ shows the counterpressure which is maximum (i.e. the term vanishes) when $W_2 = c_2(t) + p_2(t) + r_2(t)$ because, in this case, there is no more room for people coming from node 1. This bottleneck coupling is nonlinear, as shown on Figure 3.7.

As we consider only transitional dynamics where $q_k(t) = b_k(t) = 0$, Equation (3.19) can be written

$$
\eta_{c,1,2}\, c_1(t)(W_2 - V_2(t))
\tag{3.20}
$$

3.3.2 Fixed Points of the Three-Node System

The fixed-point research allows us to identify the point of equilibrium towards which the system tends during the transitional period. This equilibrium point highlights the primordial role of the size of both nodes 2 and 3 in the context of evacuation dynamics. It is important to note that the mathematically calculated equilibrium point does not

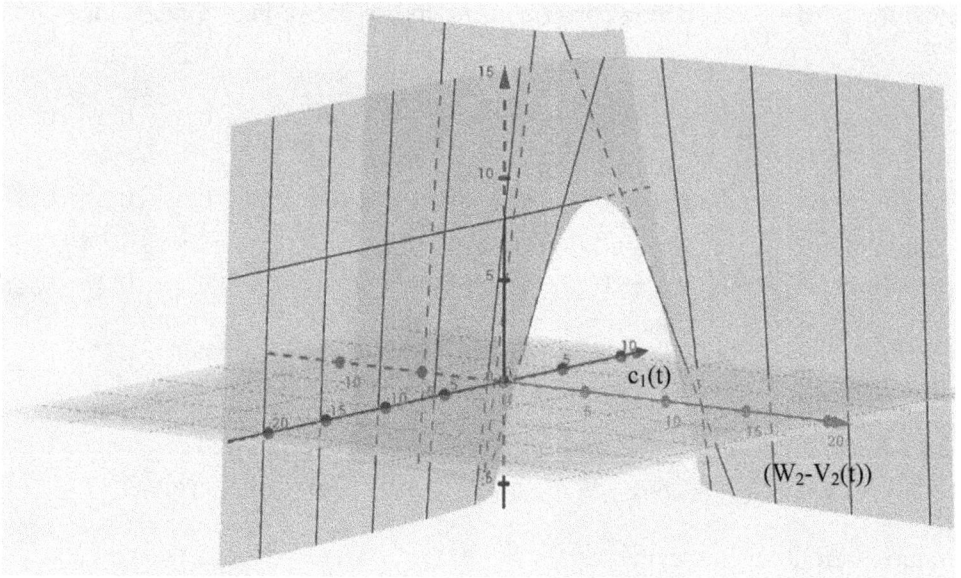

FIGURE 3.7
Graph of the bottleneck coupling function.

necessarily correspond to the equilibrium situation sought by crisis management personnel. The population stranded in the initial place (node 1) remains very vulnerable to the terrorist threat.

To stay closer to reality, we further assume that people in a reflex situation are stunned and paralyzed. They cannot rush from one node to another as indicated in Figure 3.8. Therefore, the transitions between r_1, r_2, r_3 are forbidden. People in this stunned state can only change their behavior (from r_k to p_k or c_k).

To achieve this goal Equation (3.17) is simply modified to vanish parameters $\eta_{r,i,j}$, $i = 1, 2$, $j = i+1$.

The fixed point $(r_1^*, c_1^*, p_1^*, r_2^*, c_2^*, p_2^*, r_3^*, c_3^*, p_3^*)$ of the system toward which the solution of Equation (3.17) converges, is straightforwardly computed.

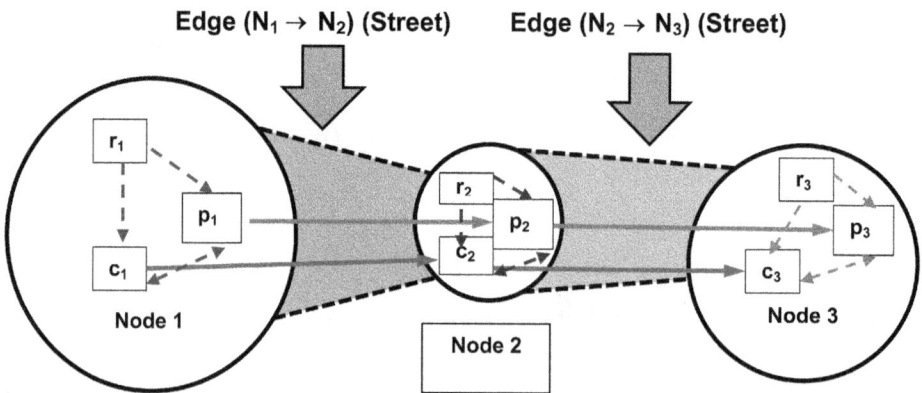

FIGURE 3.8
Three-node network: people in the reflex behavior state are stranded in their original node.

$$
\begin{cases}
r_1^* = 0 \\
c_1^* = 0 \\
p_1^* = 0 \\
r_2^* = 0 \\
c_2^* = 0 \\
p_2^* = 0 \\
r_3^* = 0 \\
c_3^* = \dfrac{C_1(V_{1,0} + V_{2,0} + V_{3,0})}{C_1 + C_2} \\
p_3^* = \dfrac{C_2(V_{1,0} + V_{2,0} + V_{3,0})}{C_1 + C_2}
\end{cases}
\qquad \text{if} \quad V_{1,0} + V_{2,0} + V_{3,0} \le W_3, \tag{3.21}
$$

$$
\begin{cases}
r_1^* = 0 \\
c_1^* = 0 \\
p_1^* = 0 \\
r_2^* = 0 \\
c_2^* = \dfrac{C_1(V_{1,0} + V_{2,0} + V_{3,0} - W_3)}{C_1 + C_2} \\
p_2^* = \dfrac{C_2(V_{1,0} + V_{2,0} + V_{3,0} - W_3)}{C_1 + C_2} \\
r_3^* = 0 \\
c_3^* = \dfrac{C_1 W_3}{C_1 + C_2} \\
p_3^* = \dfrac{C_2 W_3}{C_1 + C_2}
\end{cases}
\qquad \text{if } W_3 \le V_{1,0} + V_{2,0} + V_{3,0} \le W_2 + W_3, \tag{3.22}
$$

$$
\begin{cases}
r_1^* = 0 \\
c_1^* = \dfrac{C_1(V_{1,0} + V_{2,0} + V_{3,0} - W_2 - W_3)}{C_1 + C_2} \\
p_1^* = \dfrac{C_2(V_{1,0} + V_{2,0} + V_{3,0} - W_2 - W_3)}{C_1 + C_2} \\
r_2^* = 0 \\
c_2^* = \dfrac{C_1 W_2}{C_1 + C_2} \\
p_2^* = \dfrac{C_2 W_2}{C_1 + C_2} \\
r_3^* = 0 \\
c_3^* = \dfrac{C_1 W_3}{C_1 + C_2} \\
p_3^* = \dfrac{C_2 W_3}{C_1 + C_2}
\end{cases}
\qquad \text{if } W_2 + W_3 \le V_{1,0} + V_{2,0} + V_{3,0}, \tag{3.23}
$$

The values of this fixed point mean that:

Situation 1: if the number of people initially staying in the three nodes is less than the capacity of node 3 (i.e. $V_{1,0} + V_{2,0} + V_{3,0} \leq W_3$), after a while, both nodes 1 and 2 become empty and all the crowd has sought refuge in node 3.

Alternatively, if this number is greater than the capacity (i.e. $V_{1,0} + V_{2,0} + V_{3,0} > W_3$), then two situations can occur:

Situation 2: node 3 becomes full and the remaining people are still stranded in node 2 if $V_{1,0} + V_{2,0} + V_{3,0} \leq W_2 + W_3$

or, **Situation 3:** stranded in both nodes 1 and 2 if $V_{1,0} + V_{2,0} + V_{3,0} > W_2 + W_3$.

From Equations (3.21), (3.22), and (3.23) it is obvious that only the ratio $\dfrac{C_1}{C_2}$ is significant for the limit of solutions of Equation (3.17) because $\dfrac{c_1^*}{p_1^*} = \dfrac{c_2^*}{p_2^*} = \dfrac{c_3^*}{p_3^*} = \dfrac{C_1}{C_2}$ (when defined), instead of parameters B_1 and B_2, becomes important for the pace at which the "reservoir" of people in reflex behavior is emptied.

Of course, across the world there are different cultures, which lead to different behaviors. These behaviors can be modelled by varying parameters.

For example, if populations are not made aware of major risks and not prepared for them, it causes a panic reaction ($B_1 < B_2$); this behavior is then regulated by the ratio $\dfrac{C_1}{C_2}$.

The higher this ratio, the more the population remains in or transits to the controlled state.

In simulated situation, we choose a set of parameters that highlight a weak risk culture, while favoring the return to a controlled behavior: the values, $B_1 = 0.15$, $B_2 = 0.45$, mean that there is a weak risk culture and, $C_1 = 0.21$, $C_2 = 0.07$, mean that the panic in the crowd context is compensated by controlled reactions for a part of the population who keep self-control, notably because there is no new threat or sudden attack.

Therefore, in **Situation 1** when $V_{1,0} = 18\,000$, $V_{2,0} = 0$, $V_{3,0} = 0$, $W_2 = 1000$, $W_3 = 18\,500$ and $\eta_{c,1,2} = \eta_{c,2,3} = \eta_{p,1,2} = \eta_{p,2,3} = 0.004$, one obtains the following convergence towards the fixed point:

$$r_1^* = 0, \, c_1^* = 0, \, p_1^* = 0, \, r_2^* = 0, \, c_2^* = 0, \, p_2^* = 0,$$

$$r_3^* = 0, \, c_3^* = \frac{C_1(V_{1,0} + V_{2,0} + V_{3,0})}{C_1 + C_2}$$

$$= \frac{0.21(18\,000)}{0.21 + 0.07} = 13\,500, \, p_3^* = \frac{C_2(V_{1,0} + V_{2,0} + V_{3,0})}{C_1 + C_2} = 4\,500.$$

This value means that, initially, all the people are in node 1 and both nodes 2 and 3 are empty, but after a certain period of time, both nodes 1 and 2 are empty and everyone has reached node 3. One can see the flow of people through the two edges $(N_1 \rightarrow N_2)$; $(N_2 \rightarrow N_3)$, in Figure 3.9.

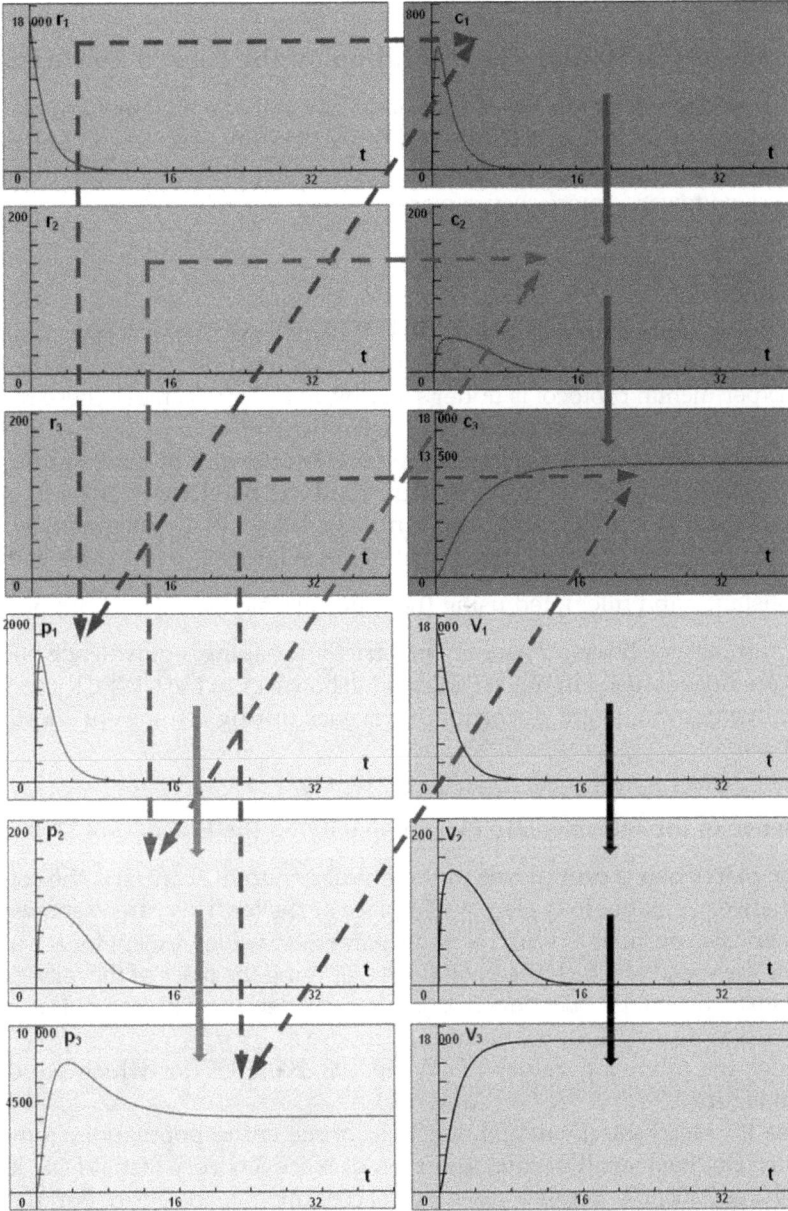

FIGURE 3.9
Situation 1: Convergence towards the fixed point (0, 0, 0, 0, 0, 0, 0, 13 500, 4 500). In this figure the change of behavioural states is symbolized by dotted arrows, and the motion between nodes by plain arrows is as displayed in Figure 3.8.

3.4 Influence of the Spatial Configuration on the Pace of Evacuation

We now seek to identify the obstacles that slow the escape of the crowd in the aftermath of the initial shock by analyzing the topology of the network of streets and places in a city. This information can potentially be used to improve the design of a city to facilitate the escape of a crowd towards more secure places.

3.4.1 Scaling the Parameters

Fixing the value of all the parameters in PCR, WPCR, or CWPCR is a very complicated task, which has not yet been done on an experimental basis. In the framework of the Com2SiCa project, an experimental protocol is under scrutiny, in order to achieve these results in the near future. However, it is important to note that only relative values between parameters are important, because there is a relationship between the unit of time and the unit used for the parameters. In other words, considering all the parameters of the adimensional Equation (3.17), (i.e. $B_i \geq 0, C_i \geq 0, \eta_{c,i,j}, \eta_{p,i,j}, \eta_{r,i,j}, i = 1, 2, j = i+1$) integrated with respect to the variable time t, it is nearly equivalent to consider such parameters multiplied by the same constant κ and integrated using the time variable $\tau = \dfrac{t}{\kappa}$ (e.g. t can be considered in seconds, minutes, or hours). There is not, strictly speaking, equivalence because such parameters are linearly used in the WPCR model; however in the CWPCR model the coupling is nonlinear and a slight distortion intervenes during a transient short period for some variables.

3.4.2 Influence of the Intermediate Place Capacity on the Evacuation Dynamics

Intermediate places play a central role in the fluidity or, to the contrary, the congestion of movement between a dangerous place and a place of shelter. This can be shown in the following numerical experiments: with the same parameter values, except for the size of node 2, we analyse the speed at which the people are emptying the place of the terrorist (node 1).

As previously described (Section 3.3.2), we consider the values of the parameters $B_1 = 0.15, B_2 = 0.45, C_1 = 0.21, C_2 = 0.07$ and we choose $\eta_{c,1,2} = \eta_{c,2,3} = \eta_{p,1,2} = \eta_{p,2,3} = 0.004$.

We consider the following values of $W_2 : 50, 100, 200, 1000$ for which we display the dynamics in Figure 3.10.

In the case $W_2 = 1000$ (black curves) the flight of the entire population from place 1 to place 2 (which can be a small square) and then to place 3 is very fast; it lasts less than 10 minutes (Figures 3.10c, f, i), because place 3 can accommodate the entire population. There is a massive influx of panicked people (Figure 3.10d), which is greater than the controlled one (Figure 3.10e), into place 2, which empties very quickly as the majority of controlled people reach the safe shelter (Figure 3.10h). However, it can be noted that a significant number of the panicked population remains in the refuge place, and this number is only slowly decreased (see bump in Figure 3.10g).

This is explained by the fact that the flight dynamics are not hindered by obstacles or bottlenecks, and the fleeing populations have not enough time to change their behavioural state.

On the other hand, if $W_2 = 50$ (red curves), the evacuation of the total population from place 1 to place 2 and then place 3 is much slower (about 35 minutes instead of less than

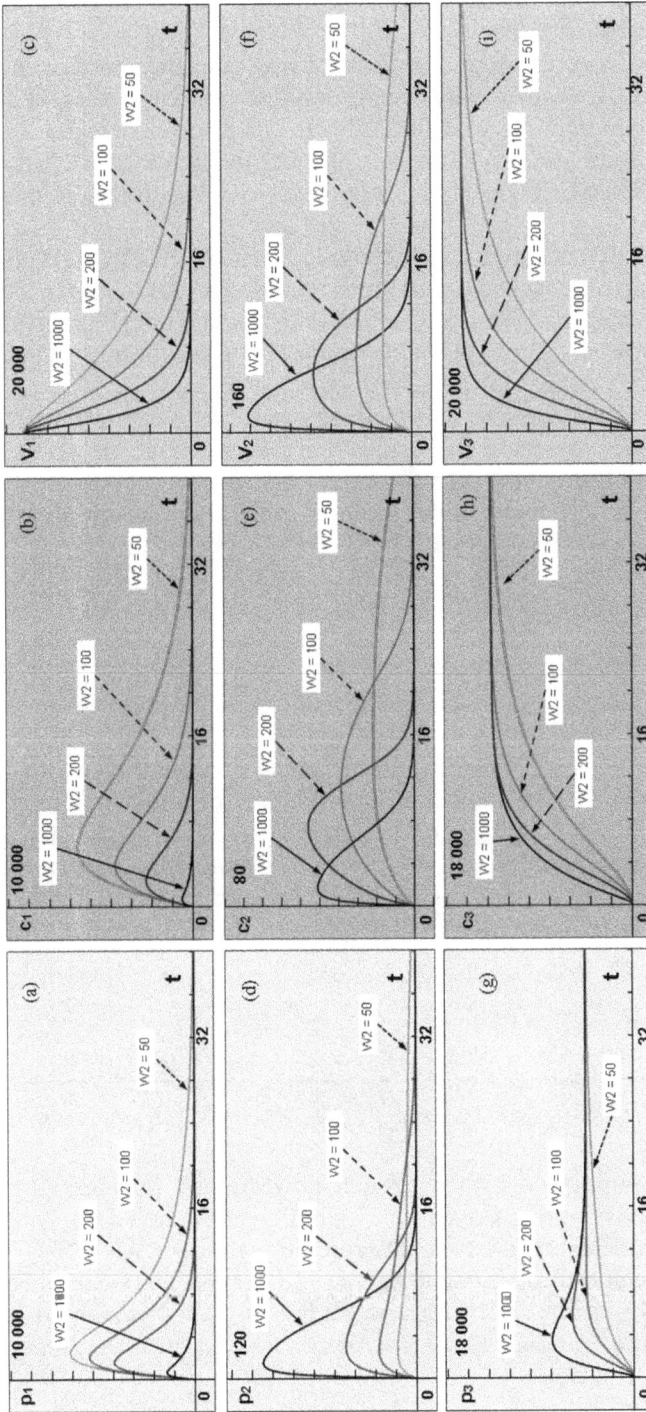

FIGURE 3.10

Convergence towards the fixed point $(0, 0, 0, 0, 0, 0, 0, 0, 13\,500, 4\,500)$ for the values of parameters $B_1 = 0.15$, $B_2 = 0.45$, $C_1 = 0.21$, $C_2 = 0.07$, $\eta_{c,1,2} = \eta_{c,2,3} = \eta_{p,1,2} = \eta_{p,2,3} = 0.004$, $V_{1,0} = 18\,000$, $V_{2,0} = 0$, $V_{3,0} = 0$, $W_3 = 18\,500$, and the values of $W_2 = 50$, $W_2 = 100$, $W_2 = 200$, $W_2 = 1000$.

10 minutes, (Figures 3.10c, f, i)). The panicked population has time to calm down in place 1, because there is no new attack (it has been assumed that there is no domino effect, $(s_i = 0, i = 1, 2)$). It is thus a population that is mainly in a state of reasoned behaviour that arrives in place 3 (Figures 3.10g, h).

It can be highlighted from this first analysis that the faster the speed of change of location (hence decreasing the vulnerability of populations), the faster this speed leads to significant flows of panic in both places 2 and 3. There is a paradox here: the fast self-safety movement of populations leads to situations of collective panic that are more difficult to manage. This fact must be taken into account by emergency services and emergency physicians.

Although the deaths that occur are not included in this CWPCR model version, one can imagine that the escape of panicked populations would give rise to more victims.

As has already been said, the simulation results shed light on the importance of the size of the intermediate places and their role in the fluidity or, to the contrary, on the congestion of movements between a dangerous place and a place of shelter.

Depending on their respective size, intermediate places will modulate the dynamics and the speed of flow of the crowds. In this sense, they become strategic places both for the planners, who must think about the organization of the area to host public events, and also for the terrorists who can use these strategic places to multiply the effect of their harmful actions by trapping the escape movements between two areas of action.

3.5 Conclusion

In this article we have developed a new model of weighted human behaviour coupled with street and place networks, in the context of an urban terrorist attack, thus improving the PCR model with bottleneck coupling and taking into account the capacity of each place and the number of people stranded in these places. The simulation results in a simple network with three nodes (places or public squares) and two edges (streets) that demonstrates the key role of the capacity of an intermediate place in the dynamics of evacuation from dangerous to safe places.

Acknowledgments

This work has been supported by the French government, through (i) the National Research Agency (ANR) under the Societal Challenge 9 "Freedom and security of Europe, its citizens and residents" with the reference number ANR-17-CE39-0008, co-financed by French Defence Procurement Agency (DGA) and The General Secretariat for Defence and National Security (SGDSN), and (ii) the UCA[JEDI] Investments in the Future project managed by the National Research Agency (ANR) with the reference number ANR-15-IDEX-01.

References

Cantin, G., Verdière, N., Lanza, V., et al. 2016. Mathematical modeling of human behaviours during catastrophic events: Stability and bifurcations. *International Journal of Bifurcation and Chaos* 26(10):1630025.

Crocq, L. 2013. *Les paniques collectives*. Paris: Odile Jacob.

Daamen, W., Hoogendoorn, S. P., Bovy, P. H. L. 2005. First-order pedestrian traffic flow theory. *Transportation Research Record: Journal of the Transportation Research Board* 1934(1):43–52.

Dauphiné, A., Provitolo, D. 2013. *Risques et catastrophes-Observer, spatialiser, comprendre, gérer*. Paris: A. Colin (2nd edition), Collection U.

Dias, C., Sarvi, M., Shiwakoti, N., Ejtemai, O., Burd, M. 2013. Investigating collective escape behaviours in complex situations. *Safety Science* 60:87–94.

Hatfield, E., Cacioppo, J. T., Rapson, R. L. 1994. *Emotional Contagion*. Cambridge: Cambridge University Press.

Kretz, T., Grünebohm, A., Schreckenberg, M. 2006. Experimental study of pedestrian flow through a bottleneck. *Journal of Statistical Mechanics: Theory and Experiment* 2006(10):10014.

Liao, W., Seyfried, A., Zhang, J., Boltes, M., Zheng, X. , Zhao, Y. 2014. Experimental study on pedestrian flow through wide bottleneck. The conference on pedestrian and evacuation dynamics 2014 (PED2014). *Transportation Research Procedia* 2:26–33.

Provitolo, D. 2005. Un exemple d'effets de dominos: La panique dans les catastrophes urbaines. *Cybergéo: Revue européenne de géographie* 328:19.

Provitolo, D., Dubos-Paillard, E., Verdière, N., et al. 2015. Les comportements humains en situation de catastrophe: De l'observation à la modélisation conceptuelle et mathématique. *Cybergeo: Européean Journal of Geography* 735:23.

Seyfried, A., Steffen, B., Klingsch, W., Boltes, M. 2005. The fundamental diagram of pedestrian movement revisited. *Journal of Statistical Mechanics: Theory and Experiment* 10:(10).

Verdière, N., Lanza, V., Charrier, R., et al. 2014. Mathematical modeling of human behaviours during catastrophic events. Paper presented at ICCSA14, 23–26 June, Le Havre, 8.

Verdière, N., Cantin, G., Provitolo, D., et al. 2015. Understanding and simulation of human behaviours in areas affected by disasters: From the observation to the conception of a mathematical model. *Global Journal of Human Social Science: H, Interdisciplinary* 15(10):10.

Zhang, X. L., Weng, W. G., Yuan, H. Y., Chen, J. G. 2013. Empirical study of a unidirectional dense crowd during a real mass event. *Physica. Part A* 392(12):2781–2791.

Zhou, J.-B., Chen, H., Yang, J., Yan, J. 2014. Pedestrian evacuation time model for urban metro hubs based on multiple video sequences data. *Mathematical Problems in Engineering* 2014:11. Article ID 843096.

4

Free Radical Processes in Medical Grade UHMWPE

M. Shah Jahan, Benjamin Walters, Afsana Sharmin, and Saghar Gomrok

CONTENTS

4.1 Introduction..61
4.2 Polymeric Free Radicals..64
4.3 Electron Spin Resonance...65
4.4 Free Radicals in Irradiated UHMWPE...66
4.5 Decay of UHMWPE Radicals as a Function of Time...........................67
4.6 Free Radicals In Vivo...70
4.7 Improvement of UHMWPE: Stabilization of Free Radical with Vitamin E...............70
4.8 Summary...74
References...74

4.1 Introduction

In arthroplasty, medical-grade biomaterials need an optimal combination of mechanical properties with appropriate design and manufacturability according to the application area. When complex forms and flexibility are required, polymeric biomaterials are used, although metals and ceramics occupy a significant segment of implant science and technology.[1, 2] Of course, being biocompatible, interactions between implants and host biological cell tissues should not induce any inflammatory response or formation of unusual tissues.[3] In total hip- and knee-joint replacement parts, for example, ultra-high-molecular-weight polyethylene (UHMWPE) is extensively used.[4, 5] It is a non-biodegradable polymer which is formed from ethylene gas with the generic chemical formula $-(C_2H_4)_n-$, where n is the degree of polymerization. With nearly 200,000 ethylene repeat units and average molecular weight up to 6 million g/mol, and density of 0.94–0.97 g/ml, UHMWPE is significantly more abrasion-resistant and wear-resistant than HDPE (high-density polyethylene).[6] The crystallinity of UHMWPE is around 30 to 40%, and crystalline regions are randomly oriented with respect to each other.[7] The UHMWPE molecules are made of very long hydrocarbon polymeric chains, which all align in the same direction. Although the intermolecular bonds are relatively weak due to Van der Waals interactions, overlaps can exist between the long molecular chains, which allow transfer of the large shear forces from molecule to molecule.[6] Approximately one million UHMWPE components are implanted yearly worldwide. The commercial production of UHMWPE started in the 1950s, and since then over 90% of UHMWPE has been used in industry.[3] The medical industry has been one of the major areas of application for UHMWPE since its first use as the bearing material in total joint replacement by Sir John Charnley in 1962.[1] Although in the recent past, the joint replacement had a life of about 15 to 20 years and was prescribed to patients above 60 years of age, now the reliability of UHMWPE in cases of orthopedic surgery has increased. Currently, many research studies are being carried out to improve the mechanical properties of medical-grade UHMWPE which can potentially increase the longevity of the joint components.[7,8] The global market size of UHMWPE was estimated at USD 1.33 billion in 2016, and Grand View Research Inc. reported that the market is expected to reach at USD 2.60 billion by 2022.[9,10]

In medical applications, however, UHMWPE is subject to ionizing radiation, in the form of electron beam (e-beam) or gamma rays, for two basic reasons: One is to crosslink the long-chain molecules, and the other is to sterilize the finished products. Although the gamma sterilization technique has several advantages over other sterilization techniques, its strong penetrating radiation breaks the bonds of polymer chains and produces highly reactive free-radical species—in the core, as well as in/at the surface—of a UHMWPE component.[11-13] Some free radicals may also get trapped for long periods of time in the crystalline region of the polyethylene matrix, and then initiate long-term oxidation reactions with oxygen as it diffuses into the polyethylene. These free radical reactions ultimately lead to increased material degradation/wear and, ultimately, joint failure.[14-17] Free-radical induced degradation may also continue when the sterilization process (of the finalized products) involves gamma-radiation that is then stored in packages containing air (oxygen).[17,18] To mitigate the oxidation and its subsequent adverse effects on the mechanical properties and wear of UHMWPE, gamma-sterilization of the joint components can be performed in vacuum-packaging or inert-gas packaging.[19] However, when implanted, *in vivo* oxidation is still possible in a body-fluid environment.[20] As for crosslinking via gamma-irradiation, to reduce *in vivo* oxidation, thermal annealing or melting of UHMWPE joint components in vacuum or in an inert gas can be used to stabilize the free radicals.[21] Nonetheless, questions remain about the optimum temperature and annealing time and their effect on material properties. Figure 4.1 outlines the impact of gamma sterilization of medical-device components in the presence or absence of oxygen.

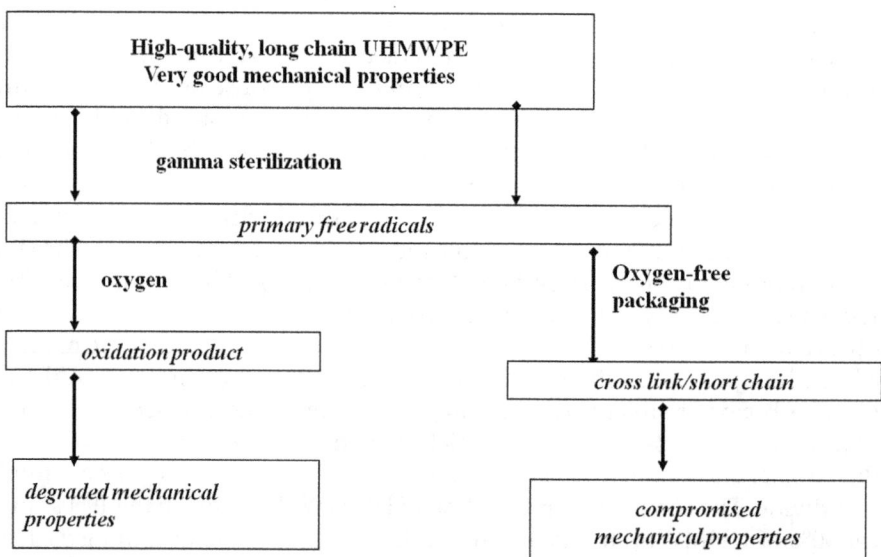

FIGURE 4.1
Effects of gamma sterilization on mechanical properties of UHMWPE.

Vitamin E (α-Tocopherol (α-T), see molecular structure in Figure 4.2) is a biocompatible and lipophilic antioxidant, and has been shown to stabilize UHMWPE, along with improving the fatigue strength, and to increase the oxidative stability and free-radical decay in UHMWPE. FTIR (Fourier-transform infrared) spectroscopy, which can be used to measure oxidation in UHMWPE, does not detect a large increase in oxidation index per ASTM F2102; as a result, the irradiated UHMWPE is considered stabilized by vitamin E. Incorporating an antioxidant such as vitamin E can, therefore, reduce the need for melting following the crosslinking process, or allow lower temperatures used for annealing.[22] On the other hand, if present within UHMWPE before crosslinking, the vitamin E may inhibit crosslinking by quenching the radicals that would otherwise be the catalysts for crosslinking. Vitamin E also has no measurable effect on post-irradiation decay of the primary radicals and subsequent production of residual radicals. Still, vitamin E has also been found to quench radicals only in part at the time of radiation-crosslinking, therefore allowing crosslinking to occur, and then a partial quantity of the original vitamin E content within the UHMWPE remains for protection from the effect of subsequent radical generation which may arise from other sources following crosslinking.[12, 23-25]

A summary of research on orthopedic UHMWPE is presented in this chapter, starting far in the past and arriving in the present; this research has included many investigations of different types, treatments, additives, storage environments, and other factors, all with the common focus on free radicals. After all, free radicals can be considered the root cause of practically all implant failures related to polyethylene (and the cause of success as well). Osteolysis (resorption of bone), is often determined to be induced by particles/wear of polyethylene. Other factors, such as the size of the particles, or the quantity of the particles, or if the particles are a particular type of polyethylene or contain vitamin E or not, have all been debated over what is the "worst" condition for the body to endure—which will most likely lead to osteolysis and implant failure. All of these factors, of course, are related to wear. While wear can have other causes—such as implant placement not being in the most wear-resistant position, a component size that may be more susceptible to wear—the most common factor is oxidative embrittlement. That is, the more oxidized and brittle a piece of polyethylene becomes, the more likely it is to wear—for any condition of implant size, position etc. So, we have osteolysis, caused by particles, caused by wear, caused by embrittlement, caused by oxidation. Then, we arrive at: What causes the oxidation? Oxygen? Well, yes, without oxygen, the oxidative process does not occur, as we will discuss later in this chapter. Still, our bodies and the environments we live in are not oxygen-free, and never will be. The implants will always be exposed to oxygen and oxidative conditions. It would be nice to have a nice oxygen-free environment just where the implant is, so that it would

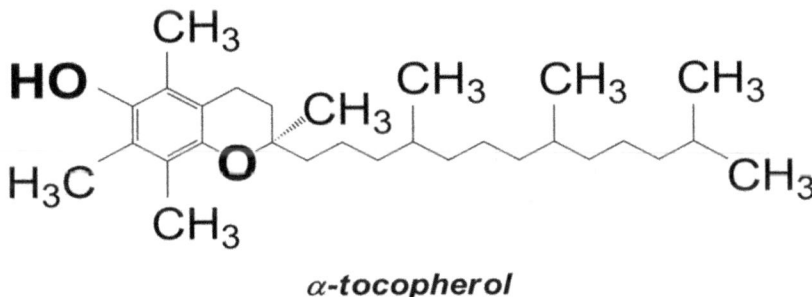

α-tocopherol

FIGURE 4.2
Molecular structure of Vitamin E (α-Tocopherol (α-T)).

not degrade for that particular reason. However, almost equivalent to that dream, but much more reachable, is to take away the free radicals. That is, without oxygen, the oxidation will be very much less likely to occur, but also without the free radicals the oxidation will be very much less likely to occur. But, just as exposure to oxygen is a part of life, so is the production of free radicals within polyethylene components of medical devices during its manufacturing process. Starting from its initial application in the 1960s, UHMWPE has gone through several stages of improvement, and this chapter highlights how free-radical research has played, and continues to play, a key role in each stage of its improvements for orthopedic applications. One primary aspect of this research has been a long-term (~20 years) targeted research study, conducted under the auspices of the Biosurface Center (NSF supported Industry-University Center for Biosurfaces (IUCRC)), under the guidance of Dr. M Shah Jahan of the Department of Physics and Materials Science at the University of Memphis, and in partnership with the State University of New York, Buffalo (SUNY Buffalo).

4.2 Polymeric Free Radicals

Figure 4.3 shows the steps of initiation, propagation, and termination of free radicals (denoted by a P• or H•, for example) of a hydrocarbon polymer (PH) as a result of irradiation with energy hν (h is Planck constant and ν is frequency of radiation). The free-radical species, denoted by (•), can be detected by the free-radical measurement technique known as Electron Spin Resonance (ESR), or Electron Paramagnetic Resonance (EPR). The non-radical or neutral species cannot be detected by ESR/EPR. One species that is particularly

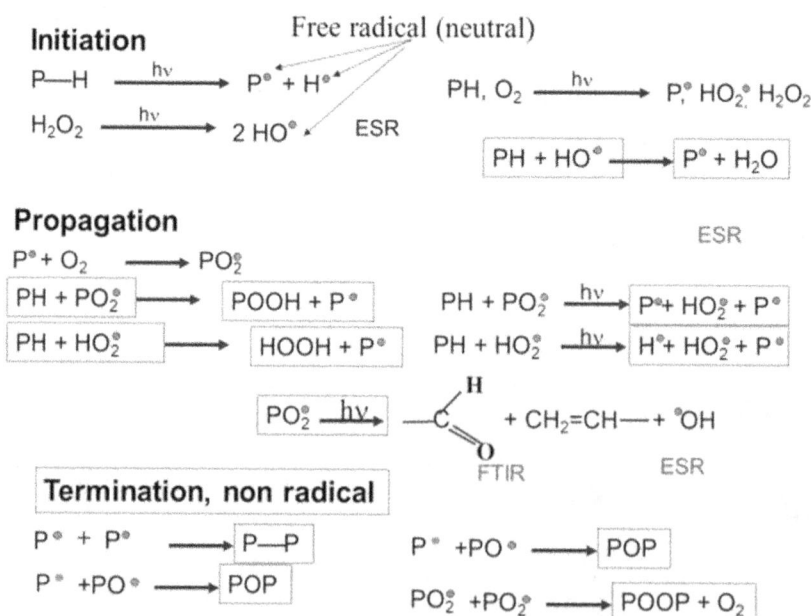

FIGURE 4.3
Formation, propagation, and termination of polymeric radicals.

important with respect to oxidation or oxidative degradation of polymers is that of the carbonyl $\left(-C^{=O}_H\right)$. The Fourier-transform infrared (FTIR) absorption band of the carbonyl group (~1720 cm^{-1}) is used to measure the oxidation index (OI) for UHMWPE per ASTM F2102.

4.3 Electron Spin Resonance

Electron Spin Resonance (ESR), also known as Electron Paramagnetic Resonance (EPR), is the only technique that can directly detect free radicals and paramagnetic species in a given material. The theory behind ESR is based on the magnetic moment of unpaired electrons in an atomic or molecular structure. When exposed to an external magnetic field, these magnetic moments either align themselves parallel or antiparallel to the field, and the energy separation between the parallel and antiparallel states is known as the Zeeman Effect. The transition between these states can occur when an absorption of energy—an exactly equal amount of microwave energy ($E = h\nu$, where h is the Planck constant and ν, the microwave frequency)—occurs. This is known as the resonance condition, and it is governed by a simple, basic formula (classical) $h\nu = g\beta H$, where g is a spectral splitting factor, β is the Bohr magneton, and H is the external magnetic field. In ESR/EPR spectroscopy, the absorption signal (ESR spectrum) is detected and displayed as a first-derivative of the absorption, with its intensity (Y axis) typically plotted as a function of external magnetic field (X-axis). A typical absorption signal, known as an ESR spectrum, is shown in Figure 4.4. The peak-to-peak height (hpp) is proportional to the number of radicals present in the test specimen. For composite spectra, the double integration method is used to determine the radical concentration (see later). A brief outline of ESR in UHMWPE can be found in the UHMWPE handbook.[24]

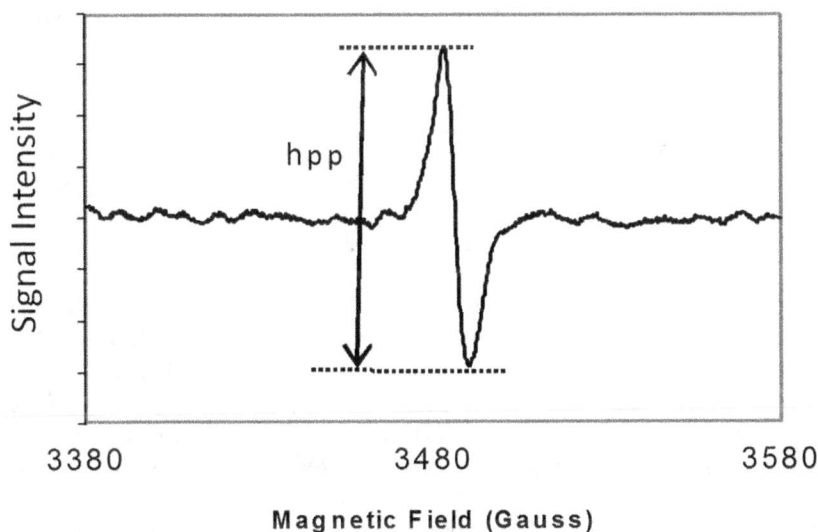

FIGURE 4.4
A first-derivative absorption signal.

Electron spin resonance data/results presented in this chapter were obtained by using X-band (~9.8 GHz) spectrometers, E4 by Varian, or EMX 300 by Bruker. Gamma irradiation using a Co-60 source was provided by an outside vendor. For X-irradiation, an in-house X-ray machine (American Instruments, Inc.) was employed.

4.4 Free Radicals in Irradiated UHMWPE

Primarily, three types of radicals are formed when PE is exposed to ionizing radiation (in the absence of oxygen). The name and chemical formula of each type are given below.

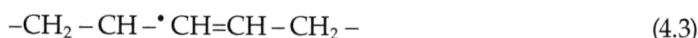

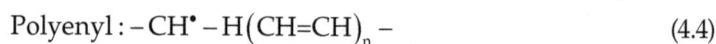

$$\text{Alkyl} : -CH_2 -^{\bullet} CH - CH_2 \tag{4.1}$$

$$\text{Allyl} : -CH_2 - CH{=}CH -^{\bullet} CH - CH_2 - \quad \text{or} \tag{4.2}$$

$$-CH_2 - CH -^{\bullet} CH{=}CH - CH_2 - \tag{4.3}$$

$$\text{Polyenyl} : -CH^{\bullet} - H(CH{=}CH)_n - \tag{4.4}$$

Figure 4.5 displays characteristic ESR spectra for each radical type; the first derivative of an absorption at resonance is plotted as a function of the resonance magnetic field (in Gauss). As mentioned earlier, the relative intensity (h_{pp}) of each line and the number of lines depend on the coupling (magnetic interaction) between the unpaired electrons (at the site of the broken molecular chain) and the nearby protons (hydrogen atoms) on the same chain.

As a function of time, a given radical decays to a non-radical species (neutral, non-detectable by ESR) when it combines with another radical and/or captures a proton (hydrogen atom). In air (presence of oxygen), a primary radical reacts with an oxygen to form a secondary or tertiary radical (oxygen-centered species); thus, it initiates an oxidation process [ATD10]. The oxygen-centered radicals and the molecular structure of each radical are listed as follows:

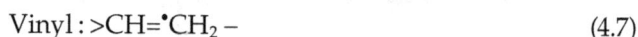

$$\text{Peroxy} : -CHO_2^{\bullet} - \tag{4.5}$$

$$\text{Alkoxy} : -CHO^{\bullet} - \tag{4.6}$$

$$\text{Vinyl} : >CH{=}^{\bullet}CH_2 - \tag{4.7}$$

An ESR spectrum, therefore, may represent a combination of any number and type of radical species depending on factors such as the irradiation environment, radiation type and dose, and temperature. For example, when UHMWPE is irradiated with gamma rays in a vacuum or an inert environment (N_2 gas, for example), a broad, 7-line spectrum, is detected (see Figure 4.6). On the other hand, when irradiated with X-rays under similar conditions, a 6-line spectrum is obtained (see Figure 4.7). The measurements are made at room temperature, in air, immediately after irradiation (X-rays) or after removing the sample from its inert-sealed package (following gamma-irradiation). For these tests, care is taken so that minimum or no oxidation occurs during measurements (i.e., limited exposure to air). This method is particularly important because it allows one to detect and analyze primarily allyl (seven-line spectrum) or alkyl (six-line spectrum) radicals before

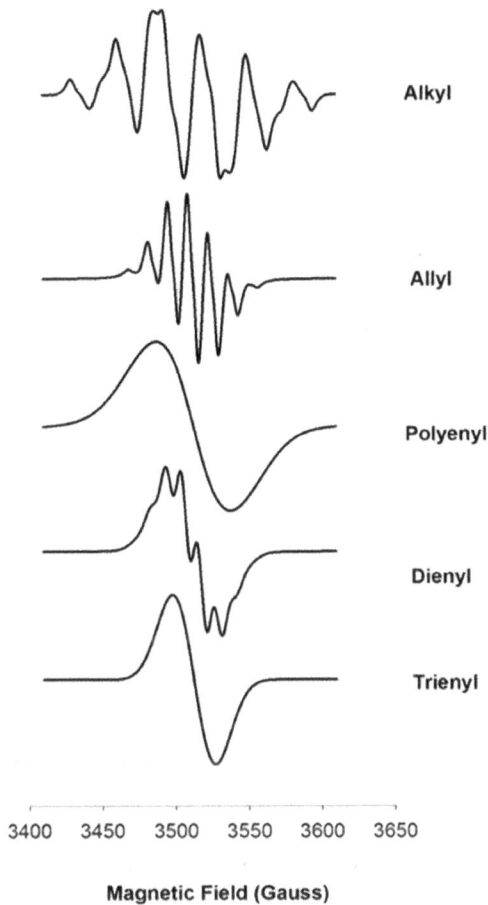

FIGURE 4.5
ESR spectra of the primary radicals in UHMWPE.

they have decayed into other radical types and convoluting the spectrum. Additionally, by employing a simulation program, the spectrum of each constituent radical can be generated to best fit the experimental spectrum (see Figures 4.3 and 4.4).

To determine the radical concentration (on a relative scale) in a given test sample, the peak-to-peak height (h_{pp}), of an absorption line (as was shown in Figure 4.4) is used. For the case of a multi-line or a composite spectrum, double-integration of the entire absorption spectrum is used. For an absolute measure (number of radicals per gram), however, the double-integration of an absorption spectrum of the test sample is compared with that of a standard material whose radical concentration is known. Our laboratory has been using a standard NIST (National Institute of Standard and Testing) reference material (Electron Paramagnetic Resonance Intensity Standard SRM 2601).

4.5 Decay of UHMWPE Radicals as a Function of Time

As stated earlier, the primary radicals in UHMWPE can decay as a function of time via a combination–recombination process in an oxygen-free environment (such as vacuum

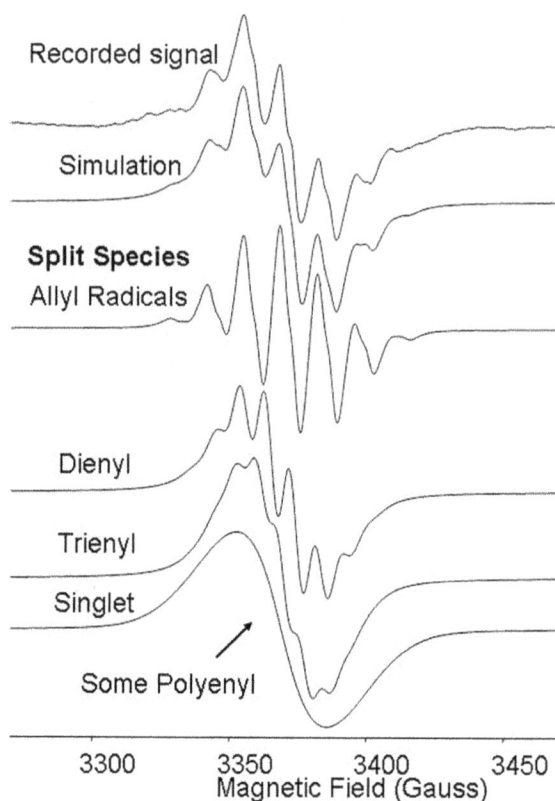

FIGURE 4.6
Simulated ESR spectrum compared with the recorded (experimental) spectrum containing primary radicals in gamma-irradiated UHMWPE.

or inert gas) and/or reacting with oxygen in an open-air environment. Figure 4.8 shows structural changes in an ESR spectrum as a function of time when UHMWPE is irradiated via X-ray at room temperature in the presence of oxygen (air). Clearly, the six-line spectrum of the primary alkyl radical is reduced to a singlet due to an oxygen-induced polyenyl radical. As is evident from the figure, the signal intensity was reduced significantly in six days because of the reduced concentration of radicals. To record the remaining spectra, therefore, the gain of the Y-scale was multiplied by 10.

Figure 4.9 shows the result of a control study performed for a period of one year. ESR samples were sealed in N2-filled or vacuum tubes (quartz), and were irradiated with gamma rays at room temperature. Following irradiation, and after recording the initial spectra of the samples, one group was placed in an oven at 75°C, and the second group at room temperature. The initial seven-line spectrum shows clear evidence of the presence of predominantly allyl radicals. After one year at room temperature, or at 75°C, the radical concentration has gone down significantly, but the radical type (allyl) has not changed; the weak shoulders among the broad outline of the derivative signal due to the allyl radicals is an indication of minimum or no oxidation reaction. This result suggests that the radicals cannot be completely eliminated or quenched by heating at 75°C; therefore, when subsequently exposed to oxygen, the radicals that remain in the sealed samples can potentially undergo oxidation and produce oxygen-induced secondary or tertiary radicals.

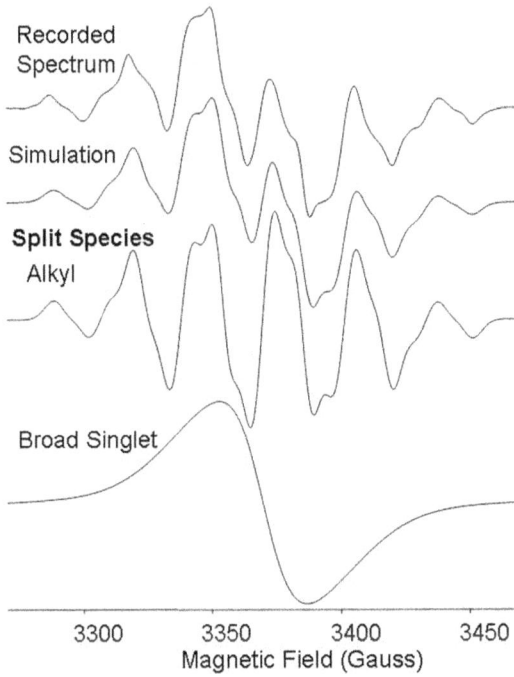

FIGURE 4.7
Simulated ESR spectrum compared with the recorded (experimental) spectrum containing primary radicals in X-irradiated UHMWPE.

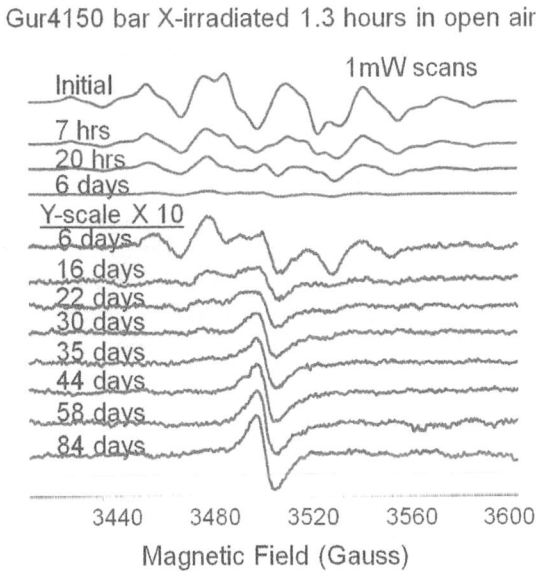

FIGURE 4.8
ESR spectra of UHMWPE radicals recorded as a function of time for 84 days in the presence of oxygen (air) following X-irradiation in air at room temperature.

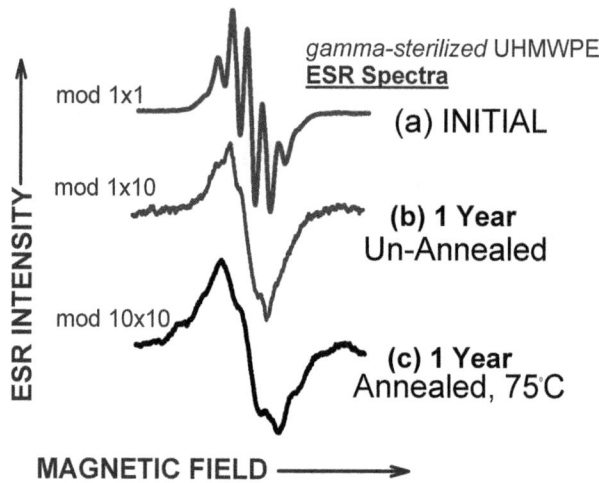

FIGURE 4.9
ESR spectra recorded after gamma-irradiation in a vacuum-sealed environment. (a) initial run with modulation amplitude (MA) 1.0 G and microwave power (mp) 1.0 mW; (b) after one year of storage at room temperature in the same sealed environment, ma = 1.0 G and mp = 10.0 mW; (c) annealed at 75°C in the same sealed environment for one year; ma = 10 G and mp = 10 mW.

4.6 Free Radicals In Vivo

Free-radical measurements were also conducted on retrieved knee- and hip-joint components that had been gamma sterilized. The finished tibial plateau insert (TPI) shown in Figure 4.10(b) was made out of a bar stock (a). The ESR samples were punched out of the control (shelf-aged) (c) and the retrieved (d) TPIs as shown by the holes in the specimens. The retrieved TPI was shelf-stored for 86 months before its use in a body for 7 months. A representative ESR signal is also shown in the figure. The signal not only shows the presence of oxygen-induced radicals (shown as a singlet—the larger feature near the middle of the spectrum), but also residual primary radicals. This result suggests that oxidation reactions due to the primary radicals continues *in vivo* for a long period of time.

Another example of the presence of long-lived radicals *in vivo* is shown in Figure 4.11. A hip cup and a schematic drawing, showing ESR samples, are presented in Figure 4.8(a). The test samples were cut out from a cup which was retrieved after six years of use. The samples were tested again 13 years later. Figure 4.8(b) shows the ESR signal intensity (Log (radical concentration/gm)) as a function of sample position. Results of this study suggest the presence of long-term radical reaction *in vivo* and outside.

4.7 Improvement of UHMWPE: Stabilization of Free Radical with Vitamin E

The results and discussions presented earlier clearly demonstrate that oxidative degradations of the UHMWPE components of the total hip- and knee-joints are primarily caused

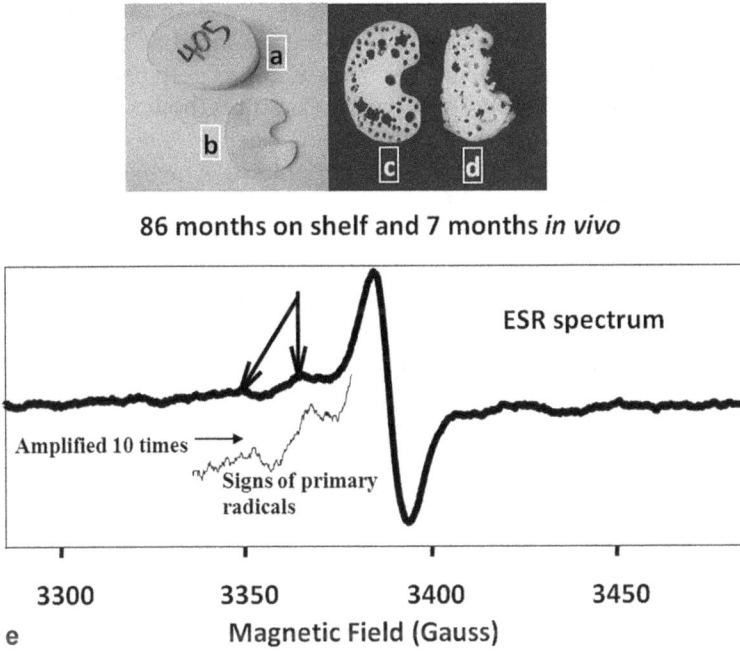

86 months on shelf and 7 months *in vivo*

FIGURE 4.10
Tibial plateau insert (TPI) as prepared for free-radical measurements via ESR. (a) bar stock from which the finished tibial plateau insert (b) was made. Samples were punched from a control (shelf-aged) (c) and retrieved (d) specimens, as shown by the holes, (e) an ESR spectrum of a retrieved specimen is also shown.

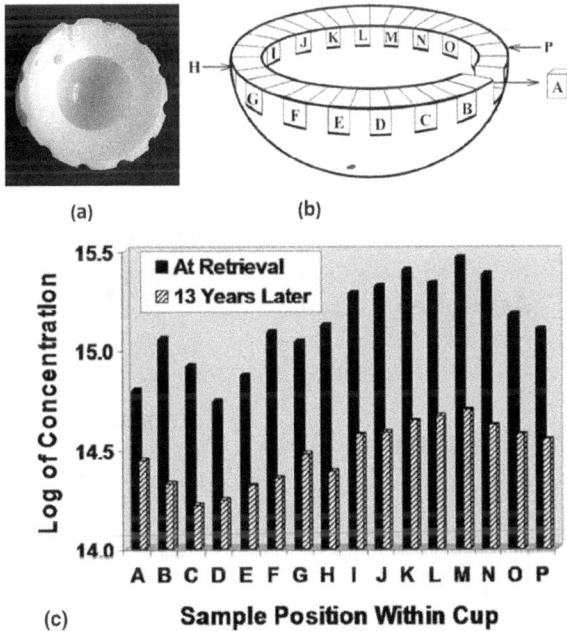

FIGURE 4.11
(a) acetabular cup; (b) example of samples cut out from a retrieved cup; (c) bar chart summarizing results of free radical measurements (Log (radicals/gram)).

by reaction of free radicals with oxygen, whether *in vivo* or in the storage environment. Therefore, as a remedy, use of vitamin E as an antioxidant has been introduced.[25] It is biocompatible and the oxidation resistance it adds to the polymer is high. The practice of blending vitamin E with UHMWPE as an antioxidant is, however, not new. In the first place, Luigi Costa reviewed how the oxidation process may be effectively blocked with an appropriate antioxidant, and recommended vitamin E, which has been proven to improve the oxidation resistance and fatigue strength of irradiated UHMWPE.[25] Vitamin E donates hydrogen to the free radicals of UHMWPE in order to stabilize them. In this process, vitamin E becomes a free radical itself (tocopheroxyl radical), although a more stable type, relative to radicals of UHMWPE. Figure 4.12 shows the structure of the vitamin E (α-Tocopherol) radical molecule and a representative ESR spectrum radical.

Figure 4.13 shows ESR spectra recorded in vitamin-E-blended UHMWPE (20% by weight α-Tocopherol) immediately after gamma irradiation and after exposure to oxygen at room temperature for 71 days. The high concentration of α-Tocopherol allows the vitamin E radicals to be monitored in the presence of the relatively strong ESR spectra of the UHMWPE radicals, which is useful to compare the radical–radical reactions. To demonstrate that the ESR signal of the E-UHMWPE was a composite one, the signals of vitamin E and

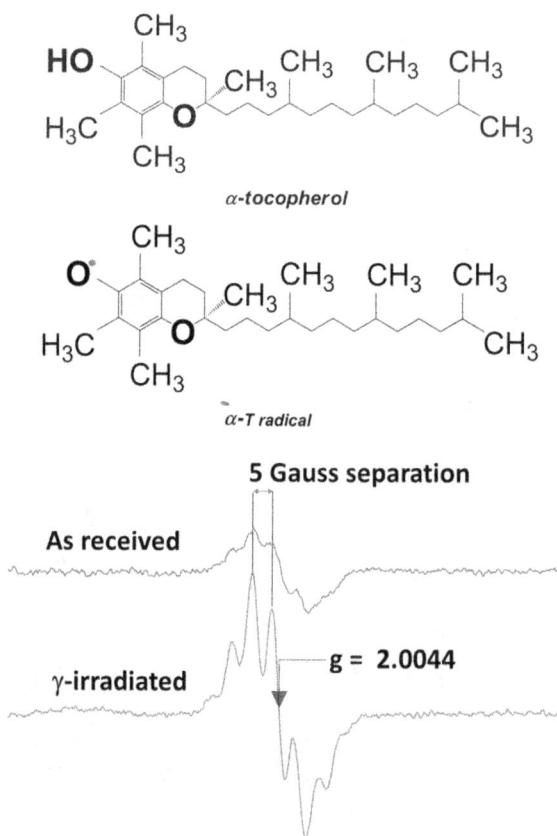

FIGURE 4.12
Molecular structure of the vitamin E (α-Tocopherol) molecule and its radical, as labeled. Also shown are two representative first-derivative ESR spectra of the α-Tocopherol radical before and after gamma-irradiation; these are of vitamin E only, as received/out of the bottle (purity 95%, Sigma Aldrich), and after being gamma-irradiated.

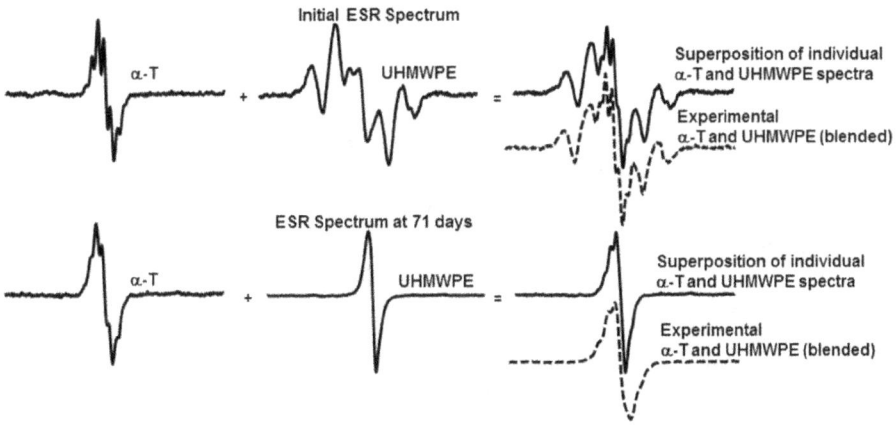

FIGURE 4.13
ESR spectra of vitamin E and UHMWPE, individually and combined, recorded immediately after irradiation (initial) and after 71 days of storage in air. Emphasized are the experimental spectra compared with the sum of Vitamin E and UHMWPE resin radicals.

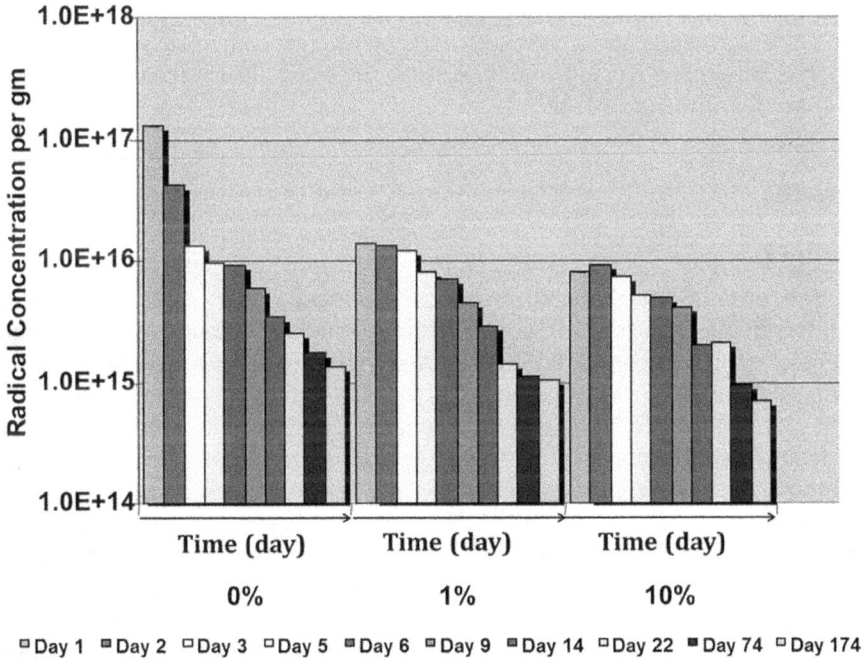

□ Day 1   ▪ Day 2   □ Day 3   □ Day 5   ▪ Day 6   □ Day 9   ▪ Day 14   □ Day 22   ▪ Day 74   □ Day 174

FIGURE 4.14
Free-radical concentration in gamma-irradiated compression molded UHMWPE containing no vitamin E (0%), 1% vitamin E, and 10% vitamin E; measurements made after opening to air.

UHMWPE, recorded separately, were added, and shown below the main spectrum with dashed lines. In a separate ESR experiment, free radical concentration as a function of % α-Tocopherol was determined (see Figure 4.14). The results of these measurements clearly demonstrate that α-Tocopherol reduces or quenches free radicals significantly. However, a measurable number of radicals do remain trapped, most likely in the crystalline region of

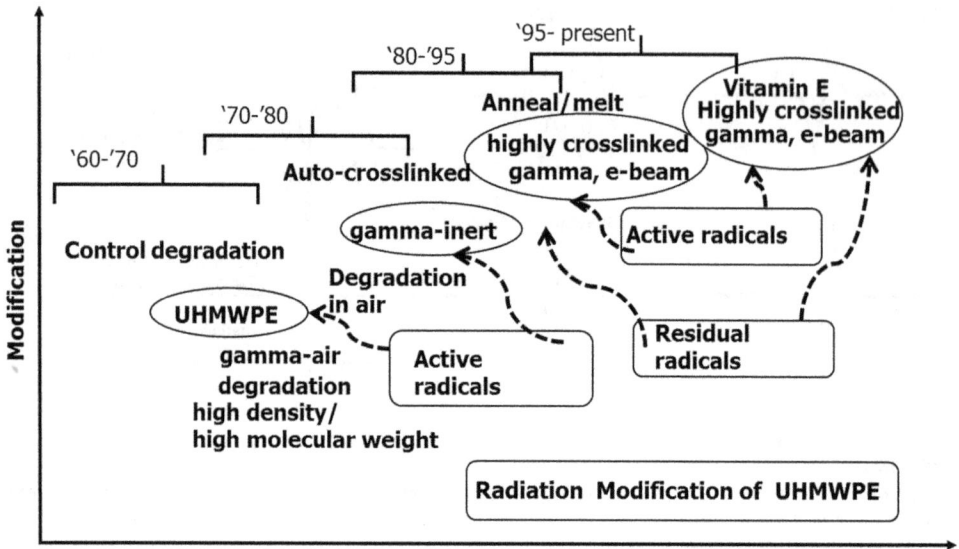

FIGURE 4.15
Improvement of medical-grade UHMWPE as a function of time period.

the polymer. And, with diffusion of oxygen into the crystalline environment, these radicals can potentially undergo oxidation.

4.8 Summary

To retain the original mechanical properties as needed in orthopedic applications, UHMWPE has undergone several stages of improvements. The flow chart in Figure 4.15 shows stages of improvement the UHMWPE has gone through since its first application in 1960. During each period of improvement, the focus of the experimental work was "how to control" the free radicals. In summary, orthopedic UHMWPE has been found to undergo oxidative degradation because of the reactions of radiation-induced free radicals with diffused or trapped oxygen. While thermal quenching of free radicals in an inert environment are found to reduce material degradation, the residual radicals are found to undergo oxidation reactions for a very long time (>25 years, in our laboratory). To combat free radicals, vitamin E has been added to UHMWPE as an antioxidant. While the experimental as well as clinical results show a significant improvement in the radical-induced degradation, ESR data show the presence of remaining free radicals at any concentration of α-Tocopherol.

References

1. A. Sáenz, E. Rivera-Muñoz, W. Brostow, V. M. Castaño, Ceramic biomaterials: An introductory overview, *Journal of Materials Education*, 21(5–6): 297–306 (1999).
2. W. Khan, E. Muntimadugu, M. Jaffe, A. J. Domb, Implantable medical devices, *Advances in Delivery Science and Technology*, 1: 33–59 (2014).

3. F. Rodríguez-González, *Introduction to Biomaterials in Orthopaedic Surgery, Biomaterials in Orthopaedic Surgery (05233G)*, ASM International (2009).
4. E. M. Brach del Prever, A. Bistolfi, P. Bracco, L. Costa, UHMWPE for arthroplasty: Past or future?, *Journal of Orthopaedics and Traumatology*, 10(1): 1–8 (2009).
5. G. L. Blunn, E. M. Brach del Preva, L. Costa, J. Fisher, M. A. Freeman, Ultra high molecular-weight polyethylene (UHMWPE) in total knee replacement: Fabrication, sterilization and wear, *The Journal of Bone and Joint Surgery*, 84(7): 946 (2002).
6. S. M. Kurtz, *The UHMWPE Handbook*, Elsevier, Burlington, MA (2014).
7. S. Affatato, B. Bordini, C. Fagnano, P. Taddei, A. Tinti, A. Toni, Effects of the sterilisation method on the wear of UHMWPE acetabular cups tested in a hip joint simulator, *Biomaterials*, 23(6): 1439–1446 (2002).
8. D. Clancy, J. Birdsall, Flies, worms and the free radical theory of ageing, *Ageing Research Reviews*, 12(1): 404–412 (2013).
9. S. H. Naidu, B. L. Bixler, M. J. R. Moulton, Radiation induced physical changes in UHMWPE implant components, *Orthopedics*, 20(2): 137–142 (1997).
10. N. Tomita, T. Kitakura, N. Onmori, Y. Ikada, E. Aoyama, Prevention of fatigue cracks in ultra-high molecular weight polyethylene joint components by the addition of vitamin E, *Journal of Biomedical Material Research: Applied Biomaterials*, 48(4): 474–478 (1999).
11. J. R. Joshi, R. P. Patel, Role of biodegradable polymers in drug deliver, *International Journal of Current Pharmaceutical Research*, 4(4): 74–81 (2012).
12. M. S. Jahan, M. Fuzail, Examination of the long-lived oxygen-induced radicals in irradiated ultra-high molecular weight polyethylene, *Nuclear Instruments and Methods in Physics Research Part B*, 265: 67–71 (2007).
13. M. D. Ridley, M. S. Jahan, Effects of packaging environments on free radicals in Υ-irradiated UHMWPE resin powder blend with vitamin E, *Journal of Biomedical Materials Research – Part A*, 88A(4): 1097–1103 (2009).
14. E. Oral, B. W. Ghali, O. K. Muratoglu, The elimination of free radicals in irradiated UHMWPEs with and without vitamin E stabilization by annealing under pressure, *Journal of Biomedical Materials Research, Part B*, 97B(1): 167–174 (2011).
15. H. Wang, L. Xu, R. Li, J. Hu, M. Wang, G. Wu, Improving the creep resistance and tensile property of UHMWPE sheet by radiation cross-linking and annealing, *Radiation Physics and Chemistry*, 125: 41–49 (2016).
16. T. Liu, A. Eyler, W. H. Zhong, Simultaneous improvements in wear resistance and mechanical properties of UHMWPE nanocomposite fabricated via a facile approach, *Materials Letters*, 177: 17–20 (2016).
17. S. Kurtz, M. Manley, A. Wang, S. J. Taylor, J. H. Dumbleton, Comparison of the properties of annealed crosslinked (crossfire) and conventional polyethylene as hip bearing materials, *Bulletin of the Hospital for Joint Disease*, 61(1–2): 17–26 (2002).
18. A. Wang, H. Zeng, S. S. Yau, A. Essner, M. Manely, J. Dumbleton, Wear, oxidation and mechanical properties of a sequentially irradiated and annealed UHMWPE in total joint replacement, *Journal of Physics. Part D*, 39(15): 3213–3219 (2006).
19. S. M. Kurtz, O. K. Muratoglu, M. Evans, A. A. Edidin, Advances in the processing, sterilization, and crosslinking of ultra-high molecular weight polyethylene for total joint arthroplasty, *Biomaterials*, 20(18): 1659–1688 (September 1999).
20. S. M. Kurtz, W. Hozack, J. Turner, J. Purtill, D. MacDonald, P. Sharkey, J. Parvizi, M. Manley, R. Rothman, Mechanical properties of retrieved highly cross-linked crossfire liners after short-term implantation, *The Journal of Arthroplasty*, 20(7): 840–849 (October 2005).
21. S. Kurtz, D. MacDonald, et al., In vivo oxidation and oxidation potential for polyethylene in total disc replacement following gamma sterilization in air and first-generation barrier packaging, *54th Annual Meeting of the Orthopedic Research Society; Poster 1324*, San Francisco, CA, 2008.
22. E. Oral, K. K. Wannomae, N. Hawkins, W. H. Harris, O. K. Muratoglu, α-Tocopherol-doped irradiated UHMWPE for high fatigue resistance and low wear, *Biomaterials*, 25(24): 5515–5522 (2004).

23. M. S. Mehmood, Hafeez Ullah, M. S. Jahan, S. Mishra, B. M. Walters, M. Ikram, The effect of high dose of gamma-irradiation on residual radicals concentration in ultra-high molecular weight polyethylene (UHMWPE) in the presence of vitamin E1, *Polymer Science, Series A*, 54(5): 343–348 (2012).

24. M. S. Jahan, ESR insights into macroradicals in UHMWPE, *UHMWPE Biomaterials Handbook*, Elsevier (2009).

25. M. S. Jahan, M. Fuzail, Examination of the long-lived, oxygen-induced radicals in irradiated ultra-high molecular weight polyethylene, *Nuclear Instruments and Methods in Physics Research, Section B (NIMB)*, 265: 67–71 (2007).

5

Inverse Problems Involving PDEs with Applications to Imaging

Taufiquar Khan

CONTENTS

5.1 Introduction...77
5.2 Electrical Impedance Tomography..78
    5.2.1 Analytical Setting for the EIT Model...79
    5.2.2 Inverse Problem in EIT...81
        5.2.2.1 Why Is the Inverse Problem Ill-Posed? 81
        5.2.2.2 Connection to DOT...82
5.3 Diffuse Optical Tomography 82
    5.3.1 Radiative Transport Equation 82
    5.3.2 DOT Model 83
5.4 Computational Aspect and the Regularization of the Ill-Posed Problem....................84
    5.4.1 Iteratively Regularized Gauss–Newton Method 86
    5.4.2 The Statistical Inverse Problem...86
        5.4.2.1 The Posterior Density 87
    5.4.3 Regularization Functions 87
        5.4.3.1 The ℓ_p Regularizations 87
        5.4.3.2 The Total Variation Regularization 87
    5.4.4 The Markov Chain Monte Carlo Method...88
5.5 Conclusion 88
References..89

5.1 Introduction

The field of inverse problem is a fairly mature area of research and was initially motivated by industrial problems [1]. The growth of this research area has been tremendous in the last two decades. It is predicted that the use of inverse problems in applications in developing countries will grow due to significant demand from the industrial sector. Therefore, inverse problems relevant to developing countries should be getting a lot more attention. For example, the application of inverse problems in biomedical imaging is relevant in developing countries. Breast cancer is the most common cancer after skin-related cancers in the US. In 2012, 232,714 women in the US were diagnosed with breast cancer, and 43,909 women died from breast cancer, whereas 144,937 women in India were diagnosed with breast cancer during 2012, and 70,218 women died from breast cancer—a significantly higher death rate than in the US [2]. This is partly due to the failure to detect breast cancer early. X-ray mammography is still the predominant method for detection. However X-ray has drawbacks in terms of exposure to harmful radiation and high false-positive rates in younger women. On the other hand, Diffuse Optical Tomography (DOT) has the potential to be a cheaper alternative, particularly useful in developing countries. However, the mathematics and computational challenges of this highly nonlinear ill-posed inverse problem still make it impossible to use it in clinical applications.

Let us describe the difference between a forward and an inverse problem. A forward problem is given by a process where with given/known model parameters q we need to predict the solution or the output/data $u(q)$. For example, q could be parameters involving a partial differential equation (PDE) and u would be the solution of the PDE. Forward problems involving PDEs have been around for hundreds of years, where mathematicians and scientists are looking for appropriate solutions given known parameters q. Many practical applications in fields such as biology, medicine, ecology, geophysics, flexible structures, as well as industry, including biomedical imaging, involve distributed parameter systems, i.e. partial differential equations (PDEs) [1] where the solution u is measured but the model parameters q are unknown. In order to validate models and/or control theoretical issues for these systems, it is often necessary to determine these model parameters q, such as coefficients, boundary terms, initial conditions, and control inputs, given a source or forcing function f and observations of the system z for which a quantitative model is sought. In general the inverse parameter estimation problem requires minimizing an error or a cost functional $J(q; z, f)$. In most cases a regularization procedure is also required to generate a well-posed minimization problem (existence, uniqueness, and stability of the inverse problem with respect to data). If an inverse problem suffers from existence, uniqueness, or stability, it is referred to as an ill-posed problem. Therefore, a typical reformulation of an ill-posed inverse parameter estimation problem requires optimizing $J(q; z, f) + \psi(q)$ where ψ is a regularization functional. Therefore for ill-posed inverse problems, one must investigating several mathematical issues: (i) the choice of appropriate error or cost functional J; (ii) the algorithm to find the parameter q in the appropriate abstract function space given an initial parameter guess q_0; (iii) the selection of the best regularization operator ψ for inverse stability.

Here is an outline of this chapter. In Section 5.2, we describe Electrical Impedance Tomography (EIT) as an example of an ill-posed inverse problem involving PDEs, and in Section 5.3, we describe Diffuse Optical Tomography (DOT) as another example of an ill-posed inverse problem. In Section 5.4, we provide details of computational challenges including regularization. In Section 5.5, we conclude with a summary of future trends in solving ill-posed inverse problems.

5.2 Electrical Impedance Tomography

The Electrical Impedance Tomography (EIT) problem involves measuring electrical voltages on the smooth boundary $\partial\Omega$ to determine the spatially varying electrical conductivity q within the bounded region $\Omega \subseteq R^d (d = 2, 3)$. We assume q is strictly positive, isotropic and bounded conductivity with no current sources inside Ω. The EIT forward problem is typically given by the following elliptic partial differential equation,

$$- \operatorname{div}(q\nabla u) = 0 \quad \text{in } \Omega, \tag{5.1}$$

where $u \in H^1(\Omega)$ is the electric potential and q is known.

In an EIT experiment, an electrical current (Neumann data) f on $\partial\Omega$ is applied and then the resulting electrical potential (Dirichlet data) g on $\partial\Omega$ is measured. The data collected then provides information and is used to approximate q from a set of EIT experiments using different input currents [3, 4, 5]. EIT can be applied to detecting ground water intrusions in geophysical applications [6], monitoring of oil and gas mixtures in oil pipelines [7], and noninvasive medical imaging [8, 9].

5.2.1 Analytical Setting for the EIT Model

We define the following Neumann and Dirichlet boundary value problems

$$- \operatorname{div}(q\nabla u) = 0 \quad \text{in } \Omega, \tag{5.2}$$

$$q\frac{\partial u}{\partial n} = f \quad \text{on } \partial\Omega,$$

and

$$- \operatorname{div}(q\nabla u) = 0 \quad \text{in } \Omega, \tag{5.3}$$

$$u = g \quad \text{on } \partial\Omega.$$

Let the conductivity $q \in Q$, an appropriate metric space. q is assumed to be bounded below and above, i.e. $0 < c_1 \le q \le c_2 < \infty$. In addition, denote by $\Gamma_D u$ the Dirichlet trace operator, i.e. the restriction of u to the boundary

$$\Gamma_D : X \to Z$$

$$u \mapsto \Gamma_D u.$$

As usual in EIT, we restore uniqueness of the solution u of the Neumann problem (2) by requiring that the Dirichlet trace $\Gamma_D u$ satisfies

$$\int_{\partial\Omega} \Gamma_D u(s)ds = 0. \tag{5.4}$$

Note that to ensure the solvability of the Neumann problem in Equation (5.2) the current f must satisfy the integrability condition, which, in the absence of a source term, reads $\int_{\partial\Omega} f(s)ds = 0$. The associated linear forward operator of the Neumann problem, which maps an input current f to the solution u, is denoted by

$$F_N^q : W \to X \tag{5.5}$$

$$f \mapsto u \text{ solves (2)}. \tag{5.6}$$

The linear operator F_D^q for the Dirichlet problem in Equation (5.3) can be defined analogously. The NtD map can be written as $\Gamma_D F_N^q$. The weak formulation of the Neumann problem in Equation (5.2) becomes

$$\int_{\Omega} q \nabla F_N^q(f) \cdot \nabla v dx = \int_{\partial \Omega} f \, \Gamma_D v ds \qquad (5.7)$$

for a suitable set of test functions v. The integral on the boundary should be understood in the sense of duality pairing, i.e. $f \in H^{-1/2}(\partial \Omega)$ and $\Gamma_D v \in H^{1/2}(\partial \Omega)$ yield $\int_{\partial \Omega} f \, \Gamma_D v ds = \langle f, \Gamma_D v \rangle_{H^{-1/2} \times H^{1/2}}$.

There are several natural choices for the spaces X, Y, and W. To this end, we introduce

$$X = \tilde{H}^1(\Omega) = \left\{ u \in L_2(\Omega) \mid \int_{\Omega} q(x) \, |\nabla u(x)|^2 \, dx < \infty, \int_{\partial \Omega} \Gamma_D \, u(s) ds = 0 \right\}. \qquad (5.8)$$

Because of Equation (5.4), the following bilinear form defines a scalar product on this space

$$\langle u, v \rangle_{\tilde{H}^1} = \int_{\Omega} q \nabla u \cdot \nabla v dx. \qquad (5.9)$$

We use the Dirichlet forward operator F_D^q as an extension operator and define the following space of functions on the boundary $\partial \Omega$

$$Z = \tilde{H}^{1/2}(\partial \Omega) = \left\{ g \in L_2(\partial \Omega) \mid \int_{\Omega} q(x) \, |\nabla F_D^q(g)(x)|^2 \, dx < \infty, \int_{\partial \Omega} g(s) ds = 0 \right\} \qquad (5.10)$$

together with its scalar product

$$\langle g, \varphi \rangle_{\tilde{H}^{1/2}} = \int_{\Omega} q(x) \, \nabla F_D^q(g)(x) \cdot \nabla F_D^q(\varphi)(x) dx. \qquad (5.11)$$

The Dirichlet-to-Neumann (DtN) operator $q \dfrac{\partial}{\partial n} F_D^q$ is well defined on $\tilde{H}^{1/2}(\partial \Omega)$, and we introduce

$$W = \tilde{H}^{-1/2}(\partial \Omega) = \left\{ f \mid f = q \frac{\partial}{\partial n} F_D^q(g), \, g \in \tilde{H}^{1/2}(\partial \Omega) \right\} \qquad (5.12)$$

together with its scalar product

$$\langle f, \psi \rangle_{\tilde{H}^{-1/2}} = \int_{\Omega} q \, \nabla F_N^q(g) \cdot \nabla F_N^q(\psi) dx. \qquad (5.13)$$

We observe that $f \in \tilde{H}^{-1/2}(\partial \Omega)$ is the Neumann trace for $u = F_D^q(g)$. This implies $-div(q\nabla u) = 0$ and the integrability condition for Neumann problems yields

$$\int_{\partial \Omega} fds = 0 \quad \text{and} \quad \tilde{H}^{-1/2}(\partial \Omega) = \left\{ f \in H^{-1/2}(\partial \Omega) \mid \int_{\partial \Omega} fds = 0 \right\}. \qquad (5.14)$$

Furthermore the natural choice for a metric space Q is as follows:

$$Q = \left\{ q \in L^{\infty}(\Omega) \mid 0 < c_1 \leq q \leq c_2 < \infty \right\}. \tag{5.15}$$

5.2.2 Inverse Problem in EIT

The forward problem uses knowledge of conductivity parameter q to find the boundary data associated with a given source. The inverse problem instead uses knowledge of the source and boundary data and finds the conductivity interior to the object. The goal is to estimate the conductivity distribution q from all pairs of current and voltage measurements. The identification of the parameter q can be formulated as the following minimization problem for the cost functional

$$J(q) = \left\| \mathscr{F}(q) - g_\delta \right\|_{L_2(\partial\Omega)}^2 \tag{5.16}$$

where g_δ approximate the exact data $g = \mathscr{F}(q)$ with the accuracy δ, i.e.

$$\left\| g - g_\delta \right\| < \delta \tag{5.17}$$

However, because of the ill-posedness of the problem, regularization is needed and one of the well-known regularizations is Tikhonov's regularization, mainly,

$$J_\alpha(q) = \left\| \mathscr{F}(q) - g_\delta(q) - g_\delta \right\|_{L_2(\partial\Omega)}^2 + \alpha \left\| q - q^* \right\|_{L_2(\Omega)}^2 \tag{5.18}$$

where λ is the regularization parameter and q^* is the prior or background parameter.

There are several other regularization approaches (see Section 5.4), for example for a particular EIT application total variation (TV) regularization functional may be most appropriate,

$$J_\alpha(q) = \left\| \mathscr{F}(q) - g^\delta \right\|_{L_2(\partial\Omega)}^2 + \alpha \left\| \nabla q \right\|_{L_1(\Omega)} \tag{5.19}$$

where α is a TV regularization parameter. The regularization parameter α is typically determined on a trial and error basis.

5.2.2.1 Why Is the Inverse Problem Ill-Posed?

The inverse problem here is ill-posed because the EIT model is an elliptic PDE and an elliptic forward operator is a highly smooth operator, which means the information about the parameter q is diffused as it travels to the boundary i.e. the boundary measurements don't have enough information about an inhomogeneity located far away from the boundary. In fact, the information from the data at the boundary about the inhomogeneity is exponentially decayed away from the boundary [10]. This means that if an inhomogeneity is closer to the boundary then the data on the boundary provides better information for a reconstruction of the inhomogeneity. If an inhomogeneity is far away from the boundary, then the data on the boundary does not provide sufficient information due to the exponential decay [10]. So the very fact that elliptic equations can handle rough parameters q and still solve the forward problem for u resulting in a very smooth solution, in turn makes the inverse problem ill-posed. Therefore, EIT is an extremely challenging inverse problem.

We will discuss more about the nonlinearity and ill-posedness of inverse problems while discussing DOT below.

5.2.2.2 Connection to DOT

EIT is a close cousin of DOT because the forward problem for both is elliptic with DOT being worse in the sense that DOT requires estimating two functions mainly $q = (D, \mu_a)$ rather than one parameter for EIT. However DOT is an important modality in practice because DOT can be used as an alternative to X-ray in detecting cancer in the breast and the brain.

5.3 Diffuse Optical Tomography

In optical imaging, low-energy visible light is used to illuminate biological tissue. The illumination of the tissue is modelled as a photon transport phenomenon. The process is described by the most widely applied equation in optical imaging, the radiative transfer, or transport equation (RTE) [11, 12]. RTE is an integro-differential equation for photon density and has spatially dependent diffusion and absorption parameters as coefficients. These coefficients are a priori unknown for a particular tissue sample of an individual who is being examined for cancer. Therefore, the problem is to infer from the measurements of the photon density on the boundary the coefficients of absorption and diffusion in the tissue. This estimate helps determine the location and size of the abnormality in the tissue.

5.3.1 Radiative Transport Equation

Let $\Omega \subset R^n$, with $n = 2, 3$ and with boundary $\partial\Omega$, $\nu(x)$ denote the outward unit normal to $\partial\Omega$ at the point $x \in \partial\Omega$, and $\Gamma_\pm$ is defined as,

$$\Gamma_\pm := \left\{ (x, s, t) \in \partial\Omega \times S^{n-1} \times [0, T], \pm\nu(x) \cdot s > 0 \right\}.$$

For example, one may assume a geometry as shown in Figure 5.1.
   Then the RTE is given by,

$$\frac{1}{c} \frac{\partial u}{\partial t}(x, s, t) + s \cdot \nabla u(x, s, t) + a(x)u(x, s, t)$$

$$- b(x) \int_{S^{n-1}} \Theta(s \cdot s')u(x, s', t)ds' = f(x, s, t) \tag{5.20}$$

together with the initial and boundary conditions,

$$u(x, s, 0) = 0 \text{ in } \Omega \times S^{n-1} \tag{5.21}$$

$$u(x, s, t) = 0 \text{ on } \Gamma_- \tag{5.22}$$

where $u(x, s, t)$ describes the density function of particles (photons) which travel in Ω at time t though the point x in the direction $s \in S^{n-1}$, unit sphere in R^n. The parameter a is

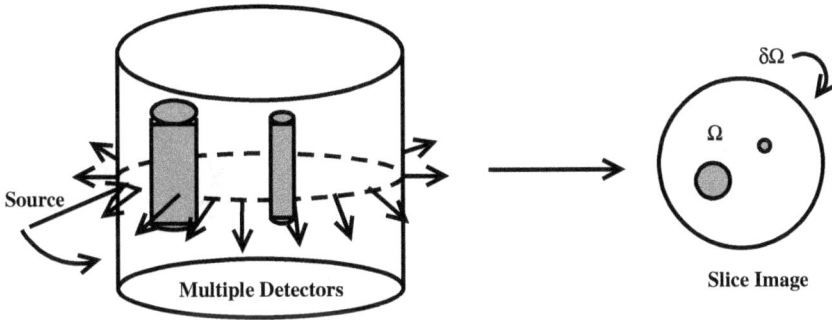

FIGURE 5.1
Tomographic imaging.

the total cross section or attenuation, and b is the scattering cross section. The difference $\mu := a - b$ has the physical meaning of an absorption cross section. The parameters a, b and μ are assumed to be real, nonnegative, and bounded functions of the position. The parameters a and b are the sought-for tissue parameters and c is the velocity of light. The function Θ is the scattering phase function characterizing the intensity of a wave incident in direction s' scattered in the direction s. It is assumed to be a real, nonnegative function and is normalized to one. The inverse problem is to recover a and b from measurements of some given functionals of the outgoing density $u_j \mid_{\Gamma_+}$ at the boundary $\partial \Gamma$ for m_s different set of source distribution f_j, $j = 1, \ldots, m_s$.

5.3.2 DOT Model

Simpler deterministic models can be derived from RTE by expanding the density u and source f in spherical harmonics and retaining a limited number of terms [13, 14, 15, 16]. Due to the prevalence of scattering, the flux is essentially isotropic a small distance away from the sources; i.e. it depends only linearly on s. Thus we may describe the process adequately by the first two moments of u. We get diffusion approximation P_0. Frequency-domain diffusion approximation can easily be obtained by Fourier transforming the time-domain equation. Furthermore, the diffusion approximation to the radiative transfer model can be written in the time independent (dc) case as in [16],

$$-\nabla \cdot D\nabla u + \mu_a u = 0. \tag{5.23}$$

The associated boundary condition is

$$u + 2D \frac{\partial u}{\partial v} = f, \ x \in \partial \Omega. \tag{5.24}$$

If we let Ω be the domain under consideration with surface $\partial \Omega$, the weak forward problem corresponding to equation (23) is to find $u \in H^1(\Omega)$ such that for all $v \in H^1(\Omega)$, the following variational equation is satisfied,

$$\int_{\Omega} D\nabla v \cdot \nabla u dx + \int_{\Omega} v \, \mu_a u dx + \int_{\partial \Omega} \frac{1}{2} vu ds = \int_{\partial \Omega} vf ds. \tag{5.25}$$

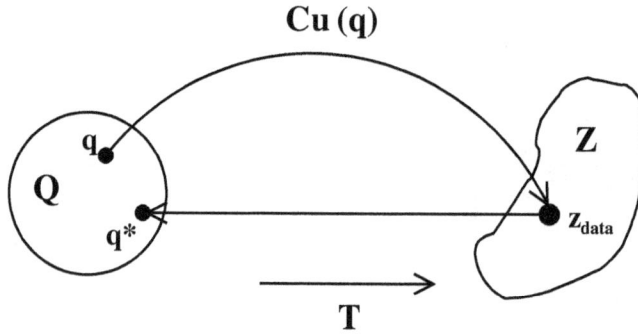

FIGURE 5.2
Nonlinear mapping.

Now, we can define the forward problem as: given sources f_j in $\partial\Omega$ and q in Q, a vector of model parameters, for example the coefficient of diffusion D and the coefficient of absorption μ_a (i.e. $q = (D, \mu_a)^T$) that belongs to a parameter set Q, find the data u on $\partial\Omega$ and the inverse problem as: given data z on $\partial\Omega$ find q. We can recast the forward problem in an abstract setting with u in an appropriate abstract space H, and f represents a source or a forcing distribution. In general, measurement of u may not be possible; only some observable part $z = \mathscr{C}u(q)$ of the actual state $u(q)$ may be measured. In this abstract setting, the objective of the inverse or parameter estimation problem is to choose a parameter q^* in Q, that minimizes an error criterion or cost functional $J(u(q), \mathscr{C}u(q), q)$ over all possible q in Q subject to $u(q)$ satisfying the diffusion approximation. A typical observation operator is,

$$\mathscr{C}^f u(q) = \left\{ -D\frac{\partial u}{\partial v}(x_i; q, f) \right\}_{i=1}^{m} \tag{5.26}$$

where x_i is in $\partial\Omega$, m is the number of measurements, and the second equality comes from the boundary condition (24). A typical Tikhonov cost functional J_λ is given as,

$$J_\lambda(q) = \frac{1}{2}\sum_{j=1}^{m_s}\sum_{i=1}^{m} w_{ij}\left|\mathscr{C}_i^{f_j}u(q) - z_i^{f_j}\right|^2 + \lambda\|q - q_0\|^2 \tag{5.27}$$

where $z_i^{f_j}$ is the measured data at the boundary for a given source f_j, w_{ij} is the weight for ij-th data, and λ is the Tikhonov regularization parameter. As shown in Figure 5.2, composing $u(q)$ and $\mathscr{C}u(q)$ we obtain the parameter-to-output mapping:

$$T : Q \to Z$$

such that $Tq = \mathscr{C}u(q)$, where Z is the space of measurments. This is the nonlinear mapping of DOT in abstract setting. The map from Q to Z is nonlinear because the solution of a partial differential equation is a nonlinear function of its coefficients $q = (D, \mu_a)$.

5.4 Computational Aspect and the Regularization of the Ill-Posed Problem

There are various approaches for solving this nonlinear ill-posed problem; we outline a few approaches in this section. In general for complex geometries, the analytic solution is intractable. Therefore, one requires numerical solutions. The finite element method (FEM) is somewhat more versatile because of its ease in complex geometries and modelling boundary effects. The FEM is a variational method used to approximate the solution by a family of finite dimensional basis functions. Then the forward problem is reduced to one of linear algebra and one computes the approximate solution using FEM codes. The FEM is derived by projecting the weak form of Equation (5.25) onto a finite dimensional function space. For example, the finite dimensional function space could be the set of continuous and twice differentiable, piecewise cubic polynomials.

More precisely, for example in EIT, we can project the infinite dimensional solution space X into a finite dimensional subspace $X_K \subset X$ which means to restrict u and v above to lie in X_K rather than X. Let us assume that there exists a family of basis functions $\phi_m(x)$ for $m = 1, \cdots, K$ for X_K. Then let $u_K = \sum_{m=1}^{K} c_m \phi_m(x)$ to be the approximation to u and we want the weak formulation to be satisfied for the K test functions $\phi_k = v$ for $k = 1, \cdots, K$. Now plugging this into the weak formulation we obtain the system of equations for c_m:

$$\sum_{m=1}^{K} \left(\int_{\Omega} q \nabla \phi_m \cdot \nabla \phi_k dx \right) c_m = \int_{\partial\Omega} f \Gamma_D \phi_k ds \tag{5.28}$$

for $k = 1, \cdots, K$. If we denote

$$A_{mk} = \int_{\Omega} q \nabla \phi_m \cdot \nabla \phi_k dx$$

$$F_k = \int_{\partial\Omega} f \Gamma_D \phi_k ds$$

then we can solve for the solution $c = (c_m)_{m=1}^{K}$, which depends nonlinearly on q using the linear system $Ac = F$ where A is a $K \times K$ matrix and F is a vector of length K. We can also approximate the infinite dimensional parameter space Q by a finite dimensional subspace $Q_M \subset Q$. The solution to the finite dimensional problem is the solution q_M in Q_M that is closest to the infinite dimensional optimization problem, for example $J_\alpha(q)$. Therefore we assume that there exists a family of basis functions $\psi_k(x)$ for $k = 1, \cdots, M$ for Q_M. Then set $q_M = \sum_{k=1}^{M} q_k \psi_k(x)$ to be the approximation to q and we arrive at the finite dimensional nonlinear optimization problem for $q_M = (q_k)_{k=1}^{M}$:

$$\hat{J}_{\alpha_1, \alpha_2}(q_M) = \sum_{k=1}^{\hat{K}} |\hat{g}_k(j) - \hat{g}_k^\delta|^2 + \alpha_1 \sum_{i=1}^{M} \delta_i |q_i - q_i^*|^p + \alpha_2 \sum_{j=1}^{Z} d_j |\Delta_j q_M|. \tag{5.29}$$

where $1 \leq p \leq 2$ and δ_i's are weights, $\hat{g}_k(f) = \Gamma_D u_K^{q_M}(f)[\hat{x}_k]$ is the trace of the solution evaluated at $\hat{K}$ boundary points $\hat{x}_k$, $\hat{g}_k^\delta$ is the noisy voltage data collected at the boundary point $\hat{x}_k$, the second term is the approximation to the TV term using a nearest neighbor approximation where Z is the number of nearest neighbors, and d_j is the distance between two neighbors [18, 17, 19].

EIT is well known to suffer from a high-degree of nonlinearity and severe ill-posedness [20, 21]. Therefore, regularization is required to produce reasonable electrical impedance images. Most reconstruction methods are deterministic, such as the factorization method

[23], d-bar method [24], and variational type methods for least-squares fitting. The variational type methods uses an iterative type method of a linearized model or fully nonlinear model such as sparisty cosnstraints [20, 21], iteratively regularized Gauss–Newton method [22]. These analytical methods can be effective in determining specific conductivity, but statistical inversion methods [25] can offer an alternative approach. In [26], Kaipio et al. optimizes the current patterns based on criteria regarding functionals of the posterior covariance matrix. The Bayesian approach has also been used to study the errors from model reduction and partially unknown geometry [27, 28]. The sparsity regularization for statistical inversion enforces the ℓ_p prior to the expansion coefficients for a certain basis like the deterministic approaches for EIT [20, 21, 29].

Since we have a nonlinear forward and inverse operator, any iterative algorithm, requires computing the Jacobian of the forward operator. Under the regularity assumptions on the domain and the coefficients, the forward operator can be shown to be differentiable.

Theorem 5.1: The operator $\mathscr{F}$ which maps q to the solution $u(q) \in H$ of the forward problem with current f is Fréchet differentiable. If $\eta \in L^\infty(\Omega)$ is such that $q + \eta \in \mathcal{Q}$, then the derivative $\mathscr{F}'(q)\eta = w$ satisfies the following variational problem

$$b(w, v) = -\int_\Omega \eta \, \nabla u \nabla v \, dx \tag{5.30}$$

for all $v \in H$, where $u = \mathcal{F}(q)$. Using the theorem above, computing the Jacobian is explained in [21].

5.4.1 Iteratively Regularized Gauss–Newton Method

Suppose λ_k is a sequence of regularizing parameters [30]. A general algorithm is given by Smirnova [30] using a line search procedure with a variable step size α_k such that

$$0 < \alpha_k \le 1 \tag{5.31}$$

yielding the following Iteratively Regularized Gauss–Newton (IRGN) algorithm

$$q_{k+1} = q_k - \alpha_k (\mathscr{F}'(q_k)^T \mathscr{F}'(q_k) + \lambda_k W_2)^{-1} \{\mathscr{F}'(q_k)^T (\mathscr{F}(q_k) - g_\delta) + \lambda_k W_2 (q_k - q^*)\} \tag{5.32}$$

where W_2 is a preconditioning matrix. Due to the inexact nature of g_δ, we adopt a stopping rule presented in [31] to terminate the iterations at the first index $\mathscr{K} = \mathscr{K}(\delta)$, such that

$$\left\| \mathscr{F}(q_\mathscr{K}) - g_\delta \right\|^2 \le \rho \delta < \left\| \mathscr{F}(q_k) - g_\delta \right\|^2, \quad 0 \le k \le \mathscr{K}, \quad \rho > 1 \tag{5.33}$$

The line search parameter α_k is chosen appropriately with a search direction using a backtracking strategy until either the strong or weak Wolfe conditions are satisfied [32], or a maximum number of backtracking steps have been taken.

5.4.2 The Statistical Inverse Problem

Inverse problems are typically written in terms of a minimization problem. Here, however, we can also write the finite dimensional inverse problem after discretization in terms of the posterior density of the conductivity q_M given the measurements g_K on $\partial\Omega$. In other words, if we know the density of the conductivity q_M given the measurements g_K we can obtain the expected value of the conductivity given the measurements. This estimate is a reasonable point estimate of the solution of the ill-posed inverse problem. In the statistical setting, one derives the posterior density of the finite dimensional version of the conductivity mainly q_M given the finite dimensional measurement $g_{\hat{K}}$ on $\partial\Omega$ [19, 17, 18].

5.4.2.1 The Posterior Density

The density of q_M^* is usually called prior density. This is because it contains all the information about q_M^* that we believe to be true. Here we assume that

$$\pi_{q_M^*}(q_M) \propto \chi_A(q_M) \exp\left[-\alpha R(q_M)\right], \tag{5.34}$$

where $R(\cdot)$ is a regularization function, $\alpha > 0$ a constant, and $\chi_A(q_M)$ an indicator function with $A = [0, \infty)^n$. In the following section we discuss several common choices for the regularizing function $R(\cdot)$.

5.4.3 Regularization Functions

In this section we discuss several choices for the regularization function $R(\cdot)$. We have several choices for the regularization function $R(\cdot)$ which are used in the analytical and the statistical settings.

5.4.3.1 The ℓ_p Regularizations

The idea of the ℓ_p regularization is to force the difference of the parameters q_M and the background q_M^b mainly $q_M - q_M^b$ to be sparse with respect to some basis. For example, when using EIT to reconstruct an object mainly made of concrete we would choose q_M^b to be the typical conductivity of concrete. The ℓ_p regularization $R_{\ell_p}(q_M)$ is defined as

$$R_{\ell_p}(q_M) := \sum_{i=1}^{n} c_i \mid q_M(i) - q_M^b(i) \mid^p, \tag{5.35}$$

where c_i represents weights and $0 < p \leq 2$ a constant [25]. In theory a good choice for the weights would be large values at the boundary and exponentially decreasing values toward the center of Ω. This is because the variance is smaller on the boundary than in the center [10]. The ℓ_p regularization enforces sparsity when $0 < p \leq 1$ and enforces smoothness when $p \geq 2$.

5.4.3.2 The Total Variation Regularization

The idea behind the total variation regularization is to obtain smooth images. This is meaningful in most practical applications. The total variation regularization is defined as

$$R_{TV_c}(q) := \int_{\Omega} \mid \nabla q \mid dx, \tag{5.36}$$

where q the continuous version of the parameter of interest q_M. The discrete analogue for a two-dimensional body of the total variation regularization R_{TV_c} [25] is

$$R_{TV}(q_M) := \sum_{i=1}^{h} l_i \mid \Delta_i q_M \mid, \tag{5.37}$$

where l_i is defined as the length of the edge corresponding to the *ith* adjacent pixel and

$$\Delta_i = (0, 0, ..., 0, 1_{a_{(1,i)}}, 0, ..., 0, -1_{a_{(2,i)}}, 0, ..., 0), \tag{5.38}$$

with $a = (a_{(j,i)})_{i=1, j \in \{1,2\}}^{h}$ as the set containing the numbers of all adjacent pixel tuples $(a_{(1,i)}, a_{(2,i)})$.

5.4.4 The Markov Chain Monte Carlo Method

In the previous sections we discussed the posterior density with several meaningful prior densities (regularizations), and hence to obtain a good estimate for q_M^* based on the measurements $g_{\hat{k}}$. That is, the algorithm seeks to find the Bayesian estimate $E(q_M^* \mid g_{\hat{k}}) = \int\limits_{R^n} q_M \, \pi_{q_M^*}(q_M \mid g_{\hat{k}}) dq_M.$

Given that the posterior density $\pi_{q_M^*}(q_M \mid g_{\hat{k}})$ does not have a closed form, there is no direct method of finding the Bayesian estimate $E(q_M^* \mid g_{\hat{k}})$. Therefore, one uses the Markov Chain Monte Carlo Method (MCMC) to generate a large random sample $\{q_M^{(i)}\}_{i=1}^{N}$ from the posterior density $\pi_{q_M^*}(q_M \mid g_{\hat{k}})$ and then approximate the Bayesian estimate by its sample mean,

$$E(q_M^* \mid g_{\hat{k}}) = \int\limits_{R^n} q_M \, \pi_{q_M^*}(q_M \mid g_{\hat{k}}) dq_M \approx \frac{1}{N} \sum_{i=1}^{N} q_M^{(i)}. \tag{5.39}$$

Typical algorithms to generate such random samples from a posterior density are the Gibbs sampler or the Metropolis–Hastings algorithm [33, 19, 34].

5.5 Conclusion

In this chapter, we have introduced inverse problems involving PDEs for imaging applications. We described the challenges and difficulties of solving an ill-posed inverse problem. We have provided details of two very well-known examples namely EIT and DOT. A range of both deterministic and statistical regularization approaches have been exposed and summarized in this chapter. The computational approaches have been discussed without any rigorous analysis of the convergence rates. The future challenges in this area can be overcome by reformulating the inverse question either using a regularization approach such as a combination of smoothness and sparsity or statistical approaches or weakening the inverse question itself. The progress in the area of inverse problems in imaging is expected to be in the intersection of computation, statistical, and analytical approaches to understand the input–output behavior of the inverse operator. One of the latest approaches proposed involves machine learning and training methods for inverse problems. There is already significant interest in that direction in the latest literature, however many questions about machine learning and adaptive training methods are still open and the subject of future research.

The author would like to thank the organizers of ISIAM, particularly Professor Pammy Manchanda and Professor A. H. Siddiqi, for editing this chapter and for the opportunity to attend the conference at Amritsar that brought together both pure and applied mathematicians to help provide the platform to discuss various ideas for applied math research.

References

1. Banks, H.T., & Kunisch, K. (2001). *Estimation Techniques for Distributed Parameter Systems*. Birkhauser.
2. Goodarzi, E., Moayed, L., Sohrabivafa, M., Adineh, H.A., & Khazei, Z. (2019). Epidemiology incidence and mortality of breast cancer and its association with the body mass index and human development index in the asian population. *World Cancer Research Journal, WCRJ* 6, e1323.
3. Cheney, M., Isaacson, D., & Newell, J. C. (1999). Electrical impedance tomography. *SIAM Review*, 41(1), 85–101.
4. Borcea, L. (2002). Electrical impedance tomography. *Inverse Problems*, 18(6), R99–R136.
5. Hanke, M., & Brühl, M. (2003). Recent progress in electrical impedance tomography. *Inverse Problems*, 19(6), S65–S90.
6. Daily, W., Ramirez, A., LaBrecque, D., & Nitao, J. (1992). Electrical resistivity tomography of vadose water movement. *Water Resources Research*, 28(5), 1429–1442.
7. Isaksen, O., Dico, A.S., & Hammer, E.A. A capacitance-based tomography system for interface measurement in separation vessels. *Measurement Science and Technology*, 5(10), 1262.
8. Holder, D.S. (ed.). (2005). *Electrical Impedance Tomography: Methods, History and Applications*. Institute of Physics Publishing, Bristol.
9. Bayford, R.H. (2006). Bioimpedance tomography (electrical impedance tomography). *Annual Review of Biomedical Engineering*, 8, 63–91.
10. Palamodov, V.P. (2002). Gabor analysis of the continuum model for impedance tomography. *Arkiv för Matematik*, 40(1), 169–187.
11. Chandrasekhar, R. (1950). *Radiation Transfer*. Oxford, Clarendon.
12. Ishimaru, A. (1978). *Single Scattering and Transport Theory (Wave Propogation and Scattering in Random Media I)*. Academic, New York.
13. Lewis, H.W. (1950). Multiple scattering in an infinite medium. *Physical Review*, 78, 526–529.
14. Bremmer, H. (1964). Random volume scattering. *Radiation Science Journal of Research*, 680, 967–981.
15. Arridge, S.R., & Hebden, J.C. (1997). Optical imaging in medicine: 2. Modelling and reconstruction. *Physics in Medicine and Biology*, 42, 841–853.
16. Arridge, S.R. (1999). Optical tomography in medical imaging: Topical review. *Inverse Problems*, 15, R41–R93.
17. Strauss, T. (2015). *Statistical Inverse Problems in Electrical Impedance and Diffuse Optical Tomography*. Doctoral dissertation, Clemson University.
18. Strauss, T., & Khan, T. (2015). Statistical inversion in electrical impedance tomography using mixed total variation and non-convex l_p regularization prior. *Journal of Inverse and Ill-Posed Problems*, 23(5), 529–542.

19. Strauss, T., Fan, X., Sun, S., & Khan, T. (2015). Statistical inversion of absolute permeability in single-phase darcy dlow. *Procedia Computer Science*, 51, 1188–1197.
20. Jin, B., & Maass, P. (2012). An analysis of electrical impedance tomography with applications to Tikhonov regularization. *ESAIM: Control, Optimisation and Calculus of Variations*, 18(04), 1027–1048.
21. Jin, B., Khan, T., & Maass, P. (2012). A reconstruction algorithm for electrical impedance tomography based on sparsity regularization. *International Journal for Numerical Methods in Engineering*, 89(3), 337–353.
22. Khan, T., & Smirnova, A. (2005). 1D inverse problem in diffusion based optical tomography using iteratively regularized Gauss-Newton algorithm. *Applied Mathematics and Computation*, 161(1), 149–170.
23. Kirsh, A., & Grinberg, N. (2008). *The Factorization Method for Inverse Problems*. Oxford University Press.
24. Isaacson, D., Mueller, J. L., Newell, J. C., & Siltanen, S. (2004). Reconstructions of chest phantoms by the D-bar method for electrical impedance tomography. *IEEE Transcation on Medical Imaging*, 23(7), 821–828.
25. Kaipio, J.P., Kolehmainen, V., Somersalo, E., & Vauhkonen, M. (2000). Statistical inversion and Monte Carlo sampling methods in electrical impedance tomography. *Inverse Problems*, 16(5), 1487.
26. Kaipio, J.P., Seppänen, A., Somersalo, E., & Haario, H. (2004). Posterior covariance related optimal current patterns in electrical impedance tomography. *Inverse Problems*, 20(3), 919.
27. Nissinen, A., Heikkinen, L.M., & Kaipio, J.P. (2008). The Bayesian approximation error approach for electrical impedance tomography experimental results. *Measurement Science and Technology*, 19(1), 015501.
28. Nissinen, A., Heikkinen, L.M., Kolehmainen, V., & Kaipio, J.P. (2009). Compensation of errors due to discretization, domain truncation and unknown contact impedances in electrical impedance tomography. *Measurement Science and Technology*, 20(10), 105504.
29. Jin, B., & Maass, P. (2012). Sparsity regularization for parameter identification problems. *Inverse Problems*, 28(12), 123001.
30. Smirnova, Alexandra, Renaut, Rosemary A., & Khan, Taufiquar. (2007). Convergence and application of a modified iteratively regularized Gauss-Newton algorithm. *Inverse Problems*, 23, 1547–1563.
31. Bakushinsky, A.B., & Smirnova, A. (2005). On application of generalized discrepancy principle to iterative methods for nonlinear ill-posed problems, *Numer. Func. Anal. Optim.*, 26, 35–48.
32. Nocedal, J., & Wright, S.J. (1999). *Numerical Optimization*, Springer, NY.
33. Chib, S., & Greenberg, E. (1995). Understanding the metropolis-hastings algorithm. *The American Statistician*, 49(4), 327–335.
34. Bardsley, J.M. (2012). MCMC-based image reconstruction with uncertainty quantification. *SIAM Journal on Scientific Computing*, 34(3), A1316–A1332.

6

On an Efficient Family of Steffensen-Like Methods for Finding Nonlinear Equations

Vali Torkashvand and Manochehr Kazemi

CONTENTS

6.1   Introduction 91
6.2   A Class of Without-Memory Methods 92
6.3   The New Methods with Memory 94
6.4   Numerical Examples 96
6.5   Conclusion 98
References 101

6.1 Introduction

Nonlinear equations are ubiquitous in many areas of applied mathematics and play vital roles in a number of applications such as science and engineering. Nonlinear equations are usually difficult to solve analytically, therefore a numerical method is needed. Construction of iterative methods to approximate a solution for solving nonlinear equation $f(x)=0$ is one of the most important tasks in applied mathematics. The best-known iterative method for determining the solution of this problem is Newton's method (NM) given by $x_{k+1} = x_k - \dfrac{f(x_k)}{f'(x_k)}$, which produces a sequence $\{x_n\}$ quadratically convergent to the simple root α, if the initial approximation x_0 is sufficiently close to α and f' exists near α. The key idea by Traub [16] that provides the order acceleration lies in a special form of the error relation and a convenient choice of a free parameter. We define a self-accelerating parameter, which is calculated during the iterative process using Newton's interpolating polynomial. Accelerating technique relies on information from the current and the previous iterative step, defining in this way two-point methods with memory. The rest of this chapter is organized as follows: Section 6.2 is devoted to modifications of the two steps of Guo and Qian methods [11]. Section 6.3 is concerned with developing new with-memory methods. Section 6.4 includes some numerical performances and conclusions are made in Section 6.5.

6.2 A Class of Without-Memory Methods

In 2015 Guo and Qian followed the two-point methods without memory (GQM) [11]:

$$\begin{cases} w_k = x_k + f(x_k), y_k = x_k - \dfrac{f(x_k)}{f[x_k, w_k]}, \quad k = 0, 1, 2, \ldots, \\[2ex] x_{k+1} = y_k - \dfrac{(x_k - y_k)(x_k - w_k)(y_k - w_k)f(y_k)}{(y_k - w_k)^2 f(x_k) + (x_k - w_k)(x_k - 2y_k + w_k)f(y_k) - (x_k - y_k)^2 f(w_k)}. \end{cases} \tag{6.1}$$

The next theorem states the error equations in the methods in Equation (6.1).

Theorem 6.1 *Let $I \subseteq \mathbb{R}$ be an open interval, $f : I \to \mathbb{R}$ be a scalar function which has a simple root α in the open interval I, and also the initail approximation x_0 is sufficiently close the simple zero, then the iteration method defined by (1) is of fourth-order convergence and it satisfies the error equation below:*

$$e_{k+1} = (1 + f'(\alpha))^2 c_2 (c_2^2 - c_3) e_k^4 + O(e_k^5). \tag{6.2}$$

Proof 6.1 *The proof is given in Guo and Qian [11].*

By imposing two parameters in the method described in Equation (6.1), the family of two-parameters without-memory method of the following fourth-order result (TKM4):

$$\begin{cases} w_k = x_k + \gamma f(x_k), \ y_k = x_k - \dfrac{f(x_k)}{f[x_k, w_k] + \lambda f(w_k)}, \quad k = 0, 1, 2, \ldots, \\[2ex] x_{k+1} = y_k - \dfrac{(x_k - y_k)(x_k - w_k)(y_k - w_k)f(y_k)}{(y_k - w_k)^2 f(x_k) + (x_k - w_k)(x_k - 2y_k + w_k)f(y_k) - (x_k - y_k)^2 f(w_k)}. \end{cases} \tag{6.3}$$

Where $\gamma \in \mathbb{R}$. The next theorem states the error equations in the methods of Equation (6.3).

Theorem 6.2 *Let $I \subseteq \mathbb{R}$ be an open interval, $f : I \to \mathbb{R}$ be a scalar function which has a simple root α in the open interval I, and also the initail approximation x_0 is sufficiently close the simple zero, then, the iteration method (3) has fourth-order convergence and satisfies the following error equation:*

$$e_{k+1} = (1 + f'(\alpha)\gamma)^2 (\lambda + c_2)(c_2(\lambda + c_2) - c_3) e_k^4 + O(e_k^5). \tag{6.4}$$

whererin $c_k = \dfrac{f^k(\alpha)}{k!}, \forall k \in \{2, 3, 4, \cdots\}, e_k = x_k - \alpha$ and λ, γ are two real valued free parameters.

Proof 6.2 *Using Taylor's series and symbolic computation; we can determine the asymptotic error constant of the two-step without memory family of methods (3). Now we expand $f(x_k)$ about the simple zero α. Hence, we have*

$$f(x_k) = f'(\alpha)[e_k + c_2 e_k^2 + c_3 e_k^3 + c_4 e_k^4 + O(e_k^5)], \tag{6.5}$$

and

$$w_k - \alpha = x_k - \alpha - e_k f'(\alpha)\gamma(1 + e_k(c_2 + c_3 e_k^3 + c_4 e_k^4))), \tag{6.6}$$

Using the first-order divided difference of the function $f[x,w] = \dfrac{f(x) - f(w)}{x - w}$, we have:

$$
\begin{aligned}
f[x_k, w_k] &= \frac{1}{e_k \gamma(1 + e_k(c_2 e_k(c_3 + c_4 e_k)))} (e_k^2 c_2 + e_k^3 c_3 + e_k^4 c_4 + e_k f'(\alpha)\gamma(1 \\
&\quad + e_k(c_2 e_k(c_3 + c_4 e_k))) - c_2(e_k - e_k f'(\alpha)\gamma(1 + e_k(c_2 + e_k(c_3 + e_k c_4))))^4 \\
&\quad - c_3(e_k - e_k f'(\alpha)\gamma(1 + e_k(c_2 + ek(c_3 + e_k c_4))))^4 - c_4(e_k - e_k \gamma f'(\alpha) \\
&\quad \cdot (1 + e_k(c_2 + e_k(c_3 + e_k c_4))))^4)
\end{aligned}
$$

Furthermore, it is easy to find:

$$
\begin{aligned}
\frac{f(x_k)}{f[x_k, w_k] + \lambda f(w_k)} &= (e_k f'(\alpha)(1 + e_k(c_2 + e_k c_3))) / [[[c_2 e_k^2 + e_3^k c_3 + e_k \gamma f'(\alpha) \\
&\quad \cdot (1 + e_k(c_2 + e_k c_3)) - c_2(e_k - e_k f'(\alpha)(1 + e_k(c_2 + e_k c_3)))^2 \\
&\quad - c_3(e_k - e_k \gamma f'(\alpha)(e_k f'(\alpha)(1 + e_k(c_2 + e_k c_3))^3] / e_k \gamma(1 + e_k(c_2 \\
&\quad + e_k c_3))] f'(\alpha)\lambda(e_k - e_k \gamma f'(\alpha)(1 + e_k(c_2 + e_k c_3)) + c_2(e_k \\
&\quad - e_k \gamma f'(\alpha)(1 + e_k(c_2 + c_3 e_k))^2 + c_3(e_k - e_k \gamma f'(\alpha)(1 + e_k(c_2 \\
&\quad + e_k c_3)) + c_2(e_k - e_k \gamma f'(\alpha)e_k c_3)))^3)]
\end{aligned}
$$

Using Equation (6.7) and some simplifications, we get:

$$
\begin{aligned}
y_k - \alpha &= x_k - \alpha - \frac{f(x_k)}{f[x_k, w_k] + \lambda f(w_k)} = -(-1 + f'(\alpha)\gamma)(\lambda + c_2)e_k^2 + (\lambda^2(-1 + f'(\alpha)\gamma)^2 \\
&\quad - 2f'(\alpha)\gamma)(-2 + \gamma f'(\alpha))c_2 + (\lambda + c_2) + 2c_3 + f'(\alpha)\gamma)(-3 + f'(\alpha)\gamma)c_3)e_k^3 \\
&\quad + (-\lambda^3(-1 + f'(\alpha)\gamma)^3 + \lambda(5 - f'(\alpha)\gamma(7 + f'(\alpha)\gamma(-4 - f'(\alpha)\gamma)))c_2^2 \\
&\quad + (4 - \gamma f'(\alpha)(5 + f'(\alpha)\gamma(-3 + f'(\alpha)\gamma)))c_2^3 + \lambda(-4 + f'(\alpha)\gamma(7 + f'(\alpha)\gamma(-5 \\
&\quad + f'(\alpha)\gamma)))c_3 + c_2(-\lambda^2(-1 + f'(\alpha)\gamma(3 + f'(\alpha)\gamma(-2 + f'(\alpha)\gamma) + (-7 + f'(\alpha) \\
&\quad \cdot \gamma(10 + f'(\alpha)\gamma(-7 + 2f'(\alpha)\gamma)))c_3)(3 + f'(\alpha)\gamma(-3 + f'(\alpha)\gamma))c_4)(e_k^4) + O(e_k^5)
\end{aligned}
$$

We use the following relationships:

$$\frac{(x_k - y_k)(x_k - w_k)(y_k - w_k)f(y_k)}{(y_k - w_k)^2 f(x_k) + (x_k - w_k)(x_k - 2y_k + w_k)f(y_k) - (x_k - y_k)^2 f(w_k)}$$

$$= -(-1 + f'(\alpha)\gamma)(\lambda + c_2)e_k^2 + (\lambda^2(-1 + f'(\alpha)\gamma)^2 2f'(\alpha)\gamma)(-2 + \gamma f'(\alpha)))c_2$$

$$+ (\lambda + c_2) + 2c_3 + f'(\alpha)\gamma)(-3 + f'(\alpha)\gamma)c_3)e_k^3 + (-\lambda^3(-1 + f'(\alpha)\gamma)^3$$

$$+ \lambda(3 - f'(\alpha)\gamma(3 + f'(\alpha)\gamma(-2 - f'(\alpha)\gamma)))c_2^2 + (3 - \gamma f'(\alpha)(3 + f'(\alpha)\gamma$$

$$\cdot (-2 + f'(\alpha)\gamma)))c_2^3 + \lambda(-3 + f'(\alpha)\gamma(5 + f'(\alpha)\gamma(-4 + f'(\alpha)\gamma)))c_3 + c_2(-\lambda^2$$

$$\cdot (-1 + f'(\alpha)\gamma(2 + f'(\alpha)\gamma(-1 + f'(\alpha)\gamma) + 2(-3 + f'(\alpha)\gamma(4 + f'(\alpha)\gamma$$

$$\cdot (-3 + f'(\alpha)\gamma)))c_3)(3 + f'(\alpha)\gamma(-3 + f'(\alpha)\gamma))c_4)(e_k^4) + O(e_k^5)$$

And finally by considering Equation (6.10),

$$x_{k+1} - \alpha = y_k - \alpha$$

$$- \frac{(x_k - y_k)(x_k - w_k)(y_k - w_k)f(y_k)}{(y_k - w_k)^2 f(x_k) + (x_k - w_k)(x_k - 2y_k + w_k)f(y_k) - (x_k - y_k)^2 f(w_k)}$$

$$= (1 + f'(\alpha)\gamma)^2(\lambda + c_2)(c_2(\lambda + c_2) - c_3)e_k^4 + O(e_k^5)$$

which manifests that Equation (6.3) is a two-step iterative method of optimal order of convergence of four consuming three function evaluations.

For more information on optimal methods, pay attention to the following conjecture: Kung and Traub conjecture [16]: Multipoint iteration methods without memory based on n evaluations, could achieve optimal convergence order 2^{n-1} (as a simple zero-finder).

6.3 The New Methods with Memory

In this section, we will improve the convergence rate of the family Equation (6.3) by varying the parameters γ and, λ in per full iteration. We observe form Equation (6.11) that the order of convergence of the family Equation (6.3) is four when $\gamma \neq \dfrac{-1}{f'(a)}$ and, $\lambda \neq c_2$. With the choice $\gamma = \dfrac{-1}{f'(a)}$ and, $\lambda = c_2$ it can be proved that the order of the family Equation (6.3) would be 7. For this purpose, first we put $\gamma = \gamma_k$, and, $\lambda = \lambda_k$ then, approximation of these parameters are given by:

$$\begin{cases} \gamma_k = -\dfrac{1}{f'(\alpha)} \approx -\dfrac{1}{\tilde{c}_1} = -\dfrac{1}{N_3'(x_k)} \\ \lambda_k = -c_2 \approx -\tilde{c}_2 = -\dfrac{N_4''(w_k)}{N_4'(w_k)}. \end{cases} \tag{6.7}$$

where $N_3(t), N_4(t)$, are Newton's interpolating polynomial of third and fourth degree. Therefore, we have considered the best possible choices for nodal points to obtain the maximal order. Therefore, the modified family is given by (TKM7):

$$\begin{cases} \gamma_k = -\dfrac{1}{N_3'(x_k)}, \quad \lambda_k = -\dfrac{N_4''(w_k)}{N_4'(w_k)}, \quad k=1,2,\ldots, \\[2ex] w_k = x_k + \gamma_k f(x_k), \quad y_k = x_k - \dfrac{f(x_k)}{f[x_k,w_k]+\lambda_k f(w_k)}, \quad k=0,1,2,\ldots, \\[2ex] x_{k+1} = y_k - \dfrac{(x_k-y_k)(x_k-w_k)(y_k-w_k)f(y_k)}{(y_k-w_k)^2 f(x_k)+(x_k-w_k)(x_k-2y_k+w_k)f(y_k)-(x_k-y_k)^2 f(w_k)}. \end{cases} \quad (6.8)$$

Recall that a multipoint method with memory reuses old information, for example, some entries from the previous iteration. In this section, we will improve the convergence rate of the family in Equation (6.8) using an old idea given, consisting of varying the parameters $\gamma=\gamma_k$ and $\lambda=\lambda_k$ as the iteration proceeds. We aim at accelerating convergence without imposing further functional evaluations per cycle. This means that using two function evaluations, we must increase the R-order of convergence by more than 4. Here, we attempt to prove that the R-order of convergence of a new derivative-free methodsin Equation (6.8) with memory is 6. For this purpose, we state the following lemma.

Lemma 6.1 *Let* $\gamma_k = -\dfrac{1}{N_3'(x_k)}$, *and* $\lambda_k = -\dfrac{N_4''(w_k)}{N_4'(w_k)}$ *where* $e_k = x_k-\alpha, e_{k,y}=y_k-\alpha, e_{k,w}=w_k-\alpha$, *then the following asymptotic relations hold:*

$$\begin{cases} 1+\gamma_k f'(\alpha) \sim \Psi_1 e_{k-1}e_{k-1,w}e_{k-1,y}, \\ c_2 + \lambda_k \sim \Psi_2 e_{k-1}e_{k-1,w}e_{k-1,y}, \end{cases} \quad (6.9)$$

where Ψ_1 and Ψ_2 these some asymptotic constants.
  Now, we state the convergence theorem for Equation (6.8).

Theorem 6.3 *If an initial guess x_0 is sufficiently close to the zero α of $f(x)$ and the parameter $\gamma=\gamma_k$ and $\lambda=\lambda_k$ in the iterative scheme in Equation (6.8) is recursively calculated by the form given in Equation (6.9), then the R-order of convergence of methods with memory in Equation (6.8) is at least 7.*

Proof 6.3 *First we assume that the R-order of convergence of sequence x_k, w_k, y_k is at least r, p, and q, respectively.*
  Hence:

$$e_{k+1} \sim e_k^r \sim e_{k-1}^{r^2}, \quad (6.10)$$

$$e_{k,w} \sim e_k^p \sim e_{k-1}^{rp}, \quad (6.11)$$

and

$$e_{k,y} \sim e_k^q \sim e_{k-1}^{rq}. \quad 6.(12)$$

By Equations (6.10), (6.11), (6.12), and Lemma (6.1), we obtain

$$1 + \gamma_k f'(\alpha) \sim e_{k-1}^{p+q+1}, \lambda_k + c_2 \sim e_{k-1}^{p+q+1} \tag{6.13}$$

On the other hand, we get

$$e_{k,w} \sim (1 + \gamma_k f'(\alpha)) e_k, \tag{6.14}$$

$$e_{k,y} \sim (1 + \gamma_k f'(\alpha))(\lambda_k + c_2) e_k^2, \tag{6.15}$$

$$e_{k+1} \sim (1 + f'(\alpha)\gamma_k)^2 (\lambda_k + c_2)(c_2(\lambda_k + c_2) - c_3) e_k^4. \tag{6.16}$$

Combining Equations (6.10)–(6.16), (6.11)–(6.14), and (6.12)–(6.15), we conclude

$$e_{k,w} \sim e_{k-1}^{(1+p+q)+r}, \tag{6.17}$$

$$e_{k,y} \sim e_{k-1}^{2(1+p+q)+2r}, \tag{6.18}$$

and

$$e_{k+1} \sim e_{k-1}^{3(1+p+q)+4r}. \tag{6.19}$$

Equating the powers of error exponents of e_{k-1} in pairs of relations in Equations (6.14)–(6.17), (6.15)–(6.18), and (6.16)–(6.19), we have

$$\begin{cases} rp - r - (p+q+1) = 0, \\ rq - 2r - 2(p+q+1) = 0, \\ r^2 - 4r - 3(p+q+1) = 0. \end{cases} \tag{6.20}$$

This system has the solution $p=2$, $q=4$, and $r=7$ which specifies the *R*-order of convergence of the derivative-free scheme with memory in Equation (6.8).

Remark 6.1 The new two-step derivative-free methods in Equation (6.8) require three function evaluations and have the order of convergence 7. Therefore, the proposed methods are of optimal order and support the Kung–Traub conjecture [16]. Hence, the efficiency index of the proposed methods is $7^{\frac{1}{3}} = 1.91293$.

6.4 Numerical Examples

In this section, we demonstrate the convergence behavior of the methods with memory (8), where γ_k and λ_k are calculated by one of the formulas in Equation (6.7). Numerical computations reported here have been carried out in a *Mathematica*10 environment. The computational order of convergence (namely, COC) [32], can be defined by

$$COC \approx \frac{\log|f(x_n)/f(x_{n-1})|}{\log|f(x_{n-1})/f(x_{n-2})|}. \tag{6.21}$$

The test functions, their simple roots, and the starting approximations are listed in Table 6.1. Table 6.2 shows the efficiency index of the proposed method with other methods. See Table 6.3 for a percentage improvement in the convergence of memory methods. Table 6.4 shows the efficiency and order of computational convergence of the proposed method with other methods on the given functions. Iter, EI, EF, and EFD stand for number of repeats, efficiency index, total number of evaluation function, and total number of evaluation derivative functions respectively.

We investigate the following methods with the proposed method:

1. Iterative methods one-step with and without memory:

   Abbasbandy (AM)[1], Newton(NM)[20], Steffensen (SM)[29], and Zheng et al. (ZLHM)[35].

2. Iterative methods two-step with and without memory:

   Cordero et al.(CHMTM)[4], Cordero et al.(CLBTM)[5], Cordero et al.(CLKTM)[6], Cordero et al.(CLTAMM)[7], Dzunic et al.(DPPM)[9], Dehghan-Hajarian (DH)

TABLE 6.1

The Test Functions

Nonlinear Function	Root	Initial Guess
$f_1(x) = x\log(1 + x\sin(x)) + e^{-1+x^2+x\cos(x)}\sin(\pi x)$	$\alpha = 0$	$x_0 = 0.3$
$f_2(x) = \sin(5x)e^x - 2$	$\alpha = 1.36$	$x_0 = 1$
$f_3(x) = 1 + \dfrac{1}{x^4} - \dfrac{1}{x} - x^2$	$\alpha = 1$	$x_0 = 1.4$
$f_4(x) = (x-2)(x^{10} + x + 2)e^{-5x}$	$\alpha = 2$	$x_0 = 2.3$
$f_5(x) = e^{x^3-x} - \cos(x^2 - 1) + x^3 + 1$	$\alpha = -1$	$x_0 = -1.3$

TABLE 6.2

Comparison Evaluation Function and Efficiency Index of the Proposed Method With Other Schemes

Without-Memory Methods	EF	EFD	COC	EI	With-Memory Methods	EF	EFD	COC	EI
AM[1]	1	2	3.000	1.4423	SecM [23]	1	0	1.680	1.6800
HPM[12]	1	2	4.000	1.5874	TrM[33]	2	0	2.410	1.5524
NM[20]	1	1	2.000	1.4142	CLBTM[5]	3	0	7.000	1.9129
OM [21]	2	1	4.000	1.5874	KKBM[14]	3	0	7.000	1.9129
CNM[3]	3	1	8.000	1.6818	DPPM[9]	4	0	11.000	1.8212
RWBM[25]	3	0	4.000	1.5874	LMNKSM[18]	4	0	12.000	1.8612
KLM[15]	2	1	4.000	1.5874	WM[34]	3	0	4.230	1.6173
ZLHM[35]	3	1	4.000	1.5874	SSSM[26]	4	0	12.000	1.8612
MM[19]	2	1	4.000	1.5874	PIDM[24]	3	0	4.562	1.6585
TM[30]	5	0	16.000	1.7411	CLKTM[6]	3	0	6.000	1.8171
DHM[8]	1	2	4.000	1.5874	PDNM[22]	3	0	4.562	1.6585
GQM[11]	5	0	16.000	1.7411	TKM[8]	3	0	7.000	1.9129

TABLE 6.3

The Percentage Improvement of the Order of Convergence of With-Memory Methods

With-Memory Methods	Number of Steps	Optimal Order	COC	Percentage Increase
TrM[33]	1	2.000	2.410	%20.5
CLKTM[5]	2	4.000	6.000	%50
KKBM[14]	2	4.000	7.000	%75
DPPM[9]	3	8.000	8.470	%5.875
DPPM[9]	3	8.000	9.000	%12.5
DPPM[9]	3	8.000	10.000	%25
DPPM[9]	3	8.000	11.000	%37.5
LMNKSM[18]	3	8.000	12.000	%50
PIDM[24]	2	4.000	4.449	%11.225
PDNM[22]	2	4.000	4.562	%14.05
SSSM[26]	3	8.000	12.000	%50
SGGM[27]	3	8.000	11.352	%41.9
SGGM[27]	3	8.000	11.657	%45.713
SGGM[27]	3	8.000	12.000	%50
TLFM[31]	2	4.000	6.000	%50
WM[34]	2	4.000	4.230	%5.75
TKM[8]	2	4.000	7.000	%75

[8], Guo-Qian(GQM)[11], Hansen-Patrick(HPM)[12], Jarratt(JM)[13], Kansal et al. (KKBM)[14], Kou-Li(KLM)[15], Petkocic et al.(PDNM)[22], Petkocic et al.(PIDM) [24], Ostrowski(OM)[21], Ren et al.(RWBM)[25], Torkashvand et al.(TLFM)[31], Truab(TM)[33], Wang(WM)[34], and Zheng et al.(ZLHM)[35].

3. Iterative methods three-step with and without memory:

   Chun-Neta(CNM)[3], Dzunic et al.(DPPM)[9], Guo-Qian(GQM)[11], Lotfi et al. (LMNKSM)[18], Kung-Truab(KTM)[16], Sharifi et al. (SSSM)[26], Sharma et al. (SGGM)[27], Soleymani (SM)[28], and Zheng et al.(ZLHM)[35].

4. Iterative methods four-step without memory:

   Geum-Kim(GKM)[10], Guo-Qian(GQM)[11], Li et al.(LMMWM)[17], Thukral(TM) [30], and Zheng et al.(ZLHM)[35].

6.5 Conclusion

In this chapter, we have presented a new family of two-step iterative methods with memory for solving nonlinear equations. Since our aim is to construct the method of higher-order convergence without additional calculation, we have used two approximations of selfcorrecting parameters, designed by Newton interpolating polynomials in the fourth-order methods to achieve higher-order convergence without any additional calculation. The R-order of convergence of the new with-memory iterative methods is increased from 4 to 6, and 7. The numerical results have been given to confirm the validity of theoretical results.

TABLE 6.4

Comparison Evaluation Function and Efficiency Index of the Proposed Method With Other Schemes

$f_1(x) = x \log(1 + x \sin(x)) + e^{-1+x^2+x\cos(x)}\sin(\pi x)$, $\alpha = 0$, $x_0 = 0.3$

Without-Memory Methods	EF	Iter	COC	EI	With-Memory Methods	EF	Iter	COC	EI
AM	3	5	3.0000	1.44225	TrM, $\lambda_0 = 0.1$	2	4	2.4282	1.55827
NM	2	7	2.0000	1.41421	SecM, $x_1 = 0.6$	2	15	1.6181	1.61808
MM	3	5	4.0000	1.58740	WM, $\lambda_0 = 1$	3	5	4.2361	1.61803
CNM, $\alpha_0 = \gamma_0 = A_0 = 1$	4	3	8.0000	1.68179	PIDM $\beta_0 = 0.1$	3	5	4.4513	1.64499
GKM	5	3	16.0000	1.74110	DPPM, $\gamma_0 = -0.1$	4	3	1.0000	1.00000
RWBM, $\beta_0 = 1$	3	4	4.0000	1.58740	TKM, $\gamma_0 = \lambda_0 = 0.1$	3	5	7.0000	1.91293

$f_2(x) = \sin(5x)e^x - 2$, $\alpha = 1.36$, $x_0 = 1$

Without-Memory Methods	EF	Iter	COC	EI	With-Memory Methods	EF	Iter	COC	EI
AM	3	1000	1.0000	1.00000	TrM, $\lambda_0 = 0.1$	2	5	0.0000	0.00000
NM	2	15	2.0000	1.41421	SecM, $x_1 = 1.2$	2	17	1.6181	1.61808
MM	3	4	4.0000	1.58740	WM, $\lambda_0 = 0.1$	3	5	4.2361	1.61803
CNM, $\alpha_0 = \gamma_0 = A_0 = 1$	4	6	8.0000	1.68179	PIDM $\beta_0 = 0.1$	3	5	4.4314	1.64253
GKM	5	3	0.000	0.00000	DPPM, $\gamma_0 = -0.1$	4	3	1.0000	1.00000
RWBM, $\beta_0 = 1$	3	4	0.0000	0.00000	TKM, $\gamma_0 = \lambda_0 = 0.1$	3	4	7.0000	1.91293

$f_3(x) = 1 + \dfrac{1}{x^4} - \dfrac{1}{x} - x^2$, $\alpha = 1$, $x_0 = 1.4$

Without-Memory Methods	EF	Iter	COC	EI	With-Memory Methods	EF	Iter	COC	EI
AM	3	6	3.0000	1.44225	TrM, $\lambda_0 = 0.1$	2	4	2.4062	1.55120
NM	2	5	2.0000	1.41421	SecM, $x_1 = 1.5$	2	16	1.6181	1.61808
MM	3	4	4.0000	1.58740	WM, $\lambda_0 = 1$	3	5	4.2361	1.61804
CNM, $\alpha_0 = \gamma_0 = A_0 = 1$	4	3	8.0000	1.68179	PIDM $\beta_0 = 0.1$	3	5	4.4661	1.64681
GKM	5	3	16.0000	1.74110	DPPM, $\gamma_0 = -0.1$	4	3	10.0000	1.77828
RWBM, $\beta_0 = 1$	3	4	4.0000	1.58740	TKM, $\gamma_0 = \lambda_0 = 0.1$	3	3	7.0000	1.91293

(Continued)

TABLE 6.4 (CONTINUED)

Comparison Evaluation Function and Efficiency Index of the Proposed Method With Other Schemes

$f_4(x) = (x-2)(x^{10}+x+2)e^{-5x}$,

							$\alpha=2, x_0=2.3$			
AM	3	1500	3.0000	1.44225	TrM, $\lambda_0=0.1$	2	4	2.4681	1.57100
NM	2	5	2.0000	1.41421	SecM, $x_1=1.8$	2	17	1.6181	1.61808
MM	3	4	4.0000	1.58740	WM, $\lambda_0=1$	3	6	4.2361	1.61803
CNM, $\alpha_0=\gamma_0=A_0=1$	4	5	8.0000	1.68179	PIDM $\beta_0=0.1$	3	4	4.3019	1.62637
GKM	5	3	16.0000	1.74110	DPPM, $\gamma_0=-0.1$	4	3	1.0000	1.00000
RWBM, $\beta_0=1$	3	4	4.0000	1.58740	TKM, $\gamma_0=\lambda_0=0.1$	3	4	7.0000	1.91293

$f_5(x) = e^{x^3-x} - \cos(x^2-1) + x^3 + 1$,

							$\alpha=-1, x_0=-1.3$			
AM	3	5	3.0000	1.44225	TrM, $\lambda_0=0.1$	2	4	2.4520	1.56549
NM	2	5	2.0000	1.41421	SecM, $x_1=-1.6$	2	18	1.6181	1.61808
MM	3	4	4.0000	1.58740	WM, $\lambda_0=1$	3	5	4.2361	1.61803
CNM, $\alpha_0=\gamma_0=A_0=1$	4	3	8.0000	1.68179	PIDM $\beta_0=0.1$	3	4	4.4517	1.64504
GKM	5	3	16.0000	1.74110	DPPM, $\gamma_0=-0.1$	4	4	1.0000	1.00000
RWBM, $\beta_0=1$	3	4	4.0000	1.58740	TKM, $\gamma_0=\lambda_0=0.1$	3	4	7.0000	1.91293

References

Abbasbandy, S. 2006. Modified homotopy perturbation method for nonlinear equations and comparison with Adomian decomposition method. *Applied Mathematics and Computation* 172:431–438.

Chun, C., and Neta, B. 2014. An analysis of a new family of eighth-order optimal methods. *Applied Mathematics and Computation* 245:86–107.

Cordero, A., Hueso, J.L., Martinez, E., and Torregrosa, J.R. 2013. A new technique to obtain derivative-free optimal iterative methods for solving nonlinear equations. *Journal of Computational and Applied Mathematics* 252:95–102.

Cordero, A., Lotfi, T., Bakhtiari, P., and Torregrosa, J.R. 2014. An efficient two-parametric family with memory for nonlinear equations. *Numerical Algorithms* 68(2):323–335.

Cordero, A., Lotfi, T., Khoshandi, A., and Torregrosa, J.R. 2015. An efficient Steffensen-like iterative method with memry. *Bulletin mathematique de la Societe des Sciences Mathematiques de Roumanie Nouvelle Serie* 58(106):49–58.

Cordero, A., Lotfi, T., Torregrosa, J.R., Assari, P., and Mahdiani, K. 2016. Some new bi-accelarator two-point methods for solving nonlinear equations. *Computational and Applied Mathematics* 35(1):251–267.

Dehghan, M., and Hajarian, M. 2010. New iterative method for solving non-linear equations with fourth-order convergence. *International Journal of Computer Mathematics* 87(4):834–839.

Dzunic, J., Petkovic, M.S., and Petkovic, L.D. 2012. Three-point methods with and without memory for solving nonlinear equations. *Applied Mathematics and Computation* 218:4917–4927.

Geum, Y.H., and Kim, Y.I. 2011. A biparametric family optimally convergent sixteenth-order multipoint methods with their fourth-step weighting function as a sum of a rational and a generic two-variable function. *Journal of Computational and Applied Mathematics* 235:3178–3188.

Guo, Q.W., and Qian, Y.H. 2015. New efficient optimal derivative-free method for solving nonlinear equations. *International Journal of Mathematics and Computational Science* 1(3):102–110.

Hansen, E., and Patrick, M. 1977. A family of root finding methods. *Numerische Mathematik* 27:257–269.

Jarratt, P. 1966. Some fourth order multipoint methods for solving equations. *Mathematics of Computation* 20:434–437.

Kansal, M., Kanwar, V., and Bhatia, S. 2016. Efficient derivative-free variants of Hansen-Patrick's family with memory for solving nonlinear equations. *Numerical Algorithms* 73:1017–1036.

Kou, J., and Li, Y. 2007. The improvements of Chebyshev-Halley methods with fifth-order convergence. *Applied Mathematics and Computation* 188:143–147.

Kung, H.T., and Traub, J.F. 1974. Optimal order of one-point and multipoint iteration. *Journal of the ACM (JACM)* 21:634–651.

Li, X., Mu, C., Ma, J., and Wang, C. 2010. Sixteenth-order method for nonlinear equations. *Applied Mathematics and Computation* 215(10):3754–3758.

Lotfi, T., Mahdiani, K., Noori, Z., Khaksar Haghani, F., and Shateyi, S. 2014. On a new three-step class of methods and its acceleration for nonlinear equations. *The Scientific World Journal* 2014:134673.

Maheshwari, A.K. 2009. A fourth order iterative method for solving nonlinear equations. *Applied Mathematics and Computation* 211:383–391.

Ortega, J.M., and Rheinboldt, W.C. 1970. *Iterative solutions of nonlinear equations in several variables*, Academic Press, New York.

Ostrowski, A.M. 1960. *Solution of equations and systems of equations*, Academic Press, New York.

Petkovic, M.S., Dzunic, J., and Neta, B. 2011. Interpolatory multipoint methods with memory for solving nonlinear equations. *Applied Mathematics and Computation* 218:2533–2541.

Petkovic, M.S., Neta, B., Petkovic, L.D., and Dzunic, J. 2013. *Multipoint methods for solving nonlinear equations*, Elsevier, Amsterdam.

Petkovic, M.S., Ilic, S., and Dzunic, J. 2010. Derivative free two-point methods with and without memory for solving nonlinear equations. *Applied Mathematics and Computation* 217:1887–1895.

Ren, H., Wu, Q., and Bi, W. 2009. A class of two-step Steffensen type methods with fourth-order convergence. *Applied Mathematics and Computation* 209:206–210.

Sharifi, S., Siegmund, S., and Salimi, M. 2015. Solving nonlinear equations by a derivative-free form of the King's family with memory. *Calcolo* 53(2):201–215.

Sharma, J.R., Guha, R.K., and Gupta, P. 2012. Some efficient derivative free methods with memory for solving nonlinear equations. *Applied Mathematics and Computation* 219:699–707.

Soleymani, F. 2013. Some optimal iterative methods and their with memory variants. *Journal of the Egyptian Mathematical Society* 2(2):1–9.

Steffensen, J.F. 1933. Remarks on iteration. *Scandinavian Aktuarietidskr* 16:64–72.

Thukral, R. 2012. New sixteenth-order derivative-free methods for solving nonlinear equations. *American Journal of Computational and Applied Mathematics* 2(3):112–118.

Torkashvand, V., Lotfi, T., and Fariborzi Araghi, M.A. 2018. Efficient iterative methods with memory for solving nonlinear equations. *49th Annual Iranian Mathematics Conference*, August 2018, Tehran, Iran.

Torkashvand, V., Lotfi, T., and Fariborzi Araghi, M.A. 2019. A new family of adaptive methods with memory for solving nonlinear equations. *Mathematical Sciences* 13:1–20.

Traub, J.F. 1964. *Iterative methods for the molution of equations*, Prentice Hall, New York, NJ.

Wang, X. 2017. An Ostrowski-type method with memory using a novel self-accelerating parameter. *Journal of Computational and Applied Mathematics* 330:710–720.

Zheng, Q., Li, J., and Huang, F. 2011. An optimal Steffensen-type family for solving nonlinear equations. *Applied Mathematics and Computation* 217:9592–9597.

7

Computational Methods for Conformable Fractional Differential Equations

A. H. Siddiqi, R. C. Singh, and Santosh Kumar

CONTENTS

7.1 Introduction.. 103
7.2 Conformable Fractional Derivative... 105
7.3 Simulation of Conformable Fractional Heat Equation 106
7.4 Simulation of Conformable Fractional Derivative of Harmonic Oscillator 108
7.5 Simulation of Conformable Fractional Derivative of Shannon Wavelet and
     Scaling Function 110
7.6 Simulation of Conformable Fractional Derivative of Sine-Cosine Wavelet 112
7.7 Road Map of Future Research Work.. 113
References.. 113

7.1 Introduction

A differential equation, an equation relating a function with its derivatives, is a powerful tool to model real-world problems. It is frequently used in all sciences and different branches of engineering and technology. It is well known that first and second classical derivatives denote fundamental concepts of physics such as velocity and acceleration. The calculus dealing with positive integer derivatives was developed by Newton, Leibniz, and Gauss et al. Since the invention of integer derivatives the concept of fractional derivative like half derivative $\frac{d^{1/2}y}{dx^{1/2}}$ was known to L'Hospital as far back as 1695. Those fractional derivatives and fractional calculus including fractional deferential equations have a history of more than 300 years. Several eminent mathematicians like Liouville, Riemann, Weyl Fourier, Abel, Leibniz, Grunwald, and Letnikov have actively worked in this field. Mathematicians and scientists were curious to know the physical meaning of fractional derivatives like half derivative, πth, and imaginary derivative. This question was raised since the days of L'Hospital but for centuries no satisfactory answer could be given. During the last few decades this theme has developed rapidly. Nowadays, the fractional calculus including fractional differential equations attracts many scientist and engineers. There are several applications of this mathematical phenomena in mechanics, physics, chemistry, control theory, and so on. It is natural that many authors tried to solve the problems concerning fractional derivatives, fractional integral and fractional differential equations in MATLAB®.

A few interesting and useful MATLAB functions have been submitted to the Math Works, Inc. Until recently, fractional calculus was considered rather esoteric and without application. But in the past couple of decades there have been tremendous research activities on the application of fractional calculus to very diverse scientific fields ranging from the physics diffusion and advection phenomena, to control systems, to finance and economics. Indeed, at present, application and activities related to fractional calculus have appeared in at least the following fields: fractional control of engineering systems; advancement of calculus of variations and optimal control to fractional dynamic systems; analytical and numerical tools and techniques; fundamental explorations of the mechanical, electrical, and thermal constitutive relations and other properties of various engineering materials such as viscoelastic polymers, foams, gels, and their engineering and scientific applications; fundamental understanding of wave and diffusion phenomenon, their measurements and verifications, including applications to plasma physics (such as diffusion in Tokamak); bioengineering and biomedical applications; thermal modeling of engineering systems such as brakes and machine tools; image and signal processing.

The increasing interest in application of fractional calculus has motivated the development and investigation of numerical methods specially devised to solve fractional differential equations (FDEs). In references, Garrappa [25, 26] has explained a basic principle behind some methods for FDEs. He has developed some MATLAB routines for the solution of a wide range of FDEs. In these references he has discussed MATLAB routines for classical fractional derivatives like Caputo and Riemann In this chapter we intend to examine numerical tools and MATLAB routines for comparatively new fractional derivatives known as conformable derivatives. For conformable fractional derivatives and associated differential equations we refer to [21, 28, 31]. Differential equations containing conformable derivatives are called conformable differential equations.

In a beautiful article, Petras [46] has presented the MATLAB code for solving a partial differential equation where derivatives are taken in the sense of Caputo, Reimann, Liouville, and Riesz. In general he has described the methods for calculation of fractional derivative and fractional integrals together with methods for solution of the fractional differential equations in MATLAB. Several useful pieces of information are provided in this paper. Chi and Gao [18] observe that numerical computation of fractional derivatives is a task. However they have efficiently computed Riemann–Liouville fractional derivatives and integrals applying Fourier series and Taylor's series combined with MATLAB. Garrappa [25, 26] has presented an updated account of numerical methods and MATLAB codes for solution of fractional differential equations. It has been observed by Avct [9] that the computational results of a conformable heat equation are quite successful in showing the sub-behavior of the heat process. Fractional calculus is considered as a laboratory for special functions and integral transforms. The efficient implementation of product integration rules has been presented in this chapter. The MATLAB codes presented are available on the websites of [25, 26].

Applications of fractional derivatives and fractional integrals are discussed in [5, 6, 7, 10, 11, 14, 17, 19, 20, 21, 22, 31, 33, 34, 36, 38, 39, 40, 42, 43, 44, 45, 48, 53, 57, 58, 60, 62, 64, 65]. The concept of inverse problem is quite popular in classical calculus, differential equations, and partial differential equations, see for example [23, 24, 27, 35, 49, 51, 52, 55, 66]. However, there are only few papers devoted to inverse problem in fractional calculus and differential equations, for example [19, 35, 40, 56, 59, 60, 63, 64] and a recent Ph.D thesis submitted to Sharda University, Greater Noida India by Mr. Masood Alam. [12, 13, 14, 15] are devoted to Shannon wavelets which are useful for numerical simulation of fractional differentiable equations. Numerical methods for solution of fractional differential equations are investigated in [3, 31, 67, 68].

In the remaining six sections we present the basic properties of conformable fractional derivative (Section 7.2), simulation of conformable fractional heat equation (Section 7.3), simulation of conformable fractional harmonic oscillator (Section 7.4), simulation of conformable fractional derivative of Shannon wavelet (Section 7.5), simulation of conformable fractional derivative of sine-cosine wavelet (Section 7.6), and a road map of future research work (Section 7.7).

7.2 Conformable Fractional Derivative

The concept of conformable fractional derivative has been studied by Khalil et al. [32], Abdel-jawed [2], Abdelhakim et al. [1]. Conformable differential equations have been studied by Hammed and Khalil [28, 30], Çenesiz and Kurt [16], Avct [9] and Nurudeen et al. [43]. We state below a few useful properties of conformable fractional derivative [2, 32]. Conformable fractional derivative is denoted by $T_\alpha f$, for $0 < \alpha < 1$, of a function $f : [0, \infty) \to \mathbb{R}$ is defined as:

$$T_\alpha f(x) = \lim_{\epsilon \to 0} \frac{f(x + \epsilon\, x^{1-\alpha}) - f(x)}{\epsilon}, \quad x > 0, \tag{7.1}$$

provided the limit exists, in this case f is called α- differentiable on $(0,a)$ for some $a > 0$. Also

$$T_\alpha f(0) = \lim_{x \to 0^+} T_\alpha(x). \tag{7.2}$$

It has been observed by Abdelhakim and Machado [1] that if f is α- differentiable in the conformable sense at $x > 0$, then it must be differentiable in the classical sense at x and

$$T_\alpha f(x) = x^{1-\alpha} f'(x). \tag{7.3}$$

Various analogous properties of classical derivative including Rolle's Theorem, Chain Theorem, Gronwall inequality, integration of parts formula are studied in references mentioned above. Khalil et al. [32] have also introduced fractional integral associated with conformable fractional derivative. The following theorems containing various properties have proved very useful.

Theorem 7.1 Let $\alpha \in (0,1)$ and f, g be α- differentiable at $t > 0$. Then

(a) $T_\alpha(\lambda f + \mu g) = \lambda T_\alpha(f) + \mu T_\alpha(g), \forall\ \lambda, \mu \in \mathbb{R}$

(b) $T_\alpha(fg) = f T_\alpha(g) + g T_\alpha(f)$.

(c) $T_\alpha\left(\dfrac{f}{g}\right) = \dfrac{g T_\alpha(g) - f T_\alpha(f)}{g^2}$.

(d) $T_\alpha(t^p) = p t^{p-\alpha}\ \forall\ p \in \mathbb{R}$.

(e) $T_\alpha(\lambda) = 0$ for all constant functions $f(t) = \lambda$.

(f) $T_\alpha(e^{cx}) = cx^{1-\alpha} e^{cx}, c \in \mathbb{R}$.

(g) $T_\alpha(\sin bx) = bx^{1-\alpha} \cos bx, b \in \mathbb{R}$.

(h) $T_\alpha(\cos bx) = -bx^{1-\alpha}\sin bx,\ b \in \mathbb{R}.$

(i) $T_\alpha\left(\dfrac{1}{\alpha}t^\alpha\right) = 1.$

(j) $T_\alpha\left(\sin\dfrac{1}{\alpha}t^\alpha\right) = \cos\dfrac{1}{\alpha}t^\alpha.$

(k) $T_\alpha\left(\cos\dfrac{1}{\alpha}t^\alpha\right) = -\sin\dfrac{1}{\alpha}t^\alpha.$

(l) $T_\alpha\left(e^{\frac{1}{\alpha}t^\alpha}\right) = e^{\frac{1}{\alpha}t^\alpha}.$

Theorem 7.2 *Let f(t) be a α- differentiable function and let g(t) be also α- differentiable in the range of f'(t) for α ∈ (0,1), then*

$$T_\alpha(f \circ g)(t) = t^{1-\alpha} g'(t) f'(g(t)).$$

7.3 Simulation of Conformable Fractional Heat Equation

The general form for a one-dimensional heat equation is

$$\frac{\partial u}{\partial t} = k\frac{\partial^2 u}{\partial x^2}.$$

A heat equation has many fractional forms ,see for example Hammed and Khalil [28, 29] and AVCT et al. [9]. In this section we investigate the solution of a time fractional heat differential equation:

$$\frac{\partial^\alpha u}{\partial t^\alpha} = k\frac{\partial^2 u}{\partial x^2}, \quad 0 < x < L,\ t > 0. \tag{7.4}$$

with conditions

$$u(0,t) = 0,\ t \geq 0 \tag{7.5}$$

$$u(L,t) = 0,\ t \geq 0 \tag{7.6}$$

$$u(x,0) = f(x),\ 0 \leq x \leq L, \tag{7.7}$$

where the derivative is conformable fractional derivative and $0 < \alpha < 1$.
   The conformable fractional linear differential equations with constant coefficients

$$\frac{\partial^\alpha y}{\partial t^\alpha} \pm \mu^2 y = 0. \tag{7.8}$$

From Equation (7.3), we can obtain

$$\frac{\partial^\alpha y}{\partial t^\alpha} = t^{1-\alpha}\frac{dy}{dt}. \tag{7.9}$$

by substituting Equation (7.9) in Equation (7.8), we get

$$t^{1-\alpha} \frac{dy}{dt} \pm \mu^2 y = 0. \tag{7.10}$$

We obtain the solution of Equation (7.10)

$$y = ce^{\frac{\pm\mu^2}{\alpha}t^\alpha}. \tag{7.11}$$

Now we can use separation of variables methods for our time fractional heat equation. Let u = p(x)q(t). Substituting the time fractional heat equation, we have

$$\frac{d^\alpha q(t)}{dt^\alpha} p(x) = k \frac{d^2 p(x)}{dx^2} q(t)$$

we obtain

$$\frac{d^\alpha q(t)}{dt^\alpha} \Big/ kq(t) = \frac{d^2 p(x)}{dx^2} \Big/ p(x) = \omega$$

As a result:

$$\frac{d^\alpha q(t)}{dt^\alpha} - \omega kq(t) = 0$$

and

$$\frac{d^2 p(x)}{dx^2} - \omega p(x) = 0.$$

There are three cases for values of ω to be evaluated. $\omega = 0$, $-\mu^2$ and μ^2 with boundary conditions, we get

$$\mu = \frac{n\pi}{L} \quad \text{and} \quad p_n(x) = a_n \sin\frac{n\pi x}{L}. \tag{7.12}$$

Equations (7.8) and (7.11) give,

$$q_n(t) = b_n e^{-\left(\frac{n\pi}{L}\right)^2 \frac{k}{\alpha}t^\alpha} \tag{7.13}$$

From Equations (7.12) and (7.13), the solution of the Cauchy problem which satisfies two boundary conditions is obtained as

$$u(x,t) = \sum_{n=1}^{\infty} c_n \sin\frac{n\pi x}{L} e^{-\left(\frac{n\pi}{L}\right)^2 \frac{k}{\alpha}t^\alpha}. \tag{7.14}$$

With the help of the initial condition, we get

$$c_n = \frac{2}{L} \int_0^L f(x) \sin\left(\frac{n\pi x}{L}\right) dx. \tag{7.15}$$

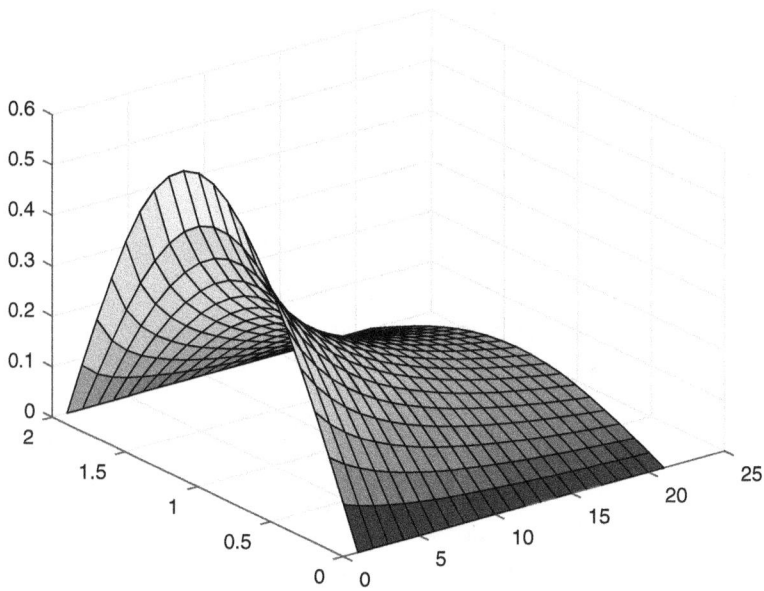

FIGURE 7.1
Conformable fractional heat equation for $\alpha = 0.5$.

Substituting Equation (7.15) in Equation (7.14), we obtain the solution

$$u(x,t) = \sum_{n=1}^{\infty} \sin\frac{n\pi x}{L} e^{-\left(\frac{n\pi}{L}\right)^2 \frac{k}{\alpha} t^{\alpha}} \left[\frac{2}{L} \int_0^L f(x) \sin\left(\frac{n\pi x}{L}\right) dx \right].$$

The MATLAB® simulation of the fractional heat equation is given in Figure 7.1.

```
MATLAB code:
[X, T]=meshgrid (0: 0.1: 2,1: 21);
alpha=0.5;
k=0.1;
L=2;
f(x)=sin(x);
u=0;
for n=1:20
fun=@(X) (sin(X).*sin((n.*pi.*X)./(L)));
format long
q=integral(fun,0,L);
u=u+sin((n.*pi.*X)./(L)).*exp(-(((n.*pi.)./(L)).^2).*(k./alpha).*T.^
    (alpha)).*(2./L).*q;
end
surf(T,X,u)
```

7.4 Simulation of Conformable Fractional Derivative of Harmonic Oscillator

The fractional harmonic oscillator is model [30] by

$$\left(m\frac{d^\alpha}{dt^\alpha}+k\right)f(t)=0, \tag{7.16}$$

where $\dfrac{d^\alpha f}{dt^\alpha}$ denotes the classical fractional derivative of order α. In this section we replace classical frictional derivative by conformable fractional derivative. Thus the equation of harmonic oscillator takes the form:

$$mt^{1-\alpha}\frac{df}{dt}+kf(t)=0. \tag{7.17}$$

The solution of this equation is

$$f=ce^{\left(\frac{-k}{m\alpha}\right)t^\alpha}$$

MATLAB simulation of conformable fractional harmonic oscillator is given in Figure 7.2. MATLAB code:

```
t=linspace(0,5);
alpha1=0.3;
c=0.5;
m=0.3;
k=0.2;
f=c.*(exp(-(k.*t.^alpha1)/(m.*alpha1)));
plot(t,f,'r')
hold on
```

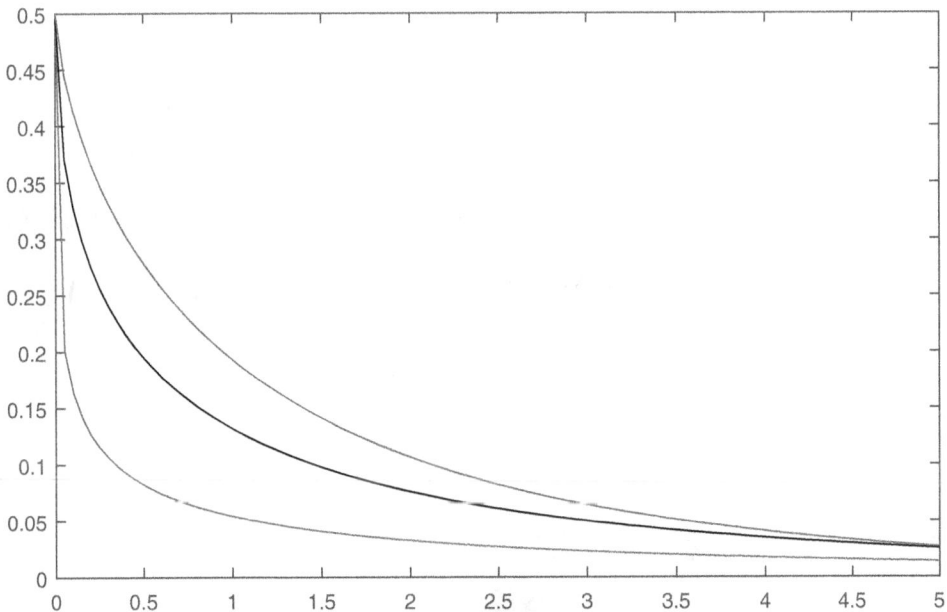

FIGURE 7.2
Conformable fractional derivative of harmonic oscillator for $\alpha=0.3$, 0.5 and 0.7.

```
alpha2=0.5;
f1=c.*(exp(-(k.*t.^alpha2)/(m.*alpha2)));
plot(t,f1,'k')
hold on
alpha3=0.7;
f2=c.*(exp(-(k.*t.^alpha3)/(m.*alpha3)));
plot(t,f2,'b')
hold off
```

7.5 Simulation of Conformable Fractional Derivative of Shannon Wavelet and Scaling Function

The Shannon wavelet theory [12, 13, 14] is based on the scaling function $\phi(x)$, also known as sinc function, and the wavelet function $\psi(x)$, respectively defined as

$$\phi(x) = \sin c \, x = \frac{\sin \pi x}{\pi x}$$

and

$$\psi(x) = \frac{\sin 2\pi \left(x - \left(\frac{1}{2} \right) \right) - \sin \pi \left(x - \left(\frac{1}{2} \right) \right)}{\pi \left(x - \left(\frac{1}{2} \right) \right)}.$$

The conformable fractional derivative of scaling function $\phi(x)$ is

$$\frac{d^\alpha \phi}{dx^\alpha} = x^{1-\alpha} \phi'(x)$$

and this implies

$$\frac{d^\alpha \phi}{dx^\alpha} = x^{1-\alpha} \left(\frac{\pi x \cos \pi x - \sin \pi x}{\pi x^2} \right).$$

The conformable fractional derivative of wavelet function $\psi(x)$ is

$$\frac{d^\alpha \psi}{dx^\alpha} = x^{1-\alpha} \psi'(x)$$

$$\frac{d^\alpha \psi}{dx^\alpha} = x^{1-\alpha} \left(\frac{\pi \left(x - \frac{1}{2} \right) \left[2\cos 2\pi \left(x - \frac{1}{2} \right) - \cos \pi \left(x - \frac{1}{2} \right) \right] + \sin \pi \left(x - \frac{1}{2} \right) - \sin 2\pi \left(x - \frac{1}{2} \right)}{\pi \left(x - \left(\frac{1}{2} \right) \right)^2} \right).$$

MATLAB simulation of conformable fractional derivative of scaling function and wavelet function is given in Figures 7.3 and 7.4.

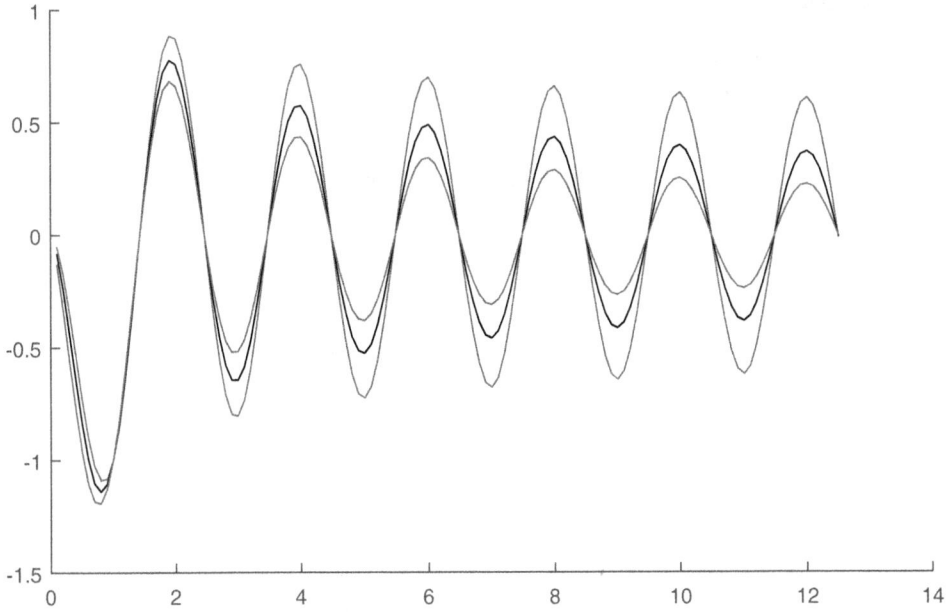

FIGURE 7.3
Conformable fractional derivative of scaling function for $\alpha = 0.2$, 0.4, and 0.6.

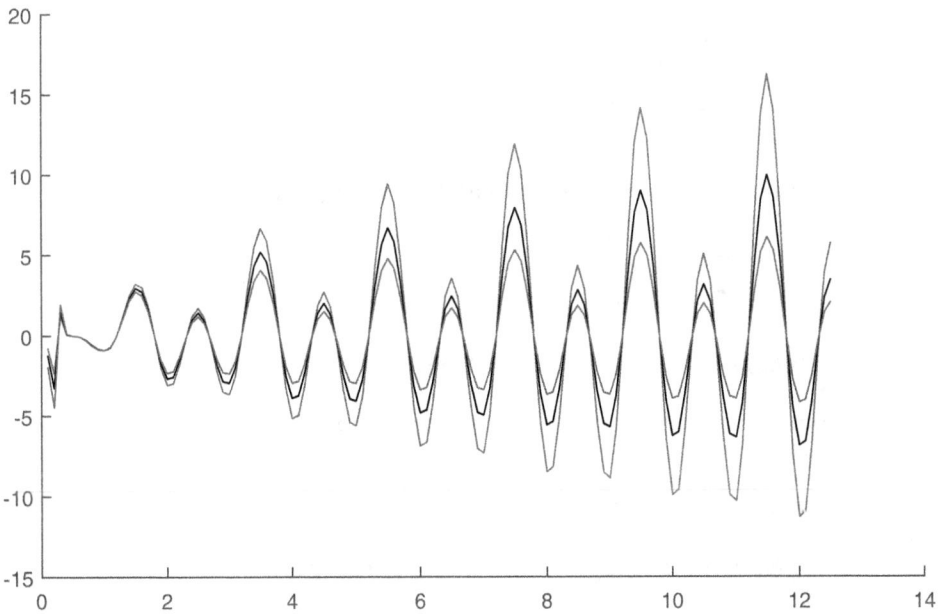

FIGURE 7.4
Conformable fractional derivative of wavelet function for $\alpha = 0.3$, 0.5 and 0.7.

7.6 Simulation of Conformable Fractional Derivative of Sine-Cosine Wavelet

Sine-cosine wavelets $\psi_{n,m}(t) = \psi(k,n,m,t)$ involve four arguments, $k = 0, 1, 2, \ldots, n = 0, \ldots, 2^k - 1$, the values m are given in Equation (7.19) and t is the normalized time. They are defined on the interval [0,1] as [41, 47, 54, 61]

$$\psi_{n,m}(t) = \begin{cases} 2^{\frac{k+1}{2}} f_m(2^k t - n), & \frac{n}{2^k} \le t \le \frac{n+1}{2^k} \\ 0, & \text{otherwise.} \end{cases} \tag{7.18}$$

where

$$f_m(t) = \begin{cases} \frac{1}{\sqrt{2}}, & m = 0 \\ \cos(2m\pi t), & m = 1, 2, \ldots L \\ \sin 2(m-L)\pi t), & m = L+1, L+2, \ldots 2L. \end{cases} \tag{7.19}$$

We used Equation (7.3) for the conformable fractional derivative

$$\frac{d^\alpha f_m}{dt^\alpha} = t^{1-\alpha} f'_m(t),$$

where

$$f'_m(t) = \begin{cases} 0, & m = 0 \\ -2m\pi \sin 2m\pi t, & m = 1, 2, \ldots L \\ 2(m-L)\pi \cos 2(m-L)\pi t, & m = L+1, L+2, \ldots 2L. \end{cases}$$

MATLAB program for conformable fractional derivative of sine wavelet:

```
t=linspace(0,1,20);
L=20;
[T, m] =meshgrid (t, 1: L);
alpha=0.5;
Y=(T.^(1-alpha)).* (cos(2.* (m - L).*pi.*T)).*(2.*(m - L).*pi);
surf(T,m,Y)
MATLAB program for conformable fractional derivative of cosine wavelet:
t=linspace(0,1,20);
L=20;
[T, m] =meshgrid (t, 1: L);
alpha=0.5;
Y=(T.^(1-alpha)).*(-sin(2.*m.*pi.*T)).*2.*m.*pi;
surf(T,m,Y)
```

Figure 7.5 represents the conformable fractional derivative of sine–cosine wavelet.

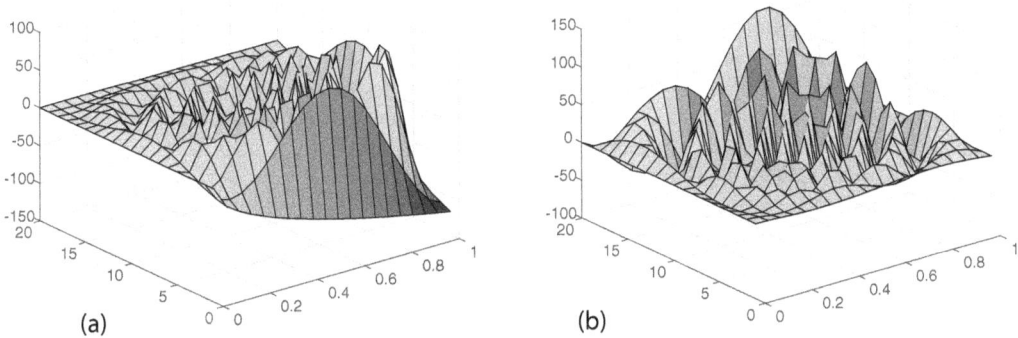

FIGURE 7.5
(a) Conformable fractional derivative of sine wavelet; (b) Conformable fractional derivative of cosine wavelet.

7.7 Road Map of Future Research Work

Adzievsk and Siddiqi [4] have studied simulation of partial differential equations using Mathematica. Similar methods could be investigate for fractional partial derivatives, particularly conformable partial differential equations. Our centre (Research Centre for Advanced Research in Applied Mathematics and Physics, Sharda University) will take up this research in the curve numerical methods for fractional analogue of partial differential equations given in [50], pp. 3–15. Harmann [30] has presented a beautiful account of applications of classical fractional derivatives like Riemann, Caputo, Liouville in several emerging areas of Physics. Conformable derivative was introduced in 2014 and its applications in several area have been studied recently. We may study the roll of conformable fractional derivative in the problems considered by Hermann. In Section 7.5 and 7.6, conformable fractional derivative of Shannon wavelet and sine-cosine wavelet have been studied. In future work we plan to study the solution of fractional differential equations and to apply these results.

References

1. Ahmad A. Abdelhakim, José A. Tenreiro Machado, A critical analysis of the conformable derivative, *Nonlinear Dynamics*, **95**(4) (2019), pp. 3063–3073.
2. T. Abdeljawad, On conformable fractional calculus, *Journal of Computational and Applied Mathematics*, **279** (2015), pp. 57–66.
3. O. Acan, D. Baleanu, A new numerical technique for solving fractional partial differential equations, *Miskole Mathematical Notes*, **19**(1) (2018), pp. 3–18.
4. K. Adzievski, A. H. Siddiqi, *Introduction to Partial Differential Equations for Scientists and Engineers Using Mathematica*, Chapman and Hall/CRC Press Taylor and Francis Group, 23 October 2013.
5. M. H. Akrami, G. H. Erjaee, Numerical solutions for fractional. Black scholes option pricing equation, *Global Analysis and Discrete Mathematics*, **1** (2016), pp. 9–14.

6. R. Almeida, What is the best fractional derivative to fit data?, arXiv:1704.00609 vl [stat. OT] 31 March 2017.

7. Saeed Amir, Saeed Umer, Sine-cosine wavelet method for fractional oscillator equations, *Mathematical Mathods in the Applied Sciences*, **42(18)** (30 July, 2019), pp. 6960–6971.

8. D. R. Anderson, D. J. Ulness, Properties of the Katumgampola fractional derivative with potential applications in quantum mechanics, *Journal of Mathematical Physics*, **56(6)** (2015), pp. 063502.

9. D. Avct, B. R. I. Eroglue, N. Ozdemir, Conformable heat equation on a radial symmetric plate, *Thermal Science*, **21(2)** (2017), pp. 819–826.

10. S. Butera, M. D. Paola, A physically based connection between fractional calculus and fractal geometry, *Annals of Physics*, **350** (2014), pp. 146–158.

11. E. Camrud, A novel approach to fractional calculus: Utilizing fractional integrals and derivatives of the Dirac delta function, *Progress in Fractional Differentiation and Applications*, **4(4)** (2018), pp. 463–478.

12. C. Cattani, Shannon wavelets theory, *Mathematical Problems in Engineering*, **2008**, Article ID 164808 (2008), 24 pages.

13. C. Cattani, Connection coefficients of Shannon wavelets, *Mathematical Modeling and Analysis*, **11(2)** (2006), pp. 117–132.

14. C. Cattani, Fractional calculus and Shannon wavelet, **2012**, Article ID 502812, 26 pages, doi:10.1155/2012/502812.

15. C. Cattani, J. J. Rushchitsky, *Wavelet and Wave Analysis as Applied to Materials with Micro or Nanostructure*, volume **74** of Series Advances in Mathematics for Applied Sciences, World Scientific Publishing, Singapore, 2007.

16. Y. Çenesiz, A. Kurt, The solutions of time and space conformable fractional heat equations with conformable Fourier transform, *Acta Universitatis Sapientiae, Mathematica*, **7(2)** (2015), pp. 130–140.

17. J. Cheng, J. Nakagawa, M. Yamamoto, T. Yamazaki, Uniqueness in an inverse problem for one-dimensional fractional diffusion equation, Inverse problem, **25(11)** (2009) 115002.

18. Chunmei Chi, Feng Gao, Simulating fractional derivatives using MATLAB, *Journal of Software*, **8(3)** (March, 2013), pp. 572–578.

19. A. O. Contreras, J. Rosales, L. M. Jimenez, J. M. Cruz-Duarte, Analysis of projectile motion in view of conformable derivative, *Open Physics*, **16(1)** (2018), pp. 581–587.

20. J. Cresson, Inverse problem of fraction calculus of variations for partial differential equations, *Communication in Nonlinear Sciences and Numerical Simulation*, **15(4)** (2010), pp. 987–996.

21. Z. L. Deng, X. M. Yang, X. L. Feng, A mollification regularization method for a fractional diffusion inverse heat conduction problem, *Mathematical Problem in Engineering*, **2013**, Article ID 109340, 9 pages.

22. A. Dixit, A. Ujlayan, Analytical solution to linear conformable fractional partial differential equations, *World Scientific News (WSN-3)*, **113** (2018), pp. 49–56.

23. Amita Garg, *Inverse Problems Related to Partial Differential Equations*, Ph.D. Thesis, Sharda University, Greater Noida, India, June 2017.

24. A. Garg, A. H. Siddiqi, Inverse estimation of 1-D heat problem with Neumann Boundary Condition by Morozov discrepancy principle, *Indian Journal of Industrial and Applied Mathematics*, **7(1)** (2016), pp. 58–64.

25. R. Garrappa, Numerical solution of fractional differential equations: A survey and a software tutorial, *Mathematics*, **6(2)**, **16**, (2018), pp. 1–23.

26. R. Garrappa, *Short Tutorial: Solving Fractional Differential Equations by MATLAB Codes*, 30 June 2014, https://www.scribd.com/document/350991589.

27. C. W. Groetsch, *Inverse Problems Activities for Undergraduate*, Washington, DC: Mathematical Association of America, 1999.

28. A. Hammad, R. Khalil, Fractional Fourier series with applications, *American Journal of Computational and Applied Mathematics*, **4(6)** (2014), pp. 187–191.

29. A. Hammad, R. Khalil, Conformable fractional heat differential equation, *International Journal of Pure and Applied Mathematics*, **94(3)** (2014), pp. 215–221.

30. R. Hermann, *Fractional Calculus: An Introduction for Physicists*, 2nd edition, World Scientific Publishing Co. Pvt. Ltd. , 2014
31. Mesut Karabacak, Muhammed Yiğider, Shannon wavelet transform for solving fractional differential-algebraic equations numerically, *International Journal of Engineering and Applied Sciences*, **3(12)** (2016), pp. 56–61.
32. R. Khalil, M. A. Horani, A. Yousef, M. Sabebah, A new definition of fractional derivative, *Journal of Computational and Applied Mathematics*, **264** (2014), pp. 65–70.
33. Yasir Khan, Naeem Faraz, Modified fractional decomposition method having integral w.r.t $d\xi^\alpha$, *Journal of King Saud University-Science*, **23(2)** (2011), pp. 157–161.
34. Anatoly A. Kilbas, Hari M. Srivastava, Juan J. Trujillo, Theory and applications of fractional differential equations, *North-Holland Mathematics Studies*, **204** (2006), pp. 1–523.
35. M. Kirane, S. A. Malik, Determination of an unknown source term and the temperature distribution for linear heat equation involving fractional derivative in time, *Applied Mathematics and Computation*, **218(1)** (2011), pp. 163–170.
36. V. Lakshmikanthan, A. S. Vatsala, Basic theory of fractional differential equations, *Nonlinear Analysis: Theory, Methods and Applications*, **69(8)** (2008), pp. 2677–2682.
37. J. T. Machado, V. Kiryakovo, F. Mainardi, Recent history of fractional calculus, *Communications in Nonlinear Science and Numerical Simulation*, **16(3)** (2011), pp. 1140–1153.
38. P. Manchanda, R. Lozi, A. H. Siddiqi, Mathematical modeling optimization, *Analytic and Numerical Solutions, Springer Nature* (2019), pp. 1–426.
39. Shuman Meng, Yujun Cui, The external solution to conformable fractional differential equations involving integral boundary conditions, *Mathematics*, **186** (2019), pp. 1–9.
40. L. Miller, M. Yamamoto, Coefficient inverse problem for a fractional diffusion equation, inverse problem for a fractional diffusion equation, *Inverse Problems*, **29(7)** (2013), p. 075013 (8pp.).
41. I. Nagma, A. H. Siddiqi, Sine-cosine wavelets approach in numerical evaluation of hankel transform for seismology, *Applied Mathematical Modelling*, **40(7–8)** (2016), pp. 4900–4907.
42. A. Neamaty, B. Agheli, R. Darzi, Solving fractional partial differential equation by using wavelet operational method, *Journal of Mathematics and in Computer Science*, **7(4)** (2013), pp. 230–240.
43. R. I. Nurudeen, F. D. Zaman, Y. F. Zakariya, Analysing the fractional heat diffusion equation solution in comparison with the new fractional derivative by decomposition method, *Malay Journal of Matematik*, **7(2)** (2019), pp. 213–222.
44. E. C. Oliveira, J. A. T. Machado, A review of definitions for fractional derivatives and integral, *Mathematical Problems in Engineering*, **2014,** Article ID 238459, 6 pages.
45. M. D. Ortigueira, J. A. T. Machado, What is fractional derivative?, *Journal of Computational Physics*, **293(15)** (2015), pp. 4–13.
46. Ivo Petras, Fractional derivatives, fractional integrals, and fractional differential equations in MATLAB, 2011. www.intechopen.com.
47. M. Razzaghi, S. Yousefi, Sine cosine wavelets operational matrix of integration and its applications in the calculus of variations, *International Journal of Systems Science*, **33(10)** (2002), pp. 805–810.
48. K. Sakamato, M. Yamamoto, Initial value/boundary value for fractional diffusion wave equations and applications to some inverse problems, *Journal of Mathematical Analysis and Applications*, **382(1)** (2011), pp. 426–447.
49. J. Shen, G. Strang, On wavelet fundamental solutions to the heat equation-heatlets, *Journal of Differential Equations*, **161(2)** (2000), pp. 403–421.
50. A. H. Siddiqi, Preface: Emerging applications of wavelet methods, *AIP Conference Proceedings*, **1463** (2012), p. 1.
51. A. H. Siddiqi, M. A. Lawati, M. Boulbrachene, *Modern Engineering Mathematics*, New York: CRC Press Taylor and Francis Group, 2017.
52. A. H. Siddiqi, R. C. Singh, G. D. V. Gowda, *Computational Science with Application*, New York: CRC Press, Taylor and Francis Group (in the press), 2019.
53. H. Sun,Y. Zhang, D. Baleanu, W. Chen, Y. Chen, A new collection of real world applications of fractional calculus in science and engineering, *Communications in Nonlinear Science and Numerical Simulation*, **64** (2018), pp. 213–231.

54. M. Tavassoli Kajani, M. Ghasemi, E. Babolian, Numerical solution of linear intgro-differential equation by using sine cosine wavelets, *Applied Mathematics and Computation*, **180**(2) (2006), pp. 569–574.

55. L. Steven, J. Brunton, N. Kutz, Data-driven science and engineering, machine learning, *Dynamical Systems and Control* (2019), pp. 1–472.

56. A. Judith, L. Wan, L. R. White, The numerical solution of inverse problems of Fourier convolution type, *Applied Mathematical Modeling*, **15**(7) (1991), pp. 359–366.

57. Faut Usta, Numerical solution of fractional elliptic PDEs by the collocation method, applications of applied mathematics, *Anais an International Journal*, **12**(1) (June 2017), pp. 470–418.

58. Wu Xiao-Er, Liang Yang-Shun, Relationship between fractal dimensions and fractional calculus, *Nonlinear Science Letter*, **1** (2017), pp. 1–10.

59. X. Xiong, H. Guo, X. Liu, An inverse problem for a fractional diffusion equation, *Journal of Computational and Applied Mathematics*, **236**(17) (2012), pp. 4474–4484.

60. S. Yang, L. Wang, S. Zhang, Conformable derivative application to non-Darcian flow in low-permeability porous media, *Applied Mathematics Letters*, **79** (2018), pp. 105–110.

61. Yanxin Wang, Tianhe Yin, Li Zhu, Sine-cosine wavelet operational matrix of fractional order integration and its applications in solving the fractional order Riccati differential equations, *Advances in Difference Equations*, **222** (2017).

62. Monika Zecova, Jan Terpak, Heat conduction modeling by using fractional order derivatives, *Applied Mathematics and Computation*, **257** (2015), pp. 265–375.

63. Y. Zhang, X. Xu, Inverse problems for a fractional diffusion equation, *Inverse Problems*, **27**(3) (2011), pp. 1–12.

64. Dazhi Zhao, Maokang Luo, General conformable fractional derivative and its physical interpretation, *Calcolo*, **54**(3) (2017), pp. 903–917.

65. H. W. Zhou, S. Yang, S. Q. Zhang, Conformable derivative approach to anomalous diffusion, *Physica. Part A*, **491** (2018), pp. 1001–1013.

66. Z. Zhao, O. Xie, Z. Meng, L. You, Determination of unknown source in the heat equation by the method of Tikhonov regularization in Hilbert scales, *Journal of Applied Mathematics and Physics*, **2** (2014), pp. 10–17.

67. Z. Zhou, X. Gao, Numerical methods for pricing American options with time fractional PDE models, *Mathematical Problems in Engineering*, **2016**, Article ID 5614950.

68. M. S. Zurigat, A. A. Momani, A. Alawneh, Analytical approximate solutions of systems of fractional algebraic-differential equations by homotopy analysis method, *Computers and Mathematics with Applications*, **59**(3) (2010), pp. 1227–1235.

8

Pan-Sharpening Using Modified Nonlocal Means-Based Guided Image Filter in NSCT Domain

Tarlok Singh and Pammy Manchanda

CONTENTS

8.1   Introduction.. 117
8.2   Proposed Method 119
    8.2.1   Nonsubsampled Contourlet Transform.. 119
        8.2.1.1   Nonsubsampled Pyramid (NSP) 120
        8.2.1.2   Nonsubsampled Directional Filter Bank (NSDFB) 122
        8.2.1.3   Combining the NSP and NSDFB 123
    8.2.2   Weighted Maps... 123
    8.2.3   Pan-Sharpening on Multi-Scale... 124
    8.2.4   Modified Non-Local Means-Based Guided Image Filter (MNLMGIF) 125
        8.2.4.1   Guided Image Filter... 125
        8.2.4.2   Nonlinear Kernel Weights 126
        8.2.4.3   Guidance Image Construction 127
8.3   Experimental Results 128
    8.3.1   Visual Analysis 129
    8.3.2   Quantitative Analysis... 131
8.4   Conclusion 132
References.. 132

8.1 Introduction

Remote sensing images are widely used for different areas ranging from mineral exploration to agricultural applications [1], but the poor quality of hyperspectral images will directly have an adverse effect on these applications [2]. Remotely sensed image pan-sharpening techniques combine a high-spatial–low-spectral–resolution image and a low-spatial–high-spectral–resolution image to attain high-quality images [3].

Satellite images are basically of two types: (a) multispectral images (MS), which have low spatial and high spectral resolution; and (b) panchromatic images with high-spatial and low-spectral resolutions as shown in Table 8.1.

The need for image pan-sharpening methods arises from the fact that satellite sensors face technical difficulties in capturing high-spatial resolution MS images. This is because detailed maps calculated through Pan images are injected to sharpen the MS images. To calculate the detailed map is the main objective of the pan-sharpening procedure.

TABLE 8.1

Multispectral vs. Panchromatic Images

Images	Spatial Resolution	Spectral Resolution
Multispectral	low	high
Panchromatic	high	low

Figure 8.1 shows the fusion effect of panchromatic and multispectral images into a pan-sharpened image.

Intensitiy hue saturation (IHS) methods begin with the transformation of RGB (Red, Green, Blue) to IHS space followed by replacing the intensity band with a Panchromatic image. Finally this new image is inversely mapped back from IHS to RGB space. This conventional IHS based method is efficient and simple to execute but it is not able to enhance an RGB-based MS image and often contains major distortions in spectral aspect, due to color changes among fused MS and the composition of re-sampled bands. Thus it misses near infrared band (NIR) which is the fourth band after RGB. [4] studies a fast pan-sharpening technique with spectral adjustment for IKONOS imagery. IKONOS is a commercial earth observing satellite collecting MS and Pan images.

In [5], two approaches are merged to form a hybrid pan-sharpening scheme: (a) local variation with contourlet transformation (CT); and (b) adaptive PCA and CT. It performs better compared to APCA+CT, APCA, PCA+CT, and PCA. Thus for multispectral satellite images, the spatial details are enhanced by introducing Pan images and extracting higher spatial details. [6] discusses a pan-sharpening method based on deep neural networks to model complex relations among independent variables by composing multiple non linear levels. Back propagation is also used for improved training as a second training phase after the pre-training phase of the complete deep neural network. In [7], sparse pan-sharpening of images based on a self-adaptive technique is studied. It makes no assumption about the spectral structure of the Pan image and inherits sparse signal reconstruction super

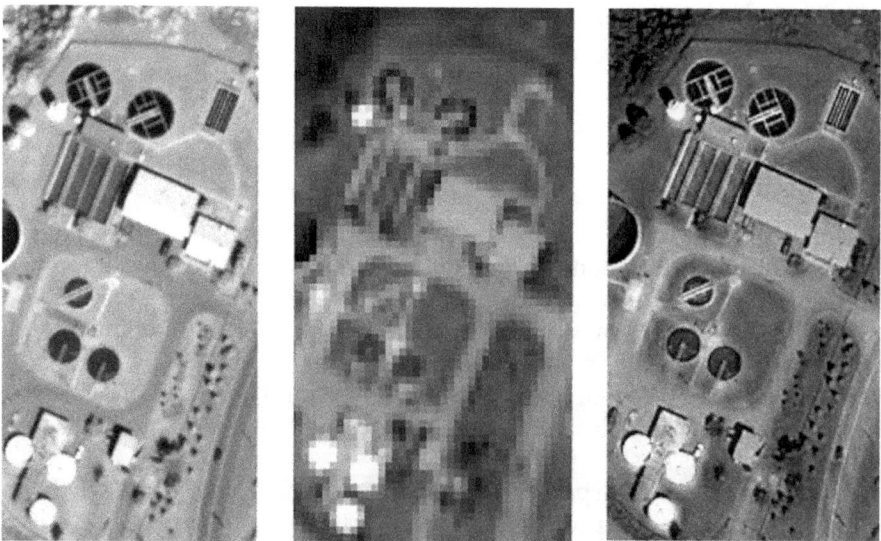

FIGURE 8.1

(Left) Panchromatic, (middle) multi-spectral, and (right) pan-sharped image.

resolution capacity; it yields less spectral distortion and higher spatial resolution. In [8], a method to remove speckle noise is discussed using generalized guided filters and Bayesian non-local means. It calculates the nonlinear weight kernel and generates guidance image with homogeniety analysis considering maximum likelihood rule and local regions.

Due to its robustness and effectiveness, multi-scale pan-sharpening based on the Laplacian pyramid decomposition is used such as shiftable complex directional pyramid [9], contrast pyramid transform [10], image decomposition [11], Laplacian pyramid with multiple features (LPMF) [12], etc. These techniques have emerged as a popular method that has shown utility in many pan-sharpening–based applications. However, the number of pyramid levels increases with the image size, which implies sophisticated data management and memory accesses, as well as additional computations [13]. Due to the loss of dependence of coefficients, most of the traditional multi-scale decomposition-based image pan-sharpening methods suffer from an inaccurate image representation. To overcome these problems, nonsubsampled shearlet transform (NSST) is used. The dependence of NSST coefficients is captured by the contextual hidden Markov model [14]. Explicit luminance and contrast masking models are combined to give the perceptual importance of each coefficient of source images produced by the dual-tree complex wavelet transform. This combined model of perceptual importance is used to select which coefficients are retained and furthermore to determine how to present the retained information in the most efficient manner as in [15]. In this chapter, we present a novel image pan-sharpening approach to overcome these limitations. The proposed method evaluates weight values of the input images and then uses the corresponding pan-sharpening map to develop the final fused image. To reduce the halo and gradient reversal artifacts further, a modified non-local means-based guided image filter (MNLMGIF) is proposed and applied on the fused image as a post-processing technique. Therefore, this method maintains all the information from both images, which is not done by most of the existing pan-sharpening methods. Moreover, the proposed technique can reduce gradient and halo artifacts problems using MNLMGIF. To justify the efficiency of the proposed method quantitatively, experiments have been carried out on remotely sensed images [16], QUICKBIRD [17], IKONOS [18], and MODIS [19] sensors.

The remainder of this chapter is organized as follows: a proposed filtering-based image pan-sharpening technique is discussed in Section 8.2. To verify the effectiveness of the proposed technique, visual and quantitative results are carried out on remote sensing images in Section 8.3. In Section 8.4 the conclusion is presented along with future directions.

8.2 Proposed Method

This section discusses the proposed methodology. A systematic procedure is adopted to evaluate the weighted map, multiscale image pan-sharpening, guidance image generation for filtering, and finally MNLMGIF for pan-sharpening is used. Figure 8.2 demonstrates the overview of graphical flow explaining the technique.

8.2.1 Nonsubsampled Contourlet Transform

Nonsubsampled Contourlet Transform (NSCT) is an extension of the well-known Contourlet Transform (CT) [20]. The multiscale decomposition feature of CT is achieved

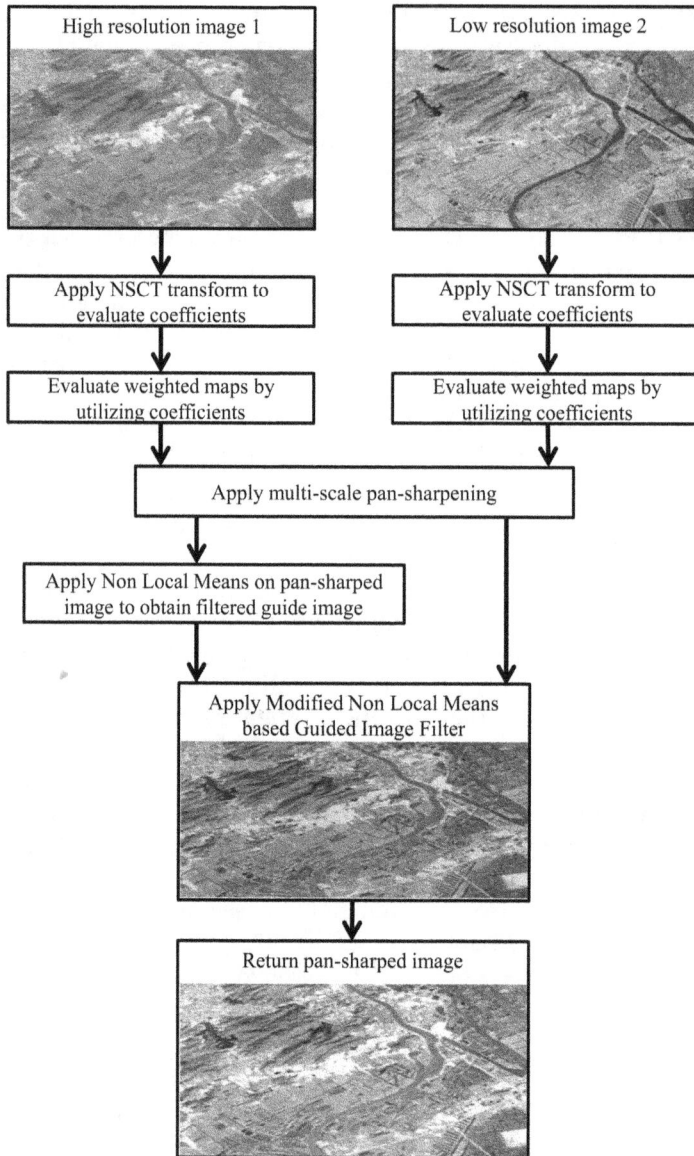

FIGURE 8.2
Graphical abstract of the proposed pan-sharpening technique.

by using Laplacian Pyramids (LPs). Directional Filter Banks (DFBs) are used to generate the directional decomposition of CT. LP and DFB utilize downsamplers and upsamplers, respectively [21]. NSCT is designed using Nonsubsampled Pyramids (NSP) and nonsubsampled DFBs to achieve shift-invariant feature (see Figure 8.3) [22].

8.2.1.1 Nonsubsampled Pyramid (NSP)

Nonsubsampled pyramid guarantees the multiscale characteristic of NSCT. NSP depends upon two-channel nonsubsampled 2D filter banks. The three stage decomposition of NSP

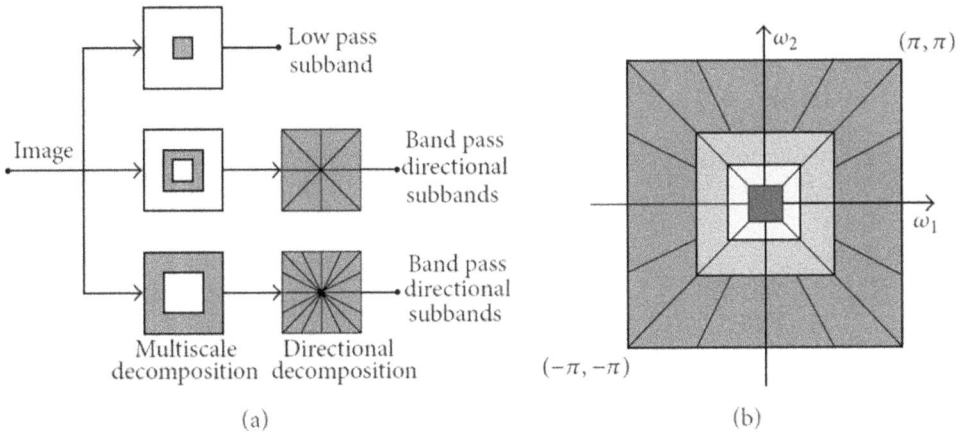

FIGURE 8.3
(a) NSFB structure that implements the NSCT; (b) idealised frequency partitioning obtained with NSFB structure (Yang et al., 2015).

is described in Figure 8.4(a). An efficient reconstruction is achieved provided the filters satisfy Bezout identity be defined as [22]:

$$H_0(z)G_0(z) + H_1(z)G_1(z) = 1 \qquad (8.1)$$

Here, $G_0(z)$ and $G_1(z)$ represent low- and high-pass reconstruction filters, respectively. $H_0(z)$ and $H_1(z)$ represent low- and high-pass decomposition filters, respectively.

The passband frequency support of the low-pass filter at *jth* level is the region $\left[-\dfrac{\pi}{2^j}, \dfrac{\pi}{2^j}\right]^2$. In the same way, the support of the high-pass filter is the complement of low-pass support

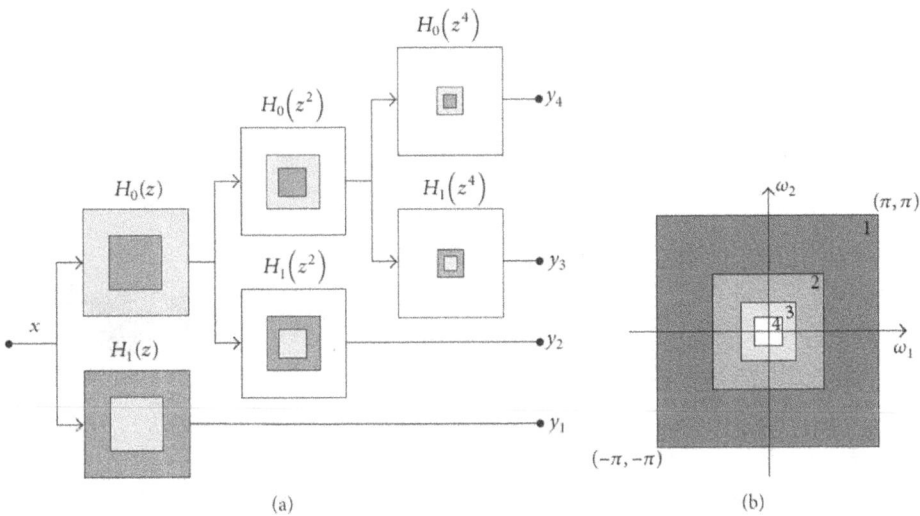

FIGURE 8.4
Nonsubsampled pyramid: (a) three-stage pyramid decomposition; (b) frequency divisions of a nonsubsampled pyramid (Yang et al., 2015).

i.e. the region $\left[-\dfrac{\pi}{2^{j-1}},\dfrac{\pi}{2^{j-1}}\right]^2 \Big/ \left[-\dfrac{\pi}{2^j},\dfrac{\pi}{2^j}\right]^2$. The equivalent filters of a *jth*-level cascading NSP can be defined as follows [22]:

$$
H_n(z) =
\begin{cases}
H_1(z^{2n-1})\displaystyle\prod_{j=0}^{n-2}H_0(z^{2j}) & 1 \le n < 2^k \\[3ex]
\displaystyle\prod_{j=0}^{n-2}H_0(z^{2j}) & n = 2^k
\end{cases}
\tag{8.2}
$$

Here, $H_0(z)$ and $H_1(z)$ represent low- and high-pass filters at the initial level, respectively. k and n represent the number of decomposition levels and the total number of decompositions of NSCT, respectively. Figure 8.4(b) shows the frequency of NSP.

8.2.1.2 Nonsubsampled Directional Filter Bank (NSDFB)

Nonsubsampled Directional Filter Bank (NSDFB) provides varying directions. It is evaluated by discarding the downsamplers and upsamplers in DFB [23]. Figure 8.5 shows a four-channel directional decomposition. The equivalent filter in each channel is defined as follows [23]:

$$
U_k(z) = U_i(z)U_j(z^Q)
\tag{8.3}
$$

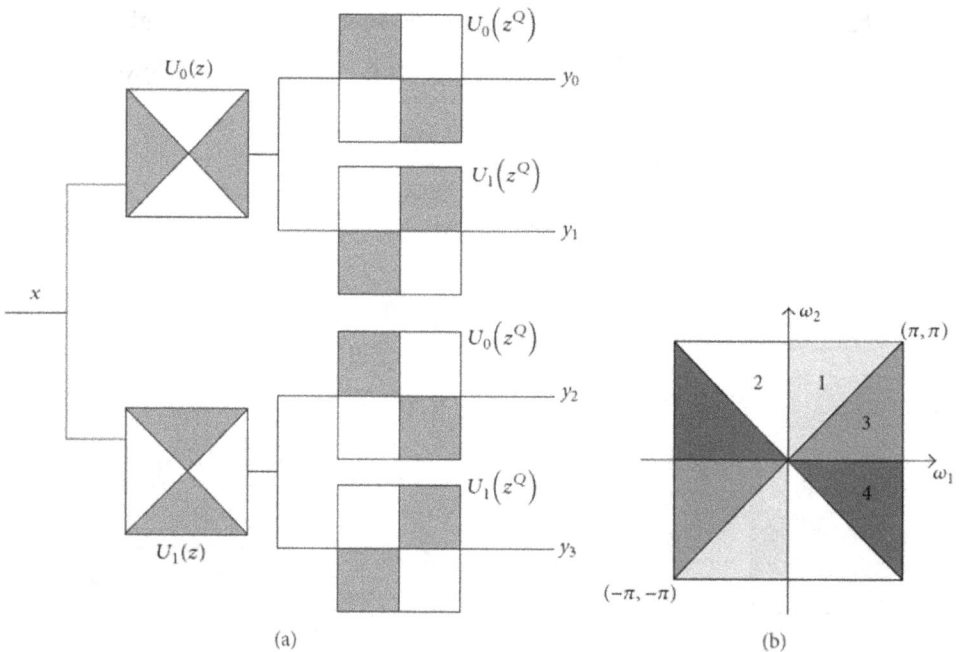

(a)   (b)

FIGURE 8.5

Four-channel NSDFB developed with two-channel fan filter bank: (a) filtering structure; (b) corresponding frequency partitioning (Yang et al., 2015).

Here, U_k, U_i, and U_j represent upsampled fan filters, which have checkerboard frequency support [20]. i and j are either 0 or 1.

8.2.1.3 Combining the NSP and NSDFB

The NSCT is achieved by integrating NSP and NSDFB (see Figure 8.3(a)). NSP contains multiscale decomposition and captures the point discontinuities. NSDFB has directional decomposition feature and links point discontinuities into linear structures [23]. This approach can be repeated continuously on low-pass subband outputs of NSP. Therefore, NSCT is appropriate for image pan-sharpening as it provides shift-invariance, multidirection, and multiresolution.

8.2.2 Weighted Maps

Image sections which consist of low light intensity and more distance (d) gets affected adversely with lesser visibility. In [24] the techniques which deal with contrast improvement on a global level are studied. The effect of d is different on the image objects depending upon its varying intensity and the contrasting color of the pixels. Now there is a family of contrast enhancement schemes which apply the same operation on the whole image and thus have this limitation in general. For example white balance, histogram equalization, and gamma correction are among popular enhancement operators. To fix this limitation, the proposed technique has three different measures based upon weight maps. They take pixels into consideration and define customized spatial functions for poor contrast regions. Thus higher values are derived for the regions with high-contrast input values.

The *luminance weight map* calculates pixel visibility and assign values proportional to their visibility. For environment affected images, color loss is calculated based upon NSP information, since they have low saturation in general. For each NSP with index n, weight is calculated as follows:

$$W_n^L = \frac{1}{\sqrt{3}}\sqrt{\text{NSP} + \text{NSDFB}} \qquad (8.4)$$

The W_n^L is an index for degradation identification caused due to d. Then $NSDFB_2$ is obtained by taking the scaled difference of luminance value of the entire image (pixels) and the average luminance. If α is a coefficient of NSDFB then

$$\text{NSDFB}_2 = \alpha(\text{NSDFB} - \text{NSDFB}_{\text{avg}}) \qquad (8.5)$$

The derived weighted luminance helps a smooth transition between derived coefficients of $NSDFB$, $NSDFB_2$, but reduces NSCT's coefficient values. Therefore, two additional weighted maps are defined based upon NSCT values.

The *chromaticity weighted map* is defined as follows:

$$W_n^C = \exp\left(\frac{(S_n - S_{\text{max}})^2}{-2\sigma^2}\right) \qquad (8.6)$$

W_n^C calculates the squared difference in saturation value and its maximum value. $\sigma = 0.3$ works well as a default value for our experiments. $S_{\text{max}} = 1$ in our case for NSCT domain. Therefore, the value is proportional to the saturation of the pixel values. Therefore, saturation regions are depicted in a better form in the outcome. The *saliency weighted map* calculates the visibility degree relative to the neighborhood areas. It basically measures the relative coefficients of a region or a segment in an NSDFB with respect to its surroundings based upon coefficient orientation and values. As discussed in [25], the weight is defined as follows:

$$W_n^S = \| \text{NSDFB}_n^f - \text{NSDFB}_n^{\text{avg}} \| \tag{8.7}$$

Here $\text{NSDFB}_n^{\text{avg}}$ is the average coefficient value of NSDFB and is of low-resolution image from where the high-frequency noise needs to be removed. It is obtained by using the high-frequency value of cut-off ($f = \pi/2.75$) into a separable binomial kernel. Therefore, the impact of these three measures is important in different aspects but the first one, which is the luminance weight map, has greater visibility impact overall. The aggregated weight map is obtained by multiplicative expression:

$$W_n = W_n^L \times W_C^n \times W_n^S \tag{8.8}$$

The normalized form of the aggregate weight guarantees that the sum will be 1 for each pixel:

$$\overline{W}_n = \frac{W_n}{\sum W_n} \tag{8.9}$$

8.2.3 Pan-Sharpening on Multi-Scale

During the image pan-sharpening, weighted inputs are computed in order to detect the features more significantly for each pixel:

$$F = \sum_n \overline{W}_n I_n \text{s.t.} \sum_n \overline{W}_n = 1 \tag{8.10}$$

Intensity scale of the resulting image is maintained because a normalized weighted map is used in the above calculation. But this equation may result in halos artifacts where the weighted maps have stronger transitions. Therefore an approach [26] which is known as pyramidal refinement is applied to get rid of this degradation. In the proposed method, I_n is actually decomposed using a Laplacian operator into a pyramid with different scales. On the parallel track, $\overline{W}_n$ is also derived from a Gaussian pyramid for each of its weighted normalized maps. Now at each respective pyramid level i (default value $i = 5$) the mixing is applied as follows:

$$F_i = \sum_n G_i(\overline{W}_n) L_i(I_n) \tag{8.11}$$

where L_i is the Laplacian operation of the input I and $G_i(\overline{W}_n)$ is the Gaussian operator on the weighted normalized map $\overline{W}_n$. Bottom-up direction is adopted for respective pyramid layers and the final restored version of I is calculated after adding all the pyramid levels using upsampling operator $U(x)$ with $x = 2^{i-1}$.

$$G(x) = \sum_i F_i(x) U(x) \tag{8.12}$$

In the proposed method, all these three weights have equal weighted contribution toward the final pan-sharpening result.

8.2.4 Modified Non-Local Means-Based Guided Image Filter (MNLMGIF)

The subsequent sections contain the discussion about guided filter, nonlinear weight kernel calculation, and guidance image formation.

8.2.4.1 Guided Image Filter

A guided filter is used for preserving the edges under the assumption about the dependence between guidance I and filtered output λ as studied in [27]. Let w_n is window with center at pixel n, (α_n, β_n) are derived through the filter input p, then

$$\lambda_i = \alpha_n \gamma_i + \beta_n, \; \forall i \in w_n \tag{8.13}$$

However, a pixel i is involved in all the overlapping windows w_n that covers i, so the value of λ_i in (13) is not identical when it is computed in different windows. A simple strategy is to average all the possible values of λ_i. So, after computing (a_n, b_n) for all windows w_n in the image, we compute the filtering output by

$$\lambda_i = \frac{1}{|w|} \sum (\alpha_n \gamma_i + \beta_n) \tag{8.14}$$

Using the least mean square idea as in [28] and letting $A_{n,j}$ and $\beta_{n,j}$ be dependent upon guidance image g, we can derive:

$$\alpha_n = \sum_n A_{n,j}(g) p_j \tag{8.15}$$

$$\beta_n = \sum_n \beta_{n,j}(g) p_j \tag{8.16}$$

Thus λ_i can be reformulated as follows:

$$\lambda_i = \frac{1}{|w|} \sum_i \sum_j [A_{n,j}(\gamma_i)\gamma_i + \beta_{n,j}g(i)] p_j \tag{8.17}$$

Let $W_{n,j}(g) = \frac{1}{|w|} \sum_n [g.A_{n,j}(\gamma_i) + \beta_{n,j}(\gamma_i)]$ be the kernel weights then

$$\lambda_i = \sum_j W_{n,j}(\gamma_i) p_j, \forall i \in w_n \tag{8.18}$$

It can be further derived that

$$\lambda_i = \sum_j W_{i,j}(\gamma_i) p_j \tag{8.19}$$

for k such that $(i, j) \in w_n$. In order to deal with the filtering issue, the linear guided filter is generalized into its nonlinear version by considering input from filtering f, i.e. speckled image, which is called GGF filter. Filtered image (d_i) can be calculated using speckled image (f), guidance image (f^g). Let $\{i, j\}$ be the pixel locations, f is available and thus $W_{i,j}$ which are nonlinear weights of kernel and construction of guidance image f^g need to be calculated.

$$d_i = \sum_j W_{i,j}(f, f^g) f_j \tag{8.20}$$

Weights can be calculated using BNLM (Bayesian Nonlocal Means) and the guidance image can be found using local region and maximum likelihood rules.

8.2.4.2 Nonlinear Kernel Weights

BNLM can be expressed as follows according to [29]. Let f_j be speckled data at jth pixel, v_i, v_j are patch vectors whose center is at (i, j) and size $N \times N$. $p(v_i|v_j)$ is a measure of similarity between v_i and v_j patches.

$$d_i = \frac{\sum p(v_i)p(v_j)f_j}{\sum p(v_i \mid v_j)p(v_j)} \tag{8.21}$$

Let r_i, r_j be radar reflectivity for speckled patches v_i, v_j to deal with similar reflectivity or almost identical-looking patches. The updated equation, after taking guidance image into consideration:

$$d_i = \frac{\sum p(v_i \mid v_j, r_i = r_j, f^g)p(v_j, r_i = r_j, f^g)v_j}{\sum p(v_i \mid v_j, r_i = r_j, f^g)p(v_j, r_i = r_j, f^g)} \tag{8.22}$$

Now conditional probability p(.) for independent speckles can be defined as:

$$p(v_i, v_j, r_i = r_j, f^g) = \Pi_n p(v_{i,n} \mid v_{j,n}, r_{i,n} = r_{j,n}, f_n^g) \tag{8.23}$$

Suppose f_m^g is independent from set of events $\{v_{i,m} \mid v_{j,m}, r_{i,m} = r_{j,m}\}$ and there is no prior knowledge about $p(v_{i,m} \mid v_{j,m})$ we have

$$p(v_{j,m}, r_{i,m} = r_{j,m}, f_g^m)p(v_{i,m} \mid v_{j,m}, r_{i,m} = rj, m, f_g^m)$$

$$= p(v_{i,m}, vj, m, r_{i,m} = r_{j,m}, f_g^m)$$

$$= p(r_{i,m} = r_{j,m}, f_g^m)p(v_{i,m}, vj, m \mid r_{i,m} = r_{j,m}, f_g^m) \tag{8.24}$$

$$\propto p(f_g^m)p(r_{i,m} = rj, m \mid f_g^m)p(vi.m, v_{j,m} \mid r_{i,m} = rj, m)$$

Suppose p_{llhd}, p_{pr} are probabilities of likelihood and prior and then we can assume

$$p_{llhd} = p(v_{i,m}, v_{j,m} \mid r_{i,m} = r_{j,m}) \tag{8.25}$$

$$p_{pr} = p(r_{i,m} = r_{i,m} \mid f_g^m) \tag{8.26}$$

Likelihood term corresponds to the fidelity of the data and prior estimates $r_{i,m} = r_{j,m}$ validity when f_g^m, i.e. guidance image-value is given. It is assumed that distribution of $p(f_g^m)$ is uniform. Thus it can be concluded that

$$p_{\text{llhd}} \propto \Theta^{2N-1}, \text{ where } \Theta = \left[\frac{v_{i,m} v_{j,m}}{v_{i,m}^2 + v_{j,m}^2} \right] \tag{8.27}$$

$$p_{\text{pr}} \propto \exp(-A.\Phi), \text{ where } \Phi = \left(\frac{| (f_{i,m}^g)^2 - (f_{j,m}^g)^2 |}{[(f_{i,m}^g)^2](f_{j,m}^g)^2} \right) \tag{8.28}$$

Using the idea of Nakagami and Rayleigh distribution, for N-look image [30] to represent image i with amplitude A:

$$p(i \mid f) = 2 \left(\frac{N}{r} \right)^N \frac{1}{\Gamma(N)} \exp\left(-\frac{Ni^2}{r} \right) i^{2N-1} \tag{8.29}$$

Let i and f be speckled and guidance images. C, D are constants for normalization to calculate these images relatively. Thus GGF kernel weight for nonlinear system becomes

$$W_{i,j}(i, f^g) = \exp\left(-\frac{1}{C} \sum_m \frac{1}{\Theta} + \frac{N}{D} \sum_m \Phi \right) \tag{8.30}$$

Proper choice of the guidance image is crucial to remove speckle noise for a nonlinear guided filter. Now analyzing Equation (8.30) it can be observed that by tuning N, C, D, one can decide about the proportional effects of i or f^g on the final filtering output through various ranges of weight kernel values.

8.2.4.3 Guidance Image Construction

It is known from the literature that guidance image plays a crucial role in image filtering, actually considered as the prior value estimation of f. An idea is proposed for construction of a guidance image from the original image which approximates the original image as closely as possible. Based upon the analysis of local-region homogeneity, variance of pixel intensities could be used for the homogeneity metric [8]. Original images are divided into various regions as described in [31] and [32] . Three variances are calculated as:

$$\zeta_1 = \sqrt{\frac{(4 / \pi - 1)}{N}} \tag{8.31}$$

$$\zeta_2 = \sqrt{3}.\zeta_1 \tag{8.32}$$

$$\zeta_3 = \frac{\sqrt{\text{var}(c_i)}}{M[\zeta_i]} \tag{8.33}$$

N is number of looks, var(c_i) is variance, $M[\zeta_i]$ is mean of pixels in the local region which is centered at c_i. Guidance image is generated by following these steps:

1. If $\zeta_3 \leq \zeta_1$ then use simple average filter to find I_g.

2. If $\zeta_1 < \zeta_3 < \zeta_2$ then compute I_g using formula $\sqrt{\dfrac{\sum_n c_{i,n}^2}{|w_i|}}$ where $c_{i,n}$ is *nth* pixel in

   local region window w_i centering at pixel c_i. $|w_i|$ is the pixel count in w_i local adaptive region w_i.

3. If $\zeta_3 \geq \zeta_2$, then I_g is same as value of c.

Finally, the filtered image from modified non-local means-based guided image filter (MNLMGIF) is obtained as follows:

$$d_i = \frac{\sum_{j \in w_i} W_{i,j}(v, f^g) v_j}{\sum_{j \in w_i} W_{i,j}(v, f^g)} \tag{8.34}$$

The normalization parameters C and D need to be tuned, which would further decide the values of p_{llhd} and $p_p r$ in the weighted kernel. This would eventually affect the final filtered image output. Patches of size 7×7 are used along with 2121 search window size. Furthermore 3×3 local regions are considered for ML filtering. Thus the other variables are set as follows: local region size = 3, search window size = 21, and patch size = 7. Moreover the number of looks N usually provide fewer looks. Figure 8.6 shows the effect of speckled images for $N = 1,2,3$. For C, D variable tuning, first C is fixed and D is varied and then vice versa. After a certain increase in the value of the variable its effect becomes negligible, and then the other parameter is tuned in a similar fashion.

8.3 Experimental Results

To achieve the objectives of this chapter, we have designed existing pan-sharpening techniques (i.e., CT, LPMF, NSST) and the proposed technique in MATLAB® 2013*a* software.

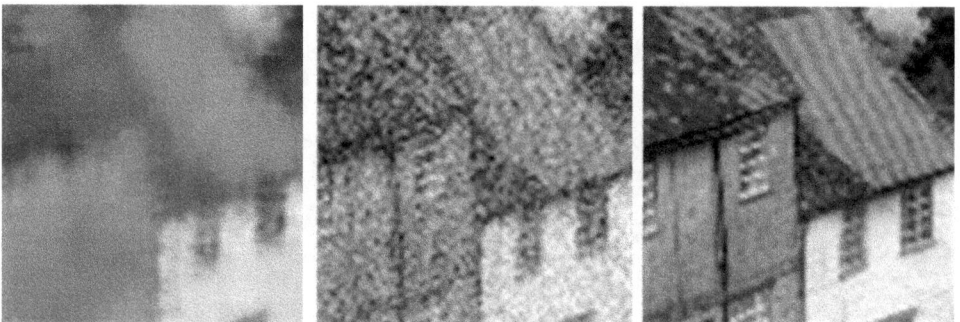

FIGURE 8.6
Filtering effects of changing the values of looks $N = 1,2,3$.

Five benchmark remote sensing images are used for performance evaluation. However, the proposed technique is not limited to these five images alone. The benchmark images are from QUICKBIRD [17], IKONOS [18], and MODIS [19] sensors. The subsequent section contains the details of the visual and the quantitative analyses.

8.3.1 Visual Analysis

Figures 8.7(a), 8.8(a) and 8.7(b), 8.8(b) show the high-spatial–resolution panchromatic images and lower-spatial–resolution images respectively of the IKONOS [18] and MODIS [19] sensors respectively. Tables 8.2 and 8.3 depict the pan-sharped images obtained from

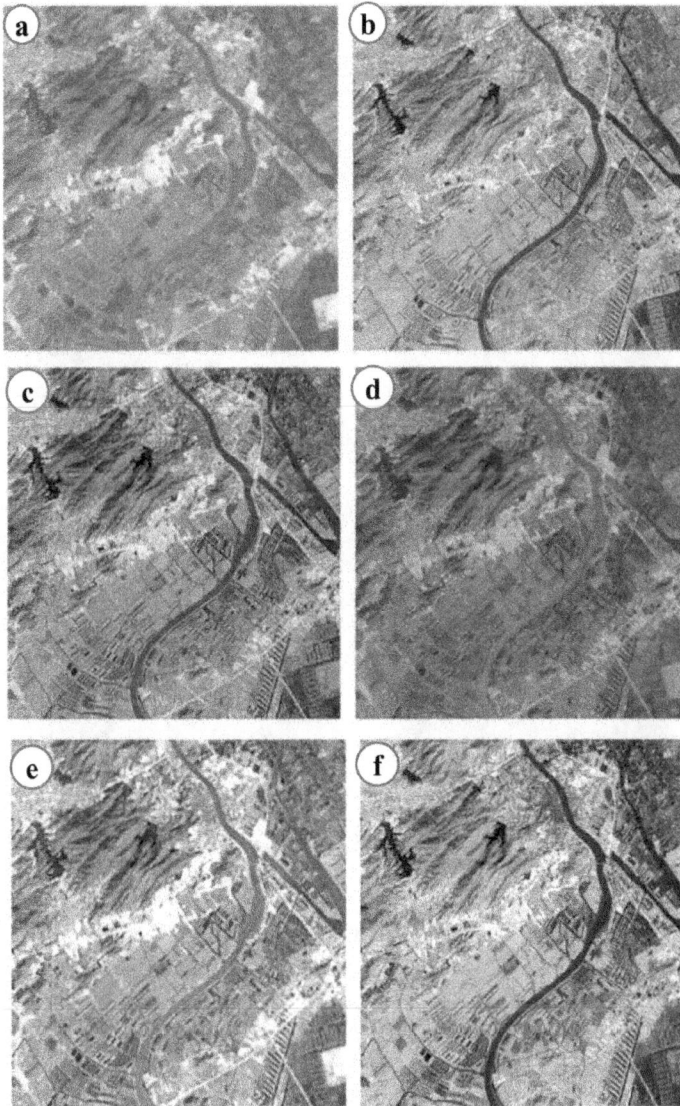

FIGURE 8.7
Experimental results of remote sensing image 1: (a) high spatial resolution panchromatic image, (b) lower spatial resolution image, (c) CT, (d) LPMF, (e) NSST, and (f) proposed.

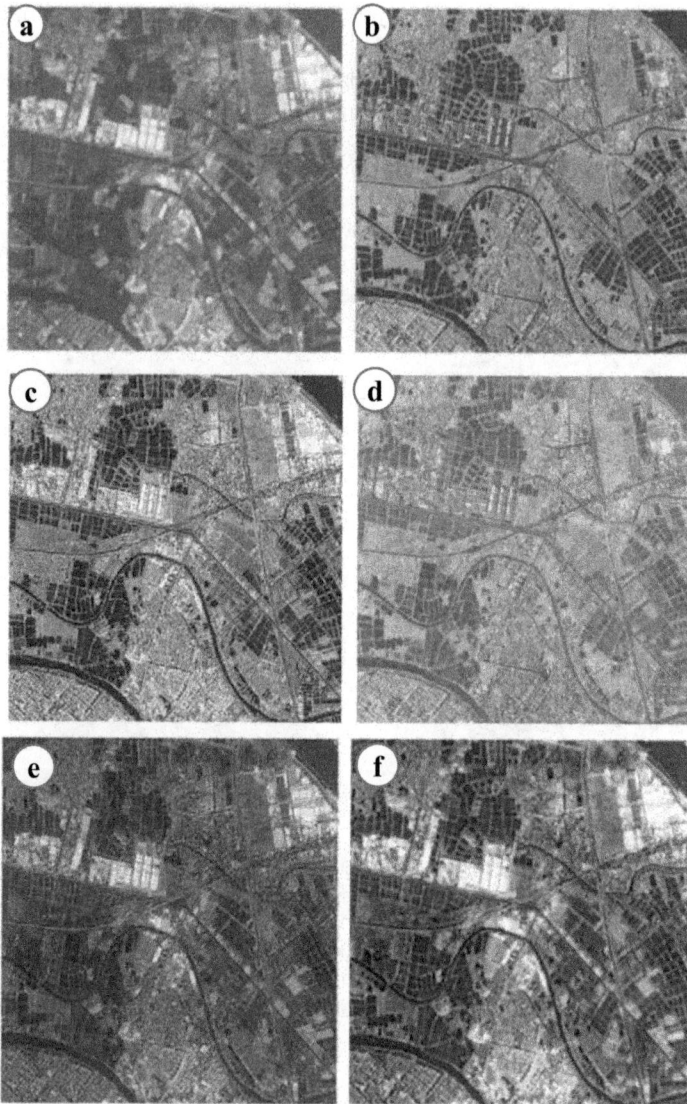

FIGURE 8.8
Experimental results of remote sensing image 2: (a) high spatial resolution panchromatic image, (b) lower spatial resolution image, (c) CT, (d) LPMF, (e) NSST, and (f) proposed.

TABLE 8.2

Entropy Analysis

Method	Image 1	Image 2	Image 3	Image 4	Image 5
CT	7.012	7.132	7.102	7.272	7.376
LPMF	7.481	7.450	7.581	7.651	7.427
NSST	7.794	7.816	7.768	7.678	7.656
proposed	**7.951**	**7.894**	**7.871**	**7.928**	**7.981**

TABLE 8.3

Analysis of the Fusion Factor

Method	Image 1	Image 2	Image 3	Image 4	Image 5
CT	5.671	5.358	5.053	5.981	5.355
LPMF	5.956	5.378	5.133	5.555	5.335
NSST	6.531	6.351	5.891	5.998	6.355
proposed	**6.831**	**6.953**	**6.396**	**6.689**	**6.953**

CT, LPMF, NSST and the proposed technique, respectively. It has been observed that the proposed technique is able to conserve good spatial, spectral, and texture information as compared to the existing techniques. The proposed technique also shows fewer edges and less color distortion compared to existing pan-sharpening techniques.

8.3.2 Quantitative Analysis

These comparisons have been drawn between the existing and the proposed pan-sharpening techniques based upon the various performance metrics such as Entropy, Fusion Factor, Peak Signal-to-Noise Ratio (PSNR), and Structural Similarity index (SSIM).

Tables 8.2 through 8.5 depict the performance analysis of existing techniques (i.e. CT, LPMF, NSST) and the proposed technique.

Table 8.2 represents the entropy values that need to be maximized. From Table 8.2, it has been observed that the proposed technique has significantly more entropy values as compared to CT, LPMF, and NSST.

Table 8.3 depicts the fusion factor analysis of the pan-sharpening technique. It needs to be maximized. Therefore, Table 8.3 reveals that the proposed technique provides significantly more fusion factor values as compared to existing techniques.

Table 8.3 shows the analysis of fusion factor, which needs to be maximized. Therefore Table 8.3 clearly demonstrates that the proposed technique has a better fusion factor than

TABLE 8.4

Peak Signal to Noise Ratio Evaluation

Method	Image 1	Image 2	Image 3	Image 4	Image 5
CT	39.91	36.81	34.19	38.92	39.88
LPMF	36.44	36.94	36.48	38.84	38.19
NSST	40.23	39.46	34.49	39.18	39.93
proposed	**42.91**	**39.81**	**39.19**	**40.32**	**42.88**

TABLE 8.5

Structural Similarity Index Metric Evaluation

Method	Image 1	Image 2	Image 3	Image 4	Image 5
CT	0.891	0.912	0.919	0.971	0.897
LPMF	0.981	0.899	0.923	0.924	0.948
NSST	0.985	0.948	0.915	0.949	0.948
proposed	**0.987**	**0.962**	**0.957**	**0.964**	**0.967**

CT, LPMF, and NSST. The mean improvement of the fusion factor of the proposed approach over existing techniques is 0.69.

Tables 8.4 and 8.5 depict the analysis of *PSNR* and *SSIM*, respectively, between existing techniques and the proposed pan-sharpening techniques. These tables clearly show that the proposed technique has an ability to achieve the more realistic results compared to the existing pan-sharpening techniques.

8.4 Conclusion

Optimistic conservation of more informative blocks from low-resolution remote sensing images from the same scene into one high-resolution image is referred to as pan-sharpening. The main objective is to conserve the spatial information from the low-resolution images. The techniques of CT, LPMF, and NSST are found to be efficient among existing pan-sharpening techniques. In this chapter, a novel NSCT-based image pan-sharpening technique is designed. To reduce the halo and gradient reversal artifacts, *MNLMGIF* is also implemented on the fused image. The proposed technique is tested on five benchmark images. Subjective and objective analysis of the existing techniques (i.e. CT, LPMF, and NSST) and the proposed technique have revealed that the proposed technique provides significant spatial, spectral, and texture detail. Average improvements of the proposed technique over existing pan-sharpening techniques are 0.78, 0.58, 0.79 dB, and 0.07 for entropy, fusion factor, *PSNR*, and *SSIM*, respectively. Therefore, the proposed technique is more suitable for real-time pan-sharpening techniques.

References

1. X. Tang and L. Jiao. Fusion similarity-based reranking for sar image retrieval. *IEEE Geoscience and Remote Sensing Letters*, 14(2):242–246, 2017.
2. A. Ansari, H. Danyali, and M. S. Helfroush. Hs remote sensing image restoration using fusion with ms images by em algorithm. *IET Signal Processing*, 11(1):95–103, 2017.
3. Q. Wei, J. Bioucas-Dias, N. Dobigeon, J. Y. Tourneret, M. Chen, and S. Godsill. Multiband image fusion based on spectral unmixing. *IEEE Transactions on Geoscience and Remote Sensing*, 54(12):7236–7249, 2016.
4. Te-Ming Tu, Ping S. Huang, Chung-Ling Hung, and Chien-Ping Chang. A fast intensity-hue-saturation fusion technique with spectral adjustment for ikonos imagery. *IEEE Geoscience and Remote Sensing Letters*, 1(4):309–312, 2004.
5. R. Akula, R. Gupta, and M. R. Vimala Devi. An efficient pan sharpening technique by merging two hybrid approaches. *Procedia Engineering*, 30:535–541, 2012.
6. Wei Huang, Liang Xiao, Zhihui Wei, Hongyi Liu, and Songze Tang. A new pan-sharpening method with deep neural networks. *IEEE Geoscience and Remote Sensing Letters*, 12(5):1037–1041, 2015.
7. Xiao Xiang Zhu and Richard Bamler. A sparse image fusion algorithm with application to pan-sharpening. *IEEE Transactions on Geoscience and Remote Sensing*, 51(5):2827–2836, 2013.
8. Weiping Ni and Xinbo Gao. Despeckling of sar image using generalized guided filter with bayesian nonlocal means. *IEEE Transactions on Geoscience and Remote Sensing*, 54(1):567–579, 2016.

9. Qiang Zhang, Long Wang, Huijuan Li, and Zhaokun Ma. Similarity-based multimodality image fusion with shiftable complex directional pyramid. *Pattern Recognition Letters*, 32(13):1544–1553, 2011.

10. Hua Xu, Yan Wang, Yujing Wu, and Yunsheng Qian. Infrared and multi- type images fusion algorithm based on contrast pyramid transform. *Infrared Physics and Technology*, 78:133–146, 2016.

11. Zhaodong Liu, Yi Chai, Hongpeng Yin, Jiayi Zhou, and Zhiqin Zhu. A novel multi-focus image fusion approach based on image decomposition. *Information Fusion*, 35:102–116, 2017.

12. Jiao Du, Weisheng Li, Bin Xiao, and Qamar Nawaz. Union laplacian pyramid with multiple features for medical image fusion. *Neurocomputing*, 194:326–339, 2016.

13. C. O. Ancuti, C. Ancuti, C. De Vleeschouwer, and A. C. Bovik. Single-scale fusion: An effective approach to merging images. *IEEE Transactions on Image Processing*, 26(1):65–78, 2017.

14. X. Luo, Z. Zhang, B. Zhang, and X. Wu. Image fusion with contextual statistical similarity and nonsubsampled shearlet transform. *IEEE Sensors Journal*, 17(6):1760–1771, 2017.

15. P. Hill, M. E. Al-Mualla, and D. Bull. Perceptual image fusion using wavelets. *IEEE Transactions on Image Processing*, 26(3):1076–1088, 2017.

16. Landsat images. http://gisgeography.com/landsat-program-satellite-imagery-bands/, accessed in 2015.

17. QUICKBIRD. *Global Land Cover Facility*, July 2016.

18. IKONOS. *Ikonos Satellites*, July 2016.

19. MODIS. *Global Land Cover Facility*, July 2016.

20. G. Bhatnagar, Q. M. J. Wu, and Z. Liu. Directive contrast based multimodal medical image fusion in NSCT domain. *IEEE Transactions on Multimedia*, 15(5):1014–1024, 2013.

21. W. Xie, L. Jiao, and J. Zhao. PolSAR image classification *via* D-KSVD and NSCT-domain features extraction. *IEEE Geoscience and Remote Sensing Letters*, 13(2):227–231, 2016.

22. Y. Yang, S. Tong, S. Huang, and P. Lin. Multifocus image fusion based on NSCT and focused area detection. *IEEE Sensors Journal*, 15(5):2824–2838, 2015.

23. Arthur L. Da Cunha, Jianping Zhou, and Minh N. Do. The nonsubsampled contourlet transform: Theory, design, and applications. *IEEE Transactions on Image Processing*, 15(10):3089–3101, 2006.

24. Raanan Fattal. Single image dehazing. *ACM Transactions on Graphics (TOG)*, 27(3):72, 2008.

25. Radhakrishna Achanta, Sheila Hemami, Francisco Estrada, and Sabine Susstrunk. Frequency-tuned salient region detection. In: *IEEE Conference on Computer Vision and Pattern Recognition, 2009 (CVPR 2009)*, pages 1597–1604. IEEE, 2009.

26. Peter Burt and Edward Adelson. The laplacian pyramid as a compact image code. *IEEE Transactions on Communications*, 31(4):532–540, 1983.

27. Kaiming He, Jian Sun, and Xiaoou Tang. Guided image filtering. *IEEE Transactions on Pattern Analysis and Machine Intelligence*, 35(6):1397–1409, 2013.

28. A. Buades, B. Coll, and J.-M. Morel. A non-local algorithm for image denoising. In: *IEEE Computer Society Conference on Computer Vision and Pattern Recognition, 2005 (CVPR 2005)*, volume 2, pages 60–65. IEEE, 2005.

29. Pierrick Coupe, Pierre Hellier, Charles Kervrann, and Christian Barillot. Bayesian non local means-based speckle filtering. In: *5th IEEE International Symposium on Biomedical Imaging: From Nano to Macro, 2008. ISBI 2008*, pages 1291–1294. IEEE, 2008.

30. H. Zhong, Y. Li, and L. C. Jiao. Sar image despeckling using bayesian nonlocal means filter with sigma preselection. *IEEE Geoscience and Remote Sensing Letters*, 8(4):809–813, 2011.

31. A. Lopes, R. Touzi, and E. Nezry. Adaptive speckle filters and scene heterogeneity. *IEEE Transactions on Geoscience and Remote Sensing*, 28(6):992–1000, 1990.

32. Guang-Ting Li, Chun-Le Wang, Ping-Ping Huang, and Wei-Dong Yu. Sar image despeckling using a space-domain filter with alterable window. *IEEE Geoscience and Remote Sensing Letters*, 10(2):263–267, 2013.

9

Moser-Trudinger and Adams Inequalities

Kunnath Sandeep

CONTENTS

9.1 Introduction 135
9.2 Preliminaries 137
9.3 Moser–Trudinger 139
9.4 Adams Inequality 142
References 147

9.1 Introduction

In this chapter we will focus on the sharp Sobolev inequalities known as the Moser–Trudinger and Adams inequalities.

Recall the well-known classical Sobolev inequality which states there exists an optimal constant $S_n > 0$ such that the inequality

$$S_n \left[\int_{\mathbb{R}^n} |u|^{2^*} \, dx \right]^{\frac{2}{2^*}} \leq \int_{\mathbb{R}^n} |\nabla u|^2 \, dx \tag{9.1}$$

holds for all $u \in C_c^1(\mathbb{R}^n)$, $n \geq 3$, where $2^* = \dfrac{2n}{n-2}$. It immediately follows from this inequality that if Ω is a bounded domain in $\mathbb{R}^n$ and $1 \leq p \leq 2^*$, then there exists an optimal constant $C_{n,p,\Omega} > 0$ such that

$$C_{n,p,\Omega} \left[\int_{\Omega} |u|^p \, dx \right]^{\frac{2}{p}} \leq \int_{\Omega} |\nabla u|^2 \, dx \tag{9.2}$$

holds for all $u \in C_c^1(\Omega)$. The inequalities of Equation (9.1) and Equation (9.2) hold for all u in $D^{1,2}(\mathbb{R}^n)$ and $H_0^1(\Omega)$ respectively where $D^{1,2}(\mathbb{R}^n)$ and $H_0^1(\Omega)$ are respectively the completion of $C_c^1(\mathbb{R}^n)$ and $C_c^1(\Omega)$ with respect to the norm $\left(\int |\nabla u|^2 \, dx \right)^{\frac{1}{2}}$.

Note that Equation (9.2) is in bounded domains in $\mathbb{R}^3$, and when $n=2$ the number 2^* formally becomes ∞ but Equation (9.2) does not hold with $p=\infty$ in this case. However one can show that:

Let Ω be a bounded domain in $\mathbb{R}^2$, then for any $1 \le p < \infty$, there exists an optimal constant $C_{p,\Omega} > 0$ such that

$$C_{p,\Omega}\left[\int_\Omega |u|^p\, dx\right]^{\frac{2}{p}} \le \int_\Omega |\nabla u|^2\, dx \tag{9.3}$$

holds for all $u \in C_c^1(\Omega)$. Moreover $C_{p,\Omega} \to 0$ as $p \to \infty$.

These inequalities play an important role in the analysis of partial differential equations, especially the ones in variational form. First recall if $(X, \|.\|_X)$ and $(Y, \|.\|_Y)$ are normed linear spaces and $f: U \to Y$ for an open subset U of X then f is said to be Fréchet differentiable at $x_0 \in U$ if there exists an $L \in BL(X,Y)$, satisfying

$$\lim_{h\to 0}\frac{\|f(x_0+h)-f(x_0)-L(h)\|_Y}{\|h\|_X}=0$$

The bounded linear map L if it exists is unique and is called the Fréchet derivative of f at x_0 and is denoted by $Df(x_0)$. Also we say that x_0 is a critical of f if $Df(x_0)=0$.

Solutions of certain PDEs appear as the critical points of some functions defined on appropriate function spaces. For example one can easily see that solutions of the boundary value problem

$$\left.\begin{array}{ll} -\Delta u = f(u) & \text{in } \Omega \\ u = 0 & \text{on } \partial\Omega \end{array}\right\} \tag{9.4}$$

where f is sufficiently regular, appear as the critical points in the suitable function space of the corresponding energy

$$E(u) = \frac{1}{2}\int_\Omega |\nabla u|^2\, dx - \int_\Omega F(u)\, dx \tag{9.5}$$

where F is the primitive of f given by $F(t) = \int_0^t f(s)\, ds$. A natural space where one would define this energy is where $\int_\Omega |\nabla u|^2\, dx < \infty$. In this space if the energy has to be well defined then $\int_\Omega F(u)\, dx$ has to be finite which follows if we can control $\int_\Omega F(u)$ by $\int_\Omega |\nabla u|^2$ and the latter is finite. Note that if f has polynomial growth of appropriate order then Equation (9.2) assures that E is well defined in $H_0^1(\Omega)$.

In the two-dimensional case if f fas polynomial growth of any order, Equation (9.3) makes the energy well-defined in $H_0^1(\Omega)$. However, there are semilinear PDEs coming from geometry, physics,etc. where we have to deal with nonlinearities which are of the exponential type. Let us see two examples.

First let us consider the problem of prescribing the Gaussian curvature on a two- dimensional manifold in a conformal way. Let (M, g) be a two-dimensional Riemannian Manifold, and $\tilde{g} = e^{2u}g$ be a metric conformal to g. Suppose the Gaussian curvatures of g and $\tilde{g}$ are denoted by k_g and k_g respectively. Then the Gaussian curvatures are related by the relation

$$-\Delta_g u + k_g = k_{\tilde{g}} e^{2u} \text{ on } M$$

Thus if we want to know whether a given function $k : M \to \mathbb{R}$ is the Gaussian curvature of a conformal metric is equivalent to proving whether the PDE

$$-\Delta_g u + k_g = k e^{2u} \text{ on } M$$

has a solution. Solutions of this PDE are the critical points of the energy functional:

$$J(u) = \frac{1}{2} \int_M |\nabla_g u|^2 \, dV_g + \int_M k_g u \, dV_g - \frac{1}{2} \int_M k e^{2u} \, dV_g \tag{9.6}$$

defined on appropriate function spaces. Another equation where similar type of non-linearity arises is the mean-field equation: Let (M, g) be a two dimensional compact Riemannian manifold, then the mean-field equation under certain assumption reduces to

$$-\Delta_g u = \lambda \left(\frac{k e^u}{\displaystyle\int_M k e^u \, dV_g} - \frac{1}{\text{Vol}(M)} \right) \text{ on } M$$

where $k : M \to \mathbb{R}$ is given. The corresponding energy is:

$$J(u) = \frac{1}{2} \int_M |\nabla_g u|^2 \, dV_g + \frac{\lambda}{\text{Vol}(M)} \int_M u \, dV_g - \frac{\lambda}{2} \log \left(\int_M k e^{2u} \, dV_g \right)$$

The above examples show that if we need the energy to be defined in the natural spaces then we need to estimate terms of the form $\int_M e^{2u} \, dV_g$ using $\int_M |\nabla_g u|^2 \, dV_g$. Indeed such estimates are available and its optimal version is known as the Moser–Trudinger inequality. Higher order versions of this inequality are known as the Adams inequality. We will discuss these inequalities in detail in the next sections.

9.2 Preliminaries

In this section we will recall some basic definition and standard results.

We will denote by D^α the differential operator

$$D^\alpha = \frac{\partial^{|\alpha|}}{\partial_{x_1}^{\alpha_1} \cdots \partial_{x_n}^{\alpha_1}}$$

corresponding to a multi-index by $\alpha = (\alpha_1, \cdots, \alpha_n)$ where $|\alpha| = \alpha_1 + \cdots + \alpha_n$.

We will denote by $C_c^k(\Omega)$ the collection of all functions with compact support in Ω for which all partial derivatives of order up to k exist and all the partial derivatives are continuous. We also denote $C_c^\infty(\Omega) = \cap_{k=1}^\infty C_c^k(\Omega)$.

Definition 9.1 *Let Ω be an open set in $\mathbb{R}^n$ we define for $1 \le p \le \infty$ and $k \in \mathbb{N}$ the Sobolev spaces $W^{k,p}(\Omega)$ and $W_0^{k,p}(\Omega)$ by*

$$W^{k,p}(\Omega) = \left\{ u \in L^p(\Omega) : D^\alpha u \in L^p(\Omega), \forall \alpha \text{ with } |\alpha| \le k \right\}$$

and

$$W_0^{k,p}(\Omega) = \text{closure of } C_c^\infty(\Omega) \text{ in } W^{k,p}(\Omega)$$

where $W^{k,p}(\Omega)$ is endowed with the norm $\| \cdot \|_{k,p}$ given by

$$\| u \|_{k,p} = \sum_{|\alpha| \le k} \| D^\alpha u \|_{L^p}$$

When $p = 2$ we denote

$$W^{k,2}(\Omega) = H^k(\Omega), \ W_0^{k,2}(\Omega) = H_0^k(\Omega)$$

With these definitions we can state the Sobolev embedding theorem:

Lemma 9.1 *Let Ω be a bounded domain in $\mathbb{R}^n$ and $kp < n$, then*

$$W^{k,p}(\Omega) \longrightarrow L^q(\Omega), 1 \le q \le \frac{np}{n - kp}$$

When $kp = n$

$$W^{k,p}(\Omega) \longrightarrow L^q(\Omega), 1 \le q < \infty$$

but $W^{k,p}(\Omega) \not\longrightarrow L^\infty(\Omega)$

Let us also recall the well-known Poincaré lemma:

Lemma 9.2 *Let Ω be a bounded domain in $\mathbb{R}^n$, then there exists a constant C depending on Ω such that*

$$C\int_\Omega | u |^p \, dx \le \int_\Omega | \nabla u |^p \, dx, \quad \forall u \in W_0^{1,p}(\Omega).$$

As a corollary we get

Corollary 9.1 *Let Ω be a bounded domain in $\mathbb{R}^n$ then $\left(\int_\Omega | \nabla u |^p \, dx \right)^{\frac{1}{p}}$ is an equivalent norm in $W_0^{1,p}(\Omega)$.*

9.3 Moser–Trudinger

The Moser–Trudinger inequality deals with the case of embedding of $W_0^{1,n}(\Omega)$. We know that

$$W_0^{1,n}(\Omega) \rightarrowtail L^q(\Omega), 1 \le q < \infty$$

and the result is not true for $q = \infty$. Now the question is whether we can embed this space into a better space. To explain this first let us note that $W_0^{1,n}(\Omega) \rightarrowtail L^q(\Omega)$ if there exists $C > 0$ such that

$$\left[\int_\Omega |u|^p \, dx \right]^{\frac{n}{p}} \le C \int_\Omega |\nabla u|^n \, dx \, , u \in C_c^\infty(\Omega)$$

Equivalently

$$\sup_{u \in C_c^\infty(\Omega), \int_\Omega |\nabla u|^n \le 1} \int_\Omega |u|^p \, dx \le C, \tag{9.7}$$

Now one may ask whether the function $|u|^q$ can be replaced by some $f(u)$ of higher growth than polynomial power. An answer to this question was given by Pohožaev ([1]) and Trudinger([2]). They established an embedding of the Sobolev space $W_0^{1,n}(\Omega)$ into an Orlicz space establishing the exponential integrability of these functions.

In 1971 J. Moser ([3]) while trying to study the question of prescribing the Gaussian curvature on sphere understood the need for establishing a sharp form of the embedding obtained by Pohožaev and Trudinger. He showed that:

Theorem 9.1 *(Moser, [3])*
There exists a positive constant C_0 depending only on n such that

$$\sup_{u \in C_c^\infty(\Omega), \int_\Omega |\nabla u|^n \le 1} \int_\Omega e^{\alpha |u|^{\frac{n}{n-1}}} \, dx \le C_0 \, |\Omega|. \tag{9.8}$$

holds for all $\alpha \le \alpha_n = n \left[\omega_{n-1} \right]^{\frac{1}{n-1}}$ where Ω is a bounded domain in $\mathbb{R}^n$, and $|\Omega|$ denotes the volume of Ω and ω_{n-1} denotes the $n-1$ dimensional area of the sphere S^{n-1}. Moreover when $\alpha > \alpha_n$, the above supremum is infinite.

The above inequality for $\alpha \le \alpha_n$ was proved by using symmetrization and reducing the problem to a one-dimensional inequality. The case $\alpha < \alpha_n$ is easier comparing with the case $\alpha = \alpha_n$. The case $\alpha > \alpha_n$ is proved using the following test functions which we call as Moser functions: For $0 < l < R < \infty$, define

$$u_{R,l}(r) = \frac{1}{\sqrt{2\pi}} \left[\sqrt{\log(\frac{R}{l})} \chi_{[0,l]} + (\log(\frac{R}{l}))^{\frac{-1}{2}} \log(\frac{r}{R}) \chi_{[l,R]} \right]$$

Choose $x_0 \in \Omega$ and $R > 0$ such that the ball $B(x_0, 2R) \subset \Omega$ and define $u_{R,l}(x) = u_{R,l}(|x - x_0|)$,
then $\int_\Omega |\nabla u_{R,l}|^n = 1$ and $\int_\Omega e^{\alpha |u_{R,l}|^{\frac{n}{n-1}}} dx \to \infty$ as $l \to 0$.

The inequality of Equation (9.8) is known as the Moser–Trudinger inequality. After this seminal work there has been lot of interest on these types of inequalities including their validity in various other contexts, improvements of this inequality, and the question of the existence of extremal functions. The question of existence of extremals was answered by Carleson and Chang ([4]) and Flucher ([5]). They showed that Equation (9.8) attains an extremal function for any bounded domain in $\mathbb{R}^n$. Various improvements of the above inequality were also obtained; one of the most noted is the improvement obtained by Adimurthi–Druet ([6]). The validity of this inequality to the entire space $\mathbb{R}^n$ was also studied and various forms of it were obtained, see [7], [8], [9], [10], [11], etc.

In this chapter, our focus will be on Moser–Trudinger type inequalities in spaces of infinite measure in Euclidean space and other geometries. To start with let us recall that in [3], Moser has established the following version of Equation (9.8) on the-two dimensional Euclidean sphere:

Theorem 9.2 *(Moser, [3])*
There exists a constant $C > 0$ such that

$$\sup_{u \in C^\infty(S^2), \int_{S^2} u = 0, \int_{S^2} |\nabla u|^2 \le 1} \int_{S^2} e^{4\pi u^2} ds \le C \tag{9.9}$$

where ds is the Riemannian volume element on S^2.

Moser's main aim in establishing this inequality was to study the question of prescribing the Gaussian curvature on S^2 by finding critical points of J in Equation (9.6). With this aim he deduced from the above theorem:

Corollary 9.1 *There exists $C \in \mathbb{R}$ such that*

$$\frac{1}{2} \int_{S^2} |\nabla_g u|^2 dV_g + \frac{1}{4\pi} \int_{S^2} u \, dV_g - \frac{1}{2} \log\left(\int_{S^2} e^{2u} dV_g \right) \ge C > -\infty$$

The above corollary implies that the functional J as defined in Equation (9.6) is bounded from below and one can look for a critical point of minimum type. A sharp form of this inequality was obtained by Onofri ([12]), which states:

$$\frac{1}{|S^2|} \int_{S^2} \left[|\nabla_g u|^2 + 2u \right] dV_g - \log\left(\frac{1}{|S^2|} \int_{S^2} e^{2u} dV_g \right) \ge 0, \ u \in C^1(S^2) \tag{9.10}$$

Beckner in 1993 generalized this inequality to the general sphere S^n (see [13] for details). The inequality (Equation (9.10)) and its n dimensional version) is called the Beckner–Onofri inequality.

The Moser–Trudinger inequality on Compact Riemannian Manifold (i.e., Equation (9.9)) with a two-dimensional compact Riemannian manifold (M, g) in place of S^2) was obtained by Cherrier [14], and Fontana [15] established the sharp version of this inequality on any compact Riemannian manifold.

Since the Moser–Trudinger was known in both spaces of constant curvatures 0 and 1, we decided to study the question in the case of negative curvature, namely the hyperbolic space. We investigated the problem in the ball model of the hyperbolic space. More precisely we studied:

Let $\mathbb{D}$ be the unit open disc in $\mathbb{R}^2$, endowed with a conformal metric $g = \rho g_e$, where g_e denotes the Euclidean metric and $\rho \in C^2(\mathbb{D})$, $\rho > 0$. When does Moser–Trudinger hold in $(\mathbb{D}, g)$? i.e., When does the following inequality hold?

$$\sup_{u \in C_c^\infty(\mathbb{D}),\, \int_{\mathbb{D}} |\nabla_g u|^2 \leq 1} \int_{\mathbb{D}} \left(e^{4\pi u^2} - 1 \right) dv_g < \infty. \tag{9.11}$$

Here ∇_g, dv_g denotes respectively the gradient and volume element for the metric g. Note that since hyperbolic space has infinite measure, we can not have inequalities of the form in Equation (9.8) and hence the modified version, Equation (9.11). In a joint work with G. Mancini we proved

Theorem 9.3 *(G. Mancini, K. Sandeep [16])*
Moser–Trudinger holds in $(\mathbb{D}, g)$ if and only if $g \leq c g_h$ for some positive constant c, where g_h is the hyperbolic metric (Poincaré metric) given by

$$g_h = \sum_{i=1}^{2} \left(\frac{2}{1 - |x^2|} \right)^2 dx_i^2.$$

Existence of this inequality to higher dimensional hyperbolic space and their various improvements were established by various authors, see for example [17], [18], [19]. Another proof of Moser–Trudinger in the hyperbolic space was obtained in [20]. Moser–Trudinger was established on any Hadamard manifold in [21].

As a consequence of the above theorem we were also able to classify domains in $\mathbb{R}^2$ where inequalities of the form in Equation (9.11) hold. i.e., let Ω be an open subser of $\mathbb{R}^2$ (not necessarily bounded). We are interested in knowing for what domains the following Moser–Trudinger holds:

$$\sup_{u \in C_c^\infty(\Omega),\, \int_{\Omega} |\nabla u|^2 \leq 1} \int_{\Omega} \left(e^{4\pi u^2} - 1 \right) dx < \infty. \tag{9.12}$$

One can easily see that the following two conditions are necessary for Equation (9.12) to hold:

- the Poincaré incquality holds, i.e.,

$$\lambda_1(\Omega) := \inf_{u \in C_c^\infty(\Omega),\, u \neq 0} \frac{\int_{\Omega} |\nabla u|^2}{\int_{\Omega} |u|^2} > 0$$

- Ω should not contain balls of arbitrary large radius, i.e.,

$$IR_\Omega := \sup\{r > 0 : B(x,r) \subset \Omega, \text{ for some } x \in \Omega\} < \infty$$

As a consequence of the above theorem we showed that these conditions are equivalent when Ω is simply connected.

Theorem 9.4 *(G. Mancini, K. Sandeep [16])*

Let Ω be a simply connected open set in $\mathbb{R}^2$, then the following are equivalent:

- Moser–Trudinger inequality (Equation (9.12)) holds in Ω.
- $\lambda_1(\Omega) > 0$
- $IR_\Omega < \infty$.

The theorem is not true if we remove the simply connectedness

It is easy to show that $(i) \Rightarrow (ii) \Rightarrow (iii)$. So it remains to show that $(iii) \Rightarrow (i)$. Since Ω is simply connected, it is conformally equivalent to the unit disc and hence one can pull back the inequality to the unit disc and the pulled-back inequality becomes like Equation (9.11), then the previous theorem applies as one can show that using Kobe's one quarter theorem (iii) implies the pullback metric is bounded from above by the Poincaé metric (see [16] for details).

Even though the theorem is not true for nonsimply connected domains, one may ask whether (i) and (ii) are equivalent in the general case. The question was answered positively in [17].

9.4 Adams Inequality

Adams inequalities are the Moser–Trudinger type inequalities in higher-order Sobolev spaces $W_0^{k,p}(\Omega)$ when $kp = n$. Even though one expects a similar type inequality to hold for higher-order Sobolev spaces, it is not at all obvious how to modify the proofs of the case $k = 1$ to $k > 1$ due to the failure of Polya–Szego type inequalities for higher-order gradients ∇^k defined by

$$\nabla^k := \begin{cases} \Delta^{\frac{k}{2}}, & \text{if } k \text{ is even,} \\ \nabla\Delta^{\frac{k-1}{2}}, & \text{if } k \text{ is odd.} \end{cases} \tag{9.13}$$

In a significant work, D.R. Adams ([22]) established the sharp embedding in the case of higher order Sobolev spaces $W_0^{k,p}(\Omega)$ when $kp = n$.

Theorem 9.5 *(Adams [22])*

Let Ω be a bounded open set in $\mathbb{R}^n$ and k be a positive integer less than n, then there exists a constant $c_0 = c_0(k, n)$ such that

$$\sup_{u \in C_c^k(\Omega),\, \int_\Omega |\nabla^k u|^p \leq 1} \int_\Omega e^{\beta|u(x)|^{p'}} \, dx \leq c_0 |\Omega|, \tag{9.14}$$

for all $\beta \leq \beta_0(k,n)$ where $p = \dfrac{n}{k}$, $p' = \dfrac{p}{p-1}$,

$$
\beta_0(k,n) = \begin{cases}
\dfrac{n}{\omega_{n-1}} \left[\dfrac{\pi^{\frac{n}{2}} 2^k \Gamma\left(\dfrac{k+1}{2}\right)}{\Gamma\left(\dfrac{n-k+1}{2}\right)} \right]^{p'}, & \text{if } k \text{ is odd,} \\[6ex]
\dfrac{n}{\omega_{n-1}} \left[\dfrac{\pi^{\frac{n}{2}} 2^k \Gamma\left(\dfrac{k}{2}\right)}{\Gamma\left(\dfrac{n-k}{2}\right)} \right]^{p'}, & \text{if } k \text{ is even,}
\end{cases}
\tag{9.15}
$$

Furthermore, if $\beta > \beta_0$, then the supremum in Equation (9.14) is infinite.

Inequalities of the above form are called Adams inequalities. As mentioned earlier, even though the statements of Adams look similar to that of Moser–Trudinger, proofs are significantly different. Adams used the Riesz transform to convert it into a convolution type estimate and then use a lemma due to O'Neil for convolution operators of nonincreasing rearrangement of functions to prove the result. Subsequently Fontana in [15] obtained the following sharp version of Equation (9.14) on compact Riemannian manifolds:

Theorem 9.6 *(Fontana [15]) Let (M, g) be an n dimensional compact Riemannian manifold without boundary, and k be a positive integer less than n, then there exists a constant $c_0 = c_0(k, M)$ such that*

$$
\sup_{u \in C^k(M), \int_M u = 0, \int_\Omega |\nabla^k u|^p \leq 1} \int_M e^{\beta |u(x)|^{p'}} \, dx \leq c_0
\tag{9.16}
$$

if $\beta \leq \beta_0(k,n)$, where p, p', ∇_g^k as above where ∇_g and Δ_g are the gradient and Laplace Beltrami operators with respect to the metric g. Furthermore, if $\beta > \beta_0$, then the supremum in Equation (9.16) is infinite.

Since the Moser–Trudinger inequality holds in the hyperbolic space, a natural question is whether we have Adams inequality in the hyperbolic space. An answer was given by Fontana and Morpurgo:

Theorem 9.7 *(Fontana and Morpurgo,[23])*
Let $1 \leq k < n$ then the Adams inequality

$$
\sup_{u \in C_c^k(\mathbb{H}^n), \int_{\mathbb{H}^n} |\nabla_g^k u|^p \leq 1} \int_{\mathbb{H}^n} Exp_{[p-1]}(\beta \, | u(x) |^{p'}) \, dx < \infty,
\tag{9.17}
$$

holds in the n dimensional hyperbolic space $\mathbb{H}^n$ iff $\beta \leq \beta_0(k,n)$ where $[p - 1]$ denotes the integer part of $p - 1$ and $Exp_m(x) = e^x - \displaystyle\sum_{i=1}^m \dfrac{x^i}{i!}$

Another approach to Adams inequality in the Hyperbolic space was taken by us in [24]. Recall that the original motivation of Moser to prove the sharp version in Equation (9.8) of the Sobolev embedding was the question of prescribing the Gaussian curvature. Our approach is from this point of view. To explain the context let us recall some notions from geometry.

Let (M, g) be a Riemannian manifold of dimension n and $P_{1,g}$ be the conformal Laplacian defined by

$$P_{1,g} = -\Delta_g + \frac{n-2}{4(n-1)} R_g,$$

where R_g is the scalar curvature of the metric. We know that if $\tilde{g} = e^{2u} g$ is a metric conformal to g, then

$$P_{1,\tilde{g}}(v) = e^{-\left(\frac{n}{2}+1\right)u} P_{1,g} \left(e^{\left(\frac{n}{2}-1\right)u} v \right),$$

for all smooth functions v. Paneitz introduces a fourth order operator $P_{2,g}$ of the form

$$P_{2,g} = \Delta_g^2 + \text{lower order terms}$$

and Branson found a sixth order operator $P_{3,g}$ of the form

$$P_{3,g} = \Delta_g^3 + \text{lower order terms}$$

with the conformal invarience properties. The existence of a general conformal operator of higher-degree was obtained by Graham, Jenne, Mason, and Sparling (what is popularly known as GJMS operators). They proved

Theorem 9.8 *Let (M, g) be a Riemannian manifold of even dimension n. For $k \in \left\{1, 2, \ldots, \frac{n}{2}\right\}$ there exists a conformally invariant differential operator $P_{k,g}$ of the form*

$$P_{k,g} = \Delta_g^k + \text{lower order terms}$$

satisfying,

$$P_{k,\tilde{g}}(v) = e^{-\left(\frac{n}{2}+k\right)u} P_{k,g} \left(e^{\left(\frac{n}{2}-k\right)u} v \right)$$

for a conformal metric $\tilde{g} = e^{2u} g$.

When n is even and $k > \frac{n}{2}$, a conformally invariant operator $P_{k,g}$ with the above properties may not exist in general. For this reason $P_{\frac{n}{2},g}$ is known as the critical GJMS operator.

Observe that one can recover the scalar curvature of a metric by acting $P_{1,g}$ on the constant function. Similarly we define higher order curvatures known as Q_k curvature for a Riemannian manifold (M, g) of even dimension n. Let k be an integer satisfying $k < \frac{n}{2}$ then the Q_k curvature is defined by

$$Q_k = \frac{2(-1)^k}{n-2k} P_k(1)$$

and Q_n curvature is defined using analytic continuation in the above formulae.

Similar to Gaussian curvature, the change of $Q_{\frac{n}{2}}$ curvature under a conformal change of the metric is governed by the GJMS operator. Let $\tilde{g} = e^{2u}g$ be a conformal metric on (M, g), then $Q_{\frac{n}{2},g}$ and $Q_{\frac{n}{2},\tilde{g}}$ are related by the relation

$$P_{\frac{n}{2},g}(u) + Q_{\frac{n}{2},g} = Q_{\frac{n}{2},\tilde{g}}\, e^{nu} \tag{9.18}$$

One can easily see that solutions of the above equations are the critical points of the energy J given by

$$J(u) == \frac{1}{2}\int_M P_k(u)u \, dV_g + \int_M Q_{k,g} u \, dV_g - \frac{1}{N}\int_M Q_{k,\tilde{g}} e^{nu} \, dV_g \tag{9.19}$$

where $k = \dfrac{n}{2}$.

A natural space where one would like to pose this functional is the space where $\displaystyle\int_M P_k(u)u \, dV_g < \infty$. Then one would like to know whether we can control $\displaystyle\int_M e^{nu} \, dV_g$ in terms of $\displaystyle\int_M P_k(u)u \, dV_g$. We answered these questions positively in [24]. We will briefly discuss the main results.

First let us recall the definition of the Sobolev space $H^k(\mathbb{H}^n)$. We define the space $H^k(\mathbb{H}^n)$ as the completion of $C_c^\infty(\mathbb{H}^n)$ with respect to the norm

$$||u||_{H^k(\mathbb{H}^n)} := \left[\sum_{m=0}^{k}\int_{\mathbb{H}^n} |\nabla_g^m u|_g^2 \, dv_g\right]^{\frac{1}{2}},$$

It is well known that we have the Poincaré inequality

$$\left(\frac{n-1}{2}\right)^2 \int_{\mathbb{H}^n} u_g^2 \, dv_g \le \int_{\mathbb{H}^n} |\nabla_g u|_g^2 \, dv_g$$

holds for all $u \in H^1(\mathbb{H}^n)$. We also have

$$\int_{\mathbb{H}^n} |\nabla_g u|_g^2 \, dv_g = \int_{\mathbb{H}^n} -(\Delta_g u)u \, dv_g \le \left(\int_{\mathbb{H}^n} |\Delta_g u|^2 \, dv_g\right)^{\frac{1}{2}}\left(\int_{\mathbb{H}^n} u^2 \, dv_g\right)^{\frac{1}{2}}$$

Combining the above two inequalities and an induction argument gives us the following higher order Poincaré type inequality:

Let k, l be non-negative integers such that $l < k$, then

$$\left(\frac{N-1}{2}\right)^{2(k-l)} \int_{\mathbb{H}^n} |\nabla_g^l u|_g^2 \, dv_g \le \int_{\mathbb{H}^n} |\nabla_g^k u|_g^2 \, dv_g$$

holds for all $u \in H^k(\mathbb{H}^n)$.

As a consequence we see that the norm $\|\cdot\|_{H^k(\mathbb{H}^n)}$ defined by

$$\|u\|_{H^k(\mathbb{H}^n)} := \left[\int_{\mathbb{H}^n} |\nabla_g^k u|_g^2 \, dv_g\right]^{\frac{1}{2}}, u \in H^k(\mathbb{H}^n) \tag{9.20}$$

defines an equivalent norm in $H^k(\mathbb{H}^n)$.

Next we define $\|\cdot\|_{k,g}$ by

$$\|u\|_{k,g} := \left[\int_{\mathbb{H}^n} (P_k u)u \, dv_g\right]^{\frac{1}{2}}, \quad u \in C_c^\infty(\mathbb{H}^n),$$

then we have:

Lemma 9.2 *(Karmakar,Sandeep [24])*

Let $\|\cdot\|_{k,g}$ be as above then it defines a norm on $C_c^\infty(\mathbb{H}^n)$ for $2k \leq n$. Moreover when $n = 2k$ there exists a positive constant Θ such that

$$\frac{1}{\Theta}\|u\|_{k,g} \leq \|u\|_{H^k(\mathbb{H}^n)} \leq \Theta\|u\|_{k,g},$$

for all $u \in C_c^\infty(\mathbb{H}^n)$.

Using this norm $\|\cdot\|_{k,g}$ we established an Adams type inequality in $\mathbb{H}^n$.

Theorem 9.9 *(Karmakar, Sandeep [24])*

Let $\mathbb{H}^n$ be the n dimensional hyperbolic space with n even and $k = \frac{n}{2}$ then,

$$\sup_{u \in C_c^\infty(\mathbb{H}^n), |u|_{k,g} \leq 1} \int_{\mathbb{H}^n} \left(e^{\beta u^2} - 1\right) dv_g < +\infty \tag{9.21}$$

iff $\beta \leq \beta_0(k, n)$, where $\beta_0(k, n)$ is as before.

Recall that a version of Adams inequality on $\mathbb{H}^n$ was obtained by Fontana and Morpurgo in [23]. A natural question is how does Theorem 9.8 compare with the inequality in [23]. For $n = 4$, 6, 8 we can check that $\|u\|_{k,g}^2 < \int_{\mathbb{H}^n} |\nabla_g^k u|^2$. Thus our inequality implies the one obtained by Fontana and Morpurgo.

To complete our aim we investigated the existence of PDEs of the form in Equation (9.18) using the Adams inequality in Equation (9.18). We could show that negative $Q_{k,\tilde{g}}$ curvature which are L^2 perturbations of $Q_{k,g}$ can be prescribed. More generally we proved

Theorem 9.10 *(Karmakar, Sandeep [24])*

Suppose $Q_1 - Q_2 \in L^2(\mathbb{H}^n)$ and $Q_2 \leq 0$ then the equation

$$P_k(u) + Q_1 = Q_2 e^{nu}$$

has a solution in $H^k(\mathbb{H}^n)$.

Study of Moser–Trudinger and Adams inequalities continues to be an active area of research and many interesting results have come out after the above-described results were established, for example, some other versions were established in [25], Adams was established in the case of pinched Hadamard manifold in [26], etc.

References

1. Pohozhaev, S.I. The Sobolev imbedding in the case $pl = n$, In: *Proc. Tech. Sci. Conf. on Adv. Sci. Research 1964–1965, Mathematics Section*, pp. 158–170. Moskov. Energet. Inst., Moscow, 1965.
2. Trudinger, Neil S. On imbeddings into Orlicz spaces and some applications, *J. Math. Mech.* 17, 473–483, 1967.
3. Moser, J. A sharp form of an inequality by N. Trudinger, *Indiana Univ. Math. J.* 20, 1077–1092, 1970/71.
4. Carleson, Lennart, and Chang, Sun-Yung A. On the existence of an extremal function for an inequality of J. Moser, *Bull. Sci. Math.* 110(2), 113–127, 1986.
5. Flucher, M. Extremal functions for the Trudinger-Moser inequality in 2 dimensions, *Comment. Math. Helv.* 67, 471–497, 1992.
6. Adimurthi , and Druet, O.. Blow-up analysis in dimension 2 and a sharp form of Trudinger-Moser inequality, *Comm. Partial Diff. Equat.* 29(1–2), 295–322, 2004.
7. Cao, D. Nontrivial solution of semilinear elliptic equations with critical exponent in $\mathbb{R}^2$, *Commun. Partial Diff. Equat.* 17, 407–435, 1992.
8. Panda, R. Nontrivial solution of a quasilinear elliptic equation with critical growth in $\mathbb{R}^N$, *Proc. Indian Acad. Sci.* 105, 425–444, 1995.
9. do Ó, J.M. N-Laplacian equations in $\mathbb{R}^N$ with critical growth, *Abstr. Appl. Anal.* 2, 301–315, 1997.
10. Ruf, B. A sharp Trudinger-Moser type inequality for unbounded domains in $\mathbb{R}^2$, *J. Funct. Anal.* 219(2), 340–367, 2005.
11. Li, Y., and Ruf, B. A sharp Trudinger-Moser type inequality for unbounded domains in $\mathbb{R}^N$, *Indiana Univ. Math. J.* 57(1), 451–480, 2008.
12. Onofri, E. On the positivity of the effective action in a theory of random surfaces, *Comm. Math. Phys.* 86(3), 321–326, 1982.
13. Beckner, William. Sharp Sobolev inequalities on the sphere and the Moser-Trudinger inequality, *Ann. of Math.* 138(1), 213–242, 1993.
14. Cherrier, Pascal. Une inégalité de Sobolev sur les variétés riemanniennes, *Bull. Sci. Math.* 103(4), 353–374, 1979.
15. Fontana, Luigi. Sharp borderline Sobolev inequalities on compact Riemannian manifolds, *Comment. Math. Helv.* 68(3), 415–454, 1993.
16. Mancini, G., and Sandeep, K. Moser-Trudinger inequality on conformal discs, *Commun. Contemp. Math.* 12(6), 1055–1068, 2010.
17. Battaglia, Luca, and Mancini, Gabriele. Remarks on the Moser-Trudinger inequality, *Adv. Nonlinear Anal.* 2(4), 389–425, 2013.
18. Lu, G., and Tang, H. Best constants for Moser-Trudinger inequalities on high dimensional hyperbolic spaces, *Adv. Nonlinear Stud.* 13(4), 1035–1052, 2013.
19. Mancini, G., Sandeep, K., and Tintarev, C. Trudinger-Moser inequality in the hyperbolic space $\mathbb{H}^n$, *Adv. Nonlinear Anal.* 2(3), 309–324, 2013.
20. Adimurthi , and Tinterev, K. On a version of Trudinger-Moser inequality with Möbius shift invariance, *Calc. Var. Part. Diff. Equat.* 39(1–2), 203–212, 2010.
21. Yang, Qiaohua, Su, Dan, and Kong, Yinying. Sharp Moser-Trudinger inequalities on Riemannian manifolds with negative curvature. *Ann. Mat. Pura Appl.* 195(2), 459–471, 2016.
22. Adams, D.R. A sharp inequality of J. Moser for higher order derivatives, *Ann. Math.* 128, 385–398, 1988.

23. Fontana, L.; Morpurgo, C. A. Inequalities for Riesz subcritical potentials. *Nonlinear Anal. 192* (2020), 111662, 32 pp.
24. Karmakar, Debabrata, and Sandeep, Kunnath. Adams inequality on the hyperbolic space, *J. Funct. Anal.* 270(5), 1792–1817, 2016.
25. Li, Jungang, Lu, Guozhen, and Yang, Qiaohua. Fourier analysis and optimal Hardy-Adams inequalities on hyperbolic spaces of any even dimension, *Adv. Math.* 333, 350–385, 2018.
26. Bertrand, Jerome, and Kunnath, Sandeep. Adams inequality on pinched Hadamard manifolds, Preprint 2019, arXiv:1809.00879v2.

10

Computational Ship Hydrodynamics: Modeling and Simulation

Sashikumaar Ganesan, Bhanu Teja, and Thivin Anandh

CONTENTS

10.1 Introduction.. 149
10.2 Ship Hydrodynamics Model.. 150
    10.2.1 Equations for Rigid Body Motion... 150
    10.2.2 Governing Equations of the Fluid Flow 154
    10.2.3 Coupled Boundary and Initial Conditions 154
        10.2.3.1 Dimensionless Form.. 155
10.3 Interface Tracking Methods 156
    10.3.1 Eulerian Approaches.. 156
        10.3.1.1 Volume of Fluid Method 156
        10.3.1.2 Level Set Method... 158
        10.3.1.3 Front Tracking Method 158
    10.3.2 Lagrangian Approaches.. 160
        10.3.2.1 Arbitrary Lagrangian Eulerian Approach 160
10.4 Finite Element Scheme 161
    10.4.1 Weak Form of the Navier–Stokes Equations 161
    10.4.2 Finite Element Discretization... 163
    10.4.3 Temporal Discretization and Linearization.. 163
    10.4.4 Linearization and Solution Process... 164
10.5 Numerical Results... 166
Acknowledgment... 169
References... 169

10.1 Introduction

Computational fluid dynamics (CFD) for ship hydrodynamics has been the subject of intense research in both scientific and industrial communities over the past few decades. The ultimate goal of the computational ship hydrodynamics is to compute the behavior of flows around the ships, the motion of the ships, etc., and to study the performance of full-scale ships under real operating conditions. Full-scale simulations will reduce the dependence on the time consuming tank tests and the costs associated with it. Further, tank tests on scaled-down prototypes need not necessarily reflect the behaviors of newly designed full-scale versions of ships. Moreover experiments on full-scale models are very challenging and are not economical, and thus computational modeling is the way forward. Thanks to the constant increase in computer memory, core count, and speed, together with improvements in the numerical schemes and parallel algorithms, the computational costs are constantly decreasing. Moreover, enhanced flow visualization and force decomposition provide much richer information and may lead to a much better understanding of the flow phenomena.

The earliest important theoretical result on ship hydrodynamics by Froude and Krylov [1, 2], dating back to the end of the nineteenth century, was the first general theory on ship motion. The authors built bases for the understanding of the different physical aspects in the dynamics of an advancing vessel. In particular, Froude investigated the role of the wave component of the resistance and gave his name to the non-dimensional number that characterizes free-surface flows—the Froude number. In 1898, a thin-ship theory was proposed in [3] to describe the waves generated by an advancing ship and an analytical estimation of the wave resistance (based on the so-called Mitchell's integral) was derived. Moreover, numerical schemes based on the solution of the potential flow equation have been developed in the field of aerodynamics (see, e.g., the pioneering work by Hess and Smith [4]) and later extended to the solution of ship hydrodynamic problems [5, 6]. However, a simple irrotational and inviscid flow model has been considered in these studies.

In the last two decades several numerical schemes for the solution of the Navier–Stokes equations have successfully been developed for naval engineering problems [7, 8]. In addition, an increase in computing power has broadened the class of naval problems that could be solved, including the possibility of treating viscous, turbulent, and separated flows.

CFD approaches for free-surface and multiphase-flow problems can be classified into two main categories based on the kinematic approaches on calculating fluid flows: (i) Eulerian approach, and (ii) Lagrangian approach. In the Eulerian approaches such as Front Tracking, Level Set, Volume-of-Fluid, Immersed Boundary, etc., the model equations are solved in a fixed computational mesh, whereas the moving interfaces are tracked separately. Accurately specifying the material properties in their respective phases, incorporating interfacial forces, and accurately tracking the interface between two fluids are the main challenges in Eulerian approaches. Contrarily, in the Lagrangian approaches like the Arbitrary Lagrangian Eulerian (ALE) method, the material properties and the interfacial forces can be incorporated accurately since the interface is resolved by the computational mesh. Nevertheless, handling geometries with large deformation and incorporating topological changes are very challenging in the ALE approach. Despite several advances made in the field of CFD, tracking interfaces, incorporating interfacial forces, and accurately coupling the existing flow solvers with structure models still remains challenging.

In this chapter, the ship hydrodynamics model, a numerical scheme based on the ALE approach, solution approaches and a few results are presented.

10.2 Ship Hydrodynamics Model

10.2.1 Equations for Rigid Body Motion

Rigid body dynamics deals with the arbitrary motion of rigid bodies in space. We treat the floating structure (ship) as a rigid body and neglect all deformations in its structure. In a given time interval [0,I], consider an arbitrarily translating and rotating rigid body $\Omega_S(t) \subset \mathbb{R}^3$, where I is a given final time and $t \in [0,I]$. Assign a set of axes fixed to the body with its origin at the center of mass of the rigid body and denote it as the rotating frame (rotates along the rigid body). Next, consider another set of axes fixed to the body with its origin again at the center of mass and denote it as inertial or a non-rotating frame (translates with the body). Assign a set of axes to be fixed to the ground (inertial frame of reference) as ground-fixed axes. These three frames of references are shown in the Figure 10.1. The rotating frame of axes xyz is body-fixed and rotates with the body, and the inertial frame of axes XYZ is a body fixed but it translates with the body, whereas the reference frame of axes $X'Y'Z'$ is fixed to the ground.

Next, the motion of the rigid body is described by six degrees of freedom (6-DOF) equations, which include variation of linear and angular momentums. Let angular $bX_b(x_b, y_b, z_b)$ be the center of mass (position vector) of the body in rotation frame xyz. Further, let $\mathbf{u}_b = (u_b, v_b, w_b)$

$$u_b = \frac{dx_b}{dt}, \quad v_b = \frac{dy_b}{dt}, \quad w_b = \frac{dz_b}{dt},$$

be the translational velocity of the body. Then, the linear momentum equation is given by

$$m\frac{d\mathbf{u}_b}{dt} = \mathbf{F}, \quad \text{in } (0, I], \tag{10.1}$$

where m is the body mass and $\mathbf{F}$ is the total force acting on the body. Further, the total force consists of

$$\mathbf{F} = \mathbf{F}_{flow} + \mathbf{F}_{ext} + mg\mathbf{e}, \quad \text{in } (0, I]$$

where $\mathbf{F}_{flow}$ is the force from the surrounding fluid which is obtained by integrating the stress from the fluid on the solid boundary (see Equation (10.7)), $\mathbf{F}_{ext}$ are the other applied forces such as towing or sail force, g is the gravitational constant, and e is an unit vector in the direction of gravitational force. Moreover, we use $\mathbf{F}_{ext}=0$ in this study.

We next derive the momentum equation for the rigid body. Consider an arbitrary vector $\mathbf{b}$ defined in $\Omega_s(t)$, and it can be decomposed as

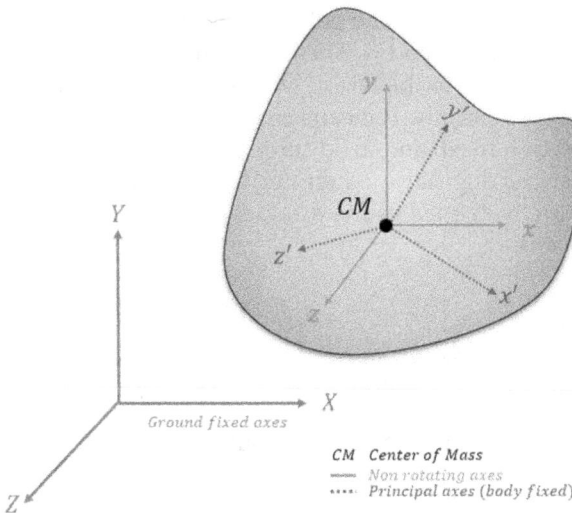

FIGURE 10.1
Frame of reference in an arbitrarly moving and rotation rigid body $\Omega_s(t)$.

$$\mathbf{b} = b_x \hat{x} + b_y \hat{y} + b_z \hat{z}$$

with respect to the body fixed axes xyz. Here, $(\hat{x}, \hat{y}, \hat{z})$ is the unit vector in the rotating frame xyz. Moreover, the vector $\mathbf{b}$ can also be decomposed as

$$\mathbf{b} = b_X \hat{X} + b_Y \hat{Y} + b_Z \hat{Z}$$

with respect to the non-rotating axes XYZ. Here, $(\hat{X}, \hat{Y}, \hat{Z})$ is the unit vector in the inertial frame XYZ. We now differentiate $\mathbf{b}$ (defined in fixed frame) with respect to time t to get

$$\frac{d\mathbf{b}}{dt} = \left[\frac{db_x}{dt} \hat{x} + \frac{db_y}{dt} \hat{y} + \frac{db_z}{dt} \hat{z} \right] + \left[b_x \frac{d\hat{x}}{dt} + b_y \frac{d\hat{y}}{dt} + b_z \frac{d\hat{z}}{dt} \right] \qquad (10.2)$$

The first part on the right hand side of Equation (10.2) can be interpreted as the instantaneous change of $\mathbf{b}$ observed from the body fixed frame xyz since the unit vectors $(\hat{x}, \hat{y}, \hat{z})$ are fixed (do not change) in the body fixed frame. The second part on the right hand side of Equation (10.2) refers to the rate of change of the unit vectors as observed from the inertial frame (non-rotating frame) XYZ. Since the origin of the inertial frame has the same instantaneous velocity as the body-fixed frame, the only way that these unit vectors change with respect to time is by a rotation about some axis passing through the common origin. We denote this rotational velocity as ω and we have

$$\frac{d\hat{x}}{dt} = \omega \times \hat{x}, \quad \frac{d\hat{y}}{dt} = \omega \times \hat{y}, \quad \frac{d\hat{z}}{dt} = \omega \times \hat{z}.$$

Hence, the second part in Equation (10.2) becomes $(\omega \times \mathbf{b})$ and Equation (10.2) reduces to

$$\frac{d\mathbf{b}}{dt} = \left(\frac{d\mathbf{b}}{dt} \right)_b + \omega \times \mathbf{b} \qquad (10.3)$$

Equation (10.3) relates the rate of change of the vector $\mathbf{b}$ between the translating axis (XYZ) and body fixed rotating axis (xyz). The subscript b indicates that the derivative is with respect to body fixed rotating frame xyz. The LHS is the rate of change of $\mathbf{b}$ in the non rotating frame XYZ, which is equal to the rate of change of $\mathbf{b}$ with respect to body fixed frame xyz plus omega cross $\mathbf{b}$. Physically, it means that the vector $\mathbf{b}$ can increase in magnitude while keeping its direction fixed, and in addition it can change its direction.

Let us now consider the angular momentum $\mathbf{L}$ (also called as rotational momentum). The angular momentum is defined as the product of moment of inertia tensor $\mathbb{I}_m$ and the angular velocity ω, that is,

$$\mathbf{L} = \mathbb{I}_m \omega, \quad \mathbb{I}_m = \begin{bmatrix} I_{xx} & -I_{xy} & -I_{xz} \\ -I_{yx} & I_{yy} & -I_{yz} \\ -I_{zx} & -I_{zy} & I_{zz} \end{bmatrix}.$$

Here, I_{xx} is the moment of inertia of the body about the x–axis defined as

$$I_{xx} = \int_{\Omega_S(t)} \rho_s \, (y^2 + z^2) dx dy dz,$$

where ρ_s is the density of the solid body. Similarly, I_{yy} and I_{zz} are moment of inertia of the body about the y–axis and z–axis, respectively, defined analogously. Further, I_{xy}, I_{xz} and I_{yz} are the product of inertia, for instance

$$I_{xy} = \int_{\Omega_S(t)} \rho_s \, xy \, dx \, dy \, dz.$$

Note that $I_{xy} = I_{yx}$, $I_{xz} = I_{zx}$ and $I_{yz} = I_{zy}$. These terms can further be simplified if we compute these quantities with respect to the body fixed axes xyz, which is also aligned with the principle axes of the rigid body. With such a reference axes (principle axes), the products of inertia I_{xy}, I_{yz} and I_{zx} become zero and I_{xx}, I_{yy} and I_{zz} become the principle moments of inertia. Hence, we have

$$\mathbf{L} = \mathbb{I}_G \omega_G = I_{xx} \omega_x \hat{x} + I_{yy} \omega_y \hat{y} + I_{zz} \omega_z \hat{z}$$

where,

$$\mathbb{I}_G = \begin{bmatrix} I_{xx} & 0 & 0 \\ 0 & I_{yy} & 0 \\ 0 & 0 & I_{zz} \end{bmatrix} \quad \text{and} \quad \omega_G = \begin{bmatrix} \omega_x \\ \omega_y \\ \omega_z \end{bmatrix}$$

Similar to Equation (10.3), the time derivative of the angular momentum $\mathbf{L}$ becomes

$$\frac{d\mathbf{L}}{dt} = \left(\frac{d\mathbf{L}}{dt} \right)_b + \omega \times \mathbf{L}.$$

We now write Equation (10.3) in component-wise and in the body-fixed frame xyz. For the first term, we get

$$\left(\frac{d\mathbf{L}}{dt} \right)_b = I_{xx} \frac{d\omega_x}{dt} \hat{x} + I_{yy} \frac{d\omega_y}{dt} \hat{y} + I_{zz} \frac{d\omega_z}{dt} \hat{z}.$$

It is true since I_{xx}, I_{yy} and I_{zz} are the principal moments of inertia and they do not change with time. Further,

$$\omega \times \mathbf{L} = [(\omega_y \omega_z (I_{zz} - I_{yy})]\hat{x} + [(\omega_z \omega_x (I_{xx} - I_{zz})]\hat{y} + [(\omega_x \omega_y (I_{yy} - I_{xx})]\hat{z}.$$

Hence, the angular momentum equation in component-wise can be written as:

$$I_{xx} \frac{d\omega_x}{dt} + (I_{zz} - I_{yy})\omega_y \omega_z = M_x ,$$

$$I_{yy} \frac{d\omega_y}{dt} + (I_{xx} - I_{zz})\omega_z \omega_x = M_y , \qquad (10.4)$$

$$I_{zz} \frac{d\omega_z}{dt} + (I_{yy} - I_{xx})\omega_x \omega_y = M_z ,$$

where $\mathbf{M} = (M_x, M_y, M_z)$ is the total external moments about the center of mass. Since these equations are defined in the body fixed frame xyz, $\dfrac{d\mathbf{L}}{dt}$ is calculated first and then projected along the principal axes. Contrarily, we cannot project **L** on the principle axis and then compute the time derivative, since these equations are valid only at an instant of time and the quantities ω and **M** would change with respect to this axes at the next time instant.

10.2.2 Governing Equations of the Fluid Flow

Let $\Omega(t) \subset \mathbb{R}^3$ be the time-dependent domain of the fluid around the ship (floating structure). We consider a free surface flow in a towing tank with a partially submerged ship in the fluid, as shown in Figure 10.2.

Further, the time-dependent boundary is defined by $\partial\Omega(t) := \Gamma_{in} \cup \Gamma_{out} \cup \Gamma_w \cup \Gamma_s \cup \Gamma_{fs}$, where Γ_{in} is the inlet boundary, Γ_{out} outlet boundary, Γ_{wall} container/tank wall, Γ_s ship-fluid interface, and Γ_{fs} is the free surface as shown in Figure 10.2. The fluid flow around the ship is described by the time-dependent incompressible Navier–Stokes equations [9]

$$\frac{\partial \mathbf{u}}{\partial t} + (\mathbf{u} \cdot \nabla)\mathbf{u} - \frac{1}{\rho}\nabla \cdot \mathbb{T}(\mathbf{u}, p) = g\mathbf{e} \quad \text{in } \Omega(t) \times [0, I],$$

$$\nabla \cdot \mathbf{u} = 0 \quad \text{in } \Omega(t) \times [0, I],$$

(10.5)

where $\mathbf{u}$, p and ρ denote the fluid velocity, pressure in the fluid and density of the fluid, respectively. Here, the stress tensor $\mathbb{T}(\mathbf{u}, p)$ and the velocity deformation tensor $\mathbb{D}(\mathbf{u})$ are given by

$$\mathbb{T}(\mathbf{u}, p) = -p\mathbb{I} + 2\mu\mathbb{D}(\mathbf{u}), \qquad \mathbb{D}(\mathbf{u}) = \frac{1}{2}\left(\nabla\mathbf{u} + \nabla\mathbf{u}^T\right), \tag{10.6}$$

where $\mathbb{I}$ and μ denote the identity tensor and the viscosity of the fluid, respectively and the superscript T is used to denote transpose.

10.2.3 Coupled Boundary and Initial Conditions

The 6-DOF Equations (10.1) and (10.4), and the Navier–Stokes Equation (10.5) are closed with the boundary and initial conditions. We consider a force-balancing boundary condition on

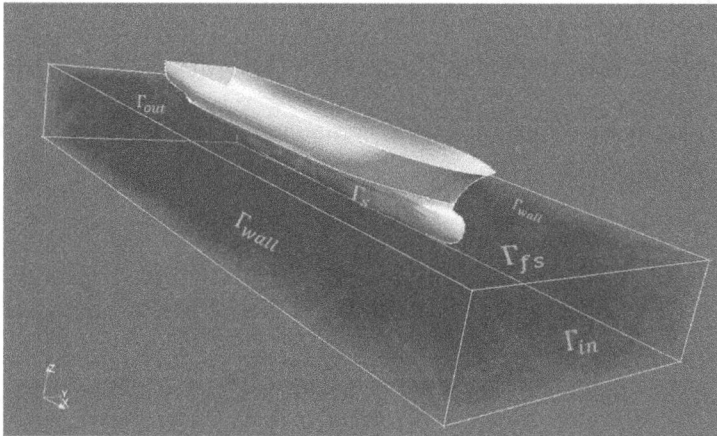

FIGURE 10.2
A computational model in ship hydrodynamics.

the free surface $\Gamma_{fs}(t)$. Moreover, a traction (surface force per unit area) condition can also be considered, where the traction can be due to wind, and in reality ocean waves are primarily caused by tidal forces. Next, we impose the slip with friction boundary condition at the fluid–solid interface $\Gamma_s(t)$, a free slip with no penetration condition on the side and bottom walls Γ_w, zero stress on the outlet boundary Γ_{out}, and a prescribed velocity on the inlet boundary Γ_{in}. The conditions can be formulated as below:

$$-\mathbb{T}(\mathbf{u},p)\cdot v = \sigma \mathcal{K} v \qquad\qquad\qquad \text{on } \Gamma_{fs}(t)\times[0,T]$$

$$\mathbf{u}\cdot v = \mathbf{u}_s\cdot v; \quad \mathbf{u}\cdot\tau = -\varepsilon_\mu(\tau\cdot\mathbb{T}\cdot v) + \eta\mathbf{u}_s\cdot\tau \qquad \text{on } \Gamma_s(t)\times[0,T]$$

$$\mathbf{u}\cdot v = 0; \quad \tau\cdot\mathbb{T}\cdot v = 0 \qquad\qquad\qquad \text{on } \Gamma_w\times[0,T]$$

$$-\mathbb{T}(\mathbf{u},p)\cdot v = 0 \qquad\qquad\qquad\qquad \text{on } \Gamma_{out}\times[0,T]$$

$$\mathbf{u} = \mathbf{u}_{in} \qquad\qquad\qquad\qquad\qquad\qquad \text{on } \Gamma_{in}\times[0,T].$$

Next, the initial conditions are given by

$$\mathbf{u}(x,0) = \mathbf{u}_0; \; \mathbf{u}_b(\mathbf{x},0) = \mathbf{u}_{b,0}; \; \omega(\mathbf{x},0) = \omega_0; \; \mathbf{u}_s(\mathbf{s},0) = \mathbf{u}_{b,0} + (\omega_0\times\mathbf{r}).$$

Here, σ, $\mathcal{K}$, $\mathbf{u}_s$, ε_μ, η, $\mathbf{u}_0$, $\mathbf{u}_{b,0}$, ω_0, $\mathbf{r}$, $\tau=(\tau_1,\tau_2)$ and v are the surface tension, curvature, total velocity (combination of translation and rotation) of the rigid body, slip coefficient, friction coefficient at the fluid-solid interface, initial fluid velocity, initial translational velocity, initial rotational velocity position vector from the center of mass in the solid, unit tangential, and outward normal with respect to fluid phase, respectively. Further, the total body force $\mathbf{F}_{flow}$ from the fluid and the total external moments $\mathbf{M}$ are given by

$$\mathbf{F}_{flow} = -\int_{\Gamma_s}\mathbb{T}\cdot v d\Gamma_s, \quad \mathbf{M}' = -\int_{\Gamma_s}\mathbf{r}\times(\mathbb{T}\cdot v)d\Gamma_s, \qquad\qquad (10.7)$$

Here, the prime on $\mathbf{M}$ indicates that the computed moment has to be projected back to principal axes xyz since we have considered rotation equations with respect to principal axes.

10.2.3.1 Dimensionless Form

To rewrite the 6-DOF Equations (10.1) and (10.4), and the Navier–Stokes Equation (10.5) in a dimensionless form, we introduce the scaling factors L and U as the characteristic length and velocity, respectively. Further, we define the non-dimensional variables

$$\tilde{x} = \frac{x}{L}, \quad \tilde{\mathbf{u}} = \frac{\mathbf{u}}{U}, \quad \tilde{\mathbf{u}}_b = \frac{\mathbf{u}_b}{U}, \quad \tilde{w} = \frac{w}{U}, \quad \tilde{t} = \frac{tU}{L}, \quad \tilde{p} = \frac{p}{\rho U^2}, \quad \tilde{\mathbb{I}} = \frac{IU}{L}$$

and the dimensionless numbers (Reynolds, Weber, Froude, and slip, respectively)

$$\text{Re} = \frac{\rho UL}{\mu_0}, \quad \text{We} = \frac{\rho U^2 L}{\sigma}, \quad \text{Fr} = \frac{U^2}{Lg}, \quad \beta = \frac{1}{\varepsilon_\mu\rho U}.$$

Using these dimensionless parameters in the governing equations and omitting the tilde afterwards, we obtain the dimensionless form of the coupled 6-DOF and Navier–Stokes model in ALE form as

$$\frac{d\mathbf{u}_b}{dt} = \frac{\rho L^3}{m} \mathbf{F}_{flow} + \frac{1}{\text{Fr}} \mathbf{e}, \qquad \text{in } (0,I]$$

$$I_{xx}\frac{d\omega_x}{dt} + (I_{zz} - I_{yy})\omega_y\omega_z = \frac{\rho}{\rho_s} M_x, \qquad \text{in } (0,I]$$

$$I_{yy}\frac{d\omega_y}{dt} + (I_{xx} - I_{zz})\omega_z\omega_x = \frac{\rho}{\rho_s} M_y, \qquad \text{in } (0,I]$$

$$I_{zz}\frac{d\omega_z}{dt} + (I_{yy} - I_{xx})\omega_x\omega_y = \frac{\rho}{\rho_s} M_z, \qquad \text{in } (0,I]$$

$$\frac{\partial \mathbf{u}}{\partial t} + (\mathbf{u} - \mathbf{w})\cdot\nabla\mathbf{u} - \nabla\cdot\mathbb{T}(\mathbf{u},p) = \frac{1}{\text{Fr}}\mathbf{e} \qquad \text{in } \Omega(t)\times(0,I],$$

$$\nabla\cdot\mathbf{u} = 0 \qquad \text{in } \Omega(t)\times(0,I], \qquad (10.8)$$

$$-\mathbb{T}(\mathbf{u},p)\cdot\nu = \frac{\mathscr{K}}{We}, \qquad \text{in } \Gamma_{fs}(t)\times[0,T]$$

$$\mathbf{u}\cdot\nu = \mathbf{u}_s\cdot\nu; \quad \mathbf{u}\cdot\tau - \eta\mathbf{u}_s\cdot\tau = -\frac{1}{\beta}(\tau\cdot\mathbb{T}\cdot\nu), \quad \text{in } \Gamma_s(t)\times[0,T]$$

$$\mathbf{u}\cdot\nu = 0; \quad \tau\cdot\mathbb{T}\cdot\nu = 0, \qquad \text{in } \Gamma_w\times[0,T]$$

$$-\mathbb{T}(\mathbf{u},p)\cdot\nu = 0, \qquad \text{in } \Gamma_{out}\times[0,T]$$

$$\mathbf{u} = \frac{\mathbf{u}_{in}}{U}, \qquad \text{in } \Gamma_{in}\times[0,T]$$

$$\mathbf{u}(\mathbf{x},0) = \frac{\mathbf{u}_0}{U}; \quad \omega_s(\mathbf{x},0) = \frac{U}{L}\omega_{s,0}.$$

Here, the dimensionless stress tensor, the total body force $\mathbf{F}_{flow}$ from the fluid and the total external moments $\mathbf{M}$ are given by

$$\mathbb{T}(\mathbf{u},p) = \frac{2}{\text{Re}}\mathbb{D}(\mathbf{u}) - p\mathbb{I}; \quad \mathbf{F}_{flow} = -\int_{\Gamma_s}\mathbb{T}\cdot\eta d\Gamma_s; \quad M' = -\int_{\Gamma_s}\mathbf{r}\times(\mathbb{T}\cdot\eta)d\Gamma_s.$$

Note that a traction forces (if known) can be used instead of surface force on the free surface.

10.3 Interface Tracking Methods

Tracking/capturing the moving or deforming interfaces and boundaries is one of the main challenges in moving boundary value problems. Several numerical methods have been proposed to track or capture the moving boundaries. We will recall a few Eulerian and Lagrangian methods briefly.

10.3.1 Eulerian Approaches

10.3.1.1 *Volume of Fluid Method*

In the volume-of-fluid method (VOF), a "volume-of-fluid function" (also called "marker function") is used to identify each fluid phase. The "volume-of-fluid function" gives the volume fraction of one of the fluids in each cell of the finite difference or finite volume or finite element mesh. Usually, rectangles and bricks are used in two- and three-dimensions, respectively.

Let us briefly discuss the volume-of-fluid method to capture the interface between two immiscible liquids, say "Liquid A" and "Liquid B," in two dimensional space as seen in Figure 10.3(a). Suppose that the domain is discretized by a rectangular grid of each side length h. Then, the volume of "Liquid B" is defined by $C_{i,j}h^2$, where $0 \leq C_{i,j} \leq 1$, $C_{i,j} \in [0,1]$ is the volume fraction of the "Liquid A" in the (i, j)th cell. Furthermore, the volume fraction of "liquid B" is given by $1 - C_{i,j}$ on the cell (i, j). If a part of the interface lies in the cell (i, j), then we have $0 < C_{i,j} < 1$. Even though the volume fraction $C_{i,j}$ is unique in the cell (i, j), the representation of the interface is not unique, see for example, Figure 10.3(b) and (c). Several interface reconstruction algorithms such as SLIC, piece wise linear approximations, have been proposed in the literature to represent the interface. See [10] for more details of these algorithms.

The volume fraction $C_{i,j}$ can be considered as an approximation of the characteristic function, which is defined as

$$C(x,y) = \begin{cases} 1 & \text{if the point } (x,y) \text{ is in "Liquid A"} \\ 0 & \text{if the point } (x,y) \text{ is in "Liquid B"} \end{cases}$$

Once the fluid velocity $\mathbf{u}$ on the background mesh is calculated at a time instance, the characteristic function $C(x, y)$ has to be recalculated by using the transport equation

$$\frac{\partial C}{\partial t} + \mathbf{u} \cdot \nabla C = 0.$$

Then the position of the interface is captured by one of the interface reconstruction algorithms using the computed volume fraction $C_{i,j}$ (Figure 10.3).

The characteristic function $C(x, y)$ is also used to include the material parameters such as density and viscosity by using

$$\rho = C\rho_A + (1-C)\rho_B, \quad \mu = C\mu_A + (1-C)\mu_B,$$

where ρ_A, ρ_B and μA, μB are the density and dynamic viscosity of the respective liquids "A" and "B."

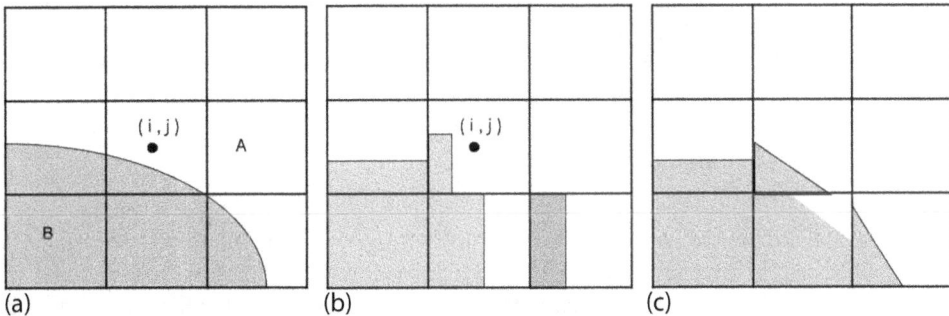

(a)   (b)   (c)

FIGURE 10.3
Interface representation in VOF based on the volume fraction in each cell: (a) original interface, (b) simple linear interface calculation (SLIC) approach, and (c) piecewise linear approximation.

The main challenge lies in the calculation of the local curvature of the interface from the volume fraction. Furthermore, guaranteeing the mass conservation separately for "Liquid A" and "Liquid B" is also very challenging in this method, see [11]. A number of recent developments, including a technique to improve the resolution of the interface with higher-order schemes [10] and local mass preserving schemes [11], have increased applicability of this method.

10.3.1.2 Level Set Method

A continuous zero-level set function represents the interface in the level set method. The level set function $\phi(x,t)$ is constructed as signed distance function in such a way that $\phi(x,t) > 0$ holds on one side of the interface and $\phi(x,t) > 0$ holds on the other side of the interface. Similarly, several interfaces can be represented by the same level set function $\phi(x,t)$.

As an example, let us consider an interface of an unit circle centre at origin at time $t = 0$. We assume that the interface inside a square domain $[-2,2]$, which is meshed by a square grid (Figure 10.4). A signed distance level set function, pictures which have positive values inside the circle and negative values outside the circle, at a time t=0 is defined by

$$\phi(x,0) = 1 - \sqrt{x^2 + y^2}.$$

Here, the zero level set function $\phi(x,0) = 0$, represents the interface. Given the fluid velocity $\mathbf{u}$ on the background mesh, the interface is advected by solving the advection equation

$$\frac{\partial \phi(x,t)}{\partial t} + \mathbf{u} \cdot \phi(x,t) = 0,$$

for a new level set function at time t≥0. The numerical error, which is obtained by discretizing the advection equation, causes a problem to guarantee mass conservation in each fluid phase. In particular, the most losses are significant on coarse meshes [10]. More accurate and robust numerical schemes (for e.g., higher order ENO,WENO) are proposed in the literature to solve this type of advection equation. In the finite element context, the mass loss can be reduced by using finer meshes. However, the computational costs would enormously increase. Alternatively, a mass correction scheme has been used in [12]. Additionally,a stabilization technique such as streamline diffusion (SDFEM) has to be used for advection equation to get a stable system [12, 13].

Moreover, it is essential that $\phi(x,t)$ remains as a signed distance function to guarantee mass conservation. In general, this property is not preserved during the advection and thus a reinitialization technique has been proposed in the literature to overcome this difficulty. Recent advances in the level set method, the natural ability to handle the topological changes of the interface, make the method attractive.

10.3.1.3 Front Tracking Method

The basic idea behind the front tracking method is to use separate front markers for the interface on a fixed mesh and modify the mesh near the front to fit the interface. The method which we discuss here is a hybrid one between the front tracking and the front capturing methods [14]. The main idea is to use stationary meshes for the fluid flow and a separate one-dimension lower moving mesh to track the interface. The front points (interface mesh points) are advected by the computed velocity, where the velocity at each front is obtained by an interpolation from the fixed background mesh.

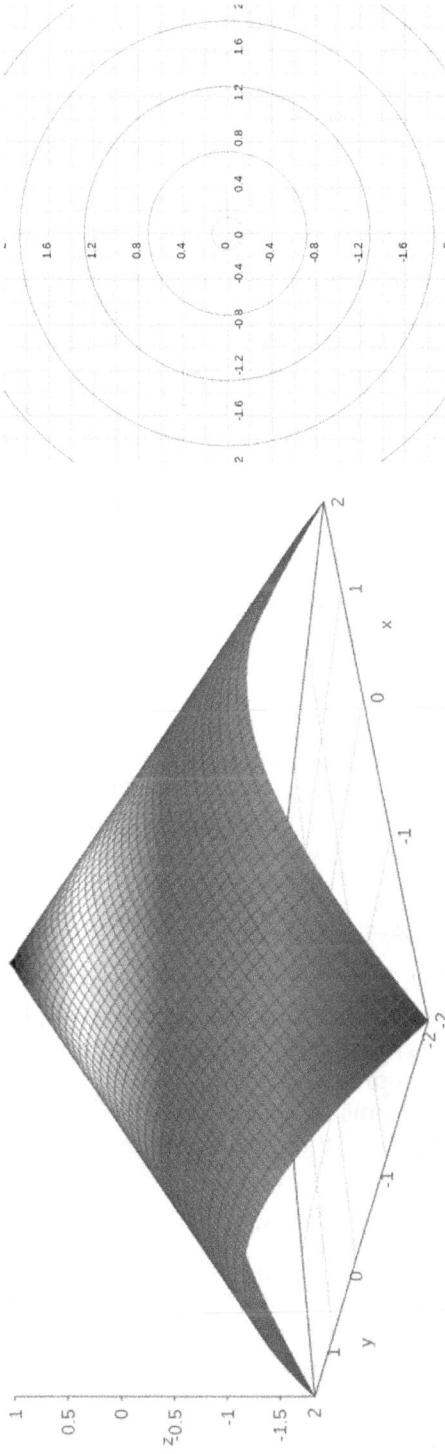

FIGURE 10.4
The level set function $\phi(x,0)$ (left) and their contours (right).

For example, in a two-space dimension, the interface is represented by a one-dimensional moving "front" consisting of connected front points χ_F. In general, smooth function (for e.g., Legendre polynomials or interpolated cubic splines) is constructed from the front points to calculate the curvature, [14, 15]. Once the fluid velocity $\mathbf{u}$ on the fixed background meshes is calculated, the velocity is interpolated at the front points. Then, the new position of the interface is obtained by the equation.

$$\frac{\partial \chi_F}{\partial t} = \mathscr{I}_{\chi_F} \mathbf{u}(\chi_F, t),$$

where $\mathscr{I}_{\chi_F}$ is the populated velocity from the background mesh to the front points χ_F. Since front points are moved in a Lagrangian manner, these points may accumulate or the solution becomes inadequate at some part of the interface. It can be addressed by adding and deleting or redistributing the front points. In front tracking method there is no need to solve any additional equation to advect the interface. However, one has to solve some algebraic equations to construct a parametrized curve, calculate the curvature, and redistribute the front points.

Since the interface is tracked explicitly, topological changes such as breaking and merging of interfaces have to be done manually. Furthermore, the interface is not advected by an interpolated velocity and generally this interpreter velocity is not divergence-free at the front points. Therefore, guaranteeing the mass conservation in each phase is a challenge in the front tracking method.

10.3.2 Lagrangian Approaches

Even though there are several advantages in computations when using Eulerian approaches, the Lagrangian approach is preferred for flows with rigid bodies. Since the interface is not deforming, interface resolved mesh in the Lagrangian approach can be used to develop efficient and accurate numerical schemes.

10.3.2.1 Arbitrary Lagrangian Eulerian Approach

An early description of the ALE approach was given in [16]. The ALE approach is quite popular for fluid structure interaction and free-surface/two-phase problems. For the application to free-surface flows in the finite element context, we refer to [17, 18, 19, 20, 21, 22, 23]. Here, we recall the ALE approach which is based on the description in [23].

Let us denote the boundary points of all interfaces by χ_F. In the ALE approach, the boundary points χ_F are advected with their corresponding velocity, as

$$\frac{d\chi_F}{dt} = \mathbf{u}(\chi_F, t)$$

and the inner points can be moved arbitrarily to preserve the mesh quality.

Let us define a family of mappings $\mathscr{A}_t$, which at each time $t \in [0, I)$ map a point (namely ALE coordinate) $\mathbf{Y} \in \hat{\Omega}$ of a reference domain $\hat{\Omega}$ onto a point (namely Eulerian coordinate) $\mathbf{X}$ of the current domain Ω_t. That is,

$$\mathscr{A}_t : \hat{\Omega} \to \Omega(t), \qquad \mathscr{A}(\mathbf{Y}) = \mathbf{X}(\mathbf{Y}, t)$$

for each $t \in [0.I]$. We assume that the mapping $\mathscr{A}_t$ is a homeomorphic function, i.e., $\mathscr{A}_t \in C^0(\hat{\Omega})$ is invertible with a continuous inverse $\mathscr{A}_t^{-1} \in C^0(\overline{\Omega(t)})$. Furthermore, we assume that the mapping

$$t \rightarrow \mathbf{X}(\mathbf{Y}, t), \qquad \mathbf{Y} \in \hat{\Omega}$$

is differentiable almost everywhere in $t \in [0, I)$. Let us consider a scalar function v on the Eulerian frame. Then, we have

$$\hat{v} := v \circ \mathscr{A}_t, \quad \hat{v} : \hat{\Omega} \times (0, I) \rightarrow \mathbb{R}, \quad \hat{v}(\mathbf{Y}, t) = v(\mathscr{A}_t(\mathbf{Y}, t),$$

which is the corresponding function on the ALE frame. Furthermore, the time derivative of v as seen from ALE frame (but expressed in Eulerian coordinates) is defined by

$$\left. \frac{\partial v}{\partial t} \right|_{\hat{\Omega}} : \Omega(t) \times (0, I) \rightarrow \mathbb{R}, \quad \left. \frac{\partial v}{\partial t} \right|_{\hat{\Omega}} (\mathbf{X}, t) = \frac{\partial \hat{v}}{\partial t}(\mathbf{Y}, t), \quad \mathbf{Y} = \mathscr{A}_t^{-1}(\mathbf{X})$$

Here, $|_{\hat{\Omega}}$ is used to indicate that the time derivative is as seen from ALE frame. Further, the time derivative as seen from Eulerian frame is indicated by $|_{\mathbf{X}}$. The domain velocity $\mathbf{w}$ is defined by

$$\mathbf{w}(\mathbf{X}, t) = \left. \frac{\partial \mathbf{X}}{\partial t} \right|_{\hat{\Omega}} .$$

Now, we apply the chain rule to the time derivative of $v \circ \mathscr{A}_t$ on the ALE frame and obtain

$$\left. \frac{\partial v}{\partial t} \right|_{\hat{\Omega}} = \left. \frac{\partial v}{\partial t} \right|_{\mathbf{X}} + \left. \frac{\partial \mathbf{X}}{\partial t} \right|_{\hat{\Omega}} \cdot \nabla_{\mathbf{X}} v = \left. \frac{\partial v}{\partial t} \right|_{\mathbf{X}} + \mathbf{w} \cdot \nabla_{\mathbf{X}} v, \tag{10.9}$$

where $\nabla_{\mathbf{X}}$ denotes the gradient with respect to the Eulerian coordinate. Using the above relation in Equation (10.9) in the momentum equation in Equation (10.5), we get

$$\left. \frac{\partial \mathbf{u}}{\partial t} \right|_{\mathbf{X}} + \mathbf{w} \cdot \nabla \mathbf{u} + (\mathbf{u} \cdot \nabla) \mathbf{u} - \nabla \cdot \mathbb{T}(\mathbf{u}, p) = \frac{1}{\mathrm{Fr}} \mathbf{e}. \tag{10.10}$$

Note that the domain velocity term is the only difference between Equations (10.5) and (10.10).

10.4 Finite Element Scheme

10.4.1 Weak Form of the Navier–Stokes Equations

Let $L^2(\hat{\Omega})$ and $H^1(\hat{\Omega})$ be the standard Sobolev spaces and $(.,.)$ be the inner product in $L^2(\Omega(t))$ and its vector/tensor-valued versions, respectively. Further, define

$$H_0^1(\hat{\Omega})^3 := \left\{ \mathbf{v} \in H^1(\hat{\Omega})^3 : \mathbf{v} \cdot \nu = 0 \quad \text{on} \quad \Gamma_{wall}(t) \cup \Gamma_s(t) \right\}.$$

We now define the functional spaces for the velocity and pressure in $\Omega(t)$ as

$$V(\Omega(t)) := \left\{ \mathbf{v} \in H^1(\Omega(t))^3 : \mathbf{v} = \hat{\mathbf{v}} \circ \mathscr{A}_t^{-1}, \hat{\mathbf{v}} \in H_0^1(\hat{\Omega})^3 \right\},$$

$$Q(\Omega(t)) := \left\{ q \in L^2(\Omega(t)) : q = \hat{q} \circ \mathscr{A}_t^{-1}, \hat{q} \in L_2(\hat{\Omega}) \right\}.$$

In order to derive the variational form of the Navier–Stokes equations, we multiply the ALE form of momentum equation in Equation (10.8) by a test function $\mathbf{v} \in V$, integrate over the computational domain $\Omega(t)$ and get

$$\int_{\Omega(t)} \frac{\partial \mathbf{u}}{\partial t} \cdot \mathbf{v}\, dx + \int_{\Omega(t)} ((\mathbf{u} - \mathbf{w}) \cdot \nabla) \mathbf{u} \cdot \mathbf{v} dx - \int_{\Omega(t)} \nabla \cdot \mathbb{T}(\mathbf{u}, p) \cdot \mathbf{v}\, dx = \frac{1}{\mathrm{Fr}} \int_{\Omega(t)} \mathbf{e} \cdot \mathbf{v}\, dx.$$

Applying integration by parts to the third term gives

$$\int_{\Omega(t)} \nabla \cdot \mathbb{T}(\mathbf{u}, p) \cdot \mathbf{v}\, dx$$

$$= \int_{\partial\Omega(t)} \mathbf{v} \cdot \mathbb{T} \cdot v ds - \int_{\Omega(t)} \mathbb{T}(\mathbf{u}, p) : \nabla \mathbf{v}\, dx$$

$$\int_{\partial\Omega(t)} \mathbf{v} \cdot \mathbb{T} \cdot v\, ds - \int_{\Omega(t)} \frac{1}{2} \mathbb{D}(\mathbf{u}) : \nabla \mathbf{v}\, dx$$

$$- \int_{\Omega(t)} \frac{1}{2} \mathbb{D}(\mathbf{u}) : \nabla \mathbf{v}^{\mathsf{T}} dx + \int_{\Omega(t)} p \nabla \cdot \mathbf{v}\, dx,$$

$$= \int_{\Gamma_{fs}} \mathbf{v} \cdot \mathbb{T} \cdot v\, ds + \int_{\Gamma_{wall}} \mathbf{v} \cdot \mathbb{T} \cdot v\, ds + \int_{\Gamma_s} \mathbf{v} \cdot \mathbb{T} \cdot v\, ds$$

$$- \int_{\Omega(t)} \mathbb{D}(\mathbf{u}) : \mathbb{D}(\mathbf{v})\, dx + \int_{\Omega(t)} p \nabla \cdot \mathbf{v}\, dx.$$

Using the vector decomposition, $\mathbf{v} = (\mathbf{v} \cdot v)v + (\mathbf{v} \cdot \tau)\tau$ and incorporating the boundary conditions results in

$$\int_{\Omega(t)} \nabla \cdot \mathbb{T}(\mathbf{u}, p) \cdot \mathbf{v}\, dx = \frac{1}{\mathrm{We}} \int_{\Gamma_{fs}} \mathscr{H} \mathbf{v} \cdot v\, ds - \frac{1}{\beta} \int_{\Gamma_s} ((\mathbf{u} - \eta \mathbf{u}_s) \cdot \tau)(\mathbf{v} \cdot \tau) ds$$

$$- \int_{\Omega(t)} \mathbb{D}(\mathbf{u}) : \mathbb{D}(\mathbf{v})\, dx \int_{\Omega(t)} p \nabla \cdot \mathbf{v}\, dx.$$

Thus the weak form of the Navier–Stokes equations in Equation (10.8) reads:
Find $\mathbf{u} \in V$ and $p \in Q$ such that for all $\mathbf{v} \in V$ and $q \in Q$

$$\left(\frac{\partial \mathbf{u}}{\partial t} \cdot \mathbf{v} \right) + a(\mathbf{u} - \mathbf{w} : \mathbf{u}, \mathbf{v}) - b(p, \mathbf{v}) + b(q, \mathbf{u}) = f(\mathbf{v}), \qquad (10.11)$$

where

$$a(\mathbf{w}:\mathbf{u},\mathbf{v}) = \int\limits_{\Omega(t)} \mathbb{D}(\mathbf{u}) : \mathbb{D}(\mathbf{v})dx + \int\limits_{\Omega(t)} (\mathbf{w}\cdot\nabla)\mathbf{u}\cdot\mathbf{v}\,dx + \frac{1}{\beta}\int\limits_{\Gamma_s} ((\mathbf{u}-\eta\mathbf{u}_s)\cdot\tau)(\mathbf{v}\cdot\tau)ds,$$

$$b(q,\mathbf{v}) = \int\limits_{\Omega(t)} q\nabla\cdot\mathbf{v}\,dx,$$

$$f(\mathbf{v}) = \frac{1}{\mathrm{Fr}}\int\limits_{\Omega(t)} \mathbf{e}\cdot\mathbf{v}\,dx + \frac{1}{\mathrm{We}}\int\limits_{\Gamma_{fs}} \mathscr{K}\,\mathbf{v}\cdot\mathbf{v}\,ds.$$

10.4.2 Finite Element Discretization

There are lot of open-source meshing tools like Gmsh [24] and Tetgen [25] to generate unstructured meshes. However, the most important task in computation is to import the mesh data and its connectivity generated by these packages into the existing data structure of the solver package. For our computations we have used Gmsh to generate mesh in'. mesh' file format which is then imported into our in-house parallel Finite Element Solver package ParMooN [26] which has built-in modules to import the mesh data structure from'. mesh' into ParMooN's mesh data structure.

Let $\{\mathscr{T}_h\}$ be a partition of the domain $\Omega(t)$ into tetrahedra or hexahedra cells. The diameter of a cell $K \in \mathscr{T}_h$ is denoted by h_K. The mesh parameter h is defined by $h = \max\{h_K|K \in \mathscr{T}_h\}$ and $\Omega_h(t) := \bigcup_{K\in\mathscr{T}_h} K$. Further, let $V_h \subset V$ and $Q_h \subset Q$ be the conforming finite dimensional subspaces on $\mathscr{T}_h$. The standard Galerkin finite element approximation of the variational problem (10.11) reads:

For given $\Omega(0)$, $\mathbf{w}$, $\mathbf{u}_{in}$ and $\mathbf{u}_0$, find $(\mathbf{u}_h, p_h) \in V_h \times Q_h$ such that

$$\left(\frac{\partial\mathbf{u}_h}{\partial t},\mathbf{v}\right)_h + a_h(\mathbf{u}_h - \mathbf{w}:\mathbf{u}_h,\mathbf{v}) - b_h(p_h,\mathbf{v}) + b_h(q,\mathbf{u}_h) = f_h(\mathbf{v}) \tag{10.12}$$

for all $(\mathbf{v},q)\in V_h \times Q_h$. Here, the subscripts h in $(\cdot,\cdot)_h$ and in the bi-/tri-linear forms denote the inner product in $\mathrm{L}^2(\Omega_h(t))$ and its vector valued versions respectively. Next, the choice of the finite element spaces for the velocity and pressure is subject to the following inf–sup condition,

$$\inf_{q_h\in Q_h}\ \sup_{v_h\in V_h} \frac{(q_h,\nabla\cdot\mathbf{v}_h)}{\|q_h\|_{Q_h}\|\mathbf{v}_h\|_{V_h}} \geq \beta_1 > 0,.$$

We can use equal-order interpolation spaces for the velocity and pressure but it will not satisfy the above inf–sup condition and a stabilization method needs to be used. Thus, in the proposed numerical scheme, we use inf–sup stable finite element pairs for the velocity and pressure (see [9] for more details).

10.4.3 Temporal Discretization and Linearization

Let $0 = t^0 < t^1 < \ldots < t^N = I$ be a decomposition of the given time interval $[0,I]$, and $\delta t = t^{n+1} - t^n$, $n = 0,\ldots,N-1$, be a uniform time step. Denote $\Omega_h^n = \Omega_h(t^n)$. For brevity, we consider the backward Euler method for temporal discretization, whereas readers are referred to [9] for an overview of other time discretization schemes in the context of finite elements.

We first transform the second-order differential equations into a system of first-order differential equations. Then the linear momentum equation becomes

$$\frac{d\mathbf{X}_b}{dt} = \mathbf{u}_b; \qquad \frac{d\mathbf{u}_b}{dt} = \frac{\rho L^3}{m} \mathbf{F}_{flow} + \frac{1}{Fr} \mathbf{e}.$$

Similarly, angular momentum equation is also transformed into a system of first-order differential equations. Let $\alpha = \alpha(\alpha_1, \alpha_2, \alpha_3)$ be the rotational angles on the body axis and $\omega = (\omega_1, \omega_2, \omega_3)$ be its rotational vector, then we have

$$\frac{d\alpha}{dt} = \omega; \qquad \mathbb{I}_m \cdot \frac{d\omega}{dt} = M - \omega \times (\mathbb{I}_m \, \omega).$$

Applying backward Euler, we get

$$\mathbf{X}_b^{n+1} = \mathbf{X}_b^n + \delta t \mathbf{u}_b^{n+1}$$

$$\mathbf{u}_b^{n+1} = \mathbf{u}_b^n + \delta t \left(\frac{\rho L^3}{m} \mathbf{F}_{flow}^{n+1} + \frac{1}{Fr} \mathbf{e} \right)$$

$$\alpha^{n+1} = \alpha^n + \delta t \omega^{n+1}$$

$$\omega^{n+1} = \omega^n + \delta t \left(\mathbb{I}_m^{n+1} \right)^{-1} \left(M^{n+1} - \omega^{n+1} \times \left(\mathbb{I}_m^{n+1} \omega^{n+1} \right) \right).$$

(10.13)

We next denote $\mathbf{u}^n = \mathbf{u}_h(\mathbf{x}, t^n)$, $p^n = p_h(\mathbf{x}, t^n)$ and $\mathbf{w}_h^n = \mathbf{w}(\mathbf{x}, t^n)$ as the approximations of the velocity, pressure and mesh velocity, respectively at time $t = t^n$, where $\mathbf{x} \in \Omega_h^n$. Applying temporal discretization to the Navier-Stokes system (Equation (10.12)) and to the 6-DOF equations in Equation (10.8), we get a sequence of stationary equations:

For given $\hat{\Omega}_h$, $\mathbf{u}_0$, $\mathbf{u}^n$, $\mathbf{w}^n$, find $\left(\mathbf{u}^{n+1}, p^{n+1} \right) \in V_h \times Q_h$ and $\mathbf{w}^{n+1} \in \mathrm{H}^1 \left(\Omega_h(t^{n+1}) \right)^3$ such that for all $(\mathbf{v}, q) \in V_h \times Q_h$

$$\left(\frac{\mathbf{u}^{n+1} - \mathbf{u}^n}{\delta t}, \mathbf{v} \right)_h + a_h(\mathbf{u}^{n+1} - \mathbf{w}^{n+1}; \mathbf{u}^{n+1}, \mathbf{v}) + -b(p^{n+1}, \mathbf{v}) = f^{n+1}(\mathbf{v})$$

(10.14)

$$b(q, \mathbf{u}^{n+1}) = 0.$$

10.4.4 Linearization and Solution Process

Monolithic or partitioned approach can be used to solve the strongly coupled discrete systems in Equation (10.13) and (10.14). In the monolithic approach, interactions between the fluid and the solid at the interface occur simultaneously and is synchronised [27, 28]. Further, the discrete systems in Equation (10.13) and (10.14) will be treated as a single system and a non-linear solver is used to solve the non-linear system. Moreover, conservation properties could be preserved [29] and the scheme will become unconditionally stable. Hence, the admissible time-step size is limited only by the required accuracy.

In a partitioned approach, the discrete solid (Equation (10.13)) and fluid (Equation (10.14)) systems are integrated in time alternatingly and the interface conditions are enforced asynchronously [30, 31]. This sequential process allows for software modularity, that is, different solvers can be used to solve Equation (10.13) and Equation (10.14) systems. Partitioned schemes require only one fluid and structure solution per time step, which can be considered as a single fluid-structure iteration. On the other hand, due to the time lag between the time integration of fluid and structure, the conservation properties of the continuum fluid-structure system are lost. Partitioned schemes are commonly energy increasing and therefore numerically unstable [31]. In practice, this introduces a restriction on the admissible time-step size.

In the proposed numerical scheme, we treat the non-linear and coupled terms semi-implicitly and apply an iteration of fixed-point type to the decoupled, linearized monolithic system. It alleviates the time-step restriction but preserves the conservation properties and allows software modularity. In a time interval (t^n, t^{n+1}), we adopt the following strategy:

Let $\mathbf{F}_{flow}^{n+1,0} = \mathbf{F}_{flow}^n$, $M^{n+1,0} = M^n$, $\mathbb{I}_m^{n+1,0} = \mathbb{I}_m^n$, $\mathbf{u}_0^{n+1} = \mathbf{u}^n$, $\mathbf{w}_0^{n+1} = \mathbf{w}^n$, then solve

$$\mathbf{X}_b^{n+1} = \mathbf{X}_b^n + \delta t \mathbf{u}_b^{n+1}$$

$$\mathbf{u}_b^{n+1} = \mathbf{u}_b^n + \delta t \left(\frac{\rho L^3}{m} \mathbf{F}_{flow}^{n+1,k} + \frac{1}{Fr} \mathbf{e} \right) \tag{10.15}$$

$$\alpha^{n+1} = \alpha^n + \delta t \omega^{n+1}$$

$$\omega^{n+1} = \omega^n + \delta t \left(\mathbb{I}_m^{n+1,k} \right)^{-1} \left(M^{n+1,k} - \omega^{n+1} \times \left(\mathbb{I}_m^{n+1,k} \cdot \omega^{n+1} \right) \right)$$

$$\left(\frac{\mathbf{u}_{k+1}^{n+1} - \mathbf{u}^n}{\delta t}, \mathbf{v} \right)_h + a_h \left(\mathbf{u}_k^{n+1} - \mathbf{w}_k^{n+1}; \mathbf{u}_{k+1}^{n+1}, \mathbf{v} \right) + -b(p^{n+1}, \mathbf{v}) = f^{n+1}(\mathbf{v}) \tag{10.16}$$

$$b(q, \mathbf{u}_{k+1}^{n+1}) = 0.$$

for $k = 0, 1, \ldots$, until the residual of the above system reduces to a threshold value or to a maximum number of prescribed steps. In this solution process, the following steps are involved in one iteration step:

- solve 6DOF Equation (10.15) in $\Omega_S(t^n)$ and compute the motion of the structural boundary
- move $\Omega(t^n)$ using linear elastic approach, see below, for the computed motion of the structural boundary
- compute the mesh velocity and restore the domain to $\Omega(t^n)$
- solve NSE (10.16) and compute structural load $\mathbf{F}_{flow}$ and $\mathbb{I}_m$ with the updated velocity and pressure.

During the iteration, we solve the linear elasticity problem to compute the mesh velocity. In the linear elastic mesh update, we first compute the displacement of the boundary vertices on the solid structure, that is,

$$\mathbf{d} = \mathbf{Z}_b^{n+1} - \mathbf{Z}_b^n, \qquad \text{where } \mathbf{Z}_b = \mathbf{X}_b \mid_{\Gamma_S}$$

that is, $\mathbf{Z}_b$ are the vertices on the boundary Γ_S. Then the displacement of inner mesh points is obtained by solving the linear elasticity equation, i.e.,

Find $Y \in H^1\left(\Omega(t^n)\right)^3$, such that

$$\nabla \cdot \mathbb{S}(\Psi) = 0 \ \text{ in } \Omega(t^n),$$

$$\Psi = \mathbf{d} \ \text{ on } \Gamma_S, \tag{10.17}$$

$$\Psi = 0 \ \text{ on } \partial\Omega(t^n) \setminus \Gamma_S,$$

where $\mathbb{S}(\Psi) = \lambda_{L1}(\nabla \cdot \Psi)\mathbb{I} + 2\lambda_{L2}\mathbb{D}(\Psi)$. Here, λ_{L1} and λ_{L2} are Lame constants, and in computations we use $\lambda_{L1} = \lambda_{L2} = 1$. Once the displacement vector Y is computed, the mesh velocity is then calculated as $\mathbf{w}_k^{n+1} = \Psi / \delta t$.

The proposed algorithms are implemented in our in-house finite element package, ParMooN [26, 32].

10.5 Numerical Results

In this section, we present a few computations results for a flow past a KVLCC2 (KRISO Very Large Container Carrier, where KRISO stands for Korea Research Institute of Ships and Ocean engineering) ship hull. Here, we considered only the fluid flow around the ship by keeping the solid structure fixed. In the considered model, the KVLCC2 hull is submerged 30 m in a fluid tank of length 500 m, width 185 m, and depth 58 m. The basic dimensions of the KVLCC2 ship hull are as follows: (i) length between perpendiculars (L_{pp}) 320 m, (ii) waterline length (L_{wl}) 325.5 m, (iii) breadth (B) 58 m, see [33] for more details. Further, the computations are performed with 146739 and 6888 DOFs for the velocity and pressure, respectively.

The plots in Figure 10.5 show the evolution of flow around the ship's hull at different instances. Further, the velocity plots are plotted on a logarithmic scale to effectively visualize the minute variations in the velocity of the flow.

Next, streamlines are most commonly used to visualize a fluid flow since it could easily represent the features like vorticity present within the fluid flow. Streamlines are imaginary lines that are drawn tangent to the velocity vector at any point along the line. Streamlines tend to group together when the flow velocity is high, whereas the streamlines are usually separated out in a low-speed flow. The plots in Figures 10.6 through 10.8 show the velocity streamlines along the axis of flow observed from various points around the ship's hull.

We could see that the streamlines tend to move upwards below the stern portion of the ship; this is due to the fact that the bottom portion on the stern side of the ship is inclined upwards toward the surface of the water. This design is similar to the design of the upper surface of an airfoil, which is done to prevent the separation of fluid flow when a flow is past an object. Separation of a fluid-flow form surface may lead to adverse pressure gradients on the rear end of the surface which significantly contributes to pressure drag, which could be observed in flow past a flat plate placed perpendicular to the direction of flow.

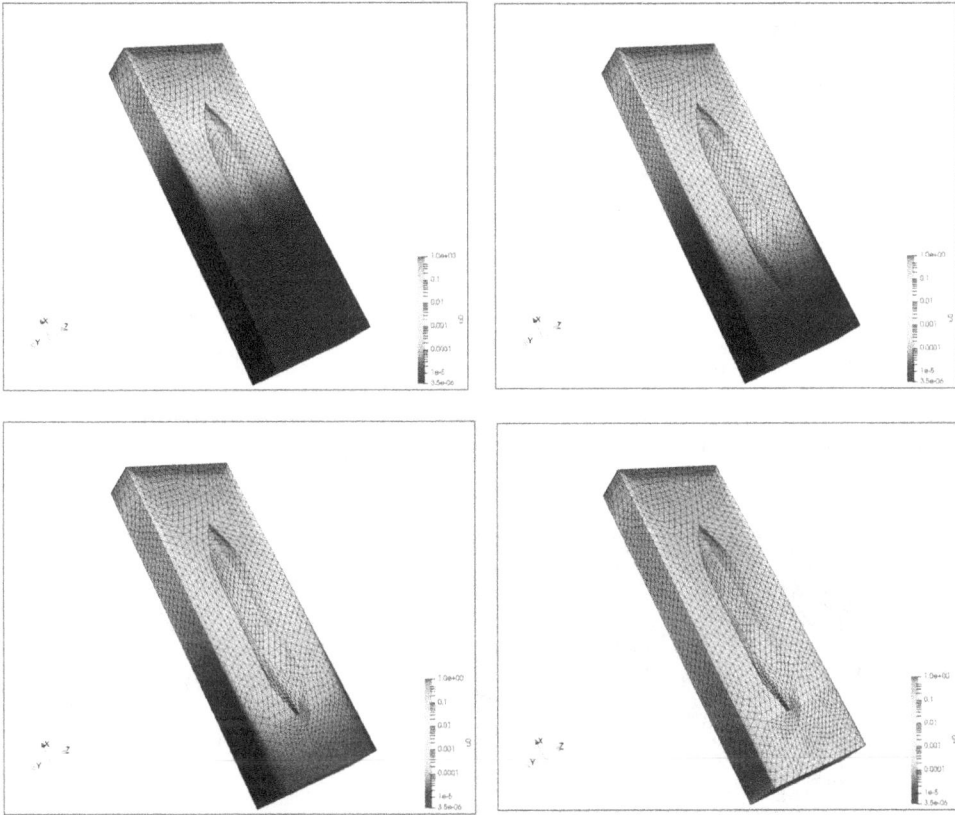

FIGURE 10.5
A computational model in ship hydrodynamics.

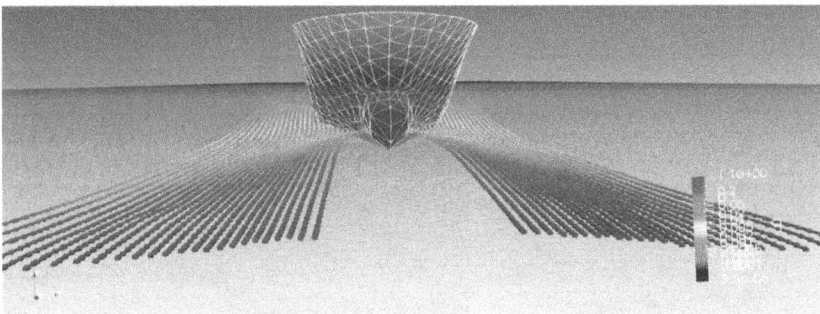

FIGURE 10.6
Velocity streamlines on the bottom portion of the ship.

From Figure 10.9, it is observed that the stress values are comparatively high at the tip of the ship. This can be explained further from the constitutive relation in Equation (10.6), the relation between fluid velocity and stress, which states that the stress tensor is proportional to the deformation of the velocity. At the tip portion of the ship, the velocity gradients will be high in that region since the flow is being obstructed, which would in turn have caused higher stress values.

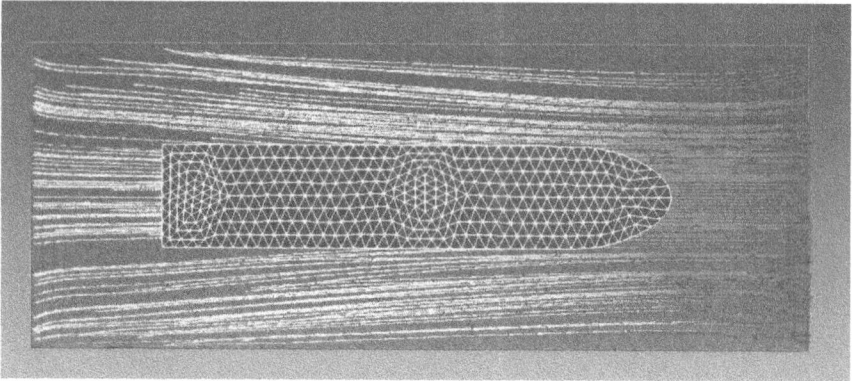

FIGURE 10.7
Streamlines of velocity around the ship's hull—top view.

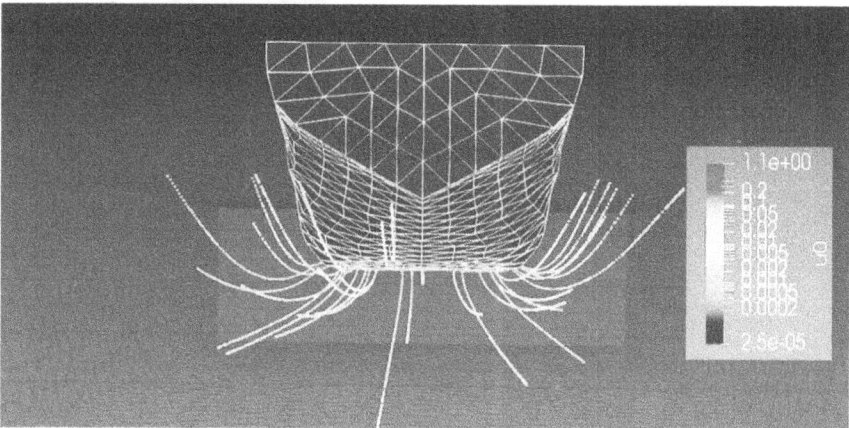

FIGURE 10.8
Streamlines visualized at the stern portion of the ship.

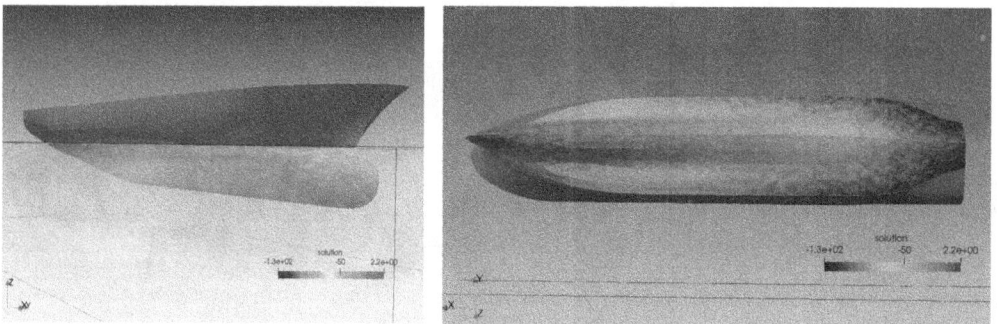

FIGURE 10.9
Visualization of stress profile on immersed portion of the ship.

Acknowledgment

This work has been supported by the Naval Research Board, DRDO, India through the grant NRB/4003/PG/368.

References

1. William Froude. On the rolling of ships. *Institute of Naval Architects*, 2:180–227, 1861.
2. A Krylov. A general theory of oscillation of ships. *Institute of Naval Architects*, 49, 1898.
3. John Henry Michell. Xi. The wave-resistance of a ship. *The London, Edinburgh, and Dublin Philosophical Magazine and Journal of Science*, 45(272):106–123, 1898.
4. John L Hess and AMO Smith. Calculation of potential flow about arbitrary bodies. *Progress in Aerospace Sciences*, 8:1–138, 1967.
5. CW Dawson. Method for solving ship wave problems. In *Proceedings of the 2nd International Conference on Numerical Ship Hydrodynamics*, 305–318, 1977.
6. S Ohring and J Telste. Numerical solutions of transient three-dimensional ship-wave problems. In *Proceedings of the Second International Conference on Numerical Ship Hydrodynamics*, University of California, Berkeley, September, 1977.
7. T Hino. Computation of viscous flows with free surface around an advancing ship. In *Proceedings of the Second Osaka International Colloquium on Viscous Fluid Dynamics in Ship and Ocean Technology*, Osaka, 1992.
8. J Farmer, Luigi Martinelli, and A Jameson. Fast multigrid method for solving incompressible hydrodynamic problems with free surfaces. *AIAA Journal*, 32(6):1175–1182, 1994.
9. S Ganesan and L Tobiska. *Finite Elements: Theory and Algorithms*. Cambridge IISc Series. Cambridge University Press, Cambridge, United Kingdom, 2017.
10. James Edward Pilliod Jr and Elbridge Gerry Puckett. Second-order accurate volume-of-fluid algorithms for tracking material interfaces. *Journal of Computational Physics*, 199(2):465–502, 2004.
11. Michael Renardy, Yuriko Renardy, and Jie Li. Numerical simulation of moving contact line problems using a volume-of-fluid method. *Journal of Computational Physics*, 171(1):243–263, 2001.
12. Sven Groß, Volker Reichelt, and Arnold Reusken. A finite element based level set method for two-phase incompressible flows. *Computing and Visualization in Science*, 9(4):239–257, 2006.
13. Anna-Karin Tornberg. *Interface Tracking Methods with Application to Multiphase Flows*. PhD thesis, Royal Institute of Technology, Department of Numerical Analysis and Computing Science, 2000.
14. Grétar Tryggvason, Bernard Bunner, Asghar Esmaeeli, Damir Juric, N Al-Rawahi, W Tauber, J Han, S Nas, and Y-J Jan. A front-tracking method for the computations of multiphase flow. *Journal of Computational Physics*, 169(2):708–759, 2001.
15. Yuriko Renardy, Stéphane Popinet, Laurent Duchemin, Michael Renardy, Stéphane Zaleski, Christophe Josserand, MA Drumright-Clarke, D Richard, Christophe Clanet, and David Quéré. Pyramidal and toroidal water drops after impact on a solid surface. *Journal of Fluid Mechanics*, 484:69–83, 2003.
16. Jean Donea. Arbitary Lagrangian-Eulerian finite elements methods. *Computational Methods in Transient Analysis*, 33(1–3), 689–723, 1983.
17. Sashikumaar Ganesan and Lutz Tobiska. A coupled arbitrary Lagrangian–Eulerian and Lagrangian method for computation of free surface flows with insoluble surfactants. *Journal of Computational Physics*, 228(8):2859–2873, 2009.

18. Sashikumaar Ganesan and Lutz Tobiska. Arbitrary Lagrangian–Eulerian finite-element method for computation of two-phase flows with soluble surfactants. *Journal of Computational Physics*, 231(9):3685–3702, 2012.

19. Sashikumaar Ganesan and Shweta Srivastava. ALE-SUPG finite element method for convection–diffusion problems in time-dependent domains: Conservative form. *Applied Mathematics and Computation*, 303:128–145, 2017.

20. Eberhard Bänsch. *Numerical Methods for the Instationary Navier-Stokes Equations with a Free Capillary Surface*. PhD thesis, Verlag nicht ermittelbar, 1998.

21. S Ganesan and L.Tobiska. Finite element simulation of a droplet impinging on a horizontal surface. In Proceedings of *Algoritmy*, vol. 2005, pp 1–11. Slovak Technical University, Bratislava, 2005.

22. Gunar Matthies. *Finite Element Methods for Free Boundary Value Problems with Capillary Surfaces*. Shaker, Aachen, Germany 2002.

23. F Nobile. *Numerical Approximation of Fluid Structure Interaction Problems with Application to Haemodynamics*. PhD thesis, École Polytechnique Fédérale de lausanne, 2001.

24. Christophe Geuzaine and Jean-François Remacle. Gmsh: A 3-d finite element mesh generator with built-in pre-and post-processing facilities. *International Journal for Numerical Methods in Engineering*, 79(11):1309–1331, 2009.

25. Hang Si. Tetgen, a delaunay-based quality tetrahedral mesh generator. *ACM Transactions on Mathematical Software (TOMS)*, 41(2):11, 2015.

26. Ulrich Wilbrandt, Clemens Bartsch, Naveed Ahmed, Najib Alia, Felix Anker, Laura Blank, Alfonso Caiazzo, Sashikumaar Ganesan, Swetlana Giere, Gunar Matthies, et al. ParMooNâ€"a modernized program package based on mapped finite elements. *Computers & Mathematics with Applications*, 74(1):74–88, 2017.

27. Patrick Le Tallec and Jean Mouro. Fluid structure interaction with large structural displacements. *Computer Methods in Applied Mechanics and Engineering*, 190(24–25):3039–3067, 2001.

28. Frederic J Blom. A monolithical fluid-structure interaction algorithm applied to the piston problem. *Computer Methods in Applied Mechanics and Engineering*, 167(3–4):369–391, 1998.

29. EH Van Brummelen, SJ Hulshoff, and R De Borst. Energy conservation under incompatibility for fluid–structure interaction problems. *Computer Methods in Applied Mechanics and Engineering*, 192(25):2727–2748, 2003.

30. Carlos A Felippa, KC Park, and Charbel Farhat. Partitioned analysis of coupled mechanical systems. *Computer Methods in Applied Mechanics and Engineering*, 190(24–25):3247–3270, 2001.

31. Serge Piperno, Charbel Farhat, and Bernard Larrouturou. Partitioned procedures for the transient solution of coupled aroelastic problems part i: Model problem, theory and two-dimensional application. *Computer Methods in Applied Mechanics and Engineering*, 124(1–2):79–112, 1995.

32. S Ganesan, V John, G Matthies, R Meesala, S Abdus, and U Wilbrandt. An object oriented parallel finite element scheme for computations of pdes: Design and implementation. In *2016 IEEE 23rd International Conference on High Performance Computing Workshops (HiPCW)*, pages 2–11, 2016.

33. Bing Jie Guo and Sverre Steen. Evaluation of added resistance of KVLCC2 in short waves. *Journal of Hydrodynamics, Series B*, 23(6):709–722, 2011.

11

DNA—Drug Interaction: Insight from Computer Simulations

Anurag Upadhyaya, Shesh Nath, and Sanjay Kumar

CONTENTS

11.1 Introduction 171
11.2 A Brief Review of DNA and Its Properties 172
11.3 Computer Simulations 175
11.4 Atomistic Simulations of DNA Rupture 179
11.5 Stability of a DNA Aptamer and Its Use as a Drug Delivery Agent 182
11.6 Stability of DNA-ICLs against Force and Its Use as an Anti-Tumor Drug 184
    11.6.1 Rupture of DNA and DNA-ICLs 186
    11.6.2 Unzipping of DNA and DNA-ICLs 188
11.7 Conclusions and Future Perspective 189
Acknowledgments 190
References 190

11.1 Introduction

Nucleic acids play crucial roles in biological processes including cell division (replication) and protein synthesis (transcription and translation). These processes happen in the normal healthy as well as in the cancerous cells. Cancerous cells are characterized by the uncontrolled cell division, which differs from the normal healthy cell division [1, 2]. The normal cells exhibit controlled cell division, where surface contact with like cells stops the cell division, a process referred to *contact inhibition*. Cancerous cells lose this ability and, as a result, they grow rapidly. Whether the cell division is mitosis or a meiosis, an exact copy of DNA replication is essential. The stability of DNA is largely affected by its length, base composition, sequence, and presence of nucleotide mismatches [3–8]. It is also strongly perturbed by changes in pH, temperature, and solvent as well as by the binding of proteins such as helicases, polymerases, or topoisomerases in living cells [7, 9, 10].

Small molecules may interact or bind with DNA [11]. Classification of such molecules depends on their mechanism of action. Broadly speaking, DNA binding molecules can be categorized in: (i) groove binders that assemble in the minor groove; (ii) intercalators that pack in between base-pairs; (iii) alkylators, which react with DNA to form DNA alkylation; and (iv) cleavage agents, which have the capability to break DNA strands. Each of these classes of molecules has a different structure and interacts with DNA differently with the primary task of stopping cancerous cell division by inhibiting DNA replication. However, it halts the cell division without distinguishing between healthy or cancerous cells [12]. As a result, it causes several other side effects e.g. nausea, diarrhea, hair loss, low blood counts, mouth sores etc. Therefore, targeted delivery of drugs into the cancer cell is a prerequisite to increase effectiveness and reduce the toxic side effects by reducing uptake into normal cells.

Recent studies have revealed that DNA aptamers can be used as agents for targeted drug delivery [13–15]. Aptamers are Guanine(G)-rich short oligonucleic acids (DNA, RNA) molecules, which can perform specific functions [16, 17]. These have been developed *in vitro* through a SELEX (Systematic Evolution of Ligands by Exponential Enrichment) process for the better understanding of the behavior of antibodies, which are produced *in vivo* or in living cells [16–18]. One of the most extensively characterized examples of the aptamer is found in telomerase at the ends of eukaryotic chromosomes, where it plays an important task in gene regulation [19, 20]. DNA loops or base-pockets consisting of Guanine can interact with small molecules and proteins and thus enhance their stability, affinity, and specificity [21]. For example, adenosine monophosphate (AMP) binds to the DNA loop (termed a binding-pocket) with eight hydrogen bonds, which enhances the stability [22–24] of the aptamer. It can be used as the target of drugs for cancer treatment [25]. Their high affinity and selectivity with target proteins make them ideal and powerful probes in biosensors and potent pharmaceuticals [26–36].

Thus, there is a need to understand the role of inter- and intra-molecular interactions involved in the drug–DNA composite. Single molecule force spectroscopy (SMFS) experiments have offered the opportunity to study these molecular forces [2, 6, 37–48]. Initially, it was thought that the interactions detected in SMFS experiments would be mostly of a mechanical nature and can be calculated by knowing the value of the applied force. However, insights gathered from these experiments revealed that the measurement of molecular interactions depends not only on the magnitude of the applied force but also on how and where the force is applied [6, 38–48]. These experiments may be used to explore the local stability arising due to the interacting sites in the drug–DNA composites.

The aim of the present chapter is to understand the conformational stability of DNA in the presence of these molecules and use them either as drugs or as a drug delivery agent. In Section 11.2, we briefly introduce DNA and its properties. Section 11.3 describes molecular dynamics (all atoms) simulations and the necessary force fields required for DNA. Section 11.4 deals with atomistic simulations of DNA rupture in the absence of drug molecules. In Section 11.5, we have extended our approach to study the stability of DNA aptamers and its use as a drug delivery agent. Section 11.6 focuses on the stability of DNA in the presence of alkylators. As discussed earlier, alkylators form covalent bonds with DNA strands, which are known as Inter Cross-Linkers (ICLs). We shall consider two drugs, namely Mitomycin C and Trabectedin, which form DNA-ICLs, and compare their stability with drug-free DNA. This chapter ends with a brief discussion and future perspectives in Section 11.7.

11.2 A Brief Review of DNA and Its Properties

Inside the cell there is a nucleus which contains the DNA (Figure 11.1). It is a molecule recognized as a "Molecule of life." Genetic information is stored in the DNA and used in the development and functioning of living organisms including many viruses [49–51].

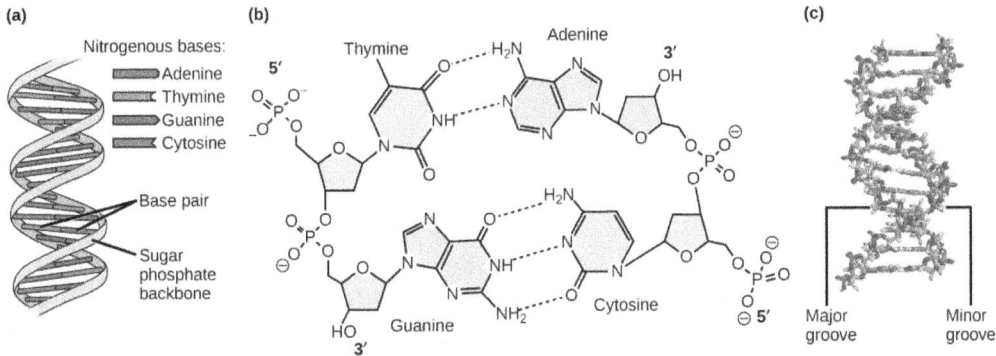

FIGURE 11.1

(a) Geometrical structure of helical DNA, (b) chemical structure of DNA, and (c) location of the major and minor groove.

A double-stranded DNA (dsDNA) is made up of two interacting single-strands of DNA (ssDNA). It consists of four different types of nucleotides, namely adenine (A), thymine (T), cytosine (C), and guanine (G). These nucleotides may be classified into two categories: (i) purine, and (ii) pyrimidine. Purine bases are adenine and guanine, while pyrimidines bases are thymine and cytosine. RNA uses uracil (U) in place of thymine. Purines are double-ring structures and pyrimidines are single-ring structures. In a dsDNA, the number of adenine is always equal to the number of thymine, and the number of guanine is always equal to the number of cytosines. Adenine and thymine form two hydrogen bonds, whereas guanine and cytosine form three hydrogen bonds (Watson–Crick base pairing). Each nucleotide is formed of three basic components: nucleo-base (nitrogenous base), a five-carbon sugar, and one phosphate group. Two nucleotides are connected by a phosphodiester bond, in which a phosphate group couples the 3′-hydroxyl group of one nucleotide to the 5′-phosphate group of the next, giving rise to directionality in the polynucleotide chain (Figure 11.1(b)). By convention, we refer to DNA sequences from the 5′- to 3′- end, which corresponds to the direction in which they are synthesized and interpreted by other molecules. The phosphate groups contain a negative charge, causing exterior of DNA an overall negative charge. This makes DNA a negative charged molecule.

From an energetic point of view, the most important contribution to the stability of the DNA helix is the intra-strand base-pair stacking interaction between adjoining nucleotide base-pairs. The base-pair stacking potential is a short-ranged attractive interaction in nature. The stacking energies of different base-pair couples are distinct. In-between the strands there are spaces known as grooves. These spaces are adjacent to the base-pairs and may provide a binding site for other molecules. Since strands are not directly parallel to each other, the grooves are unequally sized, namely major groove (22 Å wide) and minor groove (12 Å wide) [52]. The narrowness of the minor groove means that the edges of the bases are more accessible in the major groove. As a result, proteins, like transcription factors that can attach to specific sequences in double-stranded DNA, usually make links to the sides of the bases exposed in the major groove.

There are many types of conformations of dsDNA that are influenced by different factors like level of hydration, ions concentration, the presence of polyamines in the solution etc. A-DNA, B-DNA, and Z-DNA are the dsDNA conformations that are frequently used in the literature. These different conformations of the DNA helix are responsible for specific biological functions. B-DNA is the canonical right-handed DNA, which has ten bases per helical turn [53]. It is about 23.7 Å wide and extends 34 Å per ten base-pair sequence. A-DNA is the right-handed double helix with 11 base-pairs in one complete turn of the helix. In normal conditions, the distance between successive sugars or phosphates in A-DNA is found 0.6 nm. Z-DNA is the left-handed double helix in which the arrangement of bases appears in a zigzag pattern. It has 12 bases per helical twist (Figure 11.2).

There are two important biological processes, where unwinding of DNA takes place. These are DNA transcription and DNA replication processes. Transcription occurs during protein translation. It involves copying DNA into RNA. The portion of the DNA that codes for genes is transcribed or copied into messenger RNA (mRNA). DNA replication is the process of copying the DNA itself into two copies before the cell division in order to transfer genetic information from the parent cell to the daughter cells. Separation of dsDNA

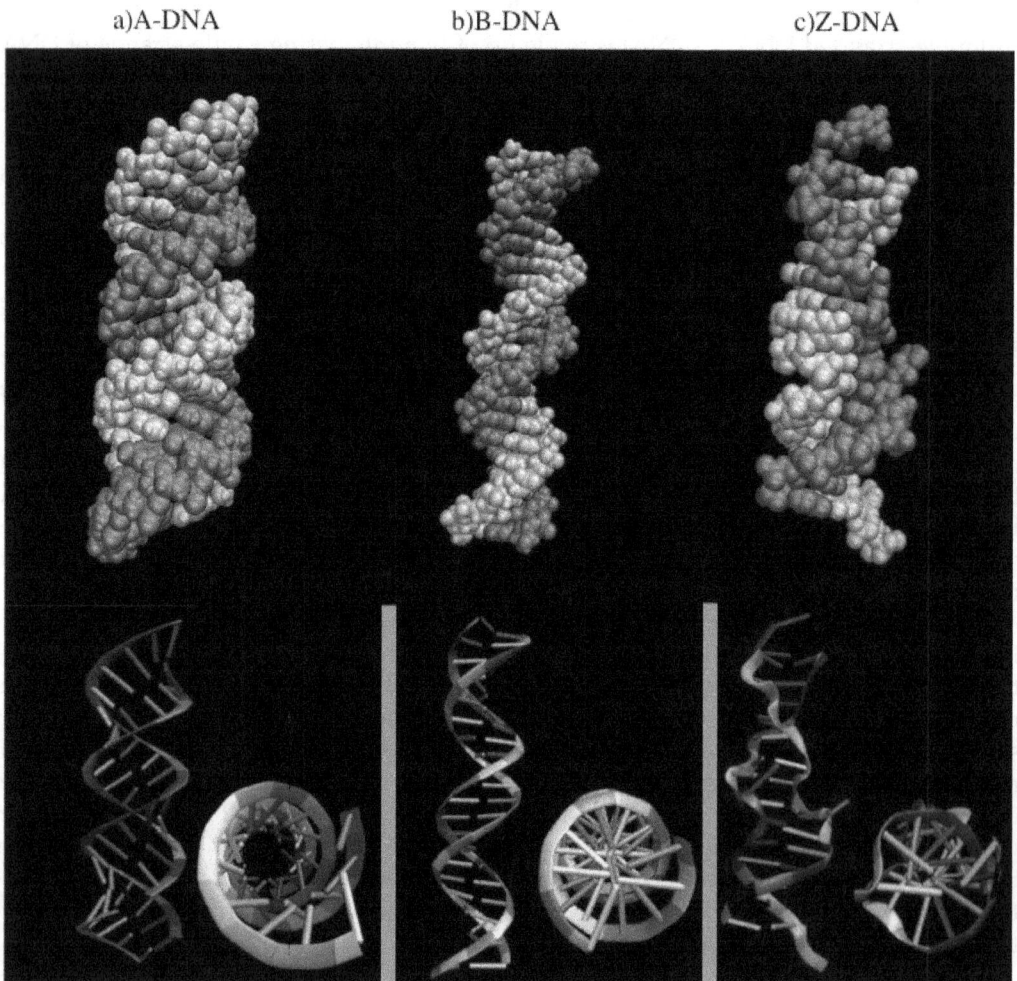

FIGURE 11.2
Atomistic representations of (a) A-DNA, (b) B-DNA and (c) Z-DNA. (d–f) represent bond view and top of view of A, B and Z-DNA.

can take place either through the heating (thermal melting) or by varying the pH value of the solvent (DNA denaturation) [54]. Thermal melting causes breaking of hydrogen bonds due to the increase in entropy and, as a result, two strands get separated. Usually, melting temperature (T_m) is defined as the temperature at which half of the base-pairs of a dsDNA gets dissociated. *In vitro*, the melting temperature was found to be in the range of 80°C and 90°C, depending on the sequence, and denaturation is obtained for the pH value greater than 9.0 [54]. Physically these two extreme conditions are not possible inside a living body. This indicates that there must be some other mechanism involved in the separation of dsDNA.

It has been realized now that these processes are mediated via certain enzymes or proteins that exert mechanical force on DNA, and, as a result, it separates [55]. Motivated by this, Bockelmamm and coworkers [39, 40] applied a force perpendicular to the helix direction i.e. at 5′–3′ ends, and measured the unzipping force ~15 pN to separate a dsDNA. Essential information about sequence and strength of the interactions involved in base pairing have been revealed.

11.3 Computer Simulations

In recent years Molecular Dynamic (MD) simulations [56, 57] have emerged as an important tool to understand the results of SMFS experiments. Such a tool has enhanced our understanding of these molecules and also predicted structures and function of these molecules. In fact, it has served as a bridge between theory and experiment. It offers so much information that is hard to quantify. Excellent reviews on molecular dynamics are in the literature, and its use in biochemistry and biophysics are numerous (see, e.g. [58] and references therein).

Before studying structural and dynamical properties of DNA through simulations, it is important to note that DNA exhibits a wide range of length and time scales over which specific processes take place. For example, local motion, which involves atomic fluctuations, side chain motion, and loop motion occurs on the length scale of 0.01 to 5 Å, and the time involved in such processes is of the order of 10^{-15} to 10^{-12} s. The motion of a helix or a sub-unit falls under the domain of rigid body motion with typical length scales between 1 and 10 Å, and the time involved in such motion is between 10^{-9} and 10^{-6} s. Large-scale motion consists of helix-coil transitions, which involve length scales of more than 10 Å and time scales from 10^{-7} to 10^{1} s. Depending upon the nature of problems, one considers either the coarse-grained description or a detailed atomistic description of the DNA.

However, in this chapter, we focus our discussion on all-atom simulations [59–61]. In fact, this chapter provides the most detailed description on the atomistic level, including the local interaction and non-bonded interactions. The latter includes the (6–12) Lenard–Jones potential, the electrostatic interaction, and the interaction with the environment. The all-atom model with the CHARMM (Chemistry at HARvard Molecular Mechanics) force field [59] was first employed by Grubmuller et al. [62]. The NAMD (NAnoscale Molecular Dynamics) software [60] is now widely used for the stretching of biomolecules by constant mechanical force and by the force with a constant loading rate. Recently, GROMACS (GROningen MAchine for Chemical Simulations) software [61] has been developed to study the mechanical unfolding in explicit water. In this chapter, we used an Assisted Model Building with Energy Refinement (AMBER) software package [63]. It is a general-purpose

molecular mechanics and dynamics program designed for the refinement of macro-molec-
ular conformations using an analytical potential energy function.

The force field describes the detail inter- and intra-molecular interactions involved
among the constituent of biomolecules [64]. The parameter of the force field is in amber
generally calculated from experiments or first-principle quantum mechanical calculations.
The force field for a system consisting of N particles can be written as the sum of the following energy terms:

$$V(r) = V_{\text{bond}} + V_{\text{angle}} + V_{\text{torsion}} + V_{\text{vdw}} + V_{\text{elec}}, \tag{11.1}$$

where $V(r)$ (Figure 11.3) denotes the total energy that is the function of position (r) of N
particles. V_{bond}, V_{angle} and V_{torsion} are the bond stretching, the bending, and the torsional
energy terms, respectively. V_{vdw} (van der Waals energy), and V_{elec} (electrostatic energy) are
non-bonded energy terms such as Lennard–Jones and Coulomb potential and calculated
between all pairs (i and j). The energy terms V_{bond}, V_{angle}, and V_{torsion} are the bonded interaction which is the sum over the sets of all bonds, angles, and dihedral angles respectively.
The most commonly used one is modeled by a simple harmonic potential function:

$$V_{\text{bond}} = \sum_{\text{bonds}} k_b (b - b_0)^2, \tag{11.2}$$

where k_b is the spring (harmonic) force constant, b is the bond length between atoms, and
b_0 is the equilibrium bond length. Similarly, the bond-angle bending between three consecutively bonded atoms (i, j, k) and the formation of an angle can also be modeled by a
simple harmonic potential function:

$$V_{\text{angle}} = \sum_{\text{angles}} k_\theta (\theta - \theta_0)^2, \tag{11.3}$$

where k_θ is the angle-bending force constant, θ is the actual bond angle between the three
consecutively bonded atoms (i, j, k), and θ_0 is the equilibrium bond angle (Figure 11.3).

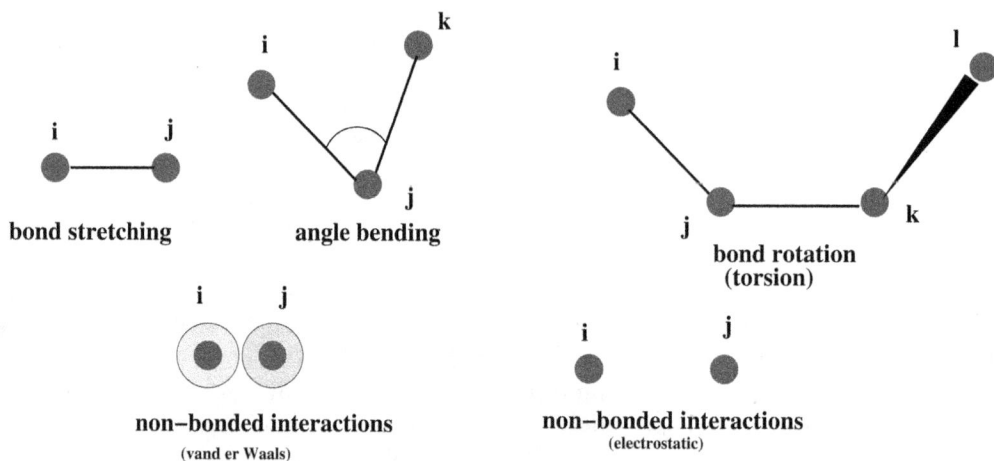

FIGURE 11.3
Schematic representations of bond-stretching, angle-bending, torsion, and non-bonded interactions terms con-
tributing to Equation (11.1).

The dihedral angle potential energy term between four atoms (i, j, k, l) is usually expressed as a cosine series expansion

$$V_{\text{torsion}} = \sum_{\text{dihedrals}} \left(\frac{V_n}{2} \right)(1 + \cos[n\phi + \delta]), \tag{11.4}$$

where V_n refers to the barrier height, ϕ is the torsion angle for the quadruple of atoms i–j–k–l, n is the multiplicity that gives the number of minima in the cosine function, and δ (the phase factor) determines which dihedral angle values correspond to these minima. V_n, n, δ are set for each type of atom quadruplet.

The non-bonded interactions are usually divided into two groups: (i) van der Waals and (ii) electrostatic interactions. The van der Waals interactions between non-bonded atoms are usually represented by a Lennard–Jones potential function:

$$V_{\text{vdw}} = \sum_{\text{nonbij}} \left(\frac{A_{ij}}{r_{ij}^{12}} \right) - \left(\frac{B_{ij}}{r_{ij}^{6}} \right) \tag{11.5}$$

where A_{ij} is the repulsive term coefficient, B_{ij} is the attractive term coefficient, and r_{ij} is the distance between the two atoms i and j.

Nucleic acids have been considered as a challenge for simulation because of the negative backbone charge and the behavior of polyelectrolytes. In simulations, particular attention should be paid to the atomic charges. The electrostatic interactions are generally described by a simple Coulomb potential function:

$$V_{\text{elec}} = \sum_{\text{nonbij}} \left(\frac{q_i q_j}{r_{ij}} \right), \tag{11.6}$$

where q_i and q_j correspond to the atomic charges of interacting atoms i and j, respectively, and r_{ij} corresponds to the distance between the two atoms.

In a typical MD simulation, an initial configuration of the system is obtained from either experimental structures or comparative modeling data, and the particles are allowed to interact under the suitable potential, say by Equation (11.1). The time evolution of a molecular system is determined by solving the following Newton's equation of motion for all atoms,

$$\mathbf{F}_i = m_i \frac{d^2 \mathbf{r}_i}{dt^2} = -\frac{dV}{d\mathbf{r}_i}, \tag{11.7}$$

where $\mathbf{F}_i$ is the force experienced by the ith particle of mass m_i due to the potential energy V of the system. The potential energy, as defined above, is a complicated function, and there is no analytic solution to the equation of motion, and therefore, one has to solve these equations numerically. They all are derived from the Taylor expansion. The expansion for the position $(\mathbf{r})$, velocity $(\mathbf{v})$, and acceleration $(\mathbf{a})$ of the particles at some instant of time t is

$$\mathbf{r}(t + \Delta t) = \mathbf{r}(t) + \mathbf{v}(t)\Delta t + \frac{1}{2}\mathbf{a}(t)\Delta t^2 + \cdots \tag{11.8}$$

$$\mathbf{v}(t + \Delta t) = \mathbf{v}(t) + \mathbf{a}(t)\Delta t + \frac{1}{2}b(t)\Delta t^2 + \cdots \tag{11.9}$$

$$\mathbf{a}(t + \Delta t) = \mathbf{a}(t) + b(t)\Delta t + \cdots, \tag{11.10}$$

where Δt is the time step used in the simulation. There are various algorithms/numerical schemes for integrating Newton's equations of motion [56, 57]. The procedure for a molecular dynamics simulation is subject to many user-defined variables. However, one should consider certain criteria in choosing these algorithms. It should conserve energy, momentum, and be computationally efficient so that long time-step integration can be performed; because the evaluation of the atomic positions is not performed on a continuous basis but at intervals of a time step ($\Delta t \sim$ femtosecond (fs)). Since this is the time scale involved in stretching the bonds with hydrogen atoms, one needs a time step of 2 fs. The algorithm that permits this increase in simulation speed is called SHAKE [56]. Application of constraints on all bond lengths and bond angles would increase the permitted time step even more, without too much loss of information.

It may be noted that the MD is a scheme for studying the time evolution of a classical system of N particles in volume V, where total energy is a constant of motion. As a result, time averages are equivalent to ensemble averages, the observables, say A, evolve according to constant energy simulation, and hence,

$$\langle A \rangle_{N,V,E} = \lim_{\tau \to \infty} \int_0^\tau A(t)dt. \tag{11.11}$$

It is possible to run many NVE simulations with different energies and the result could be used in a Boltzmann-weighted average as

$$\langle A \rangle_{NVT} = \sum_E \langle A \rangle_{NVE}\, e^{-\beta E}. \tag{11.12}$$

Conservation of energy during simulation is a tough task because of the thermal fluctuations. Therefore, one prefers to do the simulation in the constant temperature ensembles, i.e. NVT ensemble (temperature is fixed). There are different thermostats available in the literature (e.g. Berendsen thermostat [65], Langevin thermostat [66], Anderson thermostat [67], etc.). Here, we use the Langevin thermostat, where, at each time step, all particles of the system experience a random force (white noise) originating from stochastic fluctuation and have their velocities influenced by a frictional force. These two factors ensure a constant temperature during simulation. The Langevin equation of motion is described by

$$m\frac{d^2r}{dt^2} = -\zeta\frac{dr}{dt} + F + \Gamma \tag{11.13}$$

F is derived from Equation (11.7) ζ and Γ are the friction coefficient and random force, respectively. This corresponds to the implicit solvent condition, where the contribution of forces arising due to the solvent molecules is taken care of by ζ and Γ. The random force is also related to the friction coefficient by the fluctuation–dissipation [6, 56, 57] relation given by the following expression:

$$\langle \Gamma(t)\Gamma(t') \rangle = 2d\zeta T\delta(t - t') \tag{11.14}$$

Here, $\delta(t - t')$ represents the Dirac Delta function and d is dimension of space. It may be noted that the friction term ζ only influences the kinetic, but not the thermodynamic, properties [6].

These calculations [57] are sometimes carried out under vacuum conditions. A more realistic approach is to use the solvent explicitly. This is done by soaking the molecule in a box of solvent molecules. Several water models are proposed but the TIP3P water model [68] has been used quite frequently. In this water model, the water molecule is defined as a molecule with a rigid triangular geometry having a partial charge at each angle referring to the two hydrogens and the intermediate oxygen atom of a real water molecule. But the explicit description of a water molecule requires additional computational effort. Periodic Boundary Conditions (PBC) are normally employed to model the bulk solvent. In infinite PBC, the simulation box is infinitely replicated in all directions to form a lattice. Since most of the computational time is spent on evaluating the non-bonded interactions between atoms, the evaluation time step for non-bonded atom pairs can be increased to a small extent. Using a cutoff distance beyond which atoms are no longer considered to interact can significantly reduce the amount of non-bonded atom pairs. In these cutoff schemes, each particle interacts with the nearest images of the other N-1 particles or only with those minimum images contained in a sphere of radius R cutoff centered at the particle. Usually, a cutoff distance of less than half the length of the box is used. However, for long-range interactions such as electrostatic interactions, for which the range exceeds half the box, size methods such as Particle Mesh Ewald summation [69] or Ewald summation [70] are used.

11.4 Atomistic Simulations of DNA Rupture

Lee et al. [41] studied the unbinding (rupture) of dsDNA by applying a force along the helix direction i.e. at 5′–5′. The rupture force was found to be one order magnitude greater than the unzipping force. Strunz et al. [42, 43] investigated the unbinding of DNA duplex of various lengths and found that the unbinding force depends on the loading rate and sequence length. It was also found that the unbinding force is different when a force is applied at 5′–5′ and 3′–3′ of DNA [46, 48].

Hatch et al. [45] measured the rupture force of dsDNA for different sequence length. The rupture force is found to increase linearly for small sequence length and saturates for higher sequence length. This is in accordance with the earlier work of de Gennes based on the Ladder Model of DNA [71]. In another study, Chakrabarti and Nelson [72] extended the de Gennes model (nonlinear generalization of the ladder model) and studied the effects of sequence heterogeneity. Mishra et al. [73] considered a homo-sequence of DNA, where the covalent bonds and base-pairing interactions are modeled by the harmonic spring and Lennard-Jones (LJ) potential, respectively. Using Langevin dynamics (LD) simulations [73], they obtained the distribution of stretching of hydrogen bonds and the extension in the covalent bonds for a wide range of forces below the rupture, which are consistent with the Ladder model.

In this section, we shall perform atomistic simulations with the explicit solvent to study the rupture event of the base sequence used in the recent experiment [45]. More specifically, here, we are interested in the distribution of extension in the hydrogen and covalent bonds of dsDNA along the chain at the semi-microscopic level, which is experimentally difficult to obtain.

We have used AMBER10 software package [63] with all-atom (ff99SB) force field [74] to simulate the rupture event of DNA. A force routine has been added in AMBER10 to do simulation at constant force [75, 76]. In this case, the force has been applied at 5′–5′ ends

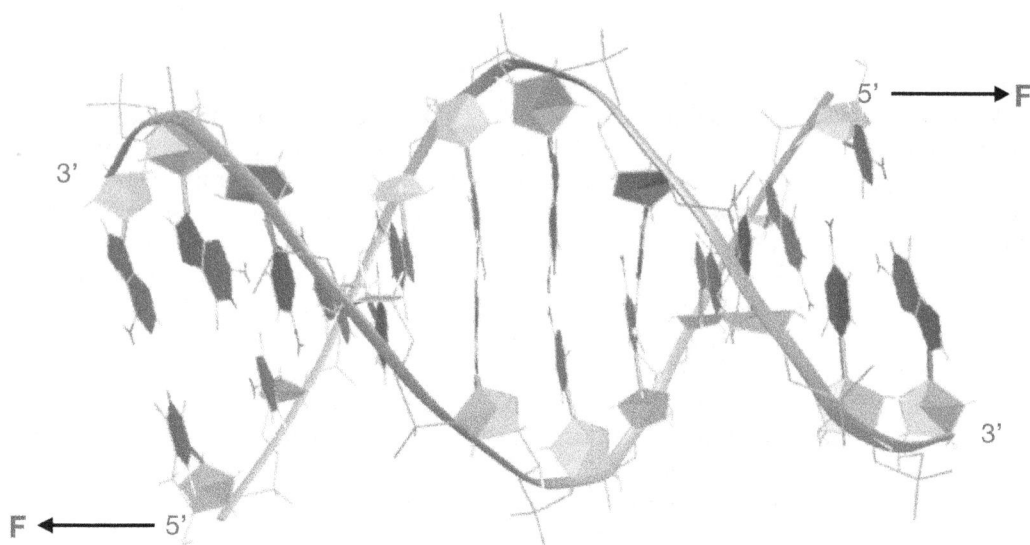

FIGURE 11.4
Schematic representation of dsDNA under the shear force applied at 5'–5' ends.

as shown in Figure 11.4. The electrostatic interactions have been calculated with Particle Mesh Ewald (PME) method [70, 77] using a cubic B-spline interpolation of order 4 and a 10^{-5} tolerance is set for the direct space sum cutoff. A real space cutoff of 10 Å is used for both the van der Waal and the electrostatic interactions.

The starting structure of the DNA duplex sequence (GTCACCTTAGAC) is built using the Nucleic Acid Builder (NAB) module of the AMBER10 suit of programs. Using the LEaP module in AMBER, we add the Na^+ (counterions) to neutralize the negative charges on the phosphate backbone group of DNA structure. This neutralized DNA structure is immersed in a water box using the TIP3P model for water [68]. We have chosen the box dimension in such a way that the ruptured DNA structure is fully inside the water box. For the 12 base-pairs sequence, we have taken the box size of $55 \times 56 \times 199$ Å^3 which contains 16690 water molecules and 22 Na^+ (counterions). The system is equilibrated at $F=0$ for 100 ps under a protocol described in Ref. [7, 8]. We carried out simulations in the isothermal-isobaric (NPT) ensemble using a time step of 1 fs. We maintain the constant pressure by isotropic position scaling [63] with a reference pressure of 1 atm and a relaxation time of 2 ps. Constant temperature was maintained at 320 K using a Langevin thermostat with a collision frequency of 1 ps^{-1}. We have used 3D periodic boundary conditions during the simulation.

To simulate the stretching of hydrogen bonds, we give sufficient time for equilibrium at a constant force. The magnitude of the applied constant force is 570 pN for the 12 base-pairs, which is sufficient enough for separating both strands of dsDNA. To have a better understanding, we have monitored the deformations in DNA at different instants of time. In Figure 11.5, we have shown some of the snapshots of the conformation (generated by visual molecular dynamics (VMD) software [78]) under constant shear force applied at 5'–5' ends at a temperature 320 K. Initially, the dsDNA remains in the zipped state as shown in Figure 11.5(a). As time passes, the dsDNA acquires the

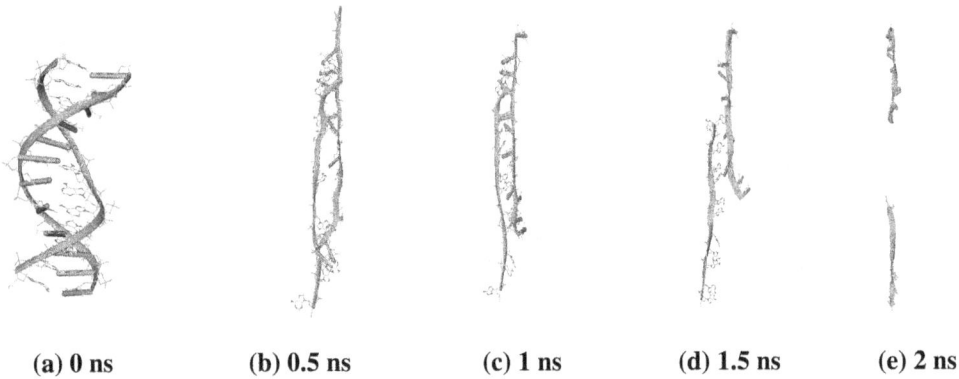

(a) 0 ns   (b) 0.5 ns   (c) 1 ns   (d) 1.5 ns   (e) 2 ns

FIGURE 11.5
Snapshots (generated using VMD software) of dsDNA ($N=12$) under constant shear force applied at $5'$–$5'$ ends at $T = 320$ K. They are taken at different times: (a) 0 ns, (b) 0.5 ns, (c) 1.0 ns, (d) 1.5 ns, and (e) 2.0 ns. We have not shown water molecules and counter-ions in the snapshots for clarity.

conformation to a ladder form (Figure 11.5(b)) and then complete rupture takes place (Figure 11.5(e)).

Figure 11.6(a) shows the variation of extension in hydrogen bond length (Δ_h) and covalent bond length (Δ_c) (i.e. deviation from their mean length) along the chain with base position. We have monitored the distance of the C4$'$ atom of complementary bases in dsDNA to measure the extension in hydrogen bond length. We have studied the system just before the rupture. Simulation results show the asymmetry in the distribution of stretching of the hydrogen bond (Figure 11.6(a)) and covalent bonds (Figure 11.6(b)).

In order to see whether this asymmetry is because of heterogeneity of the sequence, we have repeated the simulation for a designed homo-sequence

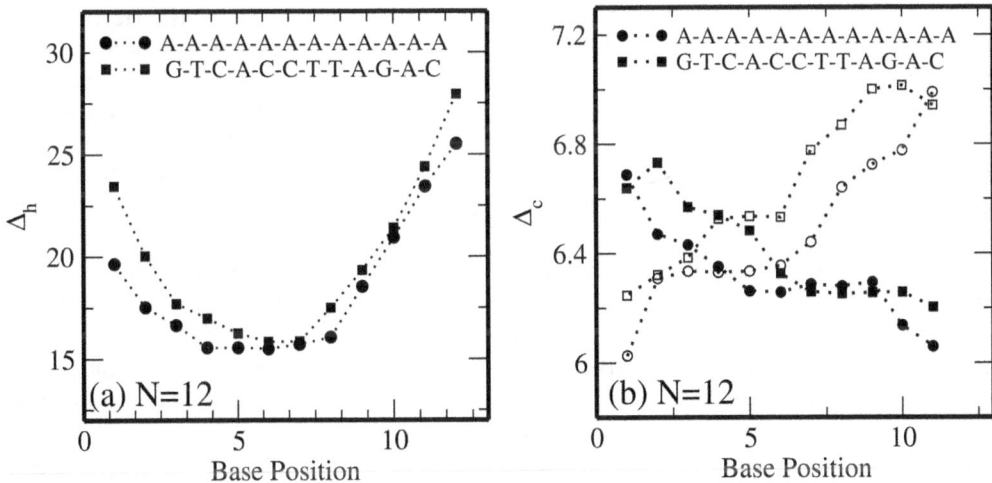

FIGURE 11.6
(a) shows the variation of extension in hydrogen bond length (Δ_h) along the chain with the base position for the chain of 12 base-pairs of the designed and experimental sequence; (b) the same as (a) but for the extension in covalent bonds (Δ_c) along the chain length. Here, open and filled symbols correspond to one strand and its complementary strand, respectively. The dotted lines are a guide for the eye.

(A-A-A-A-A-A-A-A-A-A-A-A) of the same length. The rupture force for this sequence is about 460 pN. It may be noted that the simulation carried out by Mishra et al. [73] or the analytical solution proposed by de Gennes [71], showed that distribution is symmetric for a homo-sequence chain. Surprisingly, even for a homo-sequence, one can notice that the extension in hydrogen bonds is more stretched at the pulling end which consists of thymine than the end consists of adenine in the atomistic simulations. This is due to the stacking energy involved in intra-strand bases in adenine, which is absent in the thymine. The distribution of extension in covalent bonds along the chain also shows the asymmetry.

11.5 Stability of a DNA Aptamer and Its Use as a Drug Delivery Agent

Nguyen et al. [79] measured the change in rupture force of a DNA aptamer (that forms binding pockets) with AMP and without AMP, and thereby determined the dissociation constant at a single-molecule level. In doing the experiment, they formed a DNA aptamer with two base pockets that bind two AMP molecules. These two AMP molecules formed the eight additional hydrogen bonds with the non-bonded base-pairs inside the loop and enhanced the stability of the aptamer molecule. This experiment showed that the stability of DNA aptamers increases in the presence of AMP. Papamichael et al. [80] used an aptamer-coated probe and an IgE-coated mica surface to identify specific binding areas. Efforts have also been made to determine the rupture force of aptamer binding to proteins and cells, and it was revealed that the binding pocket enhances the rupture force by many folds [81–83].

Mishra et al. [84] have studied the rupture of DNA in the presence and absence of AMP at the different base-pocket positions along the chain. They used the coarse-grained model of DNA and employed the Langevin dynamics simulations to measure the rupture force as a function of base-pocket position. They observed that the rupture force remains invariant with the binding-pocket position in the presence of AMP (Figure 11.7(a)). In the absence of AMP, the rupture force first decreases and then increases with base pocket positions, and the rupture force profile looks like a "W-shape," i.e. it has two minima (Figure 11.7(b)), where the rupture can take place at a lower force. It is desirable to know whether it is the artifact of the coarse-grained model of DNA or if it is the characteristic of the DNA aptamer.

In order to check whether this effect is a consequence of the adopted model or not, we performed the atomistic simulations with an explicit solvent. Here, we have taken a homo-sequence DNA consisting of 12 G–C bps and a varying base-pocket of 8 G-nucleotides [84]. The starting structure of the DNA duplex sequence having the base pocket is built using make-na server [85]. We used the AMBER10 software package, whose details have been discussed in the previous section. Using the LEaP module in AMBER, we add the AMP molecule (Figure 11.8(a)) in the base-pocket (Figure 11.8(b) and then the Na^+ (counter-ions) to neutralize the negative charges on the phosphate backbone group of DNA structure. This neutralized DNA aptamer structure is immersed in the water box using the TIP3P model for water. We have chosen the box dimensions in such a way that the ruptured DNA aptamer structure remains fully inside the water box. We have taken the box size of $57 \times 56 \times 183$ Å, which contains 15674 water molecules and $30Na^+$ counter-ions. Like the

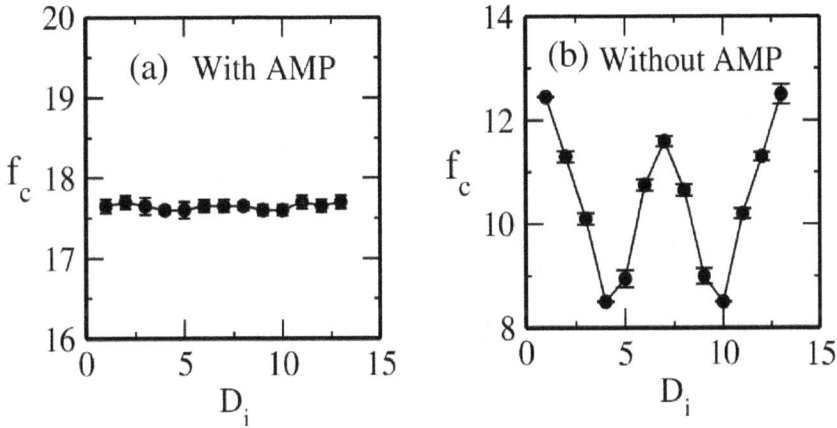

FIGURE 11.7
Variation of rupture force as a function of the base positions (a) in the presence of AMP; (b) in the absence of AMP.

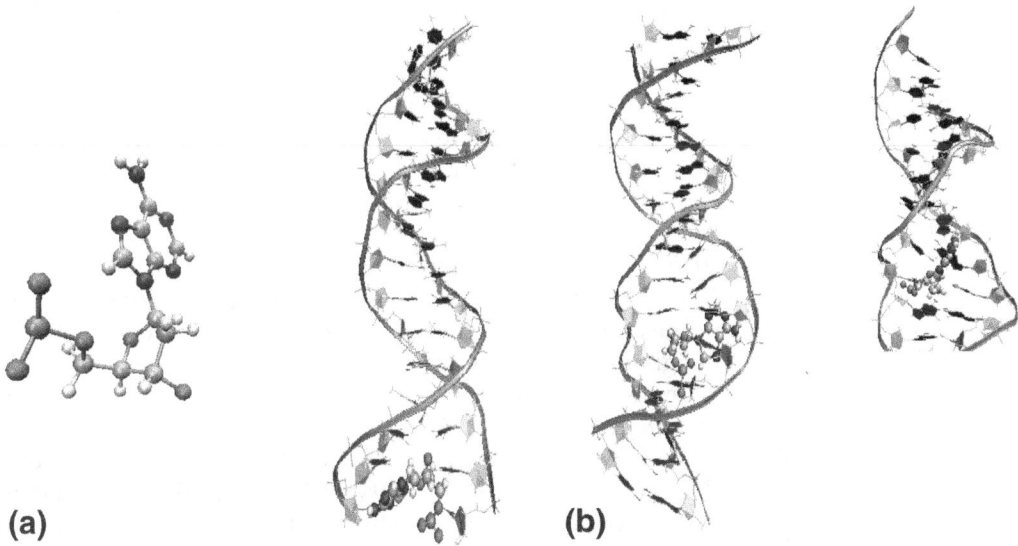

(a)   **(b)**

FIGURE 11.8
(a) The structure of the AMP molecule, which has been inserted in the base pocket; (b) Snapshots of binding of AMP molecule at three different positions ($D_i = 1, 4$ and 7) in the DNA.

DNA rupture discussed in the previous section, here we also add a force routine in the AMBER10 to do simulation at constant force [75, 76]. In this case, the force has been applied at 5′–5′-ends. The other details were same as described in the previous section. The system is equilibrated at $F = 0$ for 100 ps under a protocol described in Refs. [75, 76]. It has been ensured that AMP has bound with the base-pocket. We carried out simulations in the isothermal-isobaric (NPT) ensemble using a time step of 1 fs for 10 different realizations. The constant temperature was maintained at 300 K using a Langevin thermostat with a collision frequency of 1 ps^{-1}. We have used 3D periodic boundary conditions during the simulation.

Because of the extensive time involved in the all-atoms simulations, we restricted ourselves at three different (extreme) base-pocket positions $D_t = 1$, 4 and 7 (Figure 11.8(b)) and calculated the rupture force with 10 realizations of different seeds as a mean force [86]. The required rupture forces for these positions are 840 pN $\pm$ 20 pN, 720 pN $\pm$ 20 pN, and 830 pN $\pm$ 20 pN, indicating that the complete profile contains two minima [87]. In the presence of AMP, which interacts with the base-pocket, we find $f_c = 925$ pN $\pm$ 20 pN remains constant irrespective of binding-pocket position, as seen earlier (Figure 11.7(a)) [84]. This validates the finding of the simple coarse-grained model, which captured some essential but unexplored aspects of rupture mechanism of DNA aptamer.

11.6 Stability of DNA-ICLs against Force and Its Use as an Anti-Tumor Drug

ICL agents can form a number of adducts with DNA. Because of the chromatin structure, these adducts are not always produced randomly throughout the genome [88–92]. The three-dimensional structures of the DNA-ICLs are variable, which could influence the efficiency of ICL recognition and repair. Mitomycin C is one of the ICLs which has been used as a natural anti-tumor antibiotic (Figure 11.9). It forms adducts at the N-7 and N-2 of guanine intra-strand cross-links, and ICLs between the N-2 of guanines at d(CpG) sequences in the minor groove [89]. On the other hand, Trabectedin forms monoadducts in the minor groove (Figure 11.9) that increases the melting temperature of selected DNA oligonucleotides [90–92]. Juan et al. [93] have studied the thermal stabilization of different DNA-ICLs, e.g. Mitomycin C, mono-functional tetrahydroisoquinolines (Yondelis), Zalypsis, and PM01183. Using atomistic simulations, they analyzed the thermal denatuaration of these ICLs. However, because of a large molecular size, their simulations were limited to <200 ns at temperature 400 K; such a high temperature is unrealistic for living systems.

The aim of this section is to explore the stability of DNA-ICLs at room temperature. For this, we first choose a 15-mer sequence of TAATAACGGATTATT built by using the NAB module in the AMBER suite of programs without any ICLs agent [63]. The initial structure of dsDNA (TAATAACGGATTATT) is built by the NAB (Nucleic Acid Builder) module in the AMBER suite of programs without any ICL agent. The structure of ICLs (Trabectedin and Mitomycin C) have been taken from Protein Data Bank (PDB). The data file for ICLs has been used from the Supplementary material provided in Ref. [93]. Using the Xleap (module in the AMBER tool), we build the initial structure of dsDNA-ICLs (covalent bond between guanine of dsDNA and ICLs) and set the modified parameter (charge, atom type, etc). Finally, we build the initial coordinate file and parameter topology file, which were used in the calculation of force fields. These molecular systems are immersed in a water box using the TIP3P model for water [68]. Using the LEaP module in AMBER, 27 Na$^+$ (counter-ions) are added in the system to neutralize the negative charges on the phosphate backbone group of the DNA-ICL structure. We have chosen the box dimension in such a way that the ruptured/unzipped structure always remains inside the water box. In the case of rupture, we have taken the box size of $48 \times 55 \times 189$ Å^3, which contains 39918 atoms, while for unzipping, the box size is $188 \times 55 \times 78$ Å^3, containing 67536 atoms. Other simulation details were the same as those described in the Section 11.4.

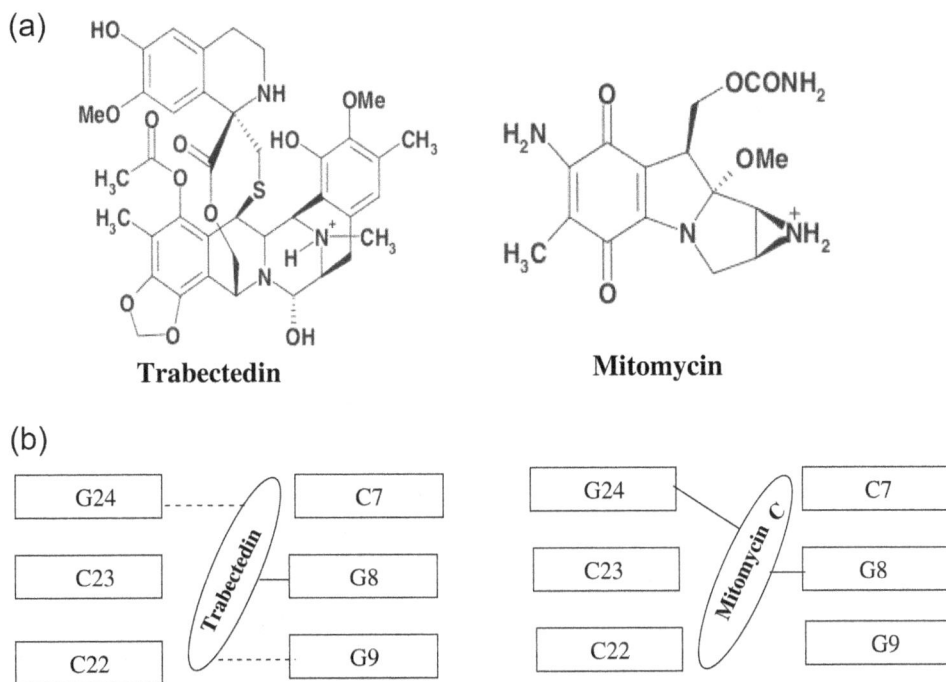

(a)

Trabectedin

Mitomycin

(b)

FIGURE 11.9

(a) The chemical structure of the N-protonated form of the drugs (Trabectedin and Mitomycin C) used as intrastrand cross-links used to study force-induced transitions in DNA-ICLs; (b) Schematic representations of binding modes of the Trabectedin and Mitomycin C to a central CGG triplet of the DNA. Base numbers correspond to their position in the 15-mer: the strand bearing the adduct = 1–15, whereas the complementary strand = 16–30. Here, the solid and dashed lines between the two strands correspond to the covalent bond and the hydrogen bond, respectively.

We have studied the force-induced separation of dsDNA in the absence and presence of ICL agents. The dynamics of the system have been studied using atomistic simulation (AMBER10 software package) with an all-atoms (parmbsc0) force field [74]. A force routine has been added in AMBER10 for simulation under constant force conditions [75, 76]. This allowed us to conduct single-molecule experiments *in silico* and observe biomolecular mechanics in action. We have carried out simulations in the isothermal-isobaric (NPT) ensemble using a time step of 1 fs. We maintain the constant pressure by isotropic position scaling with a reference pressure of 1 atm and a relaxation time of 2 ps. A constant temperature is maintained at 300 K using a Langevin thermostat with a collision frequency of 1 ps^{-1} [56].

We have used three-dimensional periodic boundary conditions during the simulation. The external force started at 0 pN and increased linearly with time steps depending on the forcing rate until the DNA separates completely. The loading rate used in these studies is kept fixed at 0.0001 pN fs^{-1}. Three-dimensional structures and trajectories were visually studied using the computer graphics program, with visual molecular dynamics (VMD) software [78]. Because of the computational limitations, we could not employ a pulling rate slower than 0.0001 pN fs^{-1}. As a result, the rupture force of DNA is seen here to be an order of magnitude higher than the one obtained in the experiments. The relationship of simulated, and measured, rupture forces has been discussed extensively by Sotomayor and Schulten [94].

In the following we discuss two cases: (a) a force is applied along the direction of the helix at 5′–5′ ends (Figure 11.10(a)) to study the rupture of DNA, (b) a force is applied perpendicular to the helical direction at 5′–3′ ends (Figure 11.10(b)) to study DNA unzipping. In order to have a better sampling of force-induced DNA strand separation, we carried out five independent MD simulations for DNA, MMC-DNA, and t-DNA at a constant loading rate (or a pulling velocity) of 0.0001 pN ps^{-1}.

11.6.1 Rupture of DNA and DNA-ICLs

We first examine rupture of the ICLs-free force-extension curve of DNA and compare it with DNA-ICLs complex. We have depicted five snapshots of ICLs-free DNA, t-DNA, and MMC-DNA at different forces. As the force increases, the DNA is stretched and undergoes a series of conformational transitions before the two strands get separated (Figure 11.11). In Figure 11.12, we plotted the force-extension (f–x) curves of 15-mer dsDNA in the absence and presence of ICLs. At low force (below 750 pN), the pathways of free DNA, t-DNA, and MMC-DNA almost overlap each other. Stretching of DNA first unwinds the DNA helix (200 pN) and then DNA acquires a planner structure like a ladder form (400 pN) before the hydrogen bonds break (600 pN). Around 750 pN, the free DNA separates completely, whereas t-DNA and MMC-DNA remain intact. At 800 pN, both strands of t-DNA and MMC-DNA remain in contact with the central CGG triplet linked with ICLs. These

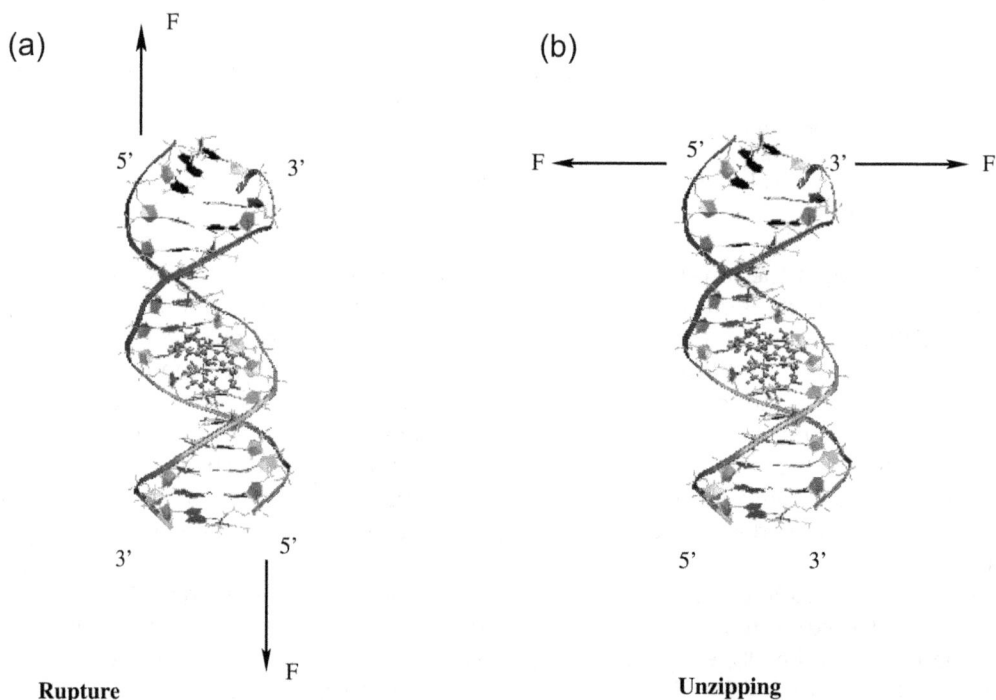

FIGURE 11.10
(a) Schematic representations of DNA-ICLs under the shear force applied at 5′–5′ (along the helix direction) ends; (b) the unzipping force applied at 5′–3′ (perpendicular to helical direction) ends.

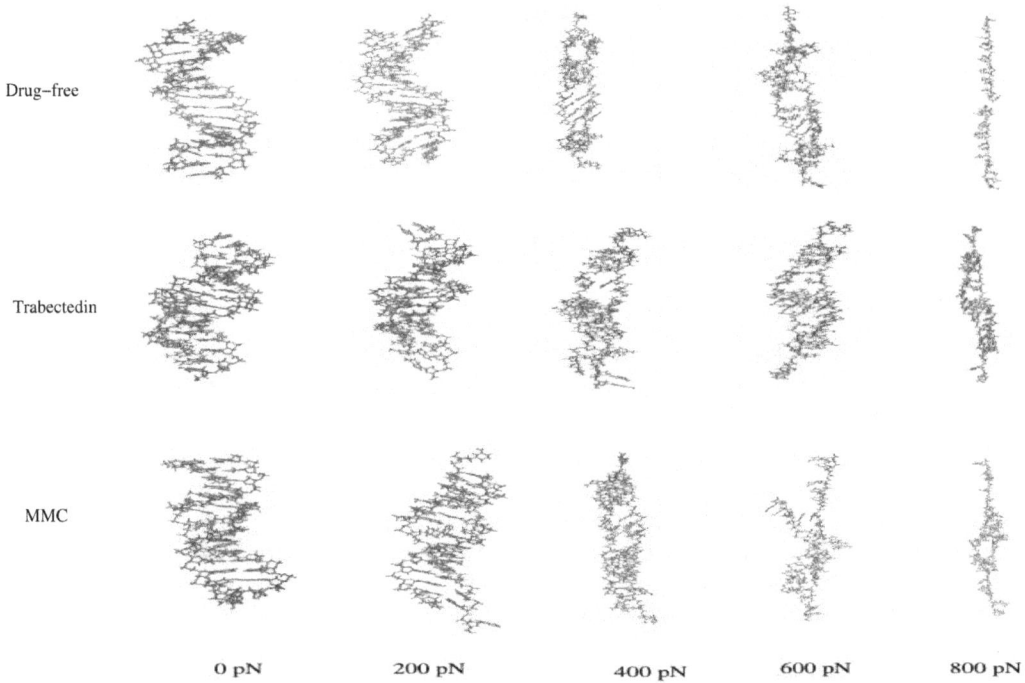

FIGURE 11.11
Instantaneous snapshots of DNA in the absence of an anti-tumor drug (top) and in the presence of anti-tumor drug Trabectedin (covalently bonded to one strand and forming an H-bond to the complementary strand) and MMC (covalently bonded to both strands) at different shearing forces applied at 5′–5′ ends. For clarity, water molecules and ions are not shown. These snapshots have been generated using VMD software.

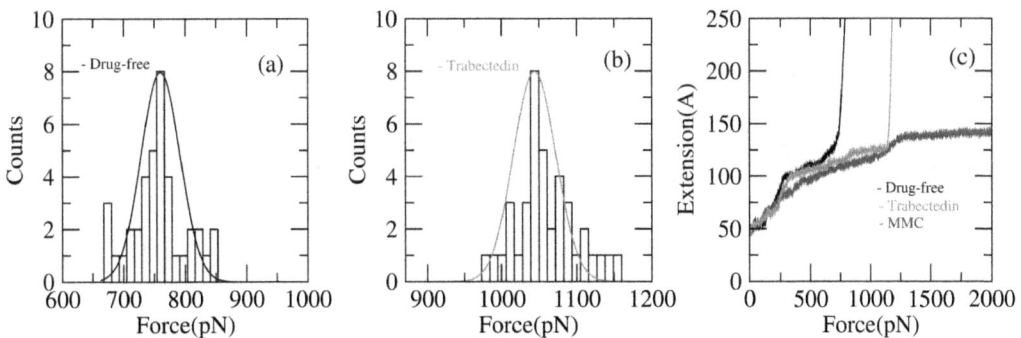

FIGURE 11.12
(a and b) Histograms of rupture forces resulting from N measurements for drug-free DNA and t-DNA, respectively. The peak in the distribution corresponds to the most probable force for the rupture; (c) Force vs. extension curves of 15-mer dsDNA in the absence and presence of anti-tumor drugs (Trabectedin and MMC). The rupture force is applied along the helical direction with a 0.0001 pN fs^{-1} pulling rate at room temperature.

figures indicate that MMC-DNA is more stable compared to t-DNA. This has also been reflected in the force-extension curve, where the extension of MMC-DNA is so large that even if there is too much force, the rupture is not obtained, whereas, t-DNA ruptured at 1044 pN (Figure 11.12(b)). This is because t-DNA is bound with a central CGG triplet with two hydrogen bonds and one covalent bond, whereas MMC-DNA is bound with a central CGG triplet with two covalent bonds. These figures clearly demonstrate that the stability of DNA-ICLs complexes has been enhanced. As a result, DNA-ICLs prevent the replication and transcription processes to cause cell death.

11.6.2 Unzipping of DNA and DNA-ICLs

In order to resolve whether the stability of MMC-DNA is greater, compared to t-DNA, because of two covalent bonds or not, we unzip the ICLs-free DNA and DNA-ICLs (MMC-DNA and t-DNA). Since unzipping force only acts at the center point of the fork, the applied force breaks only one base pair at a time and unzipping progresses gradually. We expect, when the center fork point meets the ICLs, there will be pause. However, for ICLs-free DNA, one should not expect such pause. One can see this in Figure 11.13, and that at 300 pN two strands of dsDNA are completely unzipped, whereas for DNA-ICLs, only half of the chain is unzipped, and the remaining portion of DNA remains intact. Here, one can also see that the relative extension in MMC-DNA is much larger compared to t-DNA.

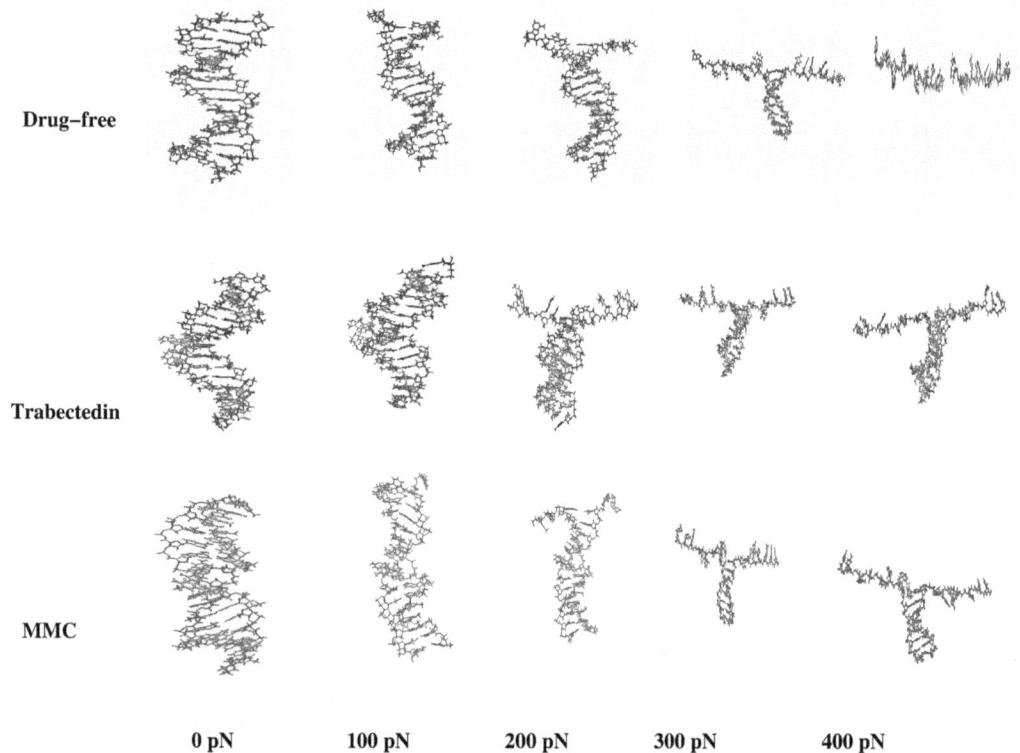

FIGURE 11.13
The same as Figure 11.11 but for different unzipping forces applied at 5′–3′ ends. One can notice that unzipping takes place at lower force compared to rupture for drug-free DNA and similarly for DNA-ICLs.

Around 400 pN, one can see that the extension in t-DNA increases because of the breaking of hydrogen bonds and at 450 pN, two strands separate. Since MMC is bound with two covalent bonds, the unzipping force was found to be as large, compared to t-DNA.

11.7 Conclusions and Future Perspective

It is pertinent to mention here that the force-extension curves of ssDNA (or RNA) consisting of similar type of nucleotides show striking differences [95–97]. It was found that the poly(T) or poly(U) show the entropic response whereas poly(A) show the plateaus arising due to the base stacking interaction. These studies indicate that the elastic constants of complimentary strands are different. However, in the analytical model [71] proposed by de Gennes and the coarse-grained molecular dynamics simulations performed by Mishra et al. [73] the same elastic constant for both strands was considered. Latter, Mishra et al. [86] performed their simulation with different elastic constants of complimentary strands and found that the extension in distribution of base-pairs are asymmetric. This also helped them to modify the Ladder model of dsDNA with different elastic constants, which also showed the asymmetry (Figure 11.14).

It was observed that the stability of the DNA aptamer depends on the base-pocket position of AMP. We have found that in the absence of AMP, the rupture force profile contains two minima while in the presence of AMP, which interacts with the base pocket; the rupture force remains constant irrespective of the binding-pocket position. This may have biological/pharmaceutical significance, because after the release of the drug molecule, the stability of carrier DNA depends on the position from where it is released.

DNA intra-strand cross links (ICLs) agents are widely used in the treatment of cancer. ICLs are thought to form a link between the same strand (intra-strand) or complimentary strand (inter-strand) and thereby increase the stability of DNA, which forbids processes like replication and transcription. As a result, cell death occurs. The atomistic simulations presented here provide an semi-microscopic description of the process of DNA separation in the presence and absence of ICLs. Force-extension curves of unzipping and rupture revealed a clear difference between the response of drug-free DNA and DNA-ICLs.

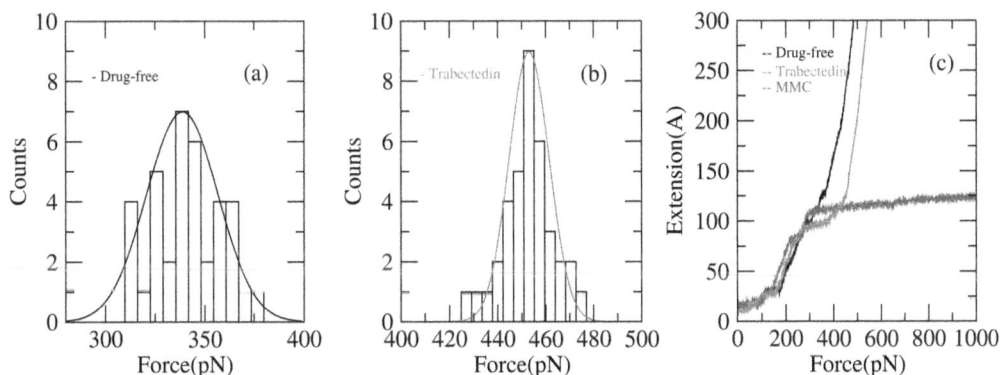

FIGURE 11.14

(a and b) The same as Figure 11.11 but for the unzipping force, which is applied perpendicular to the helical direction with a 0.0001 pN fs^{-1} pulling rate at room temperature (c) Force-extension curves for the unzipping.

In rupture, a drug-free DNA duplex and t-DNA are stretched into a zipper-like conformation followed by the strand separation. Since MMC-DNA has covalent bonds with each strand, the complete separation requires quite high force (>1500 pN) compared to t-DNA ($\approx$900 pN). For t-DNA, unzipping takes place at much lower force ($\approx$450 pN) compared to rupture. This is in agreement with previous studies, where the melting temperature of t-DNA is found to be lower as compared to MMC-DNA [93]. It is pertinent to mention that melting is defined when half of the base pairs open, and the chain can open from both ends, whereas the rupture requires complete separation of strands. Moreover, in the unzipping mode, the chain opens from the end where a force is applied. The other end remains intact. The presence of ICLs at intermediate base positions induces a pause-like event, which is not detected in melting. At this point of time our results warrant further experimental investigations to enhance our understanding about the molecular forces that govern the binding of important anti-tumor compounds to DNA. Information gathered through our simulations should be useful in designing better chemotherapeutic drugs in the future.

Acknowledgments

The financial supports from SERB, India, UGC, India SPARC, MHRD, India and DST, India are gratefully acknowledged. SK would like to thank ICTP, Trieste for his visit under Simons Associateship. This review has been compiled during his visit.

References

1. O'Connor C., 2010. *Essentials of Cell Biology*, NPG Education, Cambridge, MA.
2. Alberts B., 1994. *Molecular Biology of the Cell*, 3rd edition, Garland Press, New York.
3. Wartell R.M. and Benight A., 1985. Thermal denaturation of DNA molecules: A comparison of theory with experiment, *Phys. Rep.* 126, 67–107.
4. Goddard N.L., et al., 2000. Sequence dependent rigidity of single stranded DNA, *Phys. Rev. Lett* 85, 2400.
5. Kricka L.J., 1992. *Nonisotopic DNA Probe Techniques*, Academic Press, New York.
6. Kumar S. and Li M.S., 2010. Biomolecules under mechanical force, *Phys. Rep*, 486, 1.
7. Maiti P.K., et al., 2004. The stability of Seeman JX DNA topoisomers of paranemic crossover (PX) molecules as a function of crossover number, *Nucleic Acids Res*, 32, 6047.
8. Maiti P.K. and Bagchi B., 2006. Structure and dynamics of DNA– dendrimer complexation: Role of counterions, water and base pair sequence, *Nano Lett*, 6, 2478.
9. Mergny J.L. and Lacroix L., 2003. Analysis of thermal melting curves, *Oligonucletides*, 13, 515.
10. Stucki M., et al., 2001. A coordinated interplay:proteins with multiple functions in DNA replication, DNA repair, cell cycle/checkpoint control and transcription, *Prog. Nucleic Acid Res. Mol. Biol*, 66, 261.
11. Brown T. and Brown (Jr) T., *Nucleic Acid Book*, ATDBio Ltd, University of Southhampton, UK (https://www.atdbio.com/nucleic-acids-book).
12. Brown J.M. and Attardi L.D., 2005. The role of apoptosis in cancer development and treatment response, *Nat. Rev. Cancer*, 5, 231.
13. Ray P. and White R.R., 2010. Aptamers for targeted drug delivery, *Pharmaceuticals*, 3, 1761.

14. Zhou J, et al., 2012. Current progress of RNA aptamer-based therapeutics, *Front. Genet*, 3, 234.
15. Cerchia L., et al., 2009. Differential SELEX in human glioma cell lines, *PLOS ONE*, 4, 7971.
16. Tuerk C. and Gold L., 1990. Systematic evolution of ligands by exponential enrichment: RNA ligands to bacteriophage T4 DNA polymerase, *Science*, 249, 505.
17. Ellington A.D. and Szostak J., 1990. In vitro selection of RNA molecules that bind specific ligands, *Nature*, 346, 818.
18. O'Donoghue M.B., et al., 2012. Single-molecule atomic force microscopy on live cells compares aptamer and antibody rupture forces, *Anal. Bioanal. Chem*. 402, 3205.
19. Sarkies P., et al., 2010. Epigenetic instability due to defective replication of structured DNA, *Mol. Cell*. 40, 703.
20. Bejugam M., et al., 2007. Trisubstituted isoalloxazines as a new class of G-quadruplex binding ligands: Small molecule regulation of c-kit oncogene expression, *J. Am. Chem. Soc*. 129, 12926.
21. Keefe A.D., Pai S., and Ellington A., 2010. Aptamers as therapeutics, *Nat. Rev. Drug Discov*. 9, 537.
22. Lin C.H. and Patel D.J., 1997. Structural basis of DNA folding and recognition in an AMP-DNA aptamer complex: Distinct architectures but common recognition motifs for DNA and RNA aptamers complexed to AMP, *Chem. Biol*. 4, 817.
23. Teller C., Shimron S. and Willner I., 2009. Aptamer– dnazyme hairpins for amplified biosensing, *Anal. Chem*. 81, 9114.
24. Nonin-Lecomte S., Lin C.H. and Patel D.J., 2001. Additional hydrogen bonds and base-pair kinetics in the symmetrical AMP-DNA aptamer complex, *Biophys. J*. 81, 3422.
25. Patel D.J., Phan A.J. and Kuryavyi V., 2007. Human telomere, oncogenic promoter and 5'-UTR G-quadruplexes: Diverse higher order DNA and RNA targets for cancer therapeutics, *Nucleic Acids Res*. 35, 7429.
26. Sullenger B.A. and Gilboa E., 2002. *Nature 418*, 252.
27. Hoppe-Seyler F. and Butz K., 2000. Peptide aptamers: Powerful new tools for molecular medicine, *J. Mol. Med*. 78, 426.
28. Cho E.J., Lee J.W. and Ellington A.D., 2009. Applications of aptamers as sensors, *Annu. Rev. Ana. Chem*. 2, 241.
29. Chen J., et al., 2012. Computational lateral flow biosensor for proteins and small molecules: A new class of strip logic gates, *Anal. Chem*. 84, 6321.
30. Yangyuoru P.M., et al., 2012. Single-molecule measurements of the binding between small molecules and DNA aptamers, *Anal. Chem*. 84, 5298.
31. Zayats M., et al., 2006. Label-free and reagentless Aptamer-based sensors for small molecules, *J. Am. Chem Soc*. 128, 13666.
32. Zuo X., Xiao Y. and Plaxco K.W., 2009. High specificity, electrochemical sandwich assays based on single aptamer sequences and suitable for the direct detection of small-molecule targets in blood and other complex matrices, *J. Am. Chem Soc*. 131, 6944.
33. Olcay B., et al., 2013. Dimeric DNA aptamer complexes for high-capacity-targeted drug delivery using pH-sensitive covalent linkages, *MTNA*. 2, e107.
34. Sun H., et al., 2014. Oligonucleotide aptamers: New tools for targeted cancer therapy, *MTNA*. 3, e182.
35. Zhu G., et al., 2013. Self-assembled, aptamer-tethered DNA nanotrains for targeted transport of molecular drugs in cancer theranostics, *PNAS*. 110, 7998.
36. Li L., et al., 2014. Aptamer photoregulation in vivo, *PNAS*. 111, 17099.
37. Israelachvili J.N., 1992. *Intermolecular and Surface Forces (With Applications to Colloidal and Biological Systems)*, Academic Press, London.
38. Bustamante C., et al., 2004. Mechanical processes in biochemistry, *Annu. Rev. Biochem*. 73, 705.
39. Essevaz-Roulet B., Bockelmann U., and Heslot F., 1997. Mechanical separation of the complementary strands of dsDNA, *PNAS*. 94, 11935.
40. Bockelmann U., Essevaz-Roulet B. and Heslot F., 1997. Molecular stick-slip motion revealed by opening DNA with piconewton forces, *Phys. Rev. Lett*. 79, 4489.
41. Lee G.U., Chrisey L.A. and Colton R.J., 1994. Direct measurement of the forces between complementary strands of DNA. *Science*. 266, 771.

42. Strunz T., et al., 1999. Dynamic force spectroscopy of single DNA molecules, *PNAS*. 96, 11277.
43. Schumakovitch I., et al., 2002. Hegner temperature dependence of unbinding forces between complementary DNA strands, *Biophys J*. 82, 517.
44. Danilowicz C., et al., 2004. Measurement of the phase diagram of DNA unzipping in the temperature-force plane, *Phys. Rev. Lett.* 93, 078101.
45. Hatch K., et al., 2008. Demonstration that the shear force required to separate short double-stranded DNA does not increase significantly with sequence length for sequences longer than 25 base pairs, *Phys. Rev. E*. 78, 011920.
46. Danilowicz C., et al., 2009. The structure of DNA overstretched from the ends differs from the structure of DNA overstretched from the ends, *PNAS*. 106, 13196.
47. Küeher F., et al., 2007. Force-induced DNA slippage, *Biophys. J*. 92, 2491.
48. Naserrian-Nik A.M., Tahani M. and Karttunen M., 2013. Pulling of double-stranded DNA by atomic force microscopy: A simulation in atomistic details, *RSC Advances*. 3, 10516.
49. Saenger W., 1984. *Principles of Nucleic Acid Structure*, Springer-Verlag, New York.
50. Butler J.M., 2001. *Forensic DNA Typing: Biology and Technology Behind STR Markers*, Academic Press, London.
51. Bruce A., et al., 2002. *Molecular Biology of the Cell*, 4th edition, Garland Science, New York/London.
52. Wing R., et al., 1980. Crystal structure analysis of a complete turn of B-DNA, *Nature*. 287, 755.
53. Watson J.D. and Crick F.H.C., 1953. Molecular structure of nucleic acids: A structure for deoxyribose nucleic acid, *Nature*. 171, 737.
54. Singer M. and Berg P., 1994. *Genes and Genomes: A Changing Perspective* University Science Books.
55. Patel S.S. and Donmez I., 2006. Mechanisms of helicases, *J. Bio. Chem*. 281, 18265.
56. Frenkel D. and Smit B., 2002. *Understanding Molecular Simulation*, Academic Press, London.
57. Allen M.P. and Tildesley T.J., 1987. *Computer Simulations of Liquids*, Oxford Science, Oxford, UK.
58. Adcock S.A. and McCammon J.A., 2006. Molecular dynamics: Survey of methods for simulating the activity of proteins, *Chem. Rev*. 106, 1589.
59. Brooks B.R., et al., 1983. CHARMM: A program for macromolecular energy, minimization, and dynamics calculations, *J. Comp. Chem*. 4, 187.
60. Phillips J.C., et al., 2005. Scalable molecular dynamics with NAMD, *J. Comp. Chem*. 26, 1781.
61. Gunsteren W.F.V., et al., 1996. *Biomolecular Simulation: The Gromos96 Manual and User Guide*, vdf Hochschulverlag AG an der ETH, Zurich.
62. Grubmuller H., Heymann B. and Tavan P., 1996. Ligand binding: Molecular mechanics calculation of the streptavidin–biotin rupture force, *Science*. 271, 997.
63. Case D.A., et al., 2008. *AMBER 10*, University of California, San Francisco, CA.
64. Cheatham T.E. III and Case D.A., 2013. Twenty-five years of nucleic acid simulations, *Biopolymers*. 99, 969.
65. Berendsen H., 1991. Transport Properties Computed by Linear Response through Weak Coupling to a Bath, *Nato Science Series E: Computer Simulations in Material Science Editors: Meyer M. and Pontikis V*, p. 139. Springer, Switzerland.
66. Reichl L.E., 2004. *A Modern Course in Statistical Physics*, 2nd edition, Wiley VCH, Wenhheim, Germany.
67. Andersen H.C., 1980. Molecular dynamics simulations at constant pressure and/or temperature, *J. Chem. Phys*. 72, 2384.
68. Jorgenson W.L., 1983. Comparison of simple potential functions for simulating liquid water, *J. Chem. Phys*. 79, 926.
69. Ewald P.P., 1921. Die Berechnung optischer und elektrostatischer Gitterpotentiale, *Ann. Phys*. 64, 253.
70. Darden T., York D., Pedersen L., 1993. Particle mesh Ewald: An N:log(N) method for Ewald sums in large systems, *J. Chem Phys*. 98, 10089.
71. de Gennes P.G., 2001. Maximum pull out force on DNA hybrids force maximum de tirage sur un ADN hybride, *C. R. Acad. Sci.-Ser. IV- Phys*. 2, 1505.

72. Chakrabarti B. and Nelson D., 2009. Shear unzipping of DNA, *J. Phys. Chem. B.* 113, 3831.
73. Mishra R.K., et al., 2011. *Phys. Rev. E* . 84, 032903
74. Duan Y., et al., 2003. A point-charge force field for molecular mechanics simulations of proteins based on condensed-phase quantum mechanical calculations. *J. Comput. Chem.* 24, 1999.
75. Santosh M. and Maiti P.K., 2009. Force induced DNA melting, *J. Phys.: Condens. Matter.* 21, 034113.
76. We used the constant force routine developed by P. K. Maiti and his group.
77. Essmann U., et al., 1995. A smooth particle mesh Ewald method. *J. Chem. Phys.* 103, 8577.
78. Humphrey W., Dalke A. and Schulten K., 1996. VMD: Visual molecular dynamics, *J. Mol. Graphics.* 14, 33.
79. Nguyen T.H., et al., 2011. Measuring single small molecule binding via rupture forces of a split aptamer, *J. Am. Chem. Soc.* 133, 2025.
80. Papamichael I.K., Kreuzer P.M and Guilbault G., 2007. Viability of allergy (IgE) detection using an alternative aptamer receptor and electrochemical means , *Sens. Actuators B.* 121, 178.
81. Zlatanova J., Lindsay M.S. and Leuba H.S., 2000. Single molecule force spectroscopy in biology using the atomic force microscope, *Prog. Biophys. Mol. Biol.* 74, 37.
82. Yu J.P., et al., 2007. Energy landscape of aptamer/protein complexes studied by Single–molecule force spectroscopy, *Chem. Asian J.* 2, 284.
83. Jiang Y., et al., 2003. Specific aptamer-protein interaction studied by atomic force microscopy, *Anal. Chem.* 75, 2112.
84. Mirsha R.K., et al., 2015. Rupture of DNA aptamer: New insights from simulations, *J. Chem. Phys.* 143, 164902.
85. Shuxiang Li, Wilma K Olson, Xiang-Jun Lu. 2019. Web 3DNA 2.0 for the analysis, visualization, and modeling of 3D nucleic acid structures. *Nucleic Acids Research*, 47, W26–W34. https://doi.org/10.1093/nar/gkz394.
86. Nath S., et al., 2013. Statistical mechanics of DNA rupture: Theory and simulations, *J. Chem. Phys.* 139, 165101.
87. Here, the simulation time is about 7 ns, and hence rupture force is found to be one order higher than AFM.
88. Feuerhahn S., et al., 2011. XPF-dependent DNA breaks and RNA polymerase II arrest induced by antitumor DNA interstrand crosslinking-mimetic Alkaloids, *Chem. Biol.* 18, 988.
89. Tomasz M., 1995. Mitomycin C: Small, fast and deadly (but very selective), *Chem. Biol.* 2, 575.
90. Casado J.A., et al., 2008. Relevance of the Fanconi anemia pathway in the response of human cells to trabectedin, *Mol. Cancer Ther.* 7, 1309.
91. Marco E., et al., 2006. Further insight into the DNA recognition mechanism of trabectedin from the differential affinity of its demethylated analogue ecteinascidin ET729 for the triplet DNA binding site CGA, *J. Med. Chem.* 49, 6925.
92. D'Incalci M. and Galmarini C.M., 2010. A review of trabectedin (ET-743): A unique mechanism of action, *Mol. Cancer Ther.* 9, 2157.
93. Bueren-Calabuig J.A., et al., 2011. Temperature-induced melting of double-stranded DNA in the absence and presence of covalently bonded antitumour drugs: Insight from molecular dynamics simulations, *NAR.* 39, 8248.
94. Sotomayor M. and Schulten K., 2007. Single-molecule experiments in vitro and in silico, *Science.* 316, 1144.
95. Ke C., et al., 2007. Direct measurements of base stacking interactions in DNA by single-molecule atomic force spectroscopy, *Phys. Rev. Lett.* 99, 018302.
96. Seol Y., et al., 2007. Stretching of homopolymeric RNA reveals single-stranded helices and base-stacking, *Phys. Rev. Lett.* 98, 158103.
97. Mishra G., Giri D. and Kumar S., 2009. Stretching of a single-stranded DNA: Evidence for structural transition, *Phys. Rev. E.* 79, 031930.

12

A Graphical User Interface for Palmprint Recognition

Rohit Khokher and R. C. Singh

CONTENTS

12.1 Introduction 195
12.2 System Development 197
    12.2.1 Enrolment 197
    12.2.2 Verification 198
    12.2.3 GUI of Palmprint Recognition 198
12.3 Proposed System Functions 199
    12.3.1 Panel 1: Basic Operations 199
    12.3.2 Panel 2: Histogram 199
    12.3.3 Panel 3: Texture_F 200
    12.3.4 Panel 4: Phase_S_F 200
    12.3.5 Panel 5: Similarity_M 200
12.4 Results and Discussion 201
    12.4.1 Texture Features 201
    12.4.2 Phase and Phase Symmetry 202
    12.4.3 Similarity Measures 204
12.5 Conclusion 205
References 206

12.1 Introduction

User identification should be a requirement for many applications such as access control, aviation, online banking, e-commerce, national identity card, forensic investigations, etc. which should be safe and secure. Maintaining and managing access while protecting both the user's identity and the data has become a challenge. To resolve such an issue, biometric systems, which use distinct behavioral or physical traits of human beings, have recently become more popular because of their robustness and high accuracy. In recent decades, researchers have extensively studied several biometric features, such as, among others, fingerprints [1–4], iris [5–8], face [9–12], footprint [13–15], palmprint [16–20], etc. The palm print has been clearly noted because it is difficult to interfere with and data is easily gathered. There are two types of skin pattern on the inner surface of the palm, i.e., ridges and valleys. As a prominent component of biometric traits, the palmprint is interesting and consists of many characteristics like high articulateness, durability, and user-compatibility etc. In fact, the first use of palmprint for personal recognition was found in Chinese sales deeds in the sixteenth century [21, 22].

In the early 1950s, the first automated palmprint recognition system was made available [17]. Since then, researchers have made numerous efforts to improve the productivity and efficiency of palmprint recognition systems. In the initial stage, the ink palm was used for offline recognition [17]. The first online palmprint recognition system was proposed by Zhang et al. [18], in which 2D reduced palmprint pictures were used, taken by contact mode, to recognize an individual. It was found that principal lines and creases could be easily seen in palmprint images that were captured at 75dpi. Zhang et al. [18] also proposed the first 3D palmprint recognition system that has greater accuracy in recognizing an individual compared to a 2D system. In recent years, researchers have focused more on recognizing high-resolution palmprints. When a palmprint image is taken at 400–500dpi, features like ridges, minutiae, and pores could be extracted from it for matching purposes [22, 23]. In comparison to fingerprint, iris, and face biometric traits, palmprint has many advantages such as stability, ease of capture, and richness of features. Researchers have proposed several effective palmprint recognition methods [24, 25] that can be divided into various categories, like texture-based [26], line-based [27], based on subspace learning [28], based on correlation filters [29], based on local descriptors [30], and based on orientation coding [31]. In texture-based methods, Gabor wavelet [32], fractal dimension [33], dual-tree complex wavelets [34], discrete cosine transform [35], region covariance matrices [36], co-occurrence matrices [37], discrete orthonormal S-transform [38], and other statistical methods have been used for extracting texture features from palmprints. Line-based methods are also very important in recognizing palmprints. A method for extracting the principal line based on modified radon transformation (MFRAT) was proposed by Jia et al. [39] to recognize the palmprint. Wu et al. [40] used the first- and second-order derivatives of Gaussian function for detection of palm lines. Liu et al. [41] used isotropic responses via a circular mask for developing a wide palm-line detector. The principal component analysis (PCA) and linear discriminant analysis (LDA) [42, 43] are two subspace-learning methods that are used for palmprint recognition. In these methods 2D image data is converted into a 1D vector, which is known as the "image-as-vector." In the past decade, matrix and tensor embedding [44, 45], kernel [46], sparse learning [47], low-rank representation-based methods [48], and manifold learning [49], have also been applied for palmprint recognition. An optimal trade-off synthetic discriminant function (OTSDF) filter [50] and band-limited phase-only correlation filter (BLPOC) [51] are correlation methods that are also used for palmprint recognition. The coding methods have been found to be more accurate for recognition of an individual in terms of computing performance measures and matching speed in comparison to the other methods mentioned. Competitive code [52–54], binary orientation co-occurrence vector (BOCV) [55], robust line orientation code (RLOC) [56], and ordinal code [57, 58] are representative coding methods.

The palmprint biometric systems have good prospects for commercial development as they have achieved 100% accuracy. Graphical user interface (GUI) plays an important role in computing, whether it is web app, m-app, or any electronic or mechanical device that is used by a person in his/her daily life. Well-designed GUIs have interactive buttons that can be used by any users even when they are not trained on the particular interface. It is easier to use buttons with the same functionality rather than to write codes or commands. The motivation behind developing this GUI is to help students, young researchers, and users to understand processes like plotting of histograms, texture features extraction that includes principal lines, phase symmetry features that includes energy and convolution, and computation of similarity measures for palmprint recognition.

The rest of this chapter is organized as follows: Section 12.2 presents system development. The functionalities of the buttons of the interface are discussed in Section 12.3. Results are reported in Section 12.4. This chapter ends with the overall conclusion and future scope of GUI in Section 12.5. The MATLAB® functions have been used to develop this GUI.

12.2 System Development

A biometric system is a pattern recognition system that is used to match the features of sample images with the template images stored in a centralized database. Each biometric system comprises the following modules: i) the image acquisition module takes the biometric trait image and transmits it to the system for later processing; ii) the feature extraction module extracts the essential or discriminatory features from the captured image; iii) the matcher module matches the extracted image characteristics of the sample image to the database to obtain a match score, whereas iv) the embedded decision-making module decides whether the user is genuine or an imposter based on the matching score, and v) the database module contains the templates of the captured images. Figure 12.1 shows the functioning of a biometric system.

The process of palmprint recognition of an individual consists of following stages.

12.2.1 Enrolment

In enrolment stage, the palmprint image of an individual is acquired for feature extraction; the extracted features are stored as template in a database. A database of palmprint images has been created using the database of the blue band of the Hong Kong Polytechnic University Multispectral Palmprint database [59] and the IIT-Delhi touchless palmprint database [60]. The Multispectral palmprint images are taken from 250 individuals, comprising 195 males and 55 females, in two different sessions. The individuals were asked to provide 6 images for each palm in a session, thus the database contains 6,000 images of 500 different palms. The IIT-Delhi touchless palmprint database was acquired on the IIT-Delhi campus from July 2006 to June 2007. The database was constructed using 7 images for each palm of 235 volunteers, thus it contains 3,290 images of 470 different palms.

A centralized database of texture features, phase symmetry features, and similarity measures of 9,290 palmprint images has been created and stored for matching with the input image.

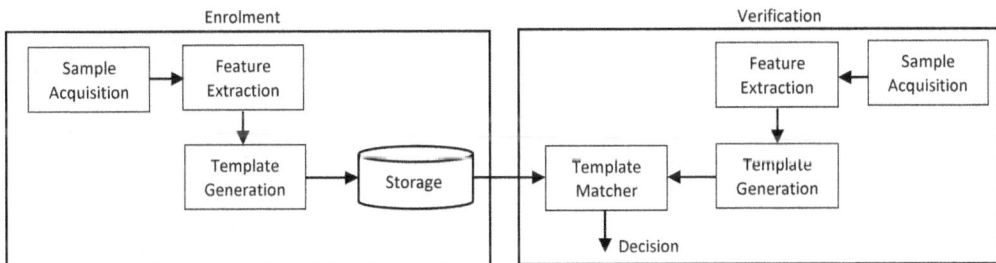

FIGURE 12.1
Functioning of biometric system.

12.2.2 Verification

At the verification stage, the system validates the identity of an individual by comparing the fingerprint image with its own palm print pattern template(s) stored in the centralized database. If a match is found with the template then the user is legitimate to access the system, otherwise, the user is fraudulent.

12.2.3 GUI of Palmprint Recognition

The GUI of palmprint recognition has been developed using MATLAB that allows steer focus and is intuitive. It consists of one axis and six panels, as shown in Figure 12.2. Axis 1 is for the sample input image. The first panel is used for performing the basic operations that contain *Load Image* and *Reset* buttons. *Load Image* button is to take the image as input for processing and the *Reset* button is used to restart the processing. Panel 2 is the histogram panel of GUI with *Histogram* button; to plot the histogram of an image, and *Histogram EQV*; to adjust the intensity of the palmprint image. The third panel is used for texture feature extraction. Principal lines have been extracted as texture features for identifying the authenticity of a user. This panel consists of *Principal Lines* and *Match Score_T* buttons. Panel 4 is used for extracting phase symmetry features. *Energy* and *Convolution* can be computed for matching purpose using this panel. Similarity measure has been computed from Panel 5, and Panel 6 is used to display numerical values of various operations.

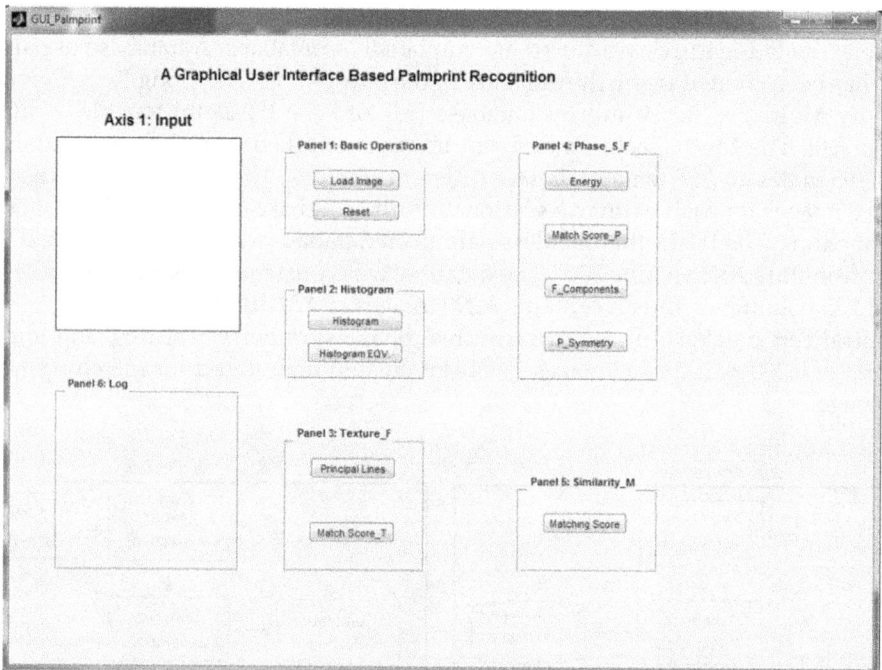

FIGURE 12.2
GUI of palmprint recognition.

12.3 Proposed System Functions

A GUI of palmprint recognition has been developed and its performance has been tested on the centralized databases. The functionalities of the palmprint GUI system are as follows.

12.3.1 Panel 1: Basic Operations

This panel contains *Load Image* and *Reset* buttons which are used to upload the input palmprint image from the database using *imread*() function and to restart the process of the system, respectively.

12.3.2 Panel 2: Histogram

Histogram and histogram equalization are techniques that modify the image by changing its intensity to the desired shape. Histogram technology maps the pixel intensity value of the input image to the output image using nonlinear and nonmonotonic transfer functions. The histogram of an image is the relative frequency of different grey levels. In a two-dimensional diagram, the x-axis shows grey levels, and the y-axis shows the number of pixels for each grey level. Using a monotonic, non-linear mapping, the histogram equalization reassigns the pixel intensity values in the input image so that the output image contains a uniform distribution of the intensities. Therefore, histogram equalization is generally used to enhance the contrast of an image [61]. Figure 12.3 shows the histogram and histogram-equalized image of a palmprint. The function *imhist*() is used to generate the histogram for the palmprint image and *histeq*() is used to equalize the histogram.

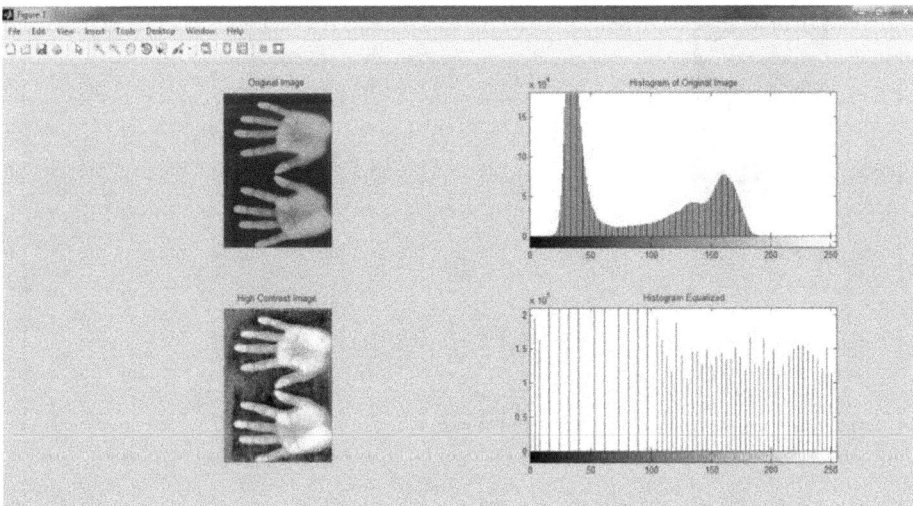

FIGURE 12.3
Histogram and histogram equalization of an image.

12.3.3 Panel 3: Texture_F

The feature of an image is defined as a function of one or more measurements, each of which specifies some quantifiable properties of the image and is computed to quantify significant characteristics of the image. Here principal lines have been extracted as a texture feature and an individual is identified using the match score of the template with the input image. A user can see the result of matching score of texture features using buttons provided on Panel 3, as shown in Figure 12.4.

12.3.4 Panel 4: Phase_S_F

The palmprint image is transformed to the frequency domain from the spatial domain to compute phase symmetry features. During the transformation of an image from one domain to another domain it can be seen that phase information is distributed across all the frequencies whereas magnitude is available at the centre [62]. Furthermore, a fixed length signal can be reconstructed at a scale factor using the phase information only. In this study, energy, phase, and phase symmetry have been computed as features that can be seen using buttons provided on Panel 4 (Figure 12.5).

12.3.5 Panel 5: Similarity_M

The biometric system using the palmprint as a biometric trait stores a user's palmprint data in the form of a template and compresses the form of an image which can be viewed as an accurate representation of the user's biometric features. The similarity measure does

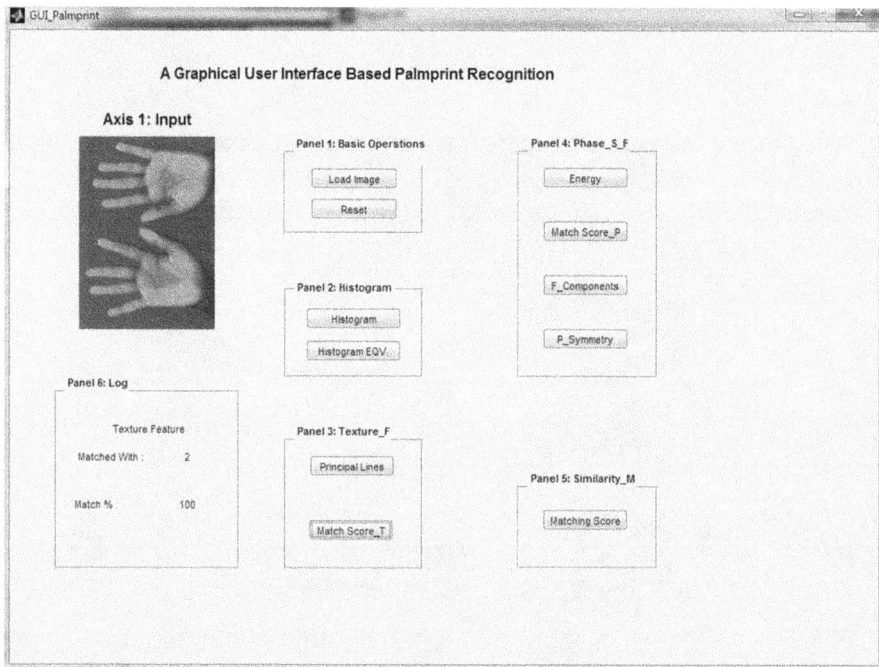

FIGURE 12.4
Matching score of texture features.

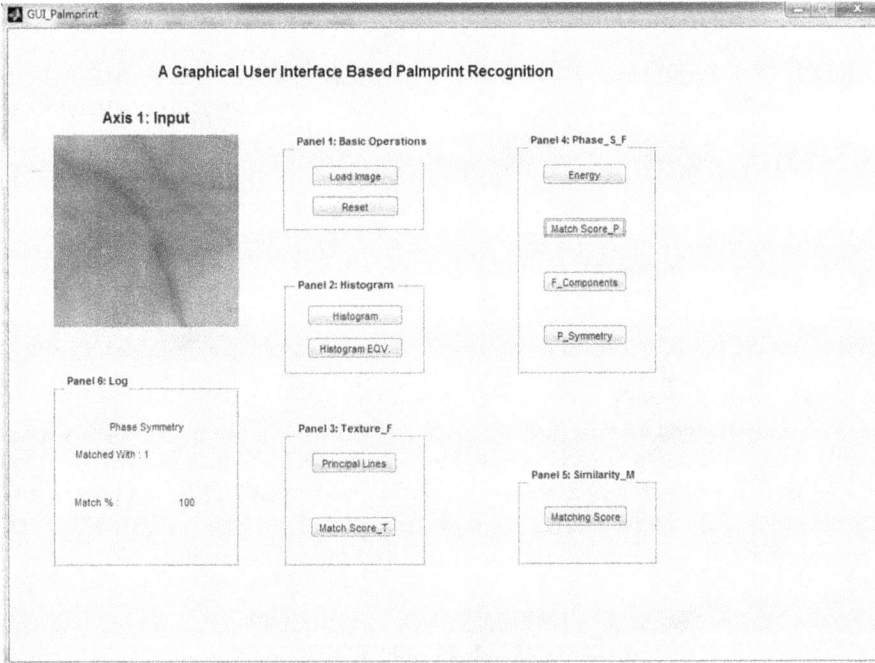

FIGURE 12.5
Matching score of phase symmetry features.

not offer the inference that two templates in the database belong to the same person, rather it is an indicator of the difficulty of recognition in comparing the two templates [63]. The similarity measure of the right and left palm can be seen using the *Matching Score* button on Panel 5.

12.4 Results and Discussion

To evaluate the performance of the GUI proposed for palmprint recognition, the following parameters have been calculated.

12.4.1 Texture Features

Feature extraction is a process of extracting the information from the image such that distinctive properties of extracted features help in differentiating between the categories of input patterns. Feature extraction is also preferred to reduce the cost of feature measurements, to increase classifier efficiency, and allows higher matching accuracy to identify an individual. The image of the palm is rich in texture features like ridges, valleys, principal lines, color etc. Thus, the use of a proper texture description scheme is expected to capture this information accurately. To extract the principal lines as texture feature of the palmprint, function *segment_palm(im)* is used to segment the image into separate palms i.e., right palm and left palm and function *palm_features(left_image, left_mask, right_image, right_mask*) is used to extract principal lines of both the palms. *Principal Lines* button is used for this operation. The output image of the operation is shown in Figure 12.6. The function *palm_matching(L_features, L_valid_corners, R_features, R_valid_corners, New_L_features, New_L_valid_corners, New_R_features, New_R_valid_corners)* is used to match the texture features of template with the input image. If a match is found, the user is authentic, and a matching percentage is displayed on the panel. *Match Score_T* button is used to compute matching percentage. In this study, 100% matching has been found between the input palmprint image and the texture features available in the centralized database.

12.4.2 Phase and Phase Symmetry

In the proposed GUI, phase symmetry features have also been computed, which are given on Panel 4. This paper proposes a method of palm identification by using phase information contained in images spread across all frequencies [62]. In addition, to reconstruct a signal within a scale factor phase information is sufficient. 2D Fourier transformation using the forward and reverse transformation function given in equations (12.1) and (12.2) is used to transform the palmprint image into the frequency domain.

$$F(a,b) = \sum_{x=0}^{M-1}\sum_{y=0}^{N-1} f(x,y) \cdot e^{-j2\pi\left(\frac{ax}{M}+\frac{by}{N}\right)} \tag{12.1}$$

$$f(x,y) = \frac{1}{MN}\sum_{p=0}^{M-1}\sum_{q=0}^{N-1} F(a,b) \cdot e^{j2\pi\left(\frac{ax}{M}+\frac{by}{N}\right)} \tag{12.2}$$

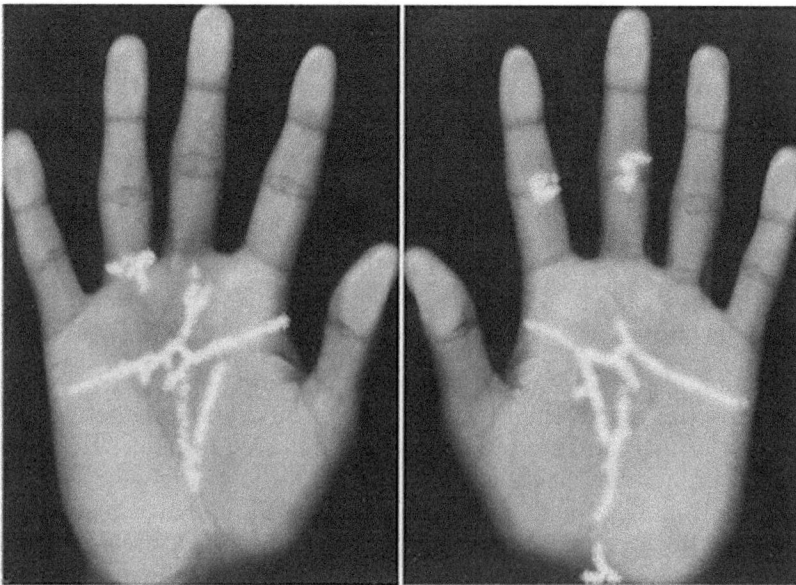

FIGURE 12.6
Texture features: Principal lines.

where $F(a, b)$ is a complex function which is represented as magnitude and phase of an image and $f(x, y)$ is the image in reality. The magnitude $|F(a,b)|$ given in Equation (12.3) depicts quantity of frequency components present in the image, and phase $\varphi(F(a,b))$ in Equation (12.4) provides information of distribution of frequency components in the image:

$$|F(a,b)| = \sqrt{\{\operatorname{Re}(F(a,b))\}^2 + \{\operatorname{Im}(F(a,b))\}^2} \tag{12.3}$$

$$\varphi(F(a,b)) = \tan^{-1}\left\{\frac{\operatorname{Im}(F(a,b))}{\operatorname{Re}(F(a,b))}\right\} \tag{12.4}$$

Figure 12.7 shows the frequency components of a palmprint image.

Symmetry is a useful feature to identify objects in an image. These objects exist at regular intervals in a structure of an image that is clearly visible in a Fourier series. The maximum and minimum points in sine waves (phase 0 or π) are symmetric points to detect symmetry phase information [64, 65]. The bandpass log-Gabor filter is used to filter the frequency domain image to evaluate phase symmetry. Field [66] suggested that instead of using a linear scale, log-Gabor filter uses a logarithmic frequency scale. A log-Gabor filter is of the form given in Equation (12.5) where w_0 is the filter's center frequency.

$$G(w) = e^{\left[\left\{-\log\left(\frac{w}{w_0}\right)^2\right\}/\left\{2\log\left(\frac{w}{w_0}\right)^2\right\}\right]} \tag{12.5}$$

To obtain the band-passed image in the frequency domain, a band-passed log-Gabor filter is applied on frequency domain images and is shown in Equation (12.6)

$$IMF = F \times G(w) \tag{12.6}$$

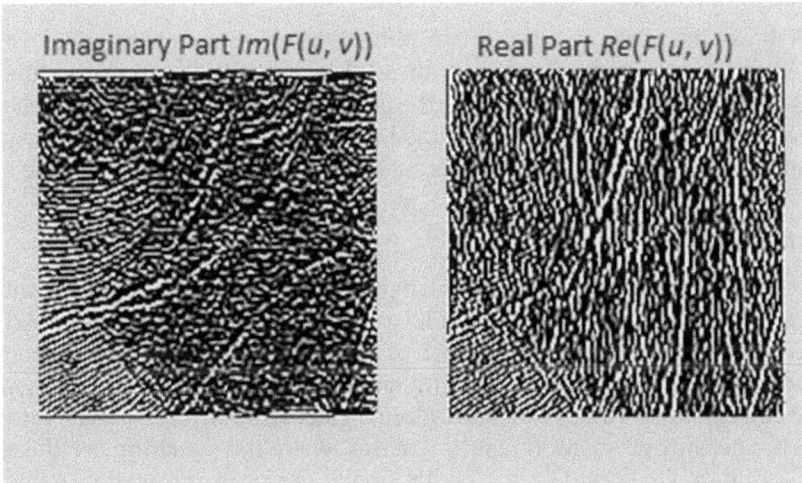

FIGURE 12.7
Frequency components of image.

The real part of the band-passed image in the spatial domain that is obtained after converting the filtered image back to the spatial domain is represented by f and is given in Equation (12.7)

$$f = \operatorname{Re}\left\{\Gamma^{-1}\left(\operatorname{IMF}\right)\right\} \tag{12.7}$$

In order to obtain the window-filtered image h shown in Equation (12.8), the log-Gabor-based adaptive window image is used where a monogenic H filter is obtained from the horizontal and vertical dimensions of the image convoluted with the band-passed image.

$$h = \Gamma^{-1}\left(\operatorname{IMF} \times H\right)\} \tag{12.8}$$

To obtain the amplitude of the filtered image Equation (12.9) is used.

$$A = \left\{\operatorname{Re}\left(h\right)\right\}^2 + \left\{\operatorname{Im}\left(h\right)\right\}^2 \tag{12.9}$$

To calculate the filtered image energy, the cumulative sum of the band-passed image and filter outputs is used and shown in Equation (12.10) on a number n of band-passed versions.

$$E = \sum_n \sqrt{f^2 + A} \tag{12.10}$$

The symmetry energy of an image is given by Equation (12.11).

$$S = \sum_n \left(f - \sqrt{A}\right) \tag{12.11}$$

The normalized phase symmetry of an image is computed using Equation (12.12).

$$P = \frac{S}{E} \tag{12.12}$$

Figure 12.8 shows the phase symmetry components of a palmprint image.

Experiments are conducted using a centralized database of 9,290 palmprint images of 485 individuals. The correlation is computed for the precision calculations using the phase energy, the training symmetry energy (the centralized database) and the test data sets (palmprint input images). The accuracy of palmprint recognition of an individual is found to be 100%.

12.4.3 Similarity Measures

A palmprint template contains spatial information about the principal lines and putative points, which are generally between 30 and 100 points. A match of six to eight putative points is usually considered enough for verification of an individual's palmprint. A biometric system must classify a pair of templates having too low or too high a similarity value that is paired with intermediate values that take a larger calculation from the matcher to classify. The similarity measure gives good results in cases where the matching is either extremely too low or too high. For a perfect match, the matching probability should be extremely high and for a mismatch, it should be extremely low. The MATLAB function that is used in this study to compute the similarity metric of left palm is *showMatchedFeatures(OLeft_Image, Left_image, L_macthed_pts1(i), L_matched_pts2(i), 'montage')* and for computing the

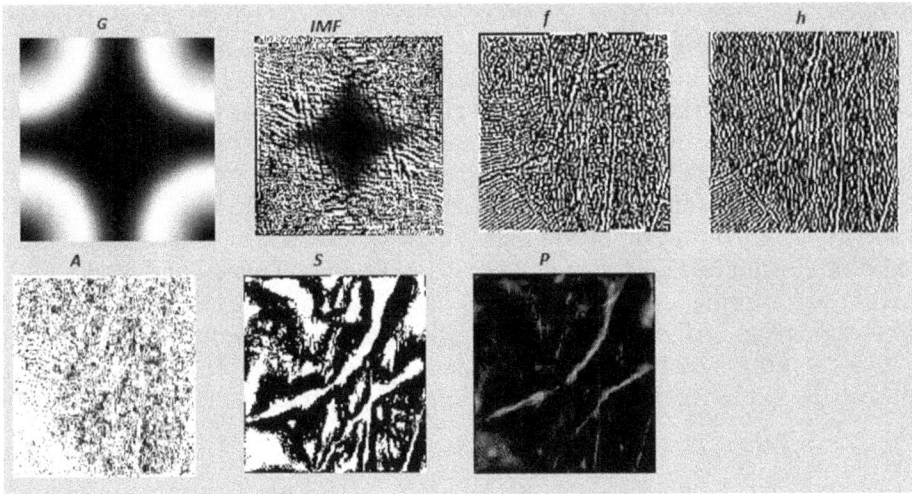

FIGURE 12.8
Phase symmetry components.

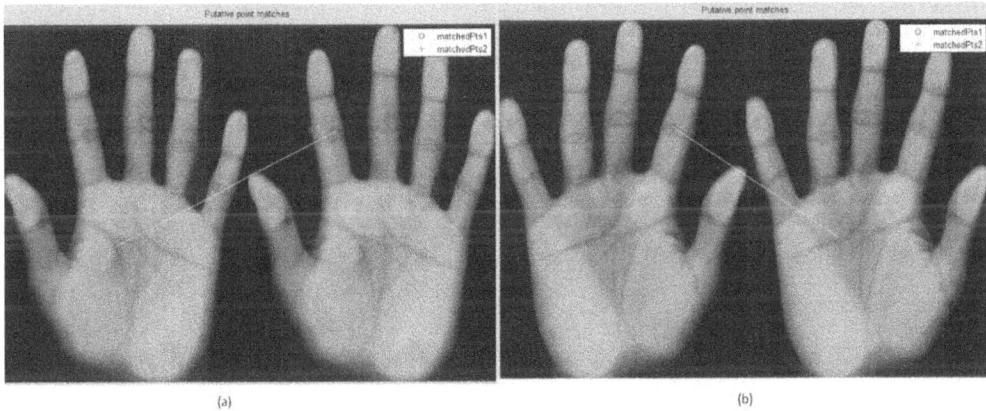

FIGURE 12.9
(a) Putative points of left palm; (b) Putative points of right palm.

matching score of the right palm function *showMatchedFeatures(ORight_Image, Right_image, R_macthed_pts1(i), R_matched_pts2(i), 'montage')* is used. Palmprint identification is found to be 100% accurate when a similarity measure score is evaluated. Figure 12.9 shows the similarity measure of the left and right palm respectively.

12.5 Conclusion

In this chapter, an attempt has been made to explore palmprint recognition through GUI using MATLAB. GUIs are much better at expressing basic operations. Information can be presented much more tersely graphically, as there are many additional features: ordering, shapes, colors, etc. The main benefit of GUIs is that they are user-friendly and thus make system operations more intuitive. The proposed system will help the people who want to understand the basic operations of palmprint recognition such as histogram, histogram equalization, texture feature extraction, phase symmetry feature extraction, and computation of similarity measure. The main advantage of this work is that a person having no prior knowledge in this area can also understand the process of palmprint recognition. In addition, GUIs are user-friendly, which makes learning more interactive and attractive. In this study, principal lines have been extracted as texture features and energy has been computed as phase symmetry feature. The similarity measure has been computed for left and right input palm images for an exact match with the template. The proposed GUI is tested on two palmprint databases discussed in Section 12.2. The results show 100% accuracy to recognize an individual using palmprint images. In the future, new algorithms can be proposed that will have faster computation speeds, feature extraction with better discrimination, and lower complexity for palmprint recognition.

References

1. Cao, K. and A. K. Jain. 2019. Automated latent fingerprint recognition. *IEEE Trans. Pattern Anal. Mach. Intell.* 2(4):788–800.
2. Jain, A. K., P. J. Flynn and A. Ross. 2007. *Handbook of Biometrics*. Springer, Boston, MA.
3. Maltoni, D., D. Maio, A. K. Jain and S. Prabhakar. 2003. *Handbook of Fingerprint Recognition*. Springer, London.
4. Ratha, N. and B. Bolle. 2004. *Automatic Fingerprint Recognition Systems*. Springer, New York.
5. Daugman, J. G. 1993. High confidence visual recognition of persons by a test of statistical independence. *IEEE Trans. Pattern Anal. Mach. Intell.* 15(11):1148–1161.
6. Daugman, J. 2004. How iris recognition works. *IEEE Trans. Circuits Syst. Video Technol.* 14(1):21–30.
7. Bowyer, K. W., K. Hollingsworth and P. J. Flynn. 2008. Image understanding for iris biometrics: A survey. *Comput. Vis. Image Underst.* 110(2):281–307.
8. Pillai, J. K., V. M. Patel, R. Chellappa and N. K. Ratha. 2011. Secure and robust iris recognition using random projections and sparse representations. *IEEE Trans. Pattern Anal. Mach. Intell.* 33(9):1877–1893.
9. Wechsler, H. 2006. *Reliable Face Recognition Methods-System Design, Implementation and Evaluation*. Springer.
10. Wright, J., A. Y. Yang, A. Ganesh, S. S. Sastry and Y. Ma. 2009. Robust face recognition via sparse representation. *IEEE Trans. Pattern Anal. Mach. Intell.* 31(2):210–227.
11. Taigman, Y., M. Yang, M. A. Ranzato and L. Wolf. 2014. DeepFace: Closing the gap to human-level performance in face verification. *Proceedings of the IEEE International Conference on Computer Vision and Pattern Recognition.* 23–28 June, Columbus, OH, 1701–1708.
12. Sun, Y., X. Wang and X. Tang. 2015. Deeply learned face representations are sparse, selective, and robust. *Proceedings of the IEEE International Conference on Computer Vision and Pattern Recognition.* 7–12 June, Boston, MA, 2892–2900.
13. Khokher, R. and R. C. Singh. 2017. Footprint-based personal recognition using Dactyloscopy technique. In: Manchanda P., Lozi R., Siddiqi A. (eds) Industrial Mathematics and Complex Systems, *Industrial and Applied Mathematics.* 207–219. Springer, Singapore.
14. King, R. R. and W. Xiaopeng. 2013. Study of biometric identification method based on naked footprint. *Int. J. Sci. Eng.* 5(2):18–24.
15. Khokher, R. and R. C. Singh. 2016. Footprint-based personal recognition using scanning technique. *Indian J. Sci. Technol.* 9(44):1–10.

16. Zhang, D. and G. Lu. 2013. *3D Biometrics–Systems and Applications*. Springer, New York.
17. Zhang, L. and H. Li. 2012. Encoding local image patterns using Riesz transforms: With applications to palmprint and finger-knuckle-print recognition. *Image Vis. Comput.* 30(12):1043–1051.
18. Zhang, D., W. Kong, J. You and M. Wong. 2003. Online palmprint identification. *IEEE Trans. Pattern Anal. Mach. Intell.* 25(9):1041–1050.
19. Jain, A. K. and J. Feng. 2009. Latent palmprint matching. *IEEE Trans. Pattern Anal. Mach. Intell.* 31(6):1032–1047.
20. Zhang, L., Y. Shen, H. Li and J. Lu. 2015. 3D palmprint identification using block wise features and collaborative representation. *IEEE Trans. Pattern Anal. Mach. Intell.* 37(8):1730–1736.
21. Cummins, H. and M. Midlo. 1961. *Fingerprints, Palms and Soles: An Introduction to Dermatoglyphics*. Dover Publications, New York.
22. Jain, A. K. and J. J. Feng. 2009. Latent palmprint matching. *IEEE Trans. Pattern Anal. Mach. Intell.* 39(6):1032–1047.
23. Dai, J. and J. Zhou. 2011. Multi-feature based high-resolution palmprint recognition. *IEEE Trans. Pattern Anal. Mach. Intell.* 33(5):945–957.
24. Kong, A., D. Zhang and M. Kamel. 2009. A survey of palmprint recognition. *Pattern Recognit.* 42(7):1408–1418.
25. Zhang, D., W. M. Zuo and F. Yue. 2012. A comparative study of palmprint recognition algorithms. *ACM Comput. Surv.* 44(1):1–37.
26. Chen, G. and W. Xie. 2007. Pattern recognition with SVM and dual-tree complex wavelets. *Image Vis. Comput.* 25(6):960–966.
27. Yong, X., F. Lunke and D. Zhang. 2015. Combining left and right palmprint images for more accurate personal identification. *IEEE Trans. Image Process.* 24(2):549–559.
28. Xu, Y., Z. Fan, M. Qiu, D. Zhang and J.-Y. Yang. 2013. A sparse representation method of bimodal biometrics and palmprint recognition experiments. *Neurocomputing* 103:164–171.
29. Zhu, Y.-H., W. Jia and L.-F. Liu. 2009. Palmprint recognition using band-limited phase-only correlation and different representations. *Emerg. Intell. Comput. Technol. Appl.* 5754:270–277. Springer.
30. Zheng, Q., A. Kumar and G. Pan. 2016. A 3D feature descriptor recovered from a single 2D palmprint image. *IEEE Trans. Pattern Anal. Mach. Intell.* 38(6):1272–1279.
31. Fei, L., Y. Xu, W. Tang and D. Zhang. 2016. Double-orientation code and nonlinear matching scheme for palmprint recognition. *Pattern Recognit.* 49:89–101.
32. Wang, X., L. Lei and M. Z. Wang. 2012. Palmprint verification based on 2D-Gabor wavelet and pulse-coupled neural network. *Knowl. Based Syst.* 27:451–455.
33. Guo, X. M., W. D. Zhou and Y. Wang. 2014. Palmprint recognition algorithm with horizontally expanded blanket dimension. *Neurocomputing* 127:152–160.
34. Chen, G. Y. and W. F. Xie. 2007. Pattern recognition with SVM and dual-tree complex wavelets. *Image Vis. Comput.* 25(6):960–966.
35. Jing, X. Y. and D. Zhang. 2004. A face and palmprint recognition approach based on discriminant DCT feature extraction. *IEEE Trans. Syst. Man, Cybern.* 34(6):2405–2415.
36. Lu, J., Y. Zhao and J. Hu. 2009. Enhanced Gabor-based region covariance matrices for palmprint recognition. *Electron. Lett.* 45(17):1–2.
37. Zhou, J. C. C., Liu Y. Zhang and, G. F. Lu. 2013. Object recognition using Gabor co-occurrence similarity. *Pattern Recognit.* 46(1):434–448.
38. Saedi, S. and N. M. Charkari. 2014. Palmprint authentication based on discrete orthonormal S-Transform. *Appl. Soft Comput.* 21:341–351.
39. Huang, D. S., W. Jia and D. Zhang. 2008. Palmprint verification based on principal lines. *Pattern Recognit.* 41(4):1316–1328.
40. Wu, X. Q., D. Zhang and K. Q. Wang. 2006. Palmline extraction and matching for personal authentication. *IEEE Trans. Syst. Man, Cybern. A* 36(5):978–987.
41. Liu, L., D. Zhang and J. You. 2007. Detecting wide lines using isotropic nonlinear filtering. *IEEE Trans. Image Process.* 16(6):1584–1595.

42. Lu, G. M., D. Zhang and K. Q. Wang. 2003. Palmprint recognition using eigen palms features. *Pattern Recognit. Lett.* 24(9–10):1463–1467.
43. Wu, X. Q., D. Zhang and K. Q. Wang. 2003. Fisher palms based palmprint recognition. *Pattern Recognit. Lett.* 24(15):2829–2838.
44. Hu, D., G. Feng and Z. Zhou. 2007. Two-dimensional locality preserving projecting (2DLPP) with its application to palmprint recognition. *Pattern Recognit.* 40(3):339–342.
45. Hu, R. X., W. Jia, D. S. Huang and Y. K. Lei. 2010. Maximum margin criterion with tensor representation. *Neurocomputing* 73(10–12):1541–1549.
46. Yang, J., A. Frangi, J. Yang, D. Zhang and J. Zhong. 2005. KPCA plus LDA: A complete kernel fisher discriminant framework for feature extraction and recognition. *IEEE Trans. Pattern Anal. Mach. Intell.* 27(2):230–244.
47. Xu, Y., Z. Z. Fan, M. N. Qiu, D. Zhang and J. Y. Yang. 2013. A sparse representation method of bimodal biometrics and palmprint recognition experiments. *Neurocomputing* 103:164–171.
48. Zhang, N. and J. Yang. 2013. Low-rank representation based discriminative projection for robust feature extraction. *Neurocomputing* 111:13–20.
49. Wu, S. S., M. M. Sun and J. Y. Yang. 2011. Stochastic neighbor projection on manifold for feature extraction. *Neurocomputing* 74(17):2780–2789.
50. Hennings-Yeomans, P., B. Kumar and M. Savvides. 2007. Palmprint classification using multiple advanced correlation filters and palm-specific segmentation. *IEEE Trans. Inf. Forensics Secur.* 2(3):613–622.
51. Ito, K., T. Aoki and H. Nakajima. 2008. A palmprint recognition algorithm using phase- only correlation. *IEICE Trans. Fundam* E91-A(4):1023–1030.
52. Kong, A. and D. Zhang. 2004. Competitive coding scheme for palmprint verification. *Proceedings of the 17th ICPR* 1:520–523.
53. Yue, F., W. M. Zuo and D. Zhang. 2009. FCM-based orientation selection for competitive code-based palmprint recogniton. *Pattern Recognit.* 42(11):2841–2849.
54. Zuo, W. M., F. Yue and D. Zhang. 2011. On accurate orientation extraction and appropriate distance measure for palmprint recognition. *Pattern Recognit.* 43(4):964–972.
55. Guo, Z. H., D. Zhang, L. Zhang and W. M. Zuo. 2009. Palmprint verification using binary orientation co-occurrence vector. *Pattern Recognit. Lett.* 30(13):1219–1227.
56. Jia, W., D. S. Huang and D. Zhang. 2008. Palmprint verification based on robust line orientation code. *Pattern Recognit.* 41(5):1504–1513.
57. Sun, Z. N., T. N. Tan, Y. H. Wang and S. Z. Li. 2005. Ordinal palmprint representation for personal identification. *Proceedings of the CVPR* 1:279–284.
58. Sun, Z. N., L. B. Wang and T. N. Tan. 2014. Ordinal feature selection for iris and palmprint recognition. *IEEE Trans. Image Process.* 23(9):3922–3934.
59. Zhang, D., Z. H. Guo, G. M. Lu, L. Zhang and W. M. Zuo. 2010. An online system of multispectral palmprint verification. *IEEE Trans. Instrum. Meas.* 59(2):480–490.
60. Kumar, A. and S. Shekhar. 2011. Personal identification using rank-level fusion. *IEEE Trans. Syst. Man. Cybern. C* 41(5):743–752.
61. Celik, T. 2012. Two-dimensional histogram equalization and contrast enhancement. *Pattern Recognit.* 45(10):3810–3824.
62. Karar, S. and R. Parekh. 2013. Palmprint recognition using phase symmetry. *Int. J. Sci. Res. Pub.* 3(4):1–6.
63. Torres-Sospedra, J., R. Montoliu, S. Trilles, O. Belmonte and J. Huerta. 2015. Comprehensive analysis of distance and similarity measures for wi-fi fingerprint in-door positioning systems. *Expert Syst. Appl.* 42(23):9263–9278.
64. Kovesi, P. 1997. Symmetry and asymmetry from local phase. *Tenth Australian Joint Conference on Artificial Intelligence.* 30 November – 04 December, Perth, Australia.
65. Jun, W., Y. Zhaoxuan and F. Dengchao. 2005. A method of image symmetry detection based on phase information. *Trans. Tianjin Univ.* 11(6):428–432.
66. Field, D. J. 1987. Relation between statistics of natural images and the response properties of cortical cells. *J. Opt. Soc. Am. Assoc.* 4(12):2379–2394.

13

Wavelet Approach for the Classification of Autism Spectrum Disorder

Noore Zahra, Hessa N. Al Eisa, Kahkashan Tabassum,
Sahar A. EI-Rahman, and Mona Jamjoom

CONTENTS

13.1 Introduction 209
13.2 Literature Survey and Background Information 210
13.3 Materials and Methods 212
    13.3.1 Wavelet Entropy 214
13.4 Implementation 215
13.5 Results and Discussion 215
    13.5.1 Wavelet Transform and Decomposition at Different Levels 215
    13.5.2 Classification of Frequency Band 215
    13.5.3 Entropy Calculation of the Frequency Bands for Different Age Groups 217
13.6 Conclusions 217
References 223

13.1 Introduction

Autism spectrum disorder (ASD) is found in approximately 1 in 59 children as determined by the Centers for Disease Control (CDC) in 2018; this leads to anxiety disorder, sleep disorder, ADHD etc. ASD is a disorder that affects communication, interaction with other people, and behaviour. It restricts a person's ability to function properly in daily life. Autism is known as a "spectrum" disorder because of its large array of variation in types and symptoms. Autism can strike at any age but mostly it is found in children of below two years of age [1].

By using models and analysis of autism spectrum disorder it is expected to get psychometric tools that will further help psychologists and pediatricians to diagnose and evaluate patients suffering with ASD. Non-linear statistical analysis provides the measurements of metabolites that facilitate an understanding of the complexities of disorder. The outcome of the analysis of ASD will help in early diagnosis of the disease, which leads to favorable results in the long run.

This chapter is dedicated to the wavelet analysis of autism that will go through the phase of multi-scale decomposition and entropy calculation of corresponding delta, theta, alpha, beta, and gamma for different age-groups of children. Results of the wavelet analysis algorithm will indicate the classification of Autism.

13.2 Literature Survey and Background Information

In order to observe brain activity, the electroencephalogram (EEG) and functional magnetic resonance imaging (fMRI) are extensively used methods, EEG being a direct and the latter an indirect method. When EEG signals are applied, the brain's electrical activity is measured and recorded as voltages. Recently researchers have produced immense evidence demonstrating autism spectrum disorder, which is a stereotyped, heterogeneous neurodevelopmental syndrome categorized by weak social communication skills and recurring activities among children, and it is proposed that EEG could be used to quantify this abnormal and digressive neural connectivity. The incredible increase in autism spectrum disorder requires methods of magnetoencephalography functional magnetic resonance imaging and electroencephalography based on neuroimaging. These methods are applicable with the help of processes of functional connectivity, segregation, and integration of information processing. The brain deficiencies/incapabilities can be solved by the application of functional connectivity [2]. This refers to the degree of compatibility of correlation or synchronization of different regions in the brain at the time of resting or performing states and the latter could be based on functional connectivity; finally the results can be combined to reflect neurological and psychiatric syndromes. An aberrant functional connectivity is applied to study and classify the cognition disorders [3, 4, 5, 6]. ASD is a neurodevelopmental disorder characterized by a typical functional integration of brain region. Brain connectivity measures may identify early ASD. Few of the earlier studies of ASD symptoms consist of local over-connectivity, global under-connectivity of aberrant functional connectivity. They presented the one core neuro-biological mechanism with applied entropy of EEG that is known to reveal unpredictable brain activities and also show the anomalies associated with it [7, 8]. The researchers McDonough et al., Ghanbari et al., and Misic et al. were able to predict that EEG complexity is useful to determine brain functional connectivity [9, 10, 11]. Li et al stated that entropy could be used to monitor functional states of the brain [12]; Piryatinska et al., Arton et al., and Tan et al. predicted the existence of neuropsychiatric and neuro-diseases such as Parkinson's disease, epilepsy, schizophrenia [13–17]. The researchers Perkins et al. and Mason et al. confirmed, after examining the studies conducted on a persons' brain connections, that a specific set of connections within regions of the brain, depending on human conditions, have abnormalities in the brain patterns, physical connections between regions sometimes referred to as structural connection or, if based on similarity of temporal characteristics in regions of brain activity, as functional connection [15, 17].

Liang et al. target entropy measures in anaesthesia [18]. At the time of perception and other active processes like cognition the temporal activities got spotted, then the task-related functional connection offers evidence that the brain regions within the network are employed in order to route and integrate information by responding to events occurred [19]. An electroencephalography test is commonly used to detect electrical brain activities and also to identify abnormal brain activities; it can predict underlying structural disorder in ASD patients [20].

Many researchers have studied the EEG signals and analyzed signals of children with autism. Kang et al. (2018) presented four entropy algorithms to discriminate autism spectrum disorder (ASD) and typical development (TD) children based on the resting-state EEG [21]. The algorithms are sample entropy, fuzzy entropy, Renyi permutation entropy, and Renyi wavelet entropy. The researchers studied the relation between age and the entropy value in ASD and TD. Also, they explored which areas in the brain are considered to differentiate the ASD children from the TD children, through 43 children with autism compared to 43 normal children aged from 4 to 8 years old. Entropy-specifically and region-specifically, the findings were that sensitivity increases with age.

Ibrahim et al. (2018) presented various EEG features extraction and classification algorithms to support ASD and epilepsy diagnosis [22]. The researchers found that the combination of discrete wavelets transform in the preprocessing stage, and that Shannon entropy and the k-nearest neighbor classifier always fulfilled the best results.

Askari et al. (2018) proposed a designed model based on EEG features and cellular neural networks to show the brain connections of autistic children. The proposed model can be used to distinguish the children with autism from the control group based on CNN and Support vector machine with an accuracy 95.1%. The findings indicate that there are considerable differences and abnormalities in the left hemisphere in the autistic children in comparison with the control children, and most abnormalities are related to the regions in frontal and parietal lobes in the autistic children [23]. Grossi et al. (2017) proposed an artificial adaptive system for features extraction in computerized EEGs and the researchers discriminate ASD children from control subjects [24]. Sharma et al. (2017) proposed an automated identification system of focal–class (FC) and non-focal–class (NFC) of EEG signals that are localized in a time–frequency domain based on orthogonal wavelet filter banks [25]. The researchers calculate different entropies from the EEG signal wavelet coefficients as discriminating features. These features are utilized to classify FC and NFC of EEG signals and ranked based on Student's t-test ranking method and then the signals are classified by Least Squares Support Vector Machine classifier. The proposed wavelet-based system fulfilled an accuracy up to 94.25%, 91.95% sensitivity, and 96.56% specificity. Faust et al. (2015) presents a wavelet algorithms which shows computer aided epilepsy diagnosis and seizure detection. The researchers found that a considerable literature review established that wavelet transform is the main method for EEG-based seizure detection, and most researchers utilize the discrete wavelet transform more than the continuous time wavelet analysis in their scientific work [26].

Duffy et al. (2013) explored the relation between Asperger's syndrome and autism utilizing EEG coherence study. The researchers got the coherence data and used principle component analysis to attenuate the frequencies. They realized that these factors could considerably distinguish neurotypical ASD from the control subjects by discriminant function analysis [27]. Ahmadlou et al. (2012) investigated the brain region's connectivity with ASD based on EEGs, Fuzzy Synchronization Likelihood, and a wavelet-chaos neural network algorithm [28]. The researchers aimed to explore the nonlinear features space for a more accurate EEG-based autism diagnosis; also, which regions in the autistic brain indicate connectivity deficits and that the deficit sub-bands that are more pronounced are determined. The findings are that the proposed method can be utilized as an effective tool for autism diagnosis, where the achieved classification accuracy is 95.5% to distinguish the autistic EEG from normal EEG. Adeli et al. proposed a wavelet-chaos neural network algorithm for EEG analysis to detect seizure and epilepsy. The methodology is applied to three EEG signals groups that include healthy, epileptic during a seizure-free interval, and epileptic during a seizure. The research ideology is that nonlinear features may not disclose considerable variations between neurological and psychiatric disorder groups and healthy groups in the entire band-limited EEGs, but may detect the variations in specific sub-bands [29].

13.3 Materials and Methods

Biomedical signals carry information of physiological events. The part of a signal related to a specific event and the analysis of a signal helps in monitoring and diagnosis of problems related with particular disease [10–13]. When an event is identified, the corresponding waveform may be analyzed in terms of its amplitude, waveshape, time duration, energy distribution, frequency contents, and so on. Event detection is an important part of bio-medical signal analysis. To reveal the pattern of electrical activity of the heart, doctors use electrocardiogram (ECG); to detect abnormal activity of muscles, electromyogram (EMG) is used; and to record the electrical activity of the brain, electroencephalogram (EEG) is widely used.

EEG rhythms or frequency bands are as follows:

Delta (δ): $0.5 \leq f < 4$ Hz; Theta (θ): $4 \leq f < 8$ Hz; Alpha (α): $8 \leq f \leq 13$ Hz; and, Beta (β): $f > 13$ Hz.

These bands make a significant contribution in the identification of brain abnormalities but it is not easy to read them or classify them without the appropriate tool. To understand them properly, computer-aided digital filter design and hidden mathematics would be required. In Figure 13.1 different lobes of the brain are shown. The occipital lobe is responsible for seeing. Alpha waves (8–13 Hz) are clearest in this lobe. The frontal lobe is responsible for consciousness and thoughts. Beta waves, which have a range greater than 13 Hz, are often found in the frontal lobe. Delta waves are the slowest waves, but with the strongest signal of highest amplitude having a frequency up to 4 Hz. This wave is common in children below one year of age. Gamma waves are of the frequency 26–100 frequencies. Theta waves are between the frequencies 4–7 Hz. These waves are called 'slow activity' and occur during sleep in young children [30]. The EEG dataset that is used here is based on the 10–20 International System as shown in Figure 13.2.

In this chapter EEG is used as an input to diagnose and classify diseases along with the computer tool MATLAB®. Wavelet transform will serve the purpose of mathematical insight of disease.

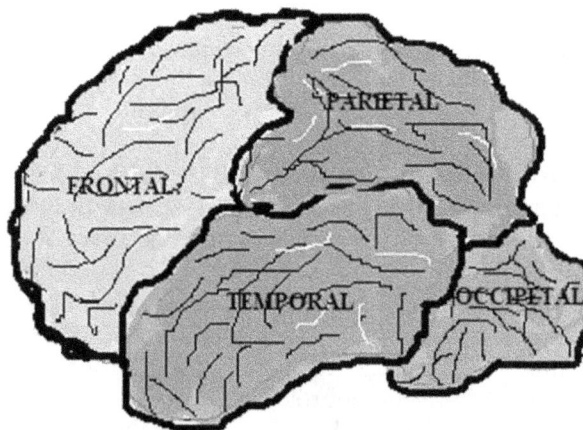

FIGURE 13.1
Anatomical region showing lobes of mammalian brain.

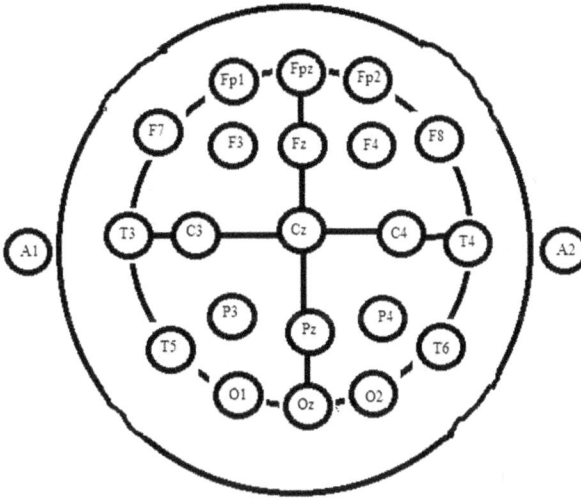

FIGURE 13.2
10–20 International System for the electrode placement of data acquisition data.

Wavelet transform is a time–frequency analysis tool widely used in signal processing. The wavelet transformation is well suited to representing various aspects of EEG signals such as trends, discontinuities, and repeated patterns where other signal processing approaches fail. Wavelet is an effective time–frequency tool for analyzing transient signals. In the wavelet transform (WT) case, WT employs a windowing technique with variable-size windows. Wavelets are mathematical functions that cut up data into different frequency components, and then study each component with a resolution matched to its scale.

In both continuous and discrete forms of wavelet analysis, the signal is decomposed into scaled and translated versions ($\psi_{a,b}(t)$) of a single function $\psi(t)$ called a mother wavelet:

$$\Psi_{a,b}(t) = \frac{1}{\sqrt{|a|}} \Psi\left(\frac{t-b}{a}\right) \tag{13.1}$$

Where a and b are the scale and translation parameters respectively $a, b \in \Re$ and $a \neq 0$.

If a signal $f(t)$ is a square integrable function of time, i.e., $f \in L^2(\Re)$ (the space of finite energy signals), then the continuous wavelet transform (CWT) of the signal is defined as

$$W(a,b) = C_{a,b} = \int_{-\infty}^{\infty} f(t) \frac{1}{\sqrt{|a|}} \psi^*\left(\frac{t-b}{a}\right) dt = \langle f, \psi_{a,b} \rangle \tag{13.2}$$

where $<.,.>$ means the inner product and the symbol "*" means complex conjugate. The factor $1/\sqrt{|a|}$ is used to normalize the energy so that it stays at the same level for different values of a and b.

The discrete wavelet transform (DWT) is obtained by discretizing the parameters a and b. We may choose $a = 2^{-j}$, $b = k2^{-j}$ with $j, k \in Z$. By substituting this in Equation (13.1) We get

$$\Psi_{j,k}(t) = 2^{j/2} \Psi(2^j t - k). \tag{13.3}$$

The DWT can be written as

$$d_{j,k} = \int_{-\infty}^{\infty} f(t) \, 2^{j/2} \Psi(2^j t - k) \, dt = \langle f, \Psi_{t,k} \rangle \qquad (13.4)$$

where $d_{j,k}$ are known as wavelet (or detailed) coefficients at scale j and location k. The wavelet coefficients $d_{j,k}$ measure the amount of fluctuation about the point $t = k2^{-j}$ with a frequency determined by the dilation index j.

Given a signal $f(t)$, its multi-resolution decomposition formula at level H is defined as

$$f(t) = \sum_{k=-\infty}^{\infty} a_{H,k} 2^{-H/2} \phi(2^{-H} t - k) + \sum_{j=-\infty}^{H} \sum_{k=-\infty}^{+\infty} d_{j,k} 2^{-j/2} \psi(2^{-j} t - k) = A_H(t) + \sum_{j=-\infty}^{H} D_j(t), \qquad (13.5)$$

$\psi(t)$ is the mother wavelet, while ϕ is a companion function, called scaling function:

$$\phi_{j,k}(t) = 2^{-j/2} \phi(2^{-j} t - k) \qquad (13.6)$$

are scaled and translated versions of the original scaling function $\phi(t)$. $a_{H,k}$ represent approximation or scaling coefficients at level H, and defined as $a_{H,k} = \langle f, \varphi_{H,k} \rangle$.

$$f(t) = \sum_{j=-\infty}^{+\infty} \sum_{k=-\infty}^{+\infty} d_{j,k} 2^{-j/2} \psi(2^{-j} t - k) \qquad (13.7)$$

Equation (13.6) expresses the synthesis of the original signal from wavelet coefficients.

By using down-sampling operations along with low-pass and high-pass filtering, signal decomposition as in Equation (13.4) can be efficiently implemented.

13.3.1 Wavelet Entropy

Wavelet entropy can be calculated from the decomposed signal [29].The orthonormal basis in $L^2(\Re)$ is family $\left\{ \Psi_{j,k}(t) = 2^{j/2} \Psi(2^j t - k) \right\}$. The energy series is given by

$$E_{j,k} = |d_{j,k}|^2 \qquad (13.8)$$

and overall energy at resolution j is

$$E_j = \sum_{k=0}^{2^{M-j}-1} |d_{j,k}|^2 \qquad (13.9)$$

The total energy for the signal can be obtained as

$$E_{tot} = |cA_i|^2 + \sum_{j=1}^{l} |d_{j,k}|^2 \qquad (13.10)$$

Relative wavelet energy at each level is calculated by

$$P_i = \frac{E_i}{E_{tot}}, \, i = 1, 2, \ldots \ldots l+1 \qquad (13.11)$$

Entropy can be defined by Shanon's entropy theory as

$$S_p = \sum_{i=1}^{l+1} P_i \ln P_i \qquad (13.12)$$

According to Renyi entropy theory, Renyi wavelet entropy is defined as

$$\text{RWE} = \frac{S_p^a}{\log N_i} \qquad (13.13)$$

13.4 Implementation

Data would be collected from the National Database for Autism Research and the Autism Centre of Excellence. The dataset was recorded according to the 10–20 International System as shown in Figure 13.2, and digitized with a sampling rate of 256 Hz, applying wavelet transforms and decomposed into different levels. These levels would be decided by a hit and trial method and db4 would be applied. Detailed study of DWT gives the information about frequency bands and finally entropy calculation would be done for ASD. In this method the EEG of a patient will be taken as input and decomposed into sublevels using wavelet transform and further differentiated into frequency bands. Entropy calculation for each band would be done and as a consequence we can classify autism for different age groups. The whole procedure is shown in Figure 13.3.

13.5 Results and Discussion

13.5.1 Wavelet Transform and Decomposition at Different Levels

In the first step DWT was applied on an EEG dataset for autism. It is observed that as the number of levels increases, frequency resolution become clearer. Starting with the range of frequencies from 0 to1200Hz after the first level, the approximation channel contains the frequencies 0 to 600 Hz and the details channels will contain frequency ranges from 600–1200 Hz. The signal is decomposed according to this procedure. It is applied on every dataset but for convenience only one result is shown here as in Figure 13.4.

13.5.2 Classification of Frequency Band

DWT provides further information about EEG sub-bands Gamma, Beta, Alpha, Theta, and Delta shown in Figures 13.5 and 13.6.

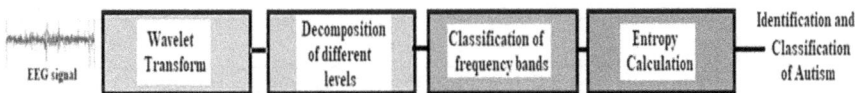

FIGURE 13.3
Methodology for identification of autism.

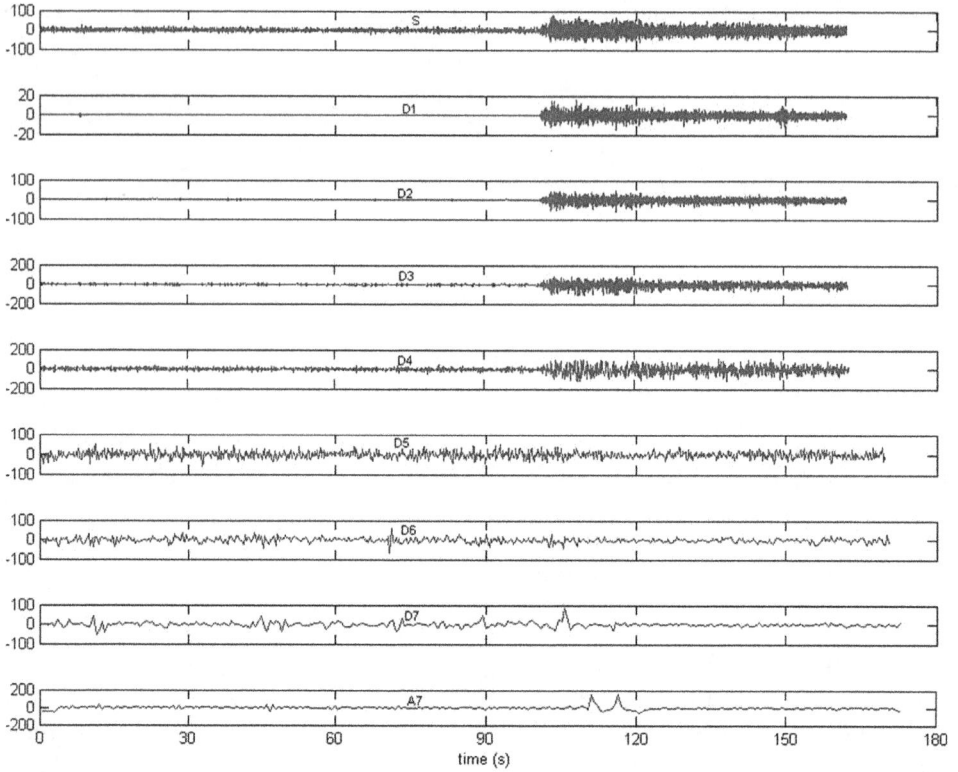

FIGURE 13.4
Wavelet decomposition using Db4 of signal S.

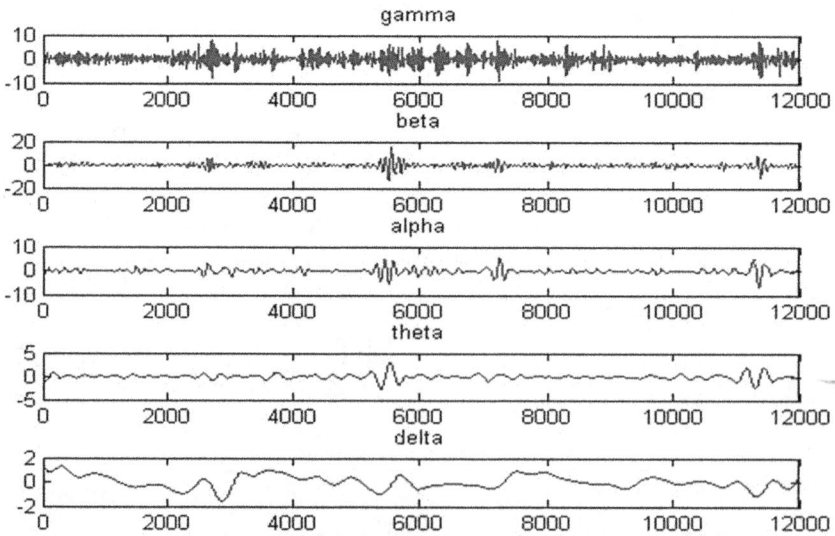

FIGURE 13.5
Frequency sub-band classification with ASD with Age Group 4.

13.5.3 Entropy Calculation of the Frequency Bands for Different Age Groups

Results shown in Figure 13.7 through 13.16 present the specific brain areas like frontal, occipital etc. which can distinguish different age groups of children by wavelet entropy method. Figures 13.6 and 13.7 represent different bands of frequencies. In Figures 13.7 through 13.16 the bar represents entropy and its height is shown by value P.

It is seen that for four-year-old children entropy calculations at the frontal lobe are (1.2–2), (1.6–2.5), (0.7–1.2), (1.1–2); temporal left lobe (1.25–1.38), (0.45–0.85), (1.25–2.0), (1.37–1.88); central lobe (1.25–2.2), (1.68–2.4), (0.65–0.8); temporal right lobe (0.75–1.25), (1.25–1.9), (0.65–0.99), (0.03–0.04); and occipital lobe (0.3–0.32), (1.6–2.4), (1.7–2.7) ,(1.35–2).

For five-year-old children, the entropy calculations for different lobes are as follows: frontal lobe are (0.00–0.00),(0.00–0.00), (0.5–0.75) (0.4–1.2); temporal left lobe (0.7–1.25), (0.10–0.12), (0.05–.1), (1.37–1.88); central lobe (0.10–0.11), (0.12–0.13), (0.1–0.2); temporal right lobe (0.25–0.35), (0.5–1.09), (0.65–1.17), (0.75–1.18); and occipital lobe (0.10–0.11), (0.7–1.2), (1.0–1.7), (1.0–1.6).

When entropy methods are applied it is observed that some specific areas show significant difference for different age groups. For the four-year-olds group, the significant amounts are concentrated on the central area for ASD children. For the five-year-olds group, there is a difference between the entropies of at the frontal and temporal left areas. It can be further revealed which entropy is more effective and which channels are more significant for different age groups affected by ASD.

13.6 Conclusions

Wavelet has been used extensively in biomedical signal processing. It has been proven as an efficient tool for classification and getting hidden information about the signal. Here

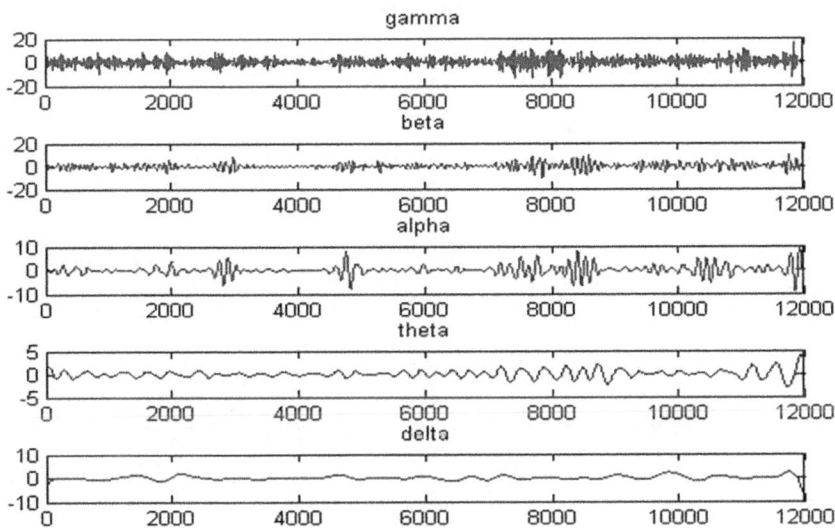

FIGURE 13.6
Frequency sub-band classifications with ASD with Age Group 5.

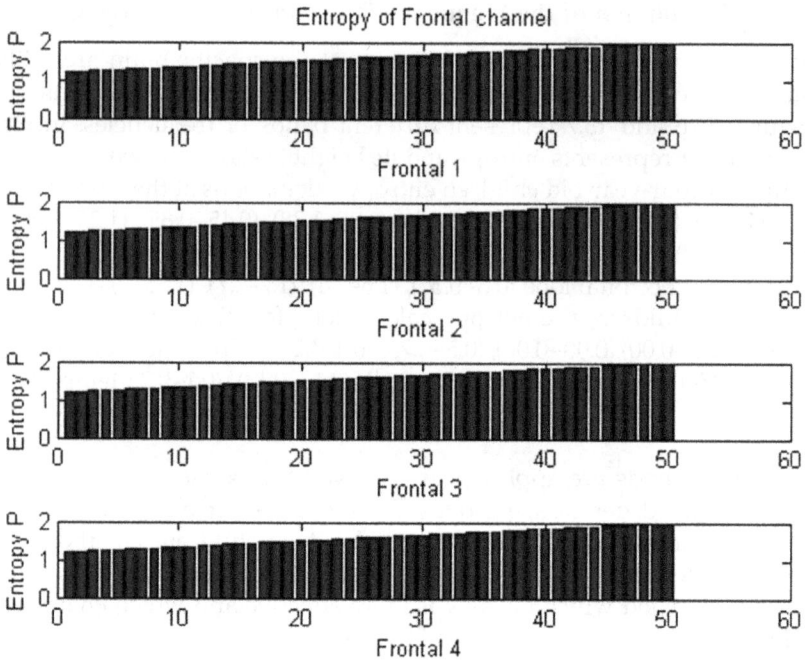

FIGURE 13.7
Accumulated entropy of frontal channels for Age Group 4.

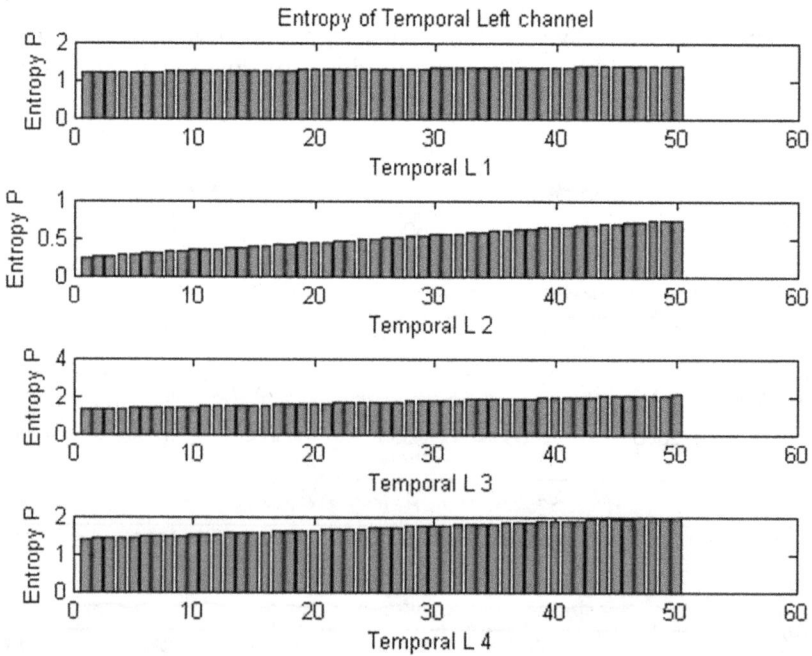

FIGURE 13.8
Accumulated entropy of temporal left channels for Age Group 4.

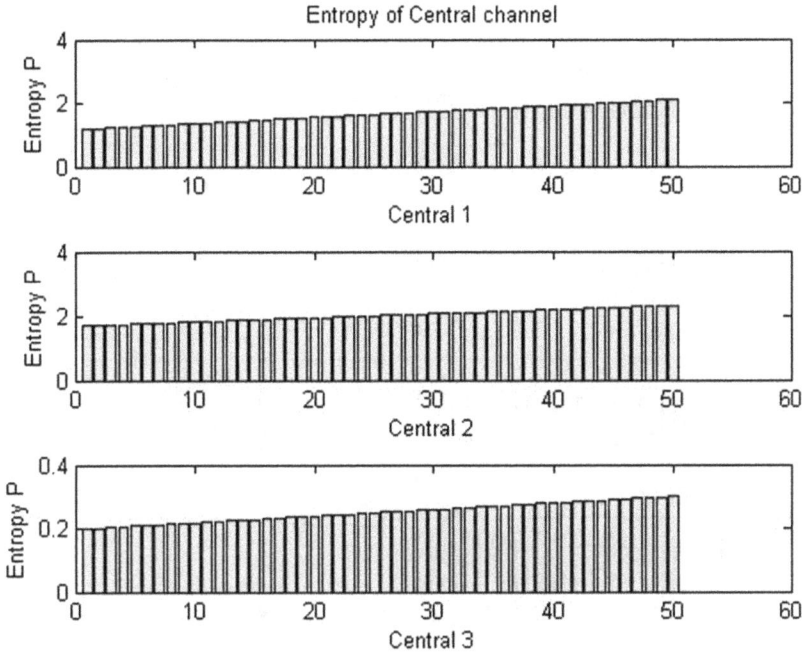

FIGURE 13.9
Accumulated entropy of central channels for Age Group 4.

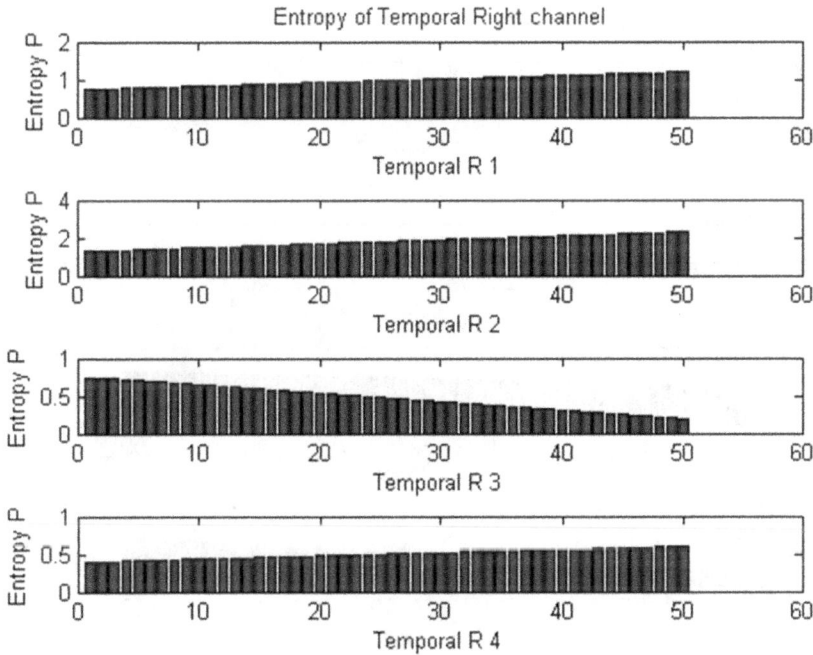

FIGURE 13.10
Accumulated entropy of temporal right channels for Age Group 4.

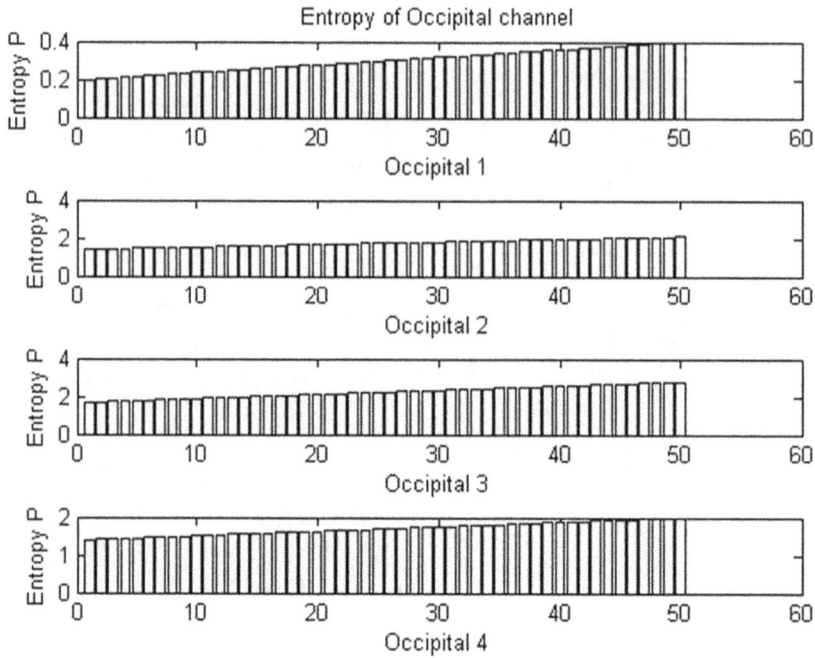

FIGURE 13.11
Accumulated entropy of occipital channels for Age Group 4.

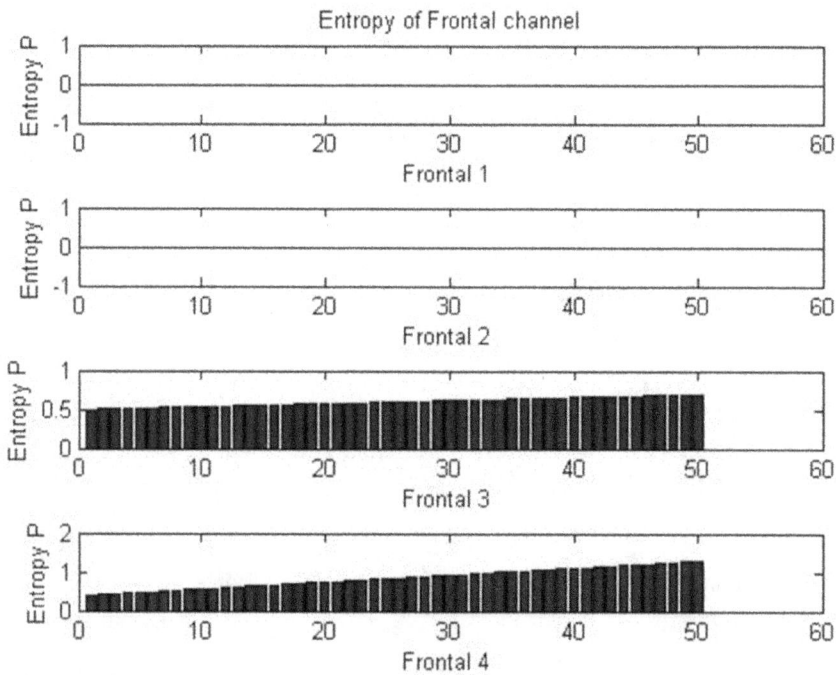

FIGURE 13.12
Accumulated entropy of frontal channels for Age Group 5.

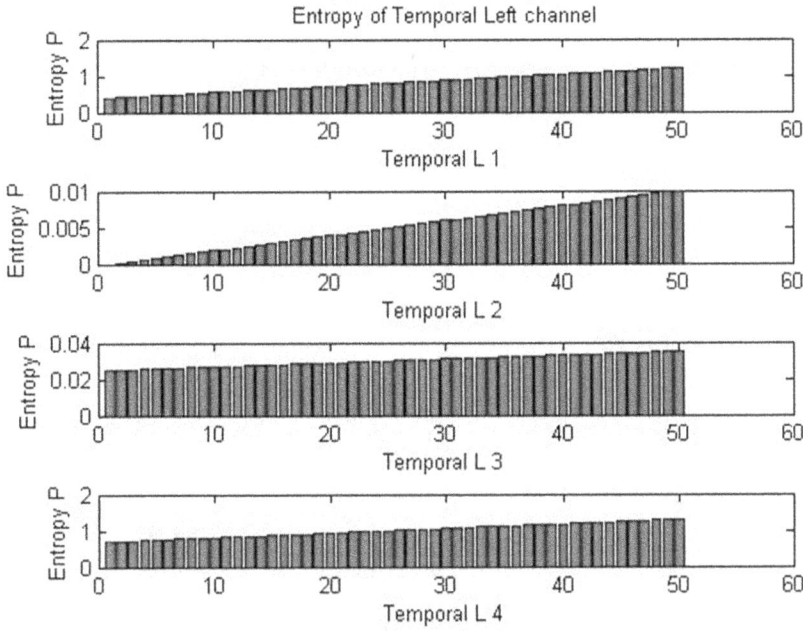

FIGURE 13.13
Accumulated entropy of temporal left channels for Age Group 5.

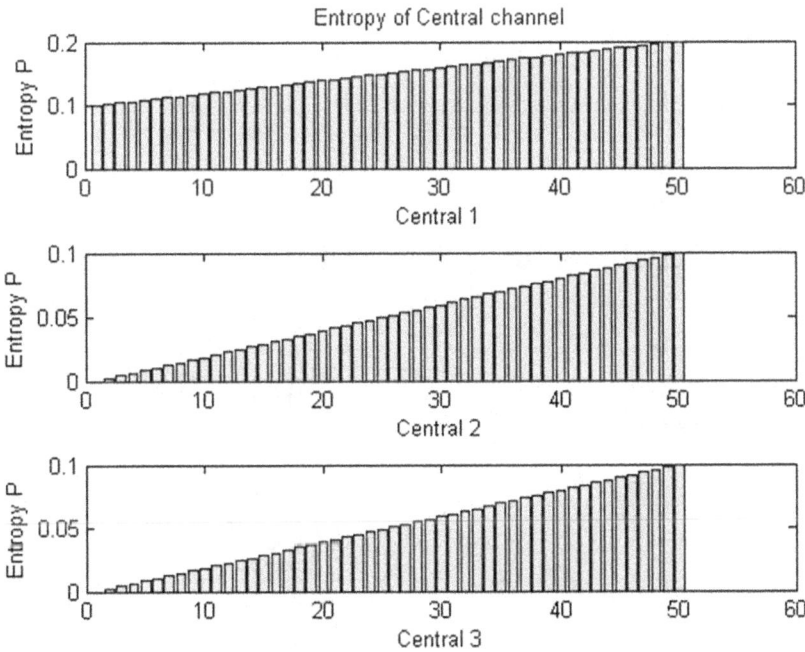

FIGURE 13.14
Accumulated entropy of central channels for Age Group 5.

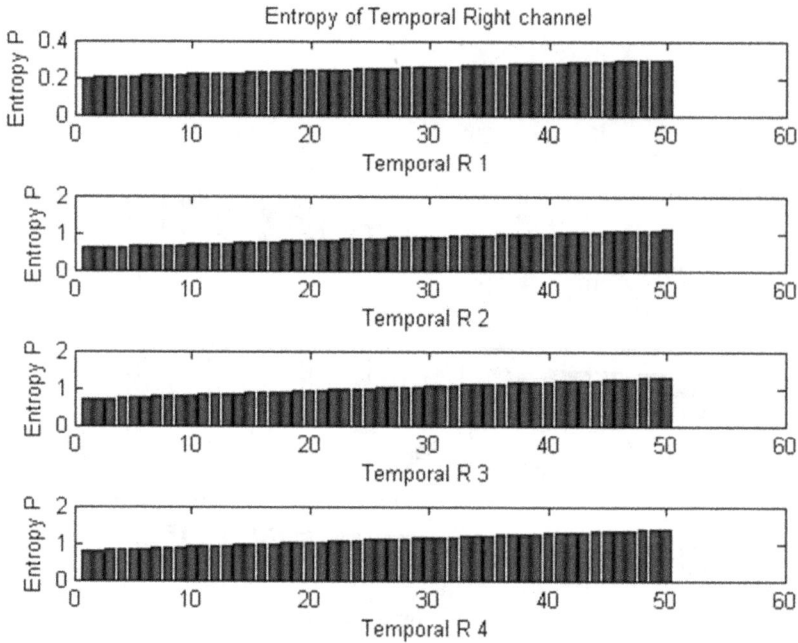

FIGURE 13.15
Accumulated entropy of temporal right channels for Age Group 5.

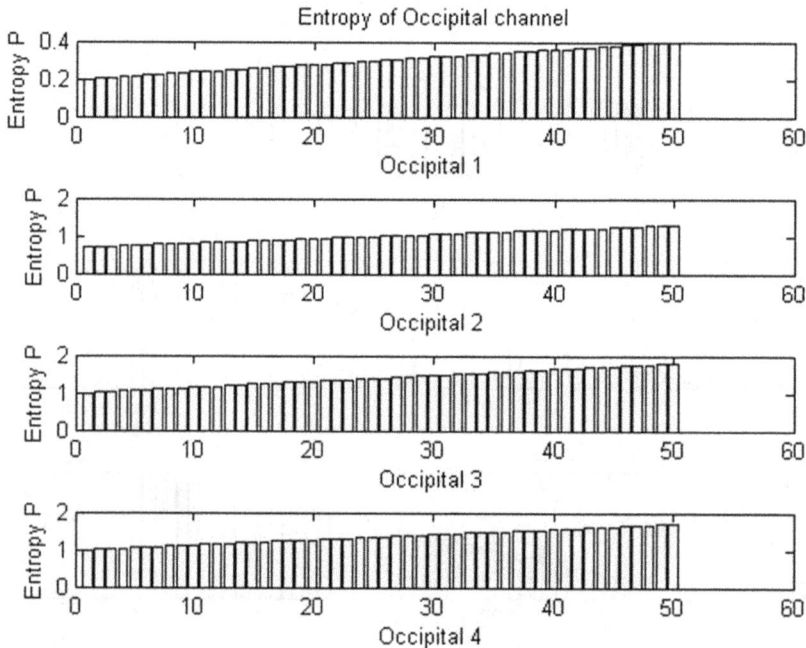

FIGURE 13.16
Accumulated entropy of occipital channels for Age Group 5.

wavelet is used for classification between different bands and further extended up to the wavelet entropy method to calculate entropy in different regions of the brain with the help of EEG signal using the 10–20 International System. It is concluded that the difference in brain regions and their wavelet entropy calculations is an efficient method to distinguish children suffering from ASD. For example, in four-year-olds there is significant difference in the central region and for five-year-olds the significant difference can be seen in the frontal and central regions of the brain. Results clearly provide evidence that the brain region changes as age increases.

References

1. Touchman Roberto. Autism and cognition within epilepsy: Social matters. *Epilepsy Currents* 2015 Jul–Aug;15(4):202–5.
2. O'Reilly C, JD Lewis, M Elsabbagh. Is functional brain connectivity atypical in autism? A systematic review of EEG and MEG studies. *PLOS ONE* 2017;12(5):e0175870.
3. Han YL, K Wang, J Jia, W Wu. Changes of EEG spectra and functional connectivity during an object-location memory task in Alzheimer's disease; *Frontiers in Behavioral Neuroscience* 31 May 2017. doi: 10.3389/fnbeh.2017.00107
4. Katarzyna J Blinowska, Franciszek Rakowski, Maciej Kaminski, Fabrizio De Vico Fallani, Claudio Del Percio, Roberta Lizio, Claudio Babiloni et al. Functional and effective brain connectivity for discrimination between Alzheimer's patients and healthy individuals: A study on resting state EEG rhythms. *Clinical Neurophysiology* 2017;128(4):667–680.
5. Lie OV, P van Mierlo. Seizure-onset mapping based on time-variant multivariate functional connectivity analysis of high-dimensional intracranial EEG: A Kalman filter approach. *Brain Topography* 2017;30(1):46–59.
6. Jun-Sang Sunwoo, Sanghun Lee, Jung-Hoon Kim, Jung-Ah Lim, Tae-Joon Kim, Jung-Ick Byun, Min Hee Jeong, Kwang Su Cha, Jeong Woo Choi, Kyung Hwan Kim, Soon-Tae Lee, Keun-Hwa Jung, Kyung-Il Park, Kon Chu, Manho Kim, Sang Kun Lee, Ki-Young Jung Altered functional connectivity in idiopathic rapid eye movement sleep behavior disorder: A resting-state EEG study. Sleep 2017 Jun 1;40(6).doi: 10.1093/sleep/zsx058.
7. Just MA, VL Cherkassky, TA Keller, NJ Minshew. Cortical activation and synchronization during sentence comprehension in high-functioning autism: Evidence of underconnectivity. *Brain* 2004;127(8):1811–21.
8. Wass S. Distortions and disconnections: Disrupted brain connectivity in autism. *Brain and Cognition* 2011;75(1):18–28.
9. McDonough IM, K Nashiro. Network complexity as a measure of information processing across resting-state networks: Evidence from the Human Connectome Project. *Frontiers in Human Neuroscience* 2014;8:409.
10. Yasser Ghanbari, Luke Bloy, J Christopher Edgar, Lisa Blaskey, Ragini Verma, Timothy PL Roberts. Joint analysis of band-specific functional connectivity and signal complexity in autism. *Journal of Autism and Developmental Disorders* 2015;45(2):444–460.
11. Misic B, VA Vakorin, T Paus, AR McIntosh. Functional embedding predicts the variability of neural activity. *Frontiers in Systems Neuroscience* 2011;5:90.
12. Li X, S Cui, LJ Voss. Using permutation entropy to measure the electroencephalographic effects of sevoflurane. *Anesthesiology* 2008;109(3):448–56.
13. Tan O, S Aydin, G Sayar, D Gursoy. EEG complexity and frequency in chronic residual schizophrenia. *Anadolu Psikiyatri Dergisi* 2016;17(5):385–92.
14. Artan NS. EEG analysis via multiscale Lempel-Ziv complexity for seizure detection. *38th Annual International Conference of the IEEE Engineering in Medicine and Biology Society (Embc)* 2016;2016:4535–538.

15. Mason RA, DL Williams, RK Kana, N Minshew, M Just. A theory of mind disruption and recruitment of the right hemisphere during narrative comprehension in autism. *Neuropsychologia* 2008;46(1):269–80.

16. Piryatinska A, B Darkhovsky, A Kaplan. Binary classification of multichannel EEG records based on the is an element of-complexity of continuous vector functions. *Compute Methods Programs Biomed* 2017;152:131–9.

17. Perkins TJ, MA Stokes, JA McGillivray, AJ Mussap, IA Cox, JJ Maller, et al. Increased left hemisphere impairment in high-functioning autism: A tract based spatial statistics study. *Psychiatry Research – Neuroimaging* 2014;224(2):119–23.

18. Zhenhu Liang, Yinghua Wang, Xue Sun, Duan Li, Logan J Voss, Jamie W Sleigh, Satoshi Hagihira, Xiaoli Li. EEG entropy measures in anesthesia. *Frontiers in Computational Neuroscience* 2015;9:16.

19. Vissers ME, MX Cohen, HM Geurts. Brain connectivity and high functioning autism: A promising path of research that needs refiened models. *Neuroscience and Biobehavioral Reviews* 2012;36(1):604–25.

20. Filipek PA, PJ Accardo, S Ashwal, GT Baranek, EH Cook, G Dawson, et al. Practice parameter: Screening and diagnosis of autism report of the quality standards sub-committee of the American Academy of Neurology and the Child Neurology Society. *Neurology* 2000;55(4):468–79.

21. Kang J, H Chen, X Li, X Li. EEG entropy analysis in autistic children. *Journal of Clinical Neuroscience* 28 Nov 2018;62:199–206. doi: 10.1016/j.jocn.2018.11.027

22. Ibrahim S, R Djemal, A Alsuwailem. Electroencephalography (EEG) signal processing for epilepsy and autism spectrum disorder diagnosis. *Biocybernetics and Biomedical Engineering* 2018;38(1):16–26. doi: 10.1016/j.bbe.2017.08.006.

23. Askari E, SK. Setarehdanb, A Sheikhanic, MR Mohammadi, M Teshnehlab, et al. Modeling the connections of brain regions in children with autism using cellular neural networks and electroencephalography analysis. *Artificial Intelligence in Medicine* 2018;89:40–50.

24. Grossi E, C Olivieri, M Buscema. Diagnosis of autism through EEG processed by advanced computational algorithms: A pilot study. *Computer Methods and Programs in Biomedicine* 2017;142:73–9. doi: 10.1016/j.cmpb.2017.02.002.

25. Sharma M, A Dhere, RB Pachori, UR Acharya. An automatic detection of focal EEG signals using new class of time–frequency localized orthogonal wavelet filter banks. *Knowledge-Based Systems* 2017;118:217–27.

26. Faust O, UR Acharya, H Adeli, A Adeli. Wavelet-based EEG processing for computer-aided seizure detection and epilepsy diagnosis. *Seizure* 2015;26:56–64. doi: 10.1016/j.seizure.2015.01.012.

27. Duffy FH, A Shankardass, GB McAnulty, H Als. The relationship of Asperger's syndrome to autism: A preliminary EEG coherence study. *BMC Medicine* 2013;11:175. doi:10.1186/1741-7015-11-175.

28. Ahmadlou M, H Adeli, A Adeli. Fuzzy synchronization likelihood-wavelet methodology for diagnosis of autism spectrum disorder. *Journal of Neuroscience Methods* 2012;211(2):203–9.

29. Adeli H, S Ghosh-Dastidar, N Dadmehr. A wavelet-chaos methodology for analysis of EEGs and EEG subbands to detect seizure and epilepsy. *IEEE Transactions on Bio-Medical Engineering* 2007;54(2):205–11.

30. www.epilepsysociety.org.uk.

14

De-Noising Raman Spectra Using Total Variation De-Noising with Iterative Clipping Algorithm

Padmesh Tripathi, Nitendra Kumar, and A. H. Siddiqi

CONTENTS

14.1 Introduction...225
14.2 Material and Method..227
    14.2.1 Material...227
    14.2.2 Total Variation (TV) De-Noising..227
    14.2.3 Algorithm for TV De-Noising...227
    14.2.4 Numerical Implementation and Performance Analysis 228
14.3 Results and Discussion 228
14.4 Conclusions...230
References..230

14.1 Introduction

Raman spectroscopy is a kind of molecular scattering spectroscopy that is characterized by frequency excursion that can be used to access information about the molecular structure of materials. It is a powerful non-destructive tool used in the estimation of chemical and physical properties of materials. It has a broad application in bio-medical, chemical, environmental, and industrial fields. The weak nature of Raman scattering results in the spectra with very low SNR can greatly influence the analytical results. There are three possible forms of noise in Raman spectra: high-frequency noise, which comes from the data acquisition electronic circuit and other sources of system variation; low-frequency noise, which arises from ambient light entering the spectrograph and fluorescence emission from the sample; and cosmic spikes, which are spurious and appear as very narrow spikes.

De-noising a signal is the process of removing the noise from the signal. Due to its applicability in many fields, it has been the center of attraction for engineers and scientists. Several techniques and methods have been used by researchers in recent years. Such a de-noising technique is based on total variation which was pioneered by Rudin et al. (1992). The generalization and extension of theories of total variation were done by Bredies et al. (2010), Hu and Jacob (2012), and Rodrigues and Wohlberg (2009). In regularization using total variation, minimization of a cost function is used to obtain the de-noising filter. All those algorithms that give a solution to the optimization problem of de-noising can be applied in the implementation of the total variation regularization. Many algorithms have been used to solve the problem. Selesnick and Bayram (2014) have developed an algorithm using the approach of Chambolle (2000), which has been used in this chapter. There are only a few reports on application of total variation for de-noising the Raman spectra. Liao et al. (2015) have used total variation to de-noise stimulated Raman spectroscopic images. Tripathi and Siddiqi (2016) have used total variation in de-noising Raman spectra solving an inverse problem.

The de-noising of Raman spectra is one of the important parts of spectral analysis. Two types of methods have been used for de-noising the Raman spectra: filtering and smoothing (Clupek et al., 2007). For filtering, the main methods used are: finite impulse response filtration, wavelet transform, and factor analysis. In finite impulse response (FIR) filtration, spectral shape is preserved, but it needs a lot of computations (Clupek et al., 2007). Wavelet transform has been established with the Fourier transform as a data-processing method in analytical fields. Several variants of wavelet transform (Ehrentreich and Summchen, 2001; Wang et al., 2006; Kumar et al., 2014; Maury and Revilla, 2015) have been used to de-noise the Raman spectra. In this method, spectral data is decomposed into the wavelet domain and reconstructed after thresholding for noise removal. In factor analysis (which is not applicable for filtering single spectrum), singular values play an important role. Singular values are the positive square roots of the eigenvalues. Using singular value decomposition, acquired spectra are projected into the orthonormal set of subspectra. Small singular values are taken as white noise. The role of large singular values is vital in maintaining the information of original spectra (Palacky et al., 2011). On the other hand, the classical Savitzky–Golay (SG) algorithm is very often used as the smoothing method to de-noise Raman spectra (Kwiatkowski et al., 2010). In this algorithm, the signal-to-noise ratio is increased without much distortion of the signal. Here, each segment of the original Raman spectrum in a small window is smoothed by fitting it to a low-degree polynomial using the least -squares. Besides these methods, Weiner estimation (Chen et al., 2013; Chen et al., 2014) has also been used for de-noising the Raman spectra.

Inverse problems deal with determining an input that produces an observed output, or determining an input that produces a desired output, often in the presence of noise. Mathematically, let $A: X \rightarrow Y$ be an operator that represents the association between the model parameter x and data y. Here X and Y are two spaces with appropriate structure. For a given input x, judging the outcome y is termed a direct problem. On the contrary, when given an observed outcome y, to judge an input x that produces it is termed an inverse problem. The solution minimizing

$$\left\| Ax - z \right\|_2^2 + \lambda \left\| Lx \right\|_1$$

is known as total variation regularized solution. Here, L is a smoothing operator that depends on application and λ is a parameter called a regularization parameter. Selection of λ plays an important role in solving an inverse problem (Wen and Chan, 2012; Langer, 2017).

In this chapter, Raman spectra of Zr doped $BaTiO_3$ (BZT) has been de-noised with the help of the total variation and the iterative clipping algorithm proposed by Selesnick and Bayram (2014). Two parameters—SNR and RMSE—are used to observe the performance of the algorithm. These are computed for different values of λ. The computed values of SNR and RMSE show that the iterative clipping algorithm with total variation is an effective algorithm to de-noise the Raman spectra.

14.2 Material and Method

In this chapter, the de-noising of Raman spectra of material Barium Zirconium Titanate is done using a method based on total variation.

14.2.1 Material

FT-Raman spectra of Zr doped $BaTiO_3$ (BZT) with composition $(x) = 0.010$ has been carried out at room temperature. The spectra has been recorded in back scattering geometry by means of a Varian (Digilab), a dedicated FT-Raman Spectrometer (Nd-YAG Laser, 1064 nm, 500 mW laser power) in the range of 100 cm^{-1}–4000 cm^{-1}.

14.2.2 Total Variation (TV) De-Noising

For a given M-point signal s(m) such that $0 \leq m \leq M$-1, the total variation of s(m) is defined by:

$$TV(s) = \sum_{m=1}^{M-1} |s(m) - s(m-1)| = \|Ds\|_1$$

where D is a matrix of order M-1 $\times$ M.

For a given measured data y(m) containing noise and represented by

$$y(m) = s(m) + \varepsilon(m), \quad m = 0, 1, 2, \ldots, M-1 \tag{14.1}$$

where s(m) is a piecewise constant signal and ε(m) is white Gaussian noise, estimation of s(m) is the inverse problem. Estimate of the signal s(m) using TV de-noising is obtained by the solution of the optimization problem

$$\arg \min_{s \in R^M} \left\{ J(s) = \frac{1}{2} \|y - s\|_2^2 + \lambda \|Dx\|_1 \right\} \tag{14.2}$$

It is worth noting here that the degree of smoothing is controlled by λ (Santamarina and Fratta, 2005).

14.2.3 Algorithm for TV De-Noising

In Equation(14.2), replacing D by a matrix A of order M $\times$ N and using the min–max property (Rockafellar, 1970; Hiriart-Urruty and Lemarechal, 1993) and the majorization-minimization method (Figueiredo et al., 2007), Selesnick and Bayram (2014) have developed the iterative clipping algorithm as:

$$x^{(i+1)} = y - \frac{\lambda}{2} A^t z^{(i)}$$

$$z^{(i+1)} = \text{clip}\left(z^{(i)} + \frac{2}{\alpha \lambda} Ax^{(i+1)}, 1 \right) \tag{14.3}$$

where *i* is the iteration index.

14.2.4 Numerical Implementation and Performance Analysis

Total variation de-noising based on Equation(14.3) is implemented by the MATLAB® program described by Selesnick and Bayram (2014). Here $\alpha=4$ is set for de-noising. In the MATLAB program, D is implemented with the *diff* command. In the algorithm, iterations are fixed at 100.

Performance of algorithm is evaluated by SNR and RMSE computed for different values of λ shown in Table 14.1.

Figure 14.1 represents the original data while Figures 14.2, 14.3, and 14.4 represent the de-noised data.

14.3 Results and Discussion

In this chapter, de-noising Raman spectra is modeled as an inverse problem with total variation and is expressed as the minimization of a non-differentiable cost function. Minimization of a non-differentiable cost function is a complex problem. It is minimized

TABLE 14.1

Computed SNR and RMSE

Values of λ	SNR	RMSE
0.1	43.3114	0.0129
0.5	35.0754	0.0332
1.0	31.6708	0.0491

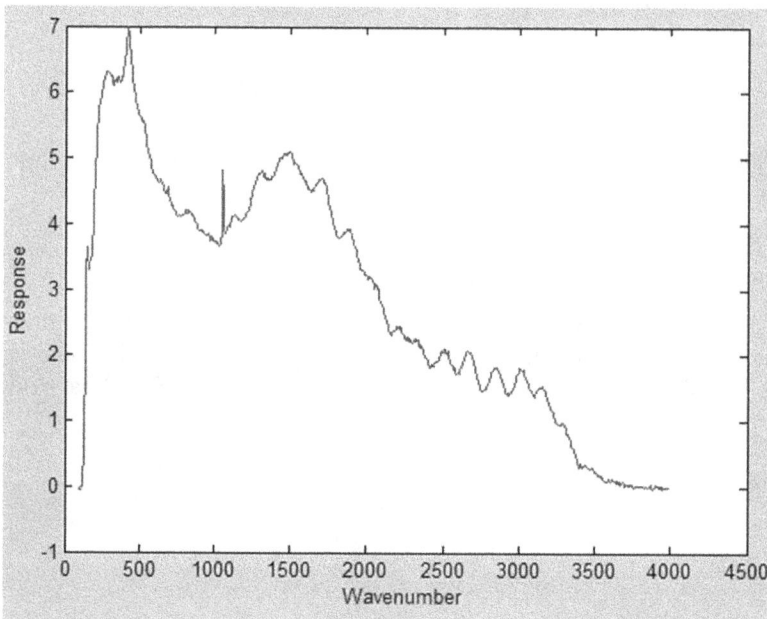

FIGURE 14.1
Original data.

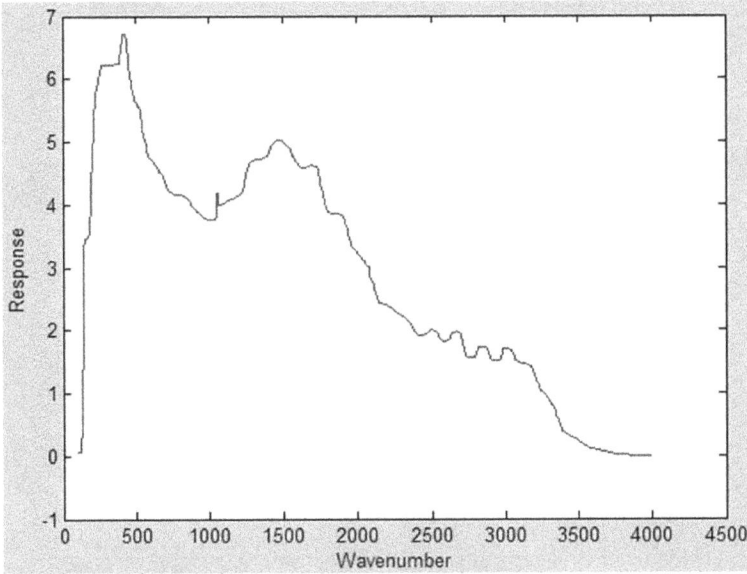

FIGURE 14.2
TV de-noising with iterative clipping algorithm for $\lambda = 0.1$.

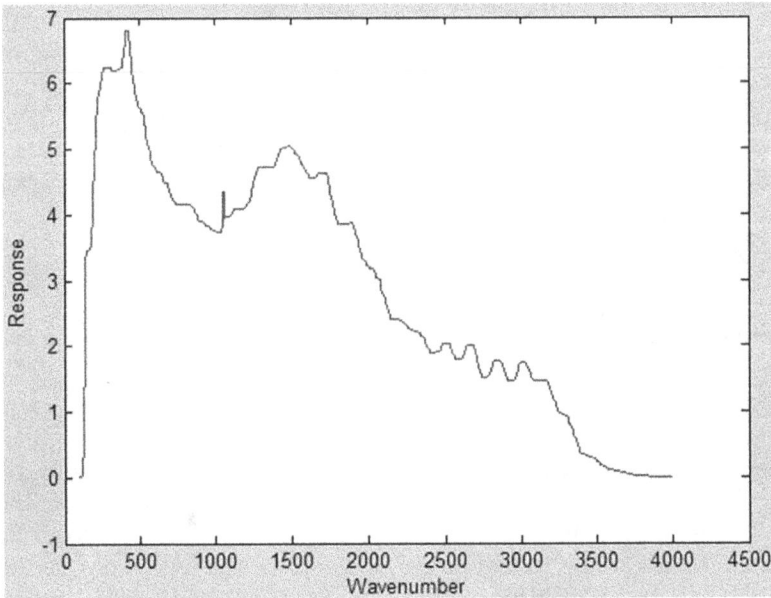

FIGURE 14.3
TV de-noising with iterative clipping algorithm for $\lambda = 0.5$.

using the iterative clipping algorithm. It has been used for de-noising the Raman spectral data of BZT (Barium Zirconium Titanate). More SNR implies more de-noising and, on the contrary, more RMSE implies less de-noising. Table 14.1 shows the values of SNR and RMSE for different values of λ. From the results obtained, we observe that iterative clipping algorithm works well for de-noising the Raman spectral data. We also observe that as

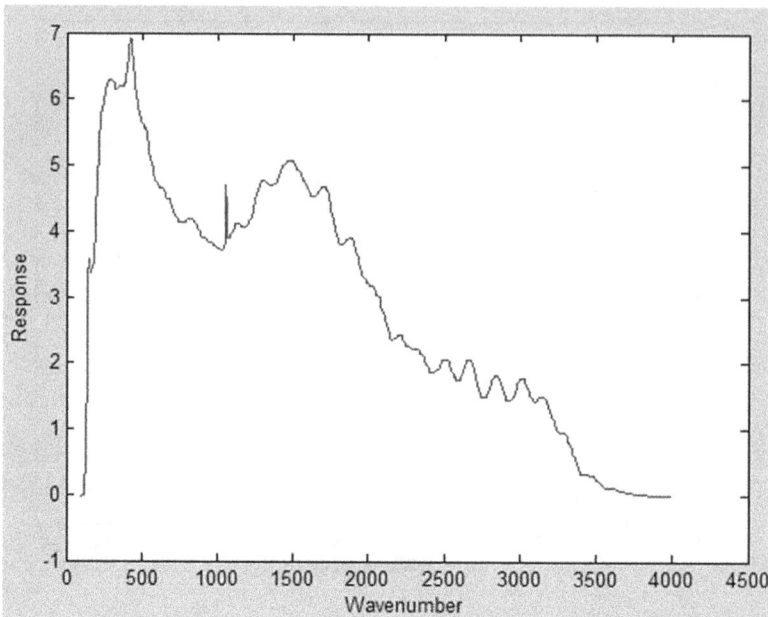

FIGURE 14.4
TV de-noising with iterative clipping algorithm for $\lambda = 1.0$.

λ increases, SNR decreases, and RMSE increases. Figure 14.1 represents the original data. Figure 14.2 represents the de-noised data with SNR = 43.3114 & RMSE = 0.0129 for $\lambda = 0.1$, Figure 14.3 represents the de-noised data with SNR = 35.0754 & RMSE = 0.0332 for $\lambda = 0.5$ and Figure 14.4 represents the de-noised data with SNR = 31.6708 & RMSE = 0.0491 for $\lambda = 1.0$.

Through de-noising, the quality of spectra is enhanced. Thus, the physical and chemical properties of material can be analyzed more accurately.

14.4 Conclusions

Increased values of signal to noise ratio for different values of regularization parameter show that iterative clipping algorithm suits for de-noising the Raman spectral data. As the values of regularization parameter increase, the SNR ratio decreases, and RMSE increases. This shows that de-noising depends on the values of the regularization parameter. Larger noise levels call for larger regularization parameter λ. This is reflected from the Table 14.1. The calculated SNR and RMSE for different values of regularization parameter show a satisfactory result.

References

Bredies, K., Kunisch, K., and Pock, T. 2010. Total generalized variation. *SIAM Journal on Imaging Sciences* 3(3): 492–526.

Chambolle, A. 2000. Total variation minimization and a class of binary MRF models. *Energy Minimization Methods in Computer Vision and Pattern Recognition*, Lecture Notes in Computer Sciences, Springer, Heidelberg, Germany, 3757, 136–52.

Chen, S., Lin, X., Yuen, C., Padmanabhan, S., Beuerman, R. W., and Liu, Q. 2014. Recovery of Raman spectra with low signal-to noise ratio using Wiener estimation. *Optics Express* 22(10): 12102–14.

Chen, S., Ong, Y. H., and Liu, Q. 2013. Fast reconstruction of Raman spectra from narrow-band measurements based on Wiener estimation. *Journal of Raman Spectroscopy* 44(6): 875–81.

Clupek, M., Matejka, P., and Volka, K. 2007. Noise reduction in Raman spectra: Finite impulse response filtration versus Savitzky–Golay smoothing. *Journal of Raman Spectroscopy* 38(9): 1174–79.

Ehrentreich, F., and Summchen, L. 2001. Spike removal and de-noising of Raman spectra by wavelet transform methods. *Analytical Chemistry* 73(17): 4364–73.

Figueiredo, M., Bioucas-Dias, J., and Nowak, R. 2007. Majorization - minimization algorithms for wavelet-based image restoration. *IEEE Transactions on Image Processing* 16(12): 2980–91.

Hiriart-Urruty, J. B., and Lemarechal, C. 1993. *Convex Analysis and Minimization Algorithms I: Fundamentals*. Berlin: Springer-Verlag.

Hu, Y., and Jacob, M. 2012. Higher degree total variation (hdtv) regularization for image recovery. *IEEE Transactions on Image Processing* 21(5): 2559–71.

Kumar, N., Siddiqi, A. H., and Alam, K. 2014. Raman spectral data de-noising based on wavelet analysis. *International Journal of Computer and Applications* 102(16): 20–2.

Kwiatkowski, A., Gnyba, M., Smulko, J., and Wierzba, P. 2010. Algorithms of chemicals detection using Raman spectra. *Metrological Measurement Systems* 17(4): 549–59.

Langer, A. 2017. Automated parameter selection for total variation minimization in image restoration. *Journal of Mathematical Imaging and Vision* 57(2): 239–68.

Liao, C. S., Choi, J. H., Zhang, D., Chan, S. H., and Cheng, J. X. 2015. De-noising stimulated Raman spectroscopic images by total variation minimization. *The Journal of Physical Chemistry C* 119(33): 19397–403.

Maury, A., and Revilla, R. I. 2015. Autocorrelation analysis combined with a wavelet transform method to detect and remove cosmic rays in a single Raman spectrum. *Applied Spectroscopy* 69(8): 984–92.

Palacky, J., Mojzes, P., and Bok, J. 2011. SVD-based method for intensity normalization, background correction and solvent subtraction in Raman spectroscopy exploiting the properties of water stretching vibrations. *Journal of Raman Spectroscopy* 42(7): 1528–39.

Rockafellar, R. T. 1970. *Convex Analysis*. Princeton: Princeton University Press.

Rodriguez, P., and Wohlberg, B. 2009. Efficient minimization method for a generalized total variation functional. *IEEE Transactions on Image Processing* 18(2): 322–32.

Rudin, L., Osher, S., and Fatemi, E. 1992. Nonlinear total variation based noise removal algorithms. *Physica D* 60(1–4): 259–68.

Santamarina, J. C., and Fratta, D. 2005. *Discrete Signals and Inverse Problems: An Introduction for Engineers and Scientists*. Chichester, NJ: Wiley and Sons.

Selesnick, I., and Bayram, I. 2014. Total-variation-filtering. http://cnx.org/contents/8NLpidGL@1/Total-Variation-Filtering (accessed December 14, 2018).

Tripathi, P., and Siddiqi, A. H. 2016. Solution of inverse problem for de-noising Raman spectral data with total variation using majorization-minimization algorithm. *International Journal of Computing Science and Mathematics* 7(3): 274–82.

Wang, Y. P., Wang, Y., and Spencer, P. 2006. Fuzzy clustering of Raman spectral imaging data with a wavelet-based noise-reduction approach. *Applied Spectroscopy* 60(7): 826–32.

Wen, Y. W., and Chan, R. H. 2012. Parameter selection for total-variation-based image restoration using discrepancy principle. *IEEE Transactions on Image Processing* 21(4): 1770–81.

15

An Efficient Identity-Based Mutual Authentication Protocol for Cloud Computing

Vinod Kumar and Musheer Ahmad

CONTENTS

15.1 Introduction...233
    15.1.1 Research Contribution 234
    15.1.2 The Organization of the Proposed Framework......................................234
15.2 Preliminaries 234
    15.2.1 Notations..234
    15.2.2 ID-Based Cryptography System 235
    15.2.3 Background of Elliptic Curve Group 235
15.3 The Proposed Protocol...236
    15.3.1 Set-Up Phase...236
    15.3.2 Extraction Phase...236
        15.3.2.1 Server private key extraction..237
    15.3.3 Mutual Authentication and Session Key Agreement Phase..................237
15.4 Security Investigation 238
15.5 Conclusion 240
References...240

15.1 Introduction

Cloud computing is an Internet-based technology which is providing resource utilization in communication system. In cloud computing the end-user is not conscious with regard to where data is stockpiled and how their data is being developed. They only admit data, proceed, and lastly, store in the cloud database. They may have the right to use the data at anytime, anywhere, if they are using internet communications. This technology is extremely flexible, scalable, and allocated in the cloud environment. In cloud computing computational sources are delivered to the client as a facility (Almorsy et al. (2016)). In the proposed framework, we followed the NIST characterization of cloud computing, a model for enabling convenient on-demand network access to a shared pool of configurable computing resources that can be rapidly provisioned and released with minimal management effort or service-provider interaction (Zhang et al. (2010)). Cloud computing technologies are divided into several services, such as: Infrastructure as a service, Platform as a service, Integration Platform as a service, Block-chain as a service, Software as a service,

Mobile backend as a service, Serverless computing, and Function as a service (Dinesha and Agrawal (2012); Jing and Jian-Jun (2010)).

In the cloud environment, outside attackers have to have a whole authentication procedure made necessary by the service producer, whenever they use the cloud facility. Frequently, a client registers through individual data, and a service supplier offers a client's acquired identifier and a validation technique in favor of client authentication once a registration is completed. In 2009, Yang and Chang (2009) suggested an ID- assisted authentication framework for mobiles which is based on ECC. In 2011, Chen et al. (2011) suggested an ECC-assisted authentication framework for cloud atmosphere and Chang and Choi (2011) a client authentication work for cloud system. After that, the user makes use of the identity and the authentication technique to work the client authentication whenever the customer starts to use cloud facilities (Nayak et al. (2012)). Mishra et al. (2013) and Khan et al. presented identity-based authentication schemes for the cloud environment; these show identity-based authentication and security in the cloud environment using different approaches. In 2017, Luo et al. presented an ID-based efficient authentication framework with a smart card applying ECC (Luo et al. (2017)). In 2018, Kumar et al. (2019b, 2018) suggested cloud-assisted authentication protocols in telecare medical information systems.

15.1.1 Research Contribution

We proposed a mutual authentication framework in the cloud computing environment. The proposed protocol has many security features and attributes. This work discusses characters such as:

- The mutual authentication property is satisfied between user and cloud server and manages security issues in cloud computing.
- Further, the proposed framework is strong against known session-specific at- tack, known-key attack, reflection attack, and unknown key-share attack, support anonymity, replay attack, no key control, PKG forward secrecy, key freshness, man-in-the-middle attack, message authentication, and off-line guessing attack.
- We evaluated the proposed framework with other presented protocols in the same domain and show that it is more secure.

15.1.2 The Organization of the Proposed Framework

The rest of this chapter is formulated as follows: Section 15.2, preliminaries; Section 15.3, the proposed protocol; Section 15.4, security evaluation; and Section 15.5, our conclusion.

15.2 Preliminaries

15.2.1 Notations

We use the followings notations/symbols in Table 15.1:

TABLE 15.1

Notations/Symbols

Symbol	Description
ECC	Elliptic curve cryptography
IBC	Identity-based cryptography system
p	A large prime
ID_i	The unique identity of entity i
$\varepsilon(F_p)$	Elliptic curve ε over a prime finite field F_p
ski	The session key generated entities i
PKG	The private key generator
U	The user
S	Cloud server
M	Master key
G	ECC-based additive group
g	Base point of G
E	Adversary
$\|\|$	Concatenation operation
Z_p^*	Additive group of order p
MASK	Mutual authentication and session key
$\oplus$	XOR-operation

15.2.2 ID-Based Cryptography System

The concept of IBC was developed by Shamir in 1984 to remove the transmission, verification, and maintenance of public-key validity. In IBC, the user submitted his/her identity to a unique trusted authority called Private Key Generator (PKG). After that, PKG provides private, part-private, and public keys to users and PKG has a secret system master key, M (Shamir (1984)).

15.2.3 Background of Elliptic Curve Group

Let p be the large prime and E indicate an elliptic curve over a prime finite field F_p, demarcated by an equation $y^2 = x^3 + cx + d$ mod q, where, $c, d \in_R F_p$ and $4c^3 + 27d^2$ mod $q \neq 0$. The additive elliptic curve group is construed as $G = \{(x,y) : x, y \in_R F_p ; (x,y) \in_R E\} \cup \{\Theta\}$, where the point Θ is identified as a point at infinity that perform as the identity element in G. The point addition in the elliptic curve is: $P = (x_P, y_P) \in_R G$ and $Q = (x_Q, y_Q) \in_R G$, where $Q \neq P$ then $P + Q = (x_i, y_i)$, where $x_i = \mu^2 - x_P - x_Q$ mod q, $y_i = (\mu(x_P - x_Q) - y_P)$ mod q, and $\mu = (y_Q - y_P)/(x_Q - x_P)$. The scalar multiplication on the group G is explained as $nP = P + P + P \ldots + P(n$ times).

- **Elliptic Curve Discrete Logarithms Problem (ECDLP):** For given $P, Q \in_R G$ find $t \in_R Z_q^*$ such that $P = tQ$, which is hard (Kumar et al. (2019a)).
- **Elliptic Curve Computational Diffie–Hellman Problem (ECCDHP):** For $a, b \in_R Z_q^*$ and g is the base point of G, given (g, ag, bg), then compute abg is hard to the group G (Khan et al. (2019)).

More information on EC, G and its application is available in Darrel et al. (2004) and Kumar et al. (2019b, 2018).

15.3 The Proposed Protocol

In this section, we discuss the proposed protocol. The architecture of the proposed protocol is shown in Figure 15.1. The presented protocol has three phase, which are as follows:

15.3.1 Set-Up Phase

PKG inputs a security value l, returns safety values, and M, for given PKG, executes the following steps:

Step 1. Inputs g of G.

Step 2. Inputs $M \in Z_p^*$ and sets public key $PK = Mg$.

Step 3. Chooses collision free one way hash functions:

$H_1: \{0, 1\}^* \times G \rightarrow Z_p^*$.

$H_2: \{0, 1\}^* \times \{0, 1\}^* \times \{0, 1\}^l \times Z_p^* \times Z_p^* \rightarrow \{0, 1\}^l$.

$H_3: \{0, 1\}^* \times \{0, 1\}^* \times G \times G \times Z_p^* \times Z_p^* \rightarrow \{0, 1\}^l$.

After that, PKG publishes systems parameters F_p, E, G, l, g, PK, H_1, H_2, H_3, and M keep safe.

15.3.2 Extraction Phase

User private key extraction
  U takes extraction as:

Step 1. U uploads her/his ID_U to PKG via secure channel.

Step 2. PKG verifies the validation of ID_U. If validation succeeds, then PKG takes a random number $x_U \in Z_p^*$.

FIGURE 15.1
Architecture of purposed protocol.

Step 3. Computes $X_U = x_U g$ and $h_U = H_1(ID_U \| X_U)$.

Step 4. With the help of M, PKG computes $Y_U = x_U + M h_U$. After that, PKG delivers $< X_U, Y_U >$ to U via a secure communication channel.

Step 5. On obtaining $< X_U, Y_U >$, U checks the condition $Y_U g = ? X_U + (H_1(ID_U \| X_U)) PK$. After that, U sets his/her public key $P_U = Y_U g$.

The flowchart of this phase is displayed in Figure 15.2.

15.3.2.1 Server private key extraction

S takes extraction as:

Step 1. S submits her/his ID_S to PKG via secure channel.

Step 2. PKG verifies the validation of ID_S. If validation succeeds, then, PKG takes a random value $x_S \in Z_p^*$.

Step 3. Computes $X_S = x_S g$ and $h_S = H_1(ID_S \| X_S)$.

Step 4. With the help of M, PKG computes $Y_S = x_S + M h_S$. After that, PKG delivers $< X_S, Y_S >$ to S key via a secure channel.

Step 5. On receiving $< X_S, Y_S >$, S checks the condition $Y_S g = ? X_S + (H_1(ID_S \| X_S)) PK$. After that, S sets his/her public key $P_U = Y_U g$.

The flowchart of this phase is displayed in Figure 15.3.

15.3.3 Mutual Authentication and Session Key Agreement Phase

Here, U and S communicate to each other and agree to a common session key. MAKS contains the following steps:

Step 1. U generates a random value $u \in Z_p^*$, computes $N_U = ug$, $ID_{Un} = ID_U \oplus H_1(ID_U \| g)$, token $T_U = H_3(ID_U \| P_U \| P_S)$ and sends message $M_1 = < N_U, ID_{Un}, T_U, u, T_1 >$ to S via public channel.

Step 2. On getting M_1, S verifies $T_2 - T_1 \leq \Delta T$. Then, S generates random value $s \in Z_p^*$, computes $N_S = sg$, $ID_U^* = ID_{Un} \oplus H_1(ID_U \| g)$ of U and $T_S = H_3(ID_U^* \| P_U \| P_S)$. Further S

User U	PKG
Sends $< I_U >$ $\cdots\cdots\cdots\cdots \rightarrow$ (via secure channel)	
	Generates $x_U \in Z_p^*$ Computes $x_U = x_U g$ Computes $h_U = H_1(ID_U \| X_U)$ Computes $Y_U = x_U + M.h_U$ Sends $< X_U, Y_U >$ $\leftarrow \cdots\cdots\cdots\cdots$ (via secure channel)
$Y_U g = ? X_U + (H_1(ID_U \| X_U)) PK$ if verified then $P_U = Y_U g$	

FIGURE 15.2
User private key extraction.

User S	PKG
Sends $< ID_S >$ ·············· → (via secure channel)	
	Generates $x_S \in Z_p^*$ Computes $x_S = x_S g$ Computes $h_S = H_1(ID_S \| X_S)$ Computes $Y_S = x_S + M.h_S$ Sends $< X_S, Y_S >$ ← ············· (via secure channel)
$Y_S g = ? X_S + (H_1(ID_S \| X_S)) PK$ if verified then $P_S = Y_S g$	

FIGURE 15.3
Server private key extraction.

verifies $T_S = ? T_U$. Further, S computes mutual authenticated code $MAC_S = H_2(ID_U^*$ $\|ID_S\|T_S\|P_S\|P_U \|uN_S)$, session key $sk_S = H_2(ID_U^* \|ID_S\|P_U \|P_S\|MAC_S\|uN_S\|T_3)$, $ID_{S1} = ID_S \oplus H_1(ID_S\|g)$ of S and sends message $M_2 = < N_S, ID_{S1}, T_S, s, MAC_S, T_3 >$ via public channel.

Step 3. On obtaining M_2, U verifies $T_4 - T_3 \leq \Delta T$. Further, U computes $ID_S^* = ID_{S1} \oplus H_1(ID_S\|g)$ of S, mutual authenticated code $MAC_S = H_2(ID_U \|ID_S^* \|T_U \|P_S\|P_U \|sN_U)$ and verifies $MAC_U = ? MAC_S$. Thus, U authenticates to S and executes session key $sk_S = H_2(ID_U \|ID_S^* \|P_U \|P_S \|MAC_U \|sN_U \|T_3)$.

Here, U and S ensure the validity of $MAC_U = MAC_S$. If successful, both U and S manage session key $sk = sk_U = sk_S$. Once the session key is established between U and S, they can communicate safely over a public communication channel. The details of this phase are shown in Figure 15.4.

15.4 Security Investigation

In this section, we have discussed security attributes and features which verify the presented protocol as:

- **Session key security:** In MAKS of this framework, U and S compute session following as:
  - S computes session key $sk_S = H_2(ID_U^* \|ID_S\|P_U \|P_S\|MAC_S\|uN_S\|T_3)$, and U executes session key $sk_S = H_2(ID_U \|ID_S^* \|P_U \|P_S\|MAC_U \|sN_U \|T_3)$. E cannot executes the session key sk_S or sk_U, where $MAC_U = H_2(ID_U \|ID_S^* \|T_U \|P_S\|P_U \|sN_U)$ and $MAC_S = H_2(ID_U^* \|ID_S\|T_S\|P_S P_U uN_S)$. MAC_S and MAC_U cannot be computed by E because of the definition hash function which is secured. Hence, the framework manages the session key.
- **Message authentication:** In MAKS, U and S verify message authentication property as:

User S	Cloud server S
$u \in Z_q^*$	
$N_U = ug$	
$ID_{U1} = ID_U \oplus H_1(ID_U\|g)$	
$T_U = H_3(ID_U\|P_U\|P_S)$	
$M_1 = <N_U, ID_{U1}, T_U, u, T_1>$	
. . . . $\longrightarrow$	
(via public channel)	
	$T_2 - T_1 \leq \Delta T$
	$s \in Z_q^*$
	$N_S = sg$
	$ID_U^* = ID_{U1} \oplus H_1(ID_U\|g)$
	$T_S = H_3(ID_U^*\|P_U\|P_S)$
	$T_S = ?T_U$
	$MAC_S = H_2(ID_U^*\|ID_S\|T_S\|P_S\|P_U\|uN_S)$
	$sk_S = H_2(ID_U^*\|ID_S\|P_U\|P_S\|MAC_S\|uN_S\|T_3)$
	$ID_{S1} = ID_S \oplus H_1(ID_S\|g)$
	$M_2 = <N_S, ID_{S1}, T_S, s, MAC_S, T_3>$
	$\longleftarrow$ . . . .
	(via public channel)
$T_4 - T_3 \leq \Delta T$	
$ID_S^* = ID_{S1} \oplus H_1(ID_S\|g)$	
$MAC_U = H_2(ID_U\|ID_S^*\|T_U\|P_S\|P_U\|sN_U)$	
$MAC_U = ?MAC_S$	
$sk_S = H_2(ID_U\|ID_S^*\|P_U\|P_S\|MAC_U\|sN_U\|T_3)$	

FIGURE 15.4
Mutual authentication and session key agreement phase.

- S receives message $M_1 = <N_U, ID_{U1}, T_U, u, T_1>$ and checks $T_2 - T_1 \leq \Delta T$ and hash function $T_S = ?T_U$.

- U receives message $M_2 = <N_S, ID_{S1}, T_S, s, MAC_S, T_3>$ and checks $T_4 - T_3 \leq \Delta T$ and hash function $MAC_U = ?MAC_S$.

If any E endeavors to change any communicated information between U and S, then U/S will recognize it. Hence, the presented protocol protests against the message authentication.

- **Known-key secrecy:** In MAKS, U computes N_U and S computes N_S. Here, $sN_U = uN_s = usg = sug$, where u and s random value taken from Z_p^*. Hence, the suggested framework manages a known-key secrecy property.

- **Known session-specific attack:** In the proposed protocol, U and S executes session key $sk = sk_U = sk_S$. It depends on MAC_S, MAC_U, ID_U, ID_S, T_S, P_S, P_U, uN_S, sN_u and T_3. These values depend on the random value u, s one-way hash function, and time-stamps. Those values are secured. Hence, the presented work is secure against this attack.

- **Man-in-the-middle attack:** In MAKS, U and S mutually authenticated each other without informing PKG. If PKG is vicious, they are trying to share incorrect information. On the other hand, to authenticate each other, U and S swap over MAC_U/MAC_S. To calculate MAC_U/MAC_S, information of the session key sk is prescribed even though sk is expected to be secret.

- **Replay attack:** In replay attack, the common countermeasures are time-stamp and random-value approaches. In the presented protocol, random-value and time-stamps are used. Hence, replay attack does not work in the presented protocol.

- **PKG forward secrecy:** If E obtains M of PKG, it means that E can compute the partial private keys of both U and S, although it is not possible to compute session

sk in MAKS. Further, session key sk contains usg which is hard to compute similar to ECCDHP in ECC.

- **Key freshness:** In this scheme, a key is fresh for every session, so a key freshness condition succeeds for each session.

- **Implicit key authentication:** In this scheme, no unauthorized party should learn the session key sk, for find it, attacker computes first $P_S = Y_S g$, $P_U = Y_U g$, $uN_S = usg$, $sN_U = sug$, $MAC_S = H_2(ID_U^* \| ID_S \| T_S \| P_S \| P_U \| uN_S)$ and time-stamps T_3. Here, usg or sug is ECCDHP on ECC and $P_S = Y_S g$, $P_U = Y_U g$ which are similar to ECDLP in ECC.

- **Support anonymity property:** In MAKS, U computed $ID_{S1} = ID_S \oplus H_1(ID_S \| ID_{U1} = ID_U \oplus H_1(ID_U \| g)$ and sends to S. On getting ID_{S1}, S computes $ID_U^* = ID_{U1} \oplus H_1(ID_U \| g)$ and uses this phase. Further, S computes $ID_{S1} = ID_S \oplus H_1(ID_S \| g)$ and sends to U. On getting ID_{S1}, U computes $ID_S^* = ID_{S1} \oplus H_1(ID_S \| g)$ and uses this phase. Hence, the presented protocol supports anonymity property.

15.5 Conclusion

This chapter has presented a secure and efficient identity-based mutual authentication protocol for cloud computing. This work has shown that it is secure and has many features and attributes such as key freshness, known session keys impersonation attack, PKG forward secrecy, replay attack, known session-specific attack, perfect forward secrecy, man-in-the-middle attack, mutual authentication, session key establishment, session key security, support anonymity property, and implicit key authentication. Further, this work proved that the framework is secured to launch safe communication channels via the unconfident channel. Future research also includes conserving the secrecy of the users' information offered to the cloud server.

References

Almorsy, M., Grundy, J., and Müller, I. (2016). An analysis of the cloud computing security problem. arXiv preprint arXiv:1609.01107.

Chang, H. and Choi, E. (2011). User authentication in cloud computing. In International Conference on Ubiquitous Computing and Multimedia Applications, pages 338–342. Springer.

Chen, T.-H., Yeh, H.-l., and Shih, W.-K. (2011). An advanced ecc dynamic id-based remote mutual authentication scheme for cloud computing. In 5th FTRA International Conference on Multimedia and Ubiquitous Engineering (MUE), 2011, pages 155–159. IEEE.

Dinesha, H. and Agrawal, V. K. (2012). Multi-level authentication technique for accessing cloud services. In International Conference on Computing, Communication and Applications (ICCCA), 2012, pages 1–4. IEEE.

Hankerson, D., Menezes, A. J., and Vanstone, S. (2005). Guide to elliptic curve cryptography. *Computing Reviews*, 46(1), 13.

Jing, X. and Jian-Jun, Z. (2010). A brief survey on the security model of cloud computing. In Ninth International Symposium on Distributed Computing and Applications to Business Engineering and Science (DCABES), 2010, pages 475–478. IEEE.

Khan, A. A., Kumar, V., and Ahmad, M. (2019). An elliptic curve cryptography based mutual authentication scheme for smart grid communications using biomet- ric approach. *Journal of King Saud University-Computer and Information Sciences*, 2019. doi:10.1016/j.jksuci.2019.04.013.

Khan, N., Kumar, V., and Kumari, A. (2014). An identity-based secure authenticated framework by using ecc in cloud computing. *International Journal of Science and Research (IJSR)*, ISSN (online), pages 2319–7064.

Kumar, V., Ahmad, M., and Kumar, P. (2019a). An identity-based authentication framework for big data security. In Proceedings of 2nd International Conference on Communication, Computing and Networking, pages 63–71. Springer.

Kumar, V., Ahmad, M., and Kumari, A. (2019b). A secure elliptic curve cryptography based mutual authentication protocol for cloud-assisted tmis. *Telematics and Informatics*, doi:10.1016/j. tele.2018.09.001.

Kumar, V., Jangirala, S., and Ahmad, M. (2018). An efficient mutual authentication framework for healthcare system in cloud computing. *Journal of Medical Systems*, 42(8): 142.

Luo, M., Zhang, Y., Khan, M. K., and He, D. (2017). A secure and efficient identity-based mu- tual authentication scheme with smart card using elliptic curve cryptography. *International Journal of Communication Systems*, 30(16), 1–10.

Mishra, D., Kumar, V., and Mukhopadhyay, S. (2013). A pairing-free identity based authenti- cation framework for cloud computing. In International Conference on Network and System Security, pages 721–727. Springer.

Nayak, S. K., Mohapatra, S., and Majhi, B. (2012). An improved mutual authentication framework for cloud computing. *International Journal of Computer Applications*, 52(5), 36–41.

Shamir, A. (1984). Identity-based cryptosystems and signature schemes. In Workshop on the Theory and Application of Cryptographic Techniques, pages 47–53. Springer.

Yang, J.-H. and Chang, C.-C. (2009). An id-based remote mutual authentication with key agree- ment scheme for mobile devices on elliptic curve cryptosystem. *Computers & Security*, 28(3–4): 138–143.

Zhang, Q., Cheng, L., and Boutaba, R. (2010). Cloud computing: state-of-the-art and research chal- lenges. *Journal of Internet Services and Applications*, 1(1): 7–18.

16

Development of a Statistical Yield Forecast Model for Rice Using Weather Variables Over Gorakhpur District of Eastern Uttar Pradesh

R. Bhatla, Rachita Tulshyan, Babita Dani, and A. Tripathi

CONTENTS

16.1 Introduction...243
16.2 Materials and Method..245
16.3 Results and Discussion 246
    16.3.1 Model Generated through SPSS 248
    16.3.2 Validation..249
    16.3.3 Crop Yield Forecast..250
16.4 Conclusions...254
Acknowledgments..254
References...254

16.1 Introduction

Rice is the staple food of the eastern and the southern parts of the country, particularly in the areas having over 1500 mm annual rainfall. Climate change has created one of the important challenges of the twenty-first century—to supply food for increasing populations, and to sustain the stressed environment, which also threatens global food security. The fourth assessment of the IPCC (Intergovernmental Panel on Climate Change) confirmed that in the last 100 years there has been an increase in atmospheric temperature of 0.74°C due to global warming and an expected increase in records of warmest year, frost, flood, heavy rainfall, etc. These changes have a tendency of posing likely threats to the ecosystem and to agricultural production all over the world. Agricultural productivity is most sensitive to climate change, as climate is the primary determinant of agricultural productivity [1].

Much work has already been done on almost all crops; however, the authors are not aware of any study conducted on rice in the Gorakhpur region of Eastern Uttar Pradesh using the statistical technique. This is the subject of this chapter. The agro- meteorological models can be used successfully for predicting, preharvest, the yield of any crop, and for policy planning to ensure food security under climate change scenarios over the regions of Central Punjab [2]. The influence of rainfall on the yield of London wheat was analyzed and a weekly weather data was developed assuming the effect of climate change and the magnitude of weather variables, composed of the terms of the polynomial function of time [3]. Studies on the impact of projected climate change on rice production in Punjab (India) have reported that the productivity of rice is adversely affected under warming climatic scenarios [4]. A study on the impacts of climatic variability on cereal productivity as shown a decrease in the yield of rice by 3% on an increase in temperature of 1°C, in the state of Punjab, India [5]. Three pre-harvest forecast models based on weather variables and weather indices were developed for Faizabad district of eastern U.P [6]. A statistical model was developed for forecasting the yield of sugarcane over Coimbatore for the period 1981–2004 [7]. A crop yield forecast model was developed for paddy and sugarcane for six districts of Gujarat, using the regression model of modified Hendrick and Scholl technique[8]. An ORYZA1 model is used to determine the relative performance of short-duration rice genotypes across the state of Tamilnadu [9]. To analyze the effect of climatic variables on the productivity of wheat in tropical and sub-tropical regions of India a WTGROWS model was developed [10]. Yield forecast models were generated for wheat and rice using weather variables and agricultural inputs on an agro-climatic zone basis [11]. Four different approaches, two on original weather variables and two on generated weather variables, were used for developing pre-harvest forecasts for wheat for Rajkot district of Gujarat [12]. A crop weather regression model is generated using real-time rice yield forecasting over Central Punjab region [13]. A wheat crop yield forecast equation for Kanpur District of Uttar Pradesh was developed by using ordinal logistic regression, probabilities obtained through ordinal logistic regression along with the years as regressors (ordinal logistic regression approach is better than discriminant function analysis) and validated using data of subsequent years [14]. In 2016, the CERES rice model was used for forecasting rice yield over different agro-climatic zones of India [15]. Various crop weather models were used to predict rice yield in India, in spite of a significant influence of solar radiation on rice yield, but none of these models used solar radiation as one of the predictors [16]. The rice yield is directly affected by the physiological processes involved in grain production due to rainfall, temperature, and solar radiation and indirectly through diseases and insects [17].

Productivity of the rice crop was estimated for Srinagar district of Kashmir Valley using past weather variables and yield records revealing that the predicted yield was in good agreement with the actual yield for the forecasted year 2011 using agro-meteorological models [18]. Under temperate conditions, the wheat yield declined up to 37.29%on an increase in temperature of 4°C while shortening the duration of wheat crop, whereas under sub-tropical conditions there was a decrease in seed yield of 32.18%[19]. A study was done under irrigated conditions to evaluate the contribution of weather to growth and yield of *kharif* maize (*Zea mays* L.), concluding that the grain yield of maize was higher during the *kharif* season if minimum temperatures ranged from 18.9 to 22.5°C while the high range of relative humidity in the early stages of the maize crop is critical [20]. A climate variability analysis and its impacts on wheat productivity over Ludhiana (Punjab) shows that the variations in crop productivity is not reliant on a single weather parameter but is a combined effect of minimum temperature, rainfall, number of rainy days, and morning time relative humidity during the growth period of the wheat crop [21]. A CERES-Rice model of DSSAT v4.6 for the rice varieties Jyothi and Kanchan over Kerala was validated and it was found that the model could predict the phenophases more accurately but more years of data are needed in order to increase the confidence level for further validation [22]. A wheat crop simulation model (CERES-Wheat) for different wheat cultivars in Varanasi was calibrated and validated for further applications (like crop production, water management, climate changes effects, and to determine the limiting biophysical factors) and for decision-making in Varanasi [23]. Using agro-meteorological variables a forecast model was generated to estimate maize yield over Jaunpur district of Eastern Uttar Pradesh which is suitable for an advance forecast of yield [24]. Calibration and validation of the

CERES-rice model was done for different rice cultivars in Navsari, hence the lowest percentage errors in grain yield and biomass production were found [25]. A statistical yield forecast model has been developed for chickpea productivity over Rajkot district of Gujarat using the combined effect of weather parameters which shows very high variations in the generated model and could be efficiently used for the pre-harvest forecast, four weeks in advance of harvesting of the crop [26]. A yield forecast of winter rice in seven districts of two agro-climatic zones of Assam (central and upper Brahmaputra valley) has been developed employing meteorological variables through combined effects which could forecast with errors of less than 1% over almost all the districts except for Golaghat (2.02%) and Tinisukia (17.61%) [27].

The objective of this chapter is to examine the effect of weather parameters on rice production, i.e. maximum and minimum temperature, rainfall, and morning and evening time relative humidity over Gorakhpur District of Eastern U.P. and to forecast rice yield five to ten weeks in advance of harvesting.

16.2 Materials and Method

The production data of the rice crop in quintal/hectare for Gorakhpur district (which lies between latitude 26°13′N and 27°29′N and longitude 83°05′E and 83°56′E and receives annual rainfall of about 1200 mm) of Eastern Uttar Pradesh for a period of 40 years, i.e. from 1971 to 2010, are collected from ICRISAT (International Crop Research Institute for the Semi-Arid Tropics). Daily weather data for this long period were collected from National Data Centre of IMD (India Meteorological Department). The five weather variables used in the study are maximum temperature (°C), minimum temperature (°C), rainfall (mm), relative morning and evening humidity in percentages. The collected daily weather data for this period were arranged in weeks for each year. According to the standard meteorological week (SMW), the rice-growing season (the period from sowing to harvesting the crop) runs from the twenty-second to the forty-first week of the year, with the weeks arranged according to the crop season. Correlation was calculated between adjusted yield and each week of weather variables to obtain the Z indices, i.e. the list of unweighted and weighted indices, and these are shown in Table 16.1 with their notations. Unweighted indices are the simple sum and weighted indices are the sum product of weather variables, their correlation and past year observed yields. Out of 40 years, 36 years were used for generating the model, i.e. 1971 to 2006. The best regression model was generated through multiple linear stepwise regression procedure using SPSS (Statistical Package for Social Sciences) [28]. The statistical analyses of variables were carried out with a probability level of 0.05 to enter and 0.1 to remove the variables. A total of five models were generated, out of which the fifth model was selected for further analysis, on the basis of coefficients of determination (R^2). Hence, with the help of this model, the yield forecast equation was generated which was then used for validation (2007 and 2008) and a pre-harvest forecast of rice yield (five and ten weeks in advance of harvest for the years 2009 and 2010). Further, Root Mean Square Error (RMSE) and Mean Bias Error (MBE) were calculated [29] as follows:

$$\text{RMSE} = \left[\sum_{i=1}^{n} (P_i - O_i)^2 / n \right]^{1/2} \tag{16.1}$$

TABLE 16.1

Notation of Unweighted and Weighted Weather Variables
Used in the Model

Indices	Notation for Unweighted Index (Z_{ij})	Notation for Weighted Index (Z_{iij})
T_{max}	Z10	Z11
T_{min}	Z20	Z21
RF	Z30	Z31
RH I	Z40	Z41
RH II	Z50	Z51
T_{max}-T_{min}	Z120	Z121
T_{max}-RF	Z130	Z131
T_{max}-RHI	Z140	Z141
T_{max}-RHII	Z150	Z151
T_{min}-RF	Z230	Z231
T_{min}-RHI	Z240	Z241
T_{min}-RHII	Z250	Z251
RF-RHI	Z340	Z341
RF-RHII	Z350	Z351
RH I-RHII	Z450	Z451

$$\text{MBE} = \sum_{i=1}^{n} [P_i - O_i]/n \tag{16.2}$$

RMSE indicate the magnitude of the average error, where "P_i" represents predicted yield, and "O_i" represents observed yield and the value of MBE is related to the magnitude of the values under investigation. Further, trend analysis was performed on all the time series data of maximum temperature, minimum temperature, morning and evening relative humidity, 24-hourly rainfall, and actual and predicted yield of sugarcane over Gorakhpur district. The actual and predicted yields were tested for significance using non-parametric Mann–Kendall rank statistics [30].

16.3 Results and Discussion

A crop yield forecast model for district-level rice production was developed over Gorakhpur region of Eastern Uttar Pradesh with the help of statistical technique, by using five weather variables, i.e. maximum and minimum temperatures, relative humidity of morning and evening time and 24-hourly rainfall, of the crop growing season, i.e. the period from sowing to harvesting of crop, and the observed yield of past years and results were discussed. The variability of weather variables (maximum and minimum temperatures, relative humidity of morning and evening time, and rainfall) are illustrated through graphs.

Interannual variation in the mean maximum temperature of the crop growing period from the year 1971 to 2010 is shown in Figure 16.1. There is a decreasing trend in the temperature at a rate of 0.044°C/year with coefficient of distribution value (R^2value) 0.3102.

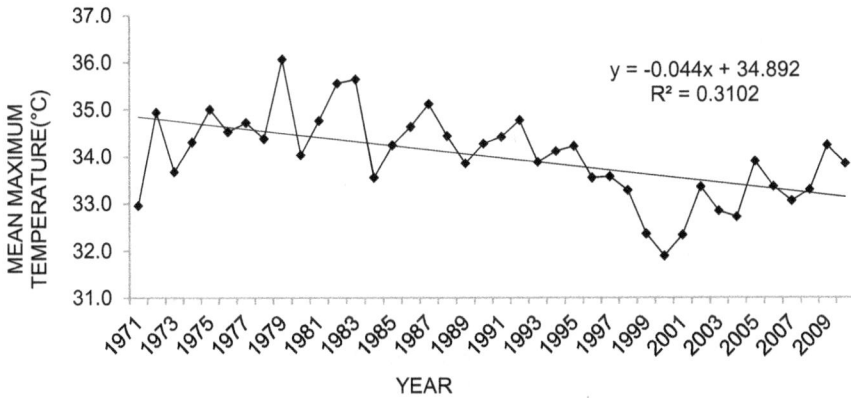

FIGURE 16.1
Variation of mean maximum temperature during the crop-growing season for the period 1971–2010.

The lowest mean maximum temperature was seen at a 31.9°C in the year 2000 while the highest mean maximum temperature was observed at a 36.1°C in the year 1979. Figure 16.2 shows interannual variation in the mean minimum temperature of the crop-growing period which represents an increasing trend at the rate of 0.025°C/year and the coefficient of distribution (R^2value) is 0.2061. The graph shows high fluctuations during 1971 to 1984, and apart from the years 1985 to 2010 almost all values seems to be equal in the entire period. The variation in the rainfall pattern of crop season of the years 1971 to 2010 is shown in Figure 16.3 which fluctuates highly. The graph represents an increasing trend of 3.595 mm/year and the coefficient of distribution value (R^2value) is 0.0244. The study shows that the maximum rainfall was found in the year 1989 which is 1622.4 mm, and minimum in the year 1982 which is 348.7 mm. Since, the graph obtained is highly fluctuating, the year 1984 and 1989 shows the sharp increase in rainfall, and shows decrease in rainfall during the year 1974, 1979, 1982, and 2002. Rainfall is a very important parameter for rice cultivation, and the graph represents an increasing trend which is good for cultivation of rice. Figure 16.4 represents the variation in morning-time relative humidity (RHI) of the crop-growing season over the period 1971–2010. The graph represents a decreasing trend at the rate of 0.079%/year with a coefficient of distribution value (R^2value) of 0.0686.

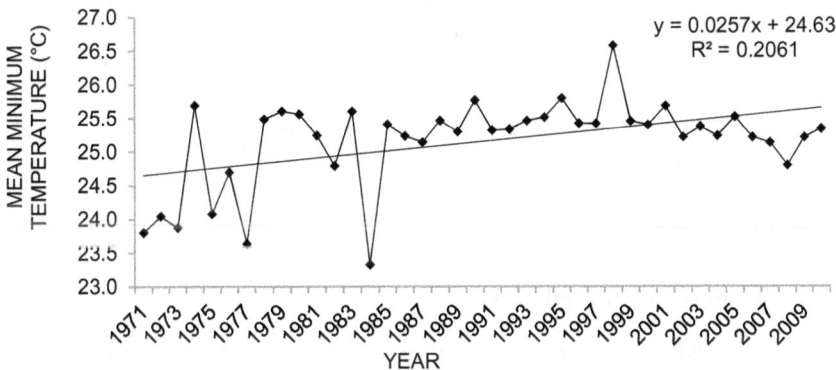

FIGURE 16.2
Variation in mean minimum temperature for the crop-growing season for the period 1971–2010.

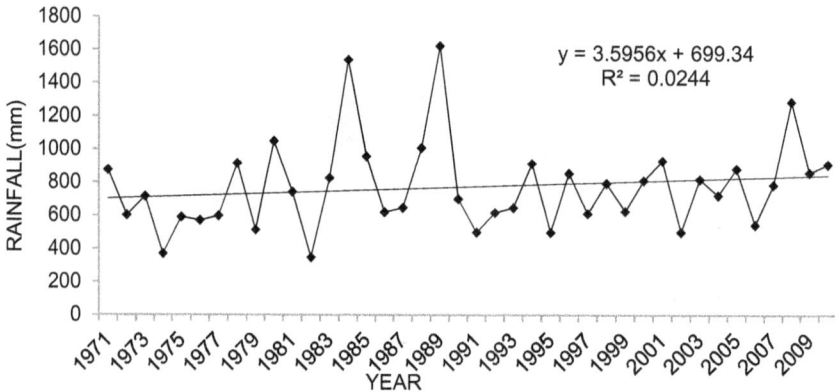

FIGURE 16.3
Variation of total rainfall for the crop-growing season for the period 1971–2010.

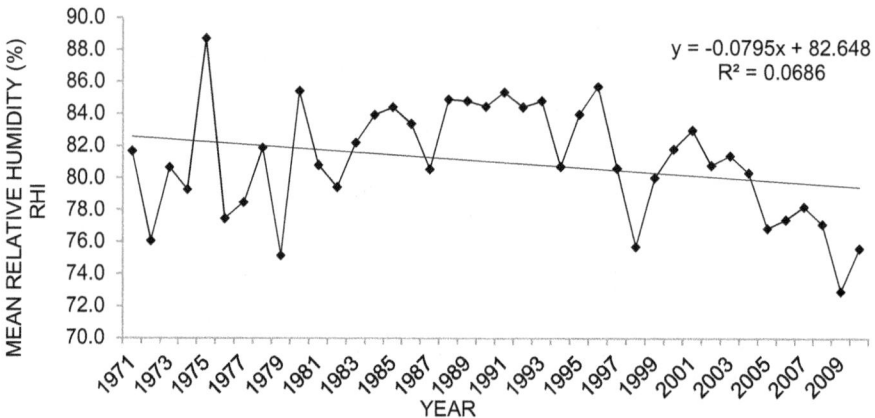

FIGURE 16.4
Variation of mean relative humidity at morning (RHI) for the crop-growing season for the period 1971–2010.

The graph shows high fluctuations throughout the year, i.e. from 1971 to 2010; the highest humidity is found in the year 1995 (88.7%/year) and the lowest humidity is found in the year 1979 (75.1%/year). It is suggested that relative humidity and rice yield are both inversely proportional to each other; if one is increasing then the other will be decreasing. Hence, the graph represents the decreasing trend with the sign of high rice yield. Figure 16.5 depicts the variation in the evening-time relative humidity (RHII) of the crop growing season over the period 1971 to 2010. There is a decreasing trend with little variations, i.e. 0.007% per year, and the coefficient of distribution (R^2value) is 0.0003. The highest humidity is found in the year 1975 (82% per year) and the lowest humidity is found in the year 1972 (62% per year).

16.3.1 Model Generated through SPSS

The model was developed by using weather variables as well as the product of two weather variables of the crop-growing season. The indices of the agro-climatic model were arranged into two categories, i.e. the first is the unweighted index, which is the simple sum

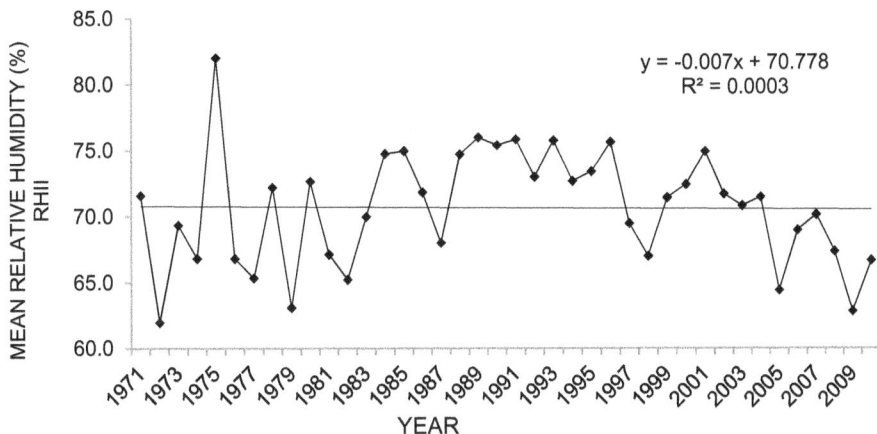

FIGURE 16.5
Variation of mean relative humidity at evening (RHII) for the crop-growing season for the period 1971–2010.

of weather variables, and the second is the weighted index, which is the sum product of weather variables and the correlation value of weather variables and yield. Multiple linear regressions were performed on these two indices to obtain the suitable regression model. Five regression models were obtained of which the fifth model was used for further analysis. The performance of a rice-yield forecast model was then validated for the years 2007 and 2008 and tested by forecasting the yield for the year 2009 and 2010. Hence, the actual and predicted yields were compared through the graph in Figure 16.6.

16.3.2 Validation

With the help of SPSS the regression model was obtained which was then validated for the years 2007 and 2008. The predicted and actual yield for the year 2007 and 2008, with estimated errors, is shown in Table 16.2. The rice-yield forecast for the district showed

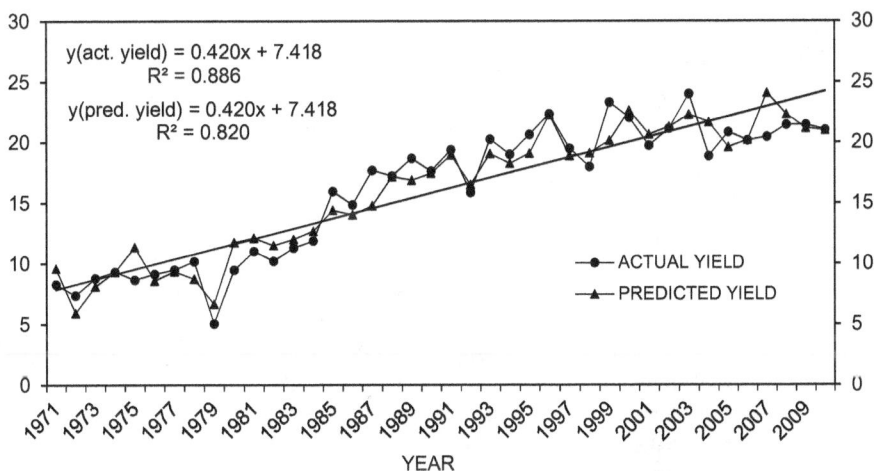

FIGURE 16.6
Comparison between actual and predicted yield over Gorakhpur District for the period 1971–2010.

TABLE 16.2

Validation of Equation for the Years 2007 and 2008

Year	Predicted Yield	Actual Yield	%Error
2007	24.1054	20.4486	15.1697
2008	22.3211	21.4498	3.9037
		Average	9.5367

consistency by producing an error below 10%. The average error calculated for the valida-tion years 2007 and 2008 was found to be 9.54% which is found to be within an acceptable limit of error. The actual and predicted yield obtained after validation for the year 2007 was 20.4486 quintal/hectare and 24.1053 quintal/hectare respectively, with an error of 15.17%, while the actual and predicted yield obtained after validation of the year 2009 was 21.4498 quintal/hectare and 22.3211 quintal/hectare respectively, with an error of 3.90%. The Root Mean Square Error (RMSE) and Mean Bias Error (MBE) in percentage obtained after generating a yield forecast model of the rice crop is 9.0894% and −0.7587% respec-tively and is shown in Table 16.3.

16.3.3 Crop Yield Forecast

A yield forecast model was developed for Gorakhpur region of Eastern Uttar Pradesh with the help of weekly weather data for obtaining highest R^2 value and lowest model error. The significant correlation (R^2) shown by the district crop yield forecast model was 0.926. The forecasting was done for the years 2009 and 2010, for five and ten weeks in advance of forecasting. The obtained predicted yields with errors are shown in Table 16.4. The fifth model was chosen for prediction of yield because it is based on more variables and associ-ated coefficients, hence supposed to give better results. Only unstandardized coefficients

TABLE 16.3

Root Mean Square Error
and Mean Bias Error

RMSE (in %)	MBE (in %)
9.0894	−0.7587

TABLE 16.4

Forecasted and Actual Yield of Rice (Five and Ten Weeks in Advance of Harvesting) and Errors in the Equation

Weeks	Years	Forecasted Yield (Quintal/Hectare)	Actual Yield (Quintal/Hectare)	Error (%)	Average Error (%)
Five Weeks in Advance of Harvesting	2009	21.9983	21.4544	2.4724	1.2995
	2010	21.0498	21.0231	0.1267	
Ten Weeks in Advance of Harvesting	2009	22.2454	21.4544	3.5556	3.5291
	2010	21.7862	21.0231	3.5026	

were used in the yield forecast equation. The final yield forecast model equation, using important independent weather variables, is:

$$Y = (-18.5407) * (\text{Constant}) + 0.4118 * T + 0.0339 * Z40$$

$$+ 0.0003 * Z351 + (-0.00009) * Z130 + (-0.00051) * Z140$$

(16.3)

R^2 = 0.926 (at 5% probability level)

where "Y" represents the predicted yield, "T" represents time, "Z40" is relative humidity at morning time, "Z351" is the sum product of rainfall, relative humidity at evening time, and their correlation values, "Z130" is the sum product of maximum temperature, rainfall, and their correlation values, and "Z140" is the sum product of maximum temperature, relative humidity at evening time, and their correlation values. These are the important parameters that came through SPSS for obtaining a yield forecast equation. The rice-yield forecasting for Gorakhpur district was made five weeks in advance for the year 2009 and 2010. The forecasted yields thus obtained with the help of the generated model, which is shown in Table 16.4, was 21.9983 and 21.04982 quintal/hectare, which is almost similar to their actual yields of 21.4544 and 21.02313 quintal/hectare, respectively for the years 2009 and 2010. The error observed was 2.47% and 0.13% for the years 2009 and 2010 respectively. Yield forecasting was made ten weeks in advance for the years 2009 and 2010 for rice crop. The forecasted yield obtained for the year 2009 and 2010 was 22.2454 and 21.78623 quintal/hectare and their actual yields were 21.45443 and 21.02313 quintal/hectare with the least errors 3.56% and 3.50% respectively. The values of forecasted and actual yields for both the years were found almost equal with least errors. In view of this discussion, comparison between the actual yield and the predicted yield over Gorakhpur District for the period 1971 to 2010 is shown in Figure 16.6 through graphical representation. The graph shows the trend of actual and predicted yield, which is increasing in both the cases. The trend lines of both the actual and the predicted yield coincide with each other. The predicted yield shows an increasing trend with a rate of 0.420 Q/Hectare/Year and R^2 value 0.820 simultaneously; the actual yield also shows an increasing trend with the rate of 0.420 Q/Hectare/Year and R^2 value 0.886, which was found to be exactly the same as the predicted yields. The actual yield and predicted yield are found to be the same in some of the years, such as in 1974 (9.3097Q/Hectare), 1977 (9.3502Q/Hectare), 1988 (17.1621Q/Hectare), 1990 (17.4416Q/Hectare), 1991 (18.9435Q/Hectare), 1996 (22.27Q/Hectare), 2000 (22.6331 Q/Hectare), 2002 (21.2728Q/Hectare), 2006 (20.1907Q/Hectare), 2009 (21.1690Q/Hectare), 2010 (21.008Q/Hectare). The highest actual and predicted yield are found in the year 2003 (23.9971Quintal/Hectare) and 2007 (24.1053 Quintal/Hectare) respectively, and the lowest actual and predicted yield are found in the same year, i.e. 1979 (5.0494Quintal/Hectare) and 1979 (6.6589Quintal/Hectare) respectively. However, Mann–Kendall rank statistics have been checked for actual and predicted yield, and both are found to be significant. Hence, it can be concluded that the yields will also follow a similar trend for the subsequent years. It can also be concluded from the above results that the generated model can give better yield predictions even five and ten weeks in advance of harvesting of various other crops like rice, wheat, sugarcane, cereals etc.

Changes in yield of rice are directly or indirectly affected by different weather variables. The impact of weather variables, i.e. maximum and minimum temperatures, relative humidity of morning and evening time, and 24-hourly rainfall on rice yield for the period

1971 to 2010, is illustrated through a graph. Figure 16.7(a) depicts the variability of mean maximum temperature with the actual and predicted yield for the period 1971 to 2010. The graph represents a decreasing trend in mean maximum temperature with an increasing trend in actual and predicted yield. The graph represents high fluctuations in temperature and, accordingly, the yields increased or decreased. It is clearly shown in the graph

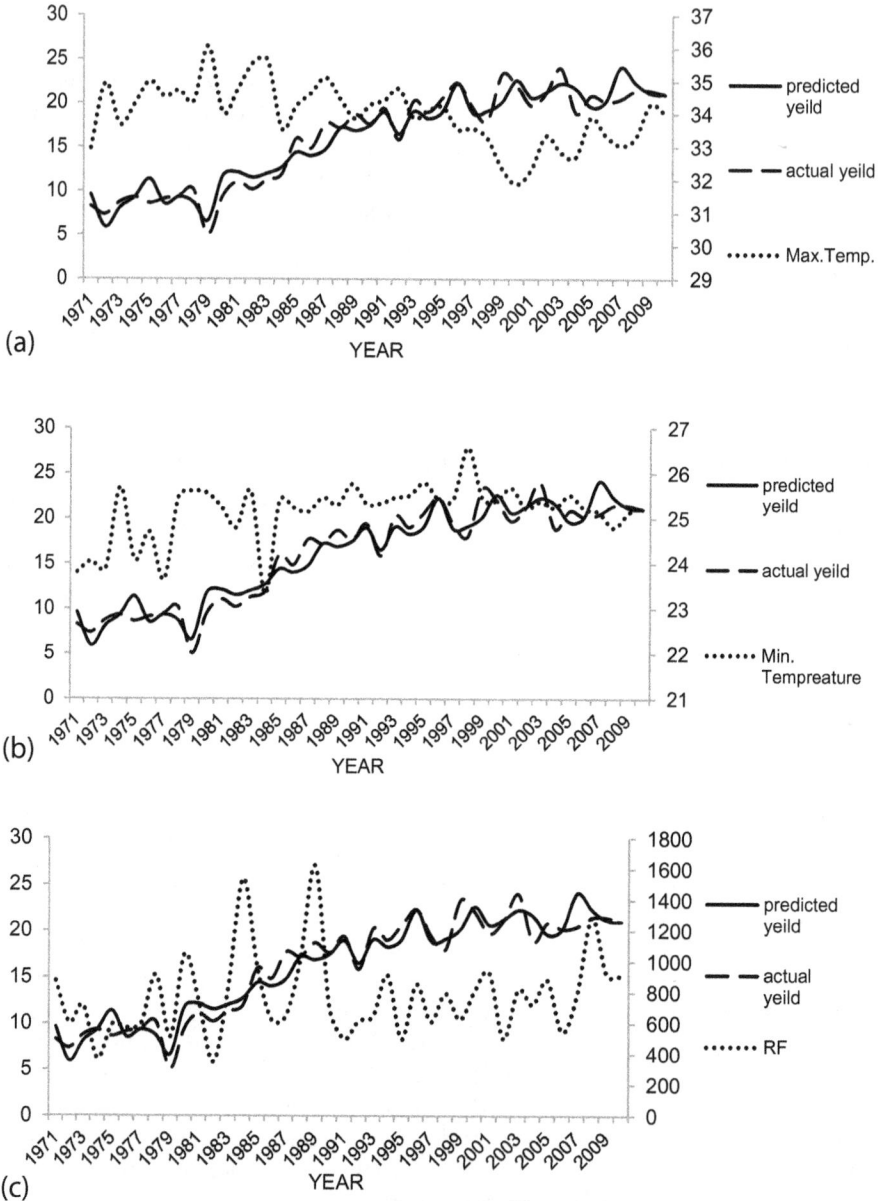

(a)

(b)

(c)

FIGURE 16.7
(a) Variation of mean maximum temperature with actual and predicted yield. (b) Variation of mean minimum temperature with actual and predicted yield. (c) Variation of rainfall with actual and predicted yield. (d) Variation of actual yield and predicted yield with relative humidity in the morning. (e) Variation of relative humidity during evening time with actual and predicted yield.

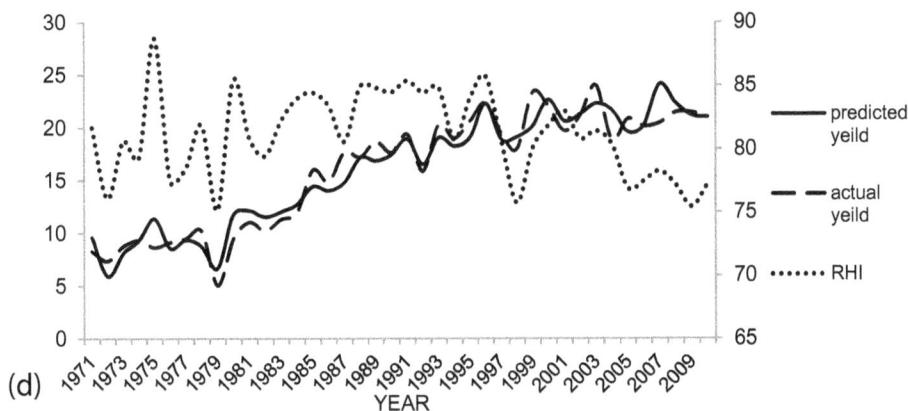

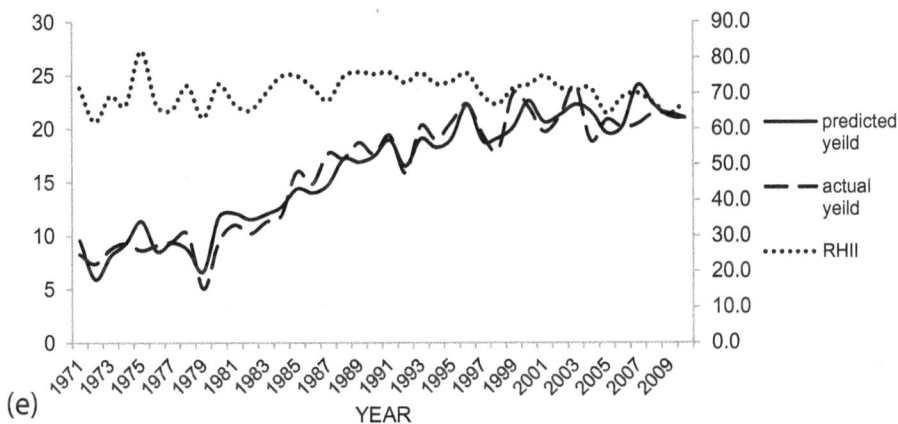

FIGURE 16.7 (CONTINUED) (a) Variation of mean maximum temperature with actual and predicted yield. (b) Variation of mean minimum temperature with actual and predicted yield. (c) Variation of rainfall with actual and predicted yield. (e) Variation of relative humidity during evening time with actual and predicted yield.

that both the yields, i.e. actual yield and predicted yield, were low when the temperature increased while both yields were high when the temperature decreased. Both are inversely proportional to each other, for example, the temperatures from the years 1971 to 1988 fluctuate more and increase but both the yields decrease; on the other hand, the temperatures decrease from the year 1993 to 2009 and yields increase. Figure 16.7(b) shows the variation of actual and predicted yield with mean minimum temperature. In the graph the trendline of actual yield, predicted yield, and mean minimum temperature shows an increasing trend, which means both the yields were directly related to the minimum temperature. Figure 16.7(c) represents the variability of actual and predicted yield with rainfall; all show an increasing trend. The yield increases with the increase in rainfall. In some years, rainfall decreases but both the yields increase, because it is not necessary that rice cultivators always use rainwater for irrigation purposes as they may also use tube wells, bore wells etc. for irrigation. Figure 16.7(d) represents the variation of actual yield and predicted yield with relative humidity in the morning. The graph shows a decreasing trend in relative humidity and an increasing trend in the actual and predicted yield, which means that rice production decreased when the morning-time relative humidity increased. From the year

1971 to 1999 the relative humidity was found to increase and yields decrease. On the other hand, from 2000 to 2009, relative humidity (RHI) decreased while the rice yield increased in both the actual and predicted yield. Figure 16.7(e) represents the variation in actual and predicted yield with relative humidity during the evening; it is known that relative humidity in the evening is higher than in the morning; the graph represents the increasing trend in both the yields and RHII. It means that the yield was found to increase, with the increase in the evening relative to humidity.

16.4 Conclusions

The study shows that the observed weather variables, i.e. maximum and minimum temperature, morning and evening time relative humidity (RHI and RHII), 24-hourly rainfall, and past observed yields of rice of previous years over Gorakhpur district, through multiple linear regression could be used to develop a yield forecast model, five, and even ten, weeks in advance of harvest. The performance of the developed yield forecast model was tested and found to be satisfactory. The average error calculated for the validation years 2007 and 2008 was found to be 9.54%, which is within an acceptable limit of error. Yield forecast for the years 2009 and 2010 five weeks in advance of harvesting shows the percentage errors of 2.47% and 0.13%, while ten weeks in advance of harvesting it shows errors of 3.56% and 3.50% respectively. The observed errors were, therefore, found to be within an acceptable limit. The yield forecast model developed in this study include independent weather variables, which indicates the prime dependence of rice yield on these weather variables that play a major role in crop growth and production, directly or indirectly. Hence, from this chapter we can say that this type of yield forecast model can efficiently be used for future yield predictions at a district level for various crops like wheat, sugarcane, maize, cereals etc.

Acknowledgments

The authors wish to express sincere thanks to India Meteorological Department and ICRISAT (International Crop Research Institute for the Semi-Arid Tropics)for providing the necessary data. One of the authors acknowledges with thanks the financial assistance in the form of a fellowship provided by the University Grants Commission for research.

References

1. Watson, R., Zinyowera, M., and Moss, R.1996. Climate change impacts, adaptation and mitigation of climate change. *Contribution of WGII to the Second Assessment Report of the IPCC*, pp. 880. Cambridge: Cambridge University Press.
2. Kingra, P. K.2016. Climate variability and impact on productivity of rice in central Punjab. *J Agromet*18(1):146–148.

3. Fisher, R. A.1924. The influence of rainfall on the yield of London wheat at Rothamsted. *Phil Trans Roy Soc B*213:89–142.

4. Mathauda, S. S., Mavi, H. S., Bhangoo B. S., and Dhaliwal, B. K.2000. Impact of projected climate change on rice production in Punjab (India). *Trop Ecol*41(1):95–98.

5. Hundal, S. S., and Kaur, P.2007. Climatic variability and its impact on cereal productivity in India. *CurrSci*92(4):506–512.

6. Pandey, K. K., Rai, V. N., Sisodia, B. V. S., Bharti, A. K., and Gairola, K. C.2013. Pre-harvest forecast models based on weather variable and weather indices for Eastern U.P. *AdvBiores*4(2):118–122.

7. Krishna, Priya, S. R., and Suresh, K. K.2009. A study on pre-harvest forecast of sugarcane yield using climatic variables. *Stat Appl*7&8(1&2):1–8.

8. Kumar, N., Pisal, R. R., Shukla, S. P., and Pandey, K. K.2016. Crop yield forecasting of paddy and sugarcane through modified Hendricks and Scholl technique for south Gujrat. *Mausam*67(2):405–410.

9. Palanisamy, S., Aggarwal, P. K., Thiyagarajan, T. M., and Ranganathan, T. B. R.1995. Simulating yield and ranking of rice genotypes in multi-location trails. In: P. K.Aggarwal, R. B.Mathews, M. J.Kropff, and H. H.Van Laar (eds.), *Application of Systems Approach in Plant Breeding, SARP Research. Proceedings*, pp. 91–95. Manila: International Rice Research Institute.

10. Aggarwal, P. K., and Kalra, N.1994. Analyzing the limitations set by climatic factors, genotypes, and water and nitrogen availability on productivity of wheat. II Climatically potential yields and optimal management strategies. *Field Crops Res*38(2):93–103.

11. Aggarwal, P. K., Talukdar, K. K., and Mall, R. K.2000. *Potential Yields of Rice-Wheat System in the Indo-Gangetic Plains of India*. Rice-Wheat Consortium Paper Series 10, pp. 16. New Delhi: Rice-Wheat Consortium for the Indo-Gangetic Plain.

12. Khistaria, M. K., Vamora, S. L., Dixit, S. K., Kalola, A. D., and Rathod, D. N.2004. Preharvest forecasting of wheat yield from weather variables in Rajkot District of Gujarat. *J Agromet*6:197–203.

13. Mallick, K., Mukherjee, J., Bal, S. K., Bhalla, S. S., and Hundal, S. S.2007. Real time rice yield forecasting over Central Punjab region using crop weather regression model. *J Agromet*9(2):158–166.

14. Kumari, V., Agrawal, R., and Kumar, A.2016. Use of ordinal logistic regression in crop yield forecasting. *Mausam*67(4):913–918.

15. Singh, P. K., Singh, K. K., Rathore, L. S., Baxla, A. K., Bhan, S. C., Gupta, A., Gohain, G. B., Balasubramanian, R., Singh, R. S., and Mall, R. K.2016. Rice (*Oryzasativa* L.) yield gap using the ceres-rice model of climate variability for different agroclimatic zones of India. *Curr Sci*110(3):405–413.

16. Kandiannan, K., Karthikeyan, R., Krishnan, R., Kailasam, C., and Balasubramanian, T. N.2002. A crop-weather model for prediction of rice (*Oryzasativa* L.) yields using an empirical-statistical technique. *J Agron Crop Sci*188(1):59–62.

17. Yoshida, S.1981. Fundamentals of rice crop science. Dissertation. International Rice Research Institute, Los Banos.

18. Mahdi, S. S., Lotus, S., Singh, G., Ahmad, L., Singh, K. N., Dar, A. L., and Bhat, A.2013. Forecast of rice (*Oryzasativa* L.) yield based on climatic parameters in Srinagar district of Kashmir Valley. *J Agromet*15(1):89–90.

19. Kour, M., Singh, K. N., Singh, M., Thakur, N. P., Kachroo, D., and Sharma, R.2013. Assessment of climate change and its impact on growth and yield of wheat under temperate and subtropical conditions. *J Agromet*15(2):142–146.

20. Rani, P. L., Sreenivas, G., and Reddy, D. R.2013. Evaluating contribution of weather to growth and yield of kharif maize (*Zea mays* L.) under irrigated conditions. *J Agromet*15(2):156–158.

21. Kingra, P. K.2016. Climate variability impacts on wheat productivity in central Punjab. *J Agromet*18(1):97–99.

22. Vysakh, A., Ajithkumar, B., and Subbarao, A. V. M.2016. Evaluation of ceres-rice model for the selected rice varieties of Kerala. *J Agromet*18(1):120–123.

23. Mall, R. K., Singh, N., and Singh, H.2016. Evaluation of ceres-wheat model for different wheat cultivars at Varanasi. *J Agromet*18(1):149–152.

24. Tripathi, A., Singh, R. S., Bhatla, R., and Kumar, A.2016. Maize yield estimation using agro-meteorological variables in Jaunpur district of Eastern Uttar Pradesh. *J Agromet*18(1):153–154.
25. Mote, B. M., and Kumar, N.2016. Calibration and validation of Ceres-rice model for different rice cultivars at Navsari. *J Agromet*18(1):155–156.
26. Marviya, P. B., Rankja, N. J., and Choddavadia, S. K.2015. Development of statistical model for forecasting chickpea productivity of Rajkot district in Gujarat state. *J Agromet*17(1):130–132.
27. Devi, M., Hussain, R., Sarma, A., Sarmah, K., and Deka, R. L.2013. Yield prediction of winter rice employing meteorological variables in central and upper Brahmaputra valley zone of Assam. *J Agromet*15 (Special Issue-I):162–166.
28. Ghosh, K., Balasubramanian, R., Bandopadhyay, S., Chattopadhyay, N., Singh, K. K., and Rathore, L. S.2014. Development of crop yield forecast models under FASAL- a case study of kharif rice in West Bengal. *J Agromet*16(1):1–8.
29. Willmott, Cort J.1982. Some comments on the evaluation of model performance. *American Meteorological Society*63(11):1309–1313.
30. WMO.1966. WMO technical note no.79, WMO no.195-TP100, pp. 64–65. Geneva: WMO.

17

Cognitive Computing Agent Systems: An Approach to Building Future Real-World Intelligent Applications

A. Chandiok, A. Prakash, A. H. Siddiqi, and D. K. Chaturvedi

CONTENTS

17.1 Introduction: Cognitive Computing-Based Intelligent Systems 257
    17.1.1 Motivation 258
    17.1.2 Literature Review 258
    17.1.3 Aim of Writing this Chapter 260
17.2 Background 260
    17.2.1 General Intelligent Agents 260
        17.2.1.1 Internal Features of Intelligent Agent 261
        17.2.1.2 External Features of Intelligent Agents 261
    17.2.2 Taxonomy of Intelligent Agent 262
17.3 Intelligent Agent Systems 263
    17.3.1 Reactive Architecture 265
17.4 Cognitive Agent System 265
    17.4.1 Symbolic Architectures 267
    17.4.2 Emergent Architectures 267
    17.4.3 Hybrid Architecture 267
17.5 Cognitive Computing Agent Systems 267
17.6 Summary 272
References 273

17.1 Introduction: Cognitive Computing-Based Intelligent Systems

The basis for an artificial agent to have human-like decision-making characteristics is due to an essential element that is known as agent architecture, the blueprint for constructing an agent having reasoning skills. Wooldridge, in 1995, discussed agent architecture as a software design that is anticipated to provide human-like rational decision-making processes (Wooldridge and Jennings, 1995). Maes defined agent architecture as a design that incorporates methods and procedures to include discrete components inside a holistic system and help these elements work together (Maes, 1991). Agent architecture is the constructing wedge for making an agent, much like building an object inside a software class. The agent architecture acts as the manager of an artificial agency as it controls the flow and representation of knowledge/information by the agent. It also decides the appropriate action of the agent based on its unique fundamental inference/reasoning tool. Therefore, researchers have developed several agent architectures based on the knowledge representation approaches and its reasoning mechanism for solving real-world problems. The research science world classifies artificial agent architectures into two broad sets: the intelligence-based single/multi-agent design, and the cognitive architecture. Agent architecture comprises reflex, model-based, goal-based, utility-based, and learning agent architectures based on their actions, while the conventional architectural designs for building explicit multi-agents include Reactive, Deliberative, BDI, and Hybrid layered architectural designs based on the concept of human information handling skills. Cognitive architecture is classified as symbolic, emergent, and hybrid cognitive, based on the ideas of the cognitive sciences.

17.1.1 Motivation

The progress of the Internet and computer technologies has increased the need for artificial agent-based distributed, dynamic, coordinated, and heterogeneous intelligent systems. Intelligent Agent skill is a novel model appropriate for evolving such structures that positions and functions dynamically under the uncertain and mixed real-world scenario. So, the aim of this chapter is to ascertain precisely what intelligent agents are. How do artificial intelligence scientists create intelligent agents? What is the progress in the area of creating artificial systems to help humanity? How does the unified architecture approach works and what are its advantages and disadvantages? Currently, there is no extensively accepted definition of what an intelligent agent is and how its research is progressing with time. In this chapter, the focus is to answer the above-mentioned question and show what points the researcher must keep in mind while developing these systems.

17.1.2 Literature Review

Architecture-based intelligent-agent system research is currently a universal methodology in the area of general artificial intelligence. Between 1990 and 2016, the creation of various intelligent and cognitive architectures took place due to the work of several researchers in creating artificial systems. Although architectural models for an artificial agent have been an important research characteristic, there is a lack of suitable state-of-the-art reviews on the subject. This review details the critical facts of previous surveys on cognitive and intelligent architecture.

Wray et al. did a pioneering survey on a mixture of intelligent and cognitive agent architectures. It explains 12 architectures like the Subsumption, Homer, Theo, Teton, X Meta-reasoning architecture, RALPH-MEA Prodigy, Atlantis, Entropy Reduction Engine, SOAR, and ICARUS. The review focuses on the comparison of intelligent agents based on architectural properties, skills of agents, environmental situation and conditions, psychological validity, efficiency, and generality. The main drawback of this website-based survey is that the comparisons are made only on the criteria and the citations given by the authors on their website, which focus only on the years up to 1991 (Wray et al., 1994).

Stollberg and Rhomberg illuminated the thorough survey on intelligent agent architecture like BDI, UPML SOAR, AI, based on constrained types of goals, goal specification language, goal description model, and goal resolution technique. The authors only focused on the feature of the goal, while not considering other intelligent agent criteria (Stollberg and Rhomberg, 2006).

Vernon et al. provided a detailed survey and compared 14 intelligent and cognitive architectures developed up to the year 2007. The author straightforwardly compares the architectures based on particular feature design and approaches (Vernon et al., 2007).

Chong et al. specify a survey on six all-time famous intelligent system architectures—openly BDI, ACT-R, SOAR, ICARUS, CLARION, and subsumption architecture. The author mines a thorough function comparison by considering a broad collection of intelligent and cognitive features like perception, learning memory, plan, goal, inference, problem solution, and design relevance to neurobiology (Chong et al., 2009). Still, the author reviews the selected architecture until the year 2009. Much new artificial agent architectures created since 2009 and old design have altered their versions. This survey did not cover these points.

Samsonovich surveys a total of 26 intelligent and cognitive architectures up until the year 2010. The review only touches the salient points and does not make any comparison between the past cognitive architectures. The review does not focus on describing the philosophies behind each design (Samonovich, 2011). Mere citations and a little description of the particulars of the existence of the architectures are not sufficient. The review needs detailed comparative research between the designs based on properties and functions common between them.

Kakkasageri tries to discourse in the survey paper regarding architectures of intelligent agents. The work gives a specific focus on BDI, BOID, Cougaar, and MicroPsi. The authors have also shown concepts in the recent applications of intelligent and cognitive agents using these designs in military, air-traffic control, etc. (Kakkasageri et al., 2012). The paper does not focus on the latest cognitive architecture and has done a review on past cognitive architecture which has already been mentioned in previous reviews.

Thagard has studied the classical approaches of agent architectural design: connectionism and a rule-based approach with significant discussion application made by researchers using these methodologies. The review highlights rule-based structures and applications based on symbolic techniques like forward chaining up to if-then, semantic network, ontology representations. The author also discusses connectionist structures like neural networks and their types. However, it does not highlight the latest hybrid and emotional artificial agent architecture which is essential to intelligent system research (Thagard, 2012).

Kajdocsi and Pozna reviews concerns in designing cognitive agent architecture based on the symbolic approach. The author focuses on architectures like SOAR, ACT-R, ACT-R/E, Icarus, LIDA, ADAPT, Prodigy (Kajdocsi and Pozna, 2014). The survey does not discuss emergent architecture, with no comparison and focus on the latest architectures.

Helie and Sun in their survey paper debate the matter of autonomy in psychological approach based cognitive architectures like SOAR, CLARION, and ACT-R. This survey specifies that a cognitive agent explicitly encompasses illustration of symbolic models, while implicitly storing sub-symbolic knowledge. The article deliberates that ACT-R performs a top-down knowledge method that progress from explicit to implicit association, and CLARION contains a bottom-up process from implicit to explicit way (Helie and Sun, 2014). However, the review argues for old architectures which are already surveyed by many researchers. It also considers only common knowledge like explicit and implicit knowledge, while no focus is given on contemporary cognitive computing techniques like machine learning, Natural Language Processing, probability analytic in the design of agents.

De Carvalho et al. in the review paper offer a novel perspective, and focus on the establishment of the enactive approach to intelligent agent design (De Carvalho et al., 2016). The article reviews sequentially different artificial system designs like cybernetics, cognitivist, connectionism, and, finally, enactive. The article stresses the new and better impact of the enactive methods in the approach of agent design. The author claims that researchers must design modern architecture based on enactive behaviour.

Lucentini and Gudwin take already reviewed cognitive agent architectures by various researchers. The authors focus on three architectures such as SOAR, CLARION, and LIDA. SOAR is mainly a cognitivist design, while, LIDA and CLARION are hybrid cognitive models merging symbolic and sub-symbolic features explicitly and implicitly (Lucentini and Gudwin, 2015). In the review, the authors make a comparison of the theoretical mechanisms of the agent architectures, and various ways they can be exploited to suggest the development of distinct applications.

The review does not aim to highlight in-depth practical intelligent and cognitive agent issues related to the unified framework design. It is, undeniably, an investigative challenge to propose enhanced human-like skills in artificial agents.

17.1.3 Aim of Writing this Chapter

In this chapter, the main aim is to cover the untouched portions in the progress of intelligent agent design which previous work has not considered. Second, it provides a comparison of intelligent and cognitive agent architectures at the functional and design level, as well as reviewing intelligent agent architecture advancement until the year 2016.

This chapter will help budding intelligent and cognitive agent researchers who are developing intelligent systems based on an architectural framework. The survey aids a thorough overview of this area, and it will offer an up-to-date theoretical literature in the area. New researchers may obtain a new prospect of progressing the development of an intelligent agent. So, the key objectives are:

1. Suggest a universal taxonomy and definition on various intelligent agents based on a physical, mental, and conscious approach.

2. Provide a state-of-the-art review on the design of artificial intelligent and cognitive agent based on the architecture approach. The review elaborates the different categories based on its architectural design, advantages, and disadvantages, removing past exploration loopholes. This chapter classifies intelligent agent architectures as reactive, deliberative BDI, and hybrid layer. On the other hand, cognitive agent architecture is categorized as symbolic, emergent, and hybrid model.

3. Give a brief review of cognitive agent systems based on new architectures progress after the year 2010. So, it leads to development on the subject, as no previous survey has focused in this way.

4. Focus on the role and applications of cognitive computing in the area of artificial intelligent and cognitive agent systems based on designs and application.

5. Finally, embrace some discussion and conclusions.

17.2 Background

17.2.1 General Intelligent Agents

Researchers define intelligent agents as the software and hardware entities that can perform in place of humans and conduct specific complex tasks or give expert decisions intelligently as shown in Figure 17.1. Intelligent agents comprise of certain internal, as well as external, characteristics which come from different software programs. The authors discuss the features as follows.

17.2.1.1 Internal Features of Intelligent Agent

An intelligent agent possesses certain inner features like autonomy, learning, inference, goal orientation, and reactive action:

- **Autonomy:** Intelligent agents can control their task. The agent can self-sense the environment and generate action based on its knowledge provided by the knowledge experts in the form of rules.
- **Learning:** An intelligent agent can learn facts and create its knowledge for developing behavior for achieving goals.
- **Inference:** An intelligent agent can provide a decision by exploiting its knowledge and choose the best action for a particular uncertain situation.
- **Goal Orientation:** Every agent has individual goals, according to the appropriate information and expertise it attempts to reach the goals.
- **Reactive Action:** Each agent has the ability to provide an immediate best reaction on perceiving information from its environment.

17.2.1.2 External Features of Intelligent Agents

Apart from the internal characteristic the agents also have few external features like mobility, cooperation, and communication.

- **Mobility:** Intelligent agents can navigate electronically through communication networks.
- **Cooperation:** For conducting specific composite tasks agents may have the skills to cooperate with other agencies and improve its self-abilities for attaining goals.
- **Communication:** Every agent can interact with its environment and other entities like humans, agents, systems, etc. to accomplish the goals.

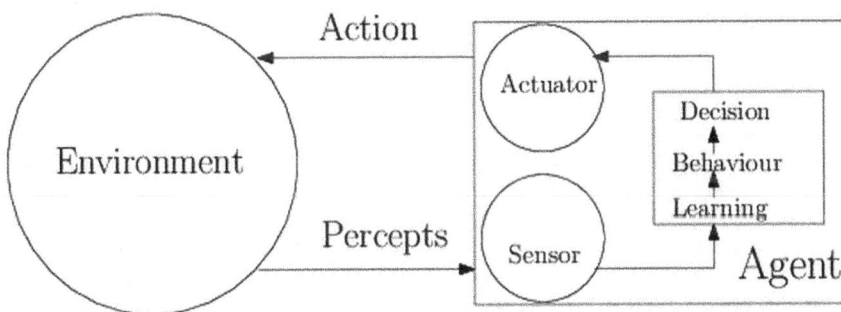

FIGURE 17.1
General intelligent agent system.

17.2.2 Taxonomy of Intelligent Agent

The research is carried out by several researchers in the field of developing and using agents for intelligent systems. Each researcher has proposed their concepts for the individual agent's model. Therefore, for removing research gaps and creating a single base for all the works in an intelligent system, offering a holistic taxonomy of agent-based on prenominal concepts is necessary, as shown in Figure 17.2.

The classification of intelligent agents is as follows:

- **Physical Agents:** Physical agents are individual biological, artificial, or a combination of both of the physical entities that can perform intelligence-based decision-making. Physical agents are further classified as:

  1. **Natural Agents:** Humans, animal, and other biological entities are examples of natural agents.

  2. **Artificial Agents:** Artificial entities have features of autonomy, learning, reactivity, and reasoning power and are the particular cases of an artificial agent that humans create using artificial technology to get help in completing specific tasks. The artificial agents are further grouped as follows:

     - **Robotic Agents:** Robots are the agents that perform mechanical actions on getting perception from an environment based on hard-coded software instructions.

     - **Intelligent Agents:** These agents perform autonomously using stored memory knowledge and a sense-making ability to complete the goal in the best way based on certain architecture models.

     - **Cognitive Agents:** Cognitive agents are software agents specifically having human-like mental intelligence using the concept of cognitive science to provide human-mind based decision-making.

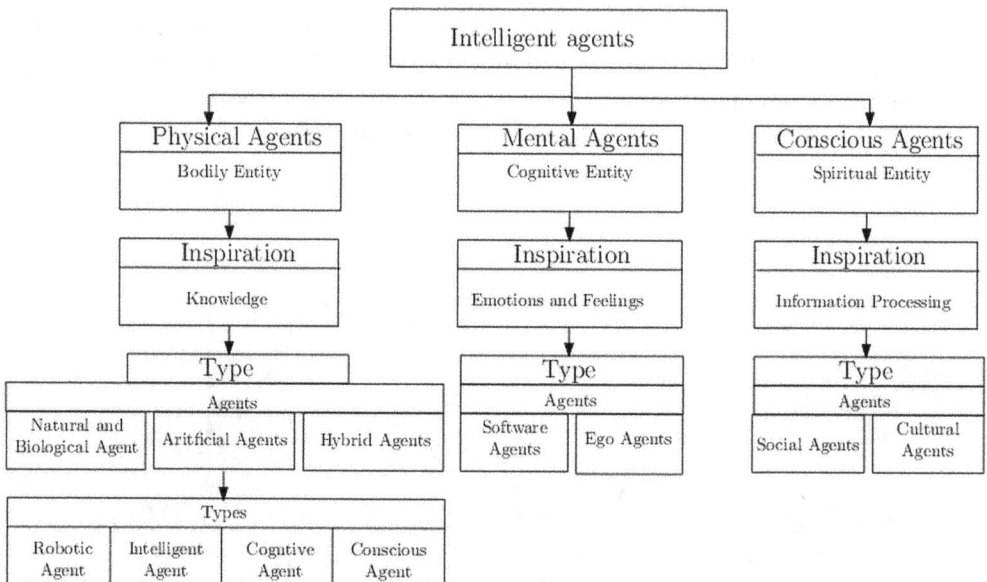

FIGURE 17.2
Taxonomy of intelligent agent system.

- **Conscious Agents:** Conscious agents use the flow of knowledge information internally and universally using natural and mental components to provide wisdom-based decision-making.

3. **Hybrid Agents:** Hybrid agents comprise simultaneous and cooperative involvement of both natural as well as artificial entities to complete the task and reach goals.

    - **Mental Agents:** Mental agents use a psychological perspective for decision-making to complete actions. An example of mental agents is the human use of intelligence, ego, emotions, and feelings. On the other hand, artificial software, cognitive agents, can be a type of mentally inspired entity that use artificially generated human reflective mental cognition.

    - **Conscious Agents:** the conscious agent is the wisdom-based agent that can perform social, cultural, and truth-based decision-making. Conscious agents handle the flow of information internally at a higher level of intelligence and socially with other actors. Conscious agents classify as:

        - **Social Agents:** A social agent can handle cooperative information and coordinate to reach the best goals. At present, modern agent-based systems involve a sole agent, for systems that consist of social agents that connect cooperatively. A social system is a system composed of numerous interacting information handling intelligent agents.

            Social actors can resolve complications, which are difficult or occasionally impossible for a single agent or a distinct colossal system to solve. Every individual agent is intelligent as it has a reasoning ability and can adapt itself to the conditions. With the ever increasing complexity in holistic and distributed approaches, the need for social agents has also increased.

        - **Cultural Agents:** A Cultural agent can handle wisdom-based information. The agent can provide truth-based results and can give a decision based on right or wrong. Self-awareness motivates this decision. This unique characteristic mainly exists in spiritual human beings.

17.3 Intelligent Agent Systems

The creation of artificial intelligent agent systems is the main aim of researchers working in Distributed Artificial Intelligence (DAI), the subdivision of AI focused on concurrent and distributed Artificial Intelligence (AI) computing. So, this in reality specifically marked in the branch of DAI, well-known as Individual and Multi-Agent (IA/MA) systems which handles the management of intelligent knowledge and behaviour between groups of autonomous, and rational artificial intelligent agents.

In earlier work, advances mostly focused on agent association, problem-solving, and management inside the individual and between multiple agents. The initial work in distributed intelligent agents was limited, given strictly abstract programmed, self-knowledge, as well as resource-unbounded. The artificial agents were operational in definite task-sharing or functioning in extremely structured job environments in which third-party programming removes inter-agent clashes. Recently, however, researchers in DAI have started to contemplate more realistic, less well-structured problem areas involving sets of distributed intelligent agents that are constrained by limited resources and abilities. So, this results in a complex and uncertain environment with bounded rationality. The artificial agent in such conditions is expected to act liberally to achieve their goals by exploiting decision methods which yield reasonable rather than ideal behavior.

So, evidently, numerous researchers from the areas of Artificial Intelligence, Distributed Artificial Intelligence, and Robotics have begun to turn their attention to the trick of designing and executing integrated architectures for artificially intelligent agents. Integrated agent architecture is a theory or model by which researchers can plan, program, and create artificially intelligent agents. So, an agent is directed characteristically to work in a dynamic, random, and frequently multi-agent world. An intelligent agent can be observed as a structured assembly of sensors, computers, and actuators as shown in Figure 17.3. In this structure, the sensors perceive the present state in the world, the computer system processes the sensory information, and the actuators yield action in the world to reach the goal.

Subsequently, variations in the environment comprehended by the agent's actuators will create a closed loop to the agent's sensors. Therefore the agent is designated as being embedded in its world. Recently researchers have proposed various integrated architectures, with each design intended to provide artificial agents with a specific level of intelligence, and autonomy. Specifically, if the artificial agent chooses actions by explicit deliberative skills with the use of an abstract symbolic knowledge model, through a search of its plan domain, or by using the anticipated utility of accessible execution methods, such agents are known as deliberative.

Otherwise, if the agent's selection of action is determined according to the situation, through pre-programming or by hardwired to execute particular action under known certain environmental conditions, then the agent is defined as non-deliberative. Also,

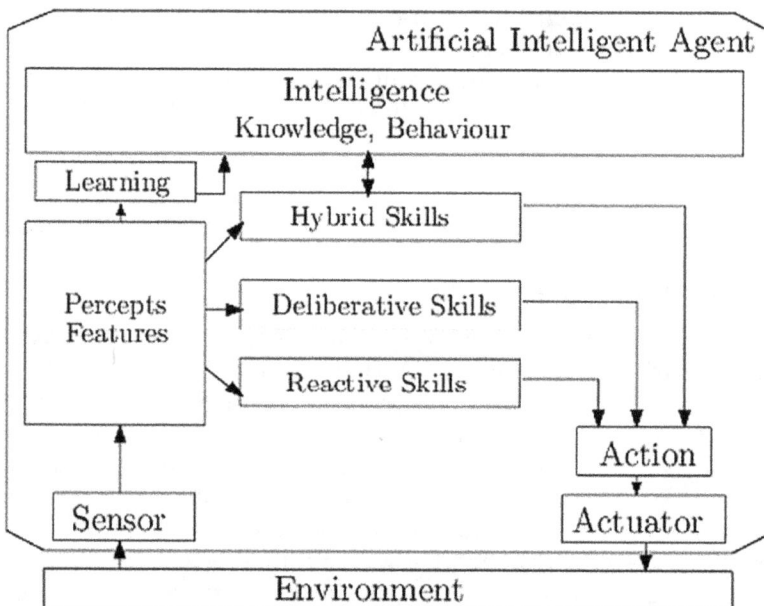

FIGURE 17.3
Artificial intelligent agent system.

individual researchers have shown interest in architectures where the selection of action is comprehended by a grouping of deliberative and non-deliberative techniques. These artificial agent architectures are known as a hybrid.

17.3.1 Reactive Architecture

Reactive agent architecture is grounded on the straight mapping of perception states into action. It does not use any fundamental symbolic knowledge model and reasoning as used in deliberative architecture. This artificial agent reacts to direct changes in the real-world environment in a perception-and-action response. The reactive architecture comprehends by using a combination of sensors and actuators, where sensor input is mapped to the actuator based on alterations in the environment.

Brooks' subsumption architecture is accepted as the most exquisite untainted reactive architecture. Brooks developed the architecture based on several assessments of various drawbacks in deliberative architecture. The percept situation mapped into an action which specifically respondes to the percept situation.

Brooks' basic idea of subsumption architecture is that agents can generate intelligence in performance without explicit knowledge illustrations and abstract inference by using symbolic artificial intelligence techniques. Intelligence in an agent is an evolving feature of complex iterative systems. Subsumption architecture is understood in finite state machines through exclusive mapping layers linked to sensors that observe the environment variations and map the individual action executed.

17.4 Cognitive Agent System

As specified in (Langley, 2004), and also shown in the proposed taxonomy, there are numerous ways and types of building artificial intelligent agents, which are the robotic agent, software-based rational (single and multi-agent) approach, cognitive architecture, and conscious computing agent approach. "A cognitive architecture approach is a blueprint for developing artificial agents system with human-like intelligence" (Anderson et al., 2004). It postulates the fundamental framework of an intelligent agent that acts like real cognitive systems by proposing human-like artificial computational software and hardware structure as shown in Figure 17.4.

So, it is not wrong to say that cognitive architectures denotes blueprints of the human mind. So, the model shows not only the behavioural properties of the system but also the structural features of the system.

Cognitive architecturel, classified as symbolic, emergent, and hybrid as shown in Table 17.1, is grounded on the basis of cognitive science. Cognitive science centers on human-level cognition, intelligence, neurology, anthropology, and psychology. Cognitive architecture began with a particular category of design acknowledged as a production system and is still progressing over time with the fantastic research work of cognitive scientists. The earliest architecture for building a cognitive agent is ACT for forming human-like behavior (Anderson et al., 2004). The production of an intelligent agent using cognitive architecture focuses on the cognition part, the simulation and modelling of human behaviour. The cognitive architecture approach is said to be different from the artificial intelligent agent architecture approach in the following ways (Langley et al., 2009).

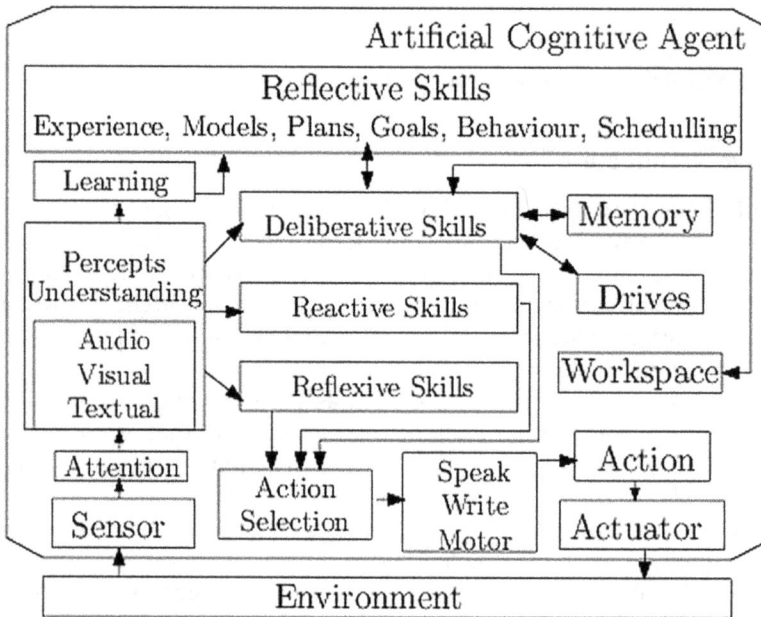

FIGURE 17.4
Cognitive agent system.

TABLE 17.1

Cognitive Architecture Taxonomy

- Symbolic architecture based on rules
- Emergent architectures based on models
- Hybrid architecture (A combination of symbolic and connectionist models)

Cognitive architecture is the layout to develop cognitive agents. It is the foundation for the agent and leads to the creation of a unified software program framework having the ability to perceive information from the environment, learn and encode knowledge, store it in the memory, and associate it with reasoning and decision-making to achieve the goal completely inside the framework. So, cognitive architecture offers the following outcomes:

1. It provides cognitive agents with robust conventions on modular knowledge representation and its processing.

2. It provides the cognitive agent with system-level mental, psychological, and human-like intelligent behavior. So, it removes the fundamental level of function design to complete particular tasks as in the case of an intelligent agent.

3. It delivers a unified cognitive system methodology in which agents can have a common base of knowledge representation and method. So, it decreases the need for constructing specific local and unique features in the agent.

4. It gives short- and long-term memories for artificially storing the cognitive agent's knowledge, experiences, beliefs, and goals.

5. It helps the cognitive agent's framework for the representation and organization of memories and for acquiring the knowledge for achieving goals. So, in a cognitive architecture, memory and learning are the two principal components. The grouping of these two modules forms a taxonomy of three broad kinds of cognitive agent architecture, explicitly symbolic, emergent, and hybrid models (Duch et al., 2008).

17.4.1 Symbolic Architectures

Symbolic archetypes track an extensive belief in Artificial Intelligence and are founded on the postulate that the mind occurs primarily to operate symbols that indicate features of the world. Specimens for most popular symbolic architectures amongst many others are SOAR, ICARUS, and GLAIR. Though symbolic architectures comprise several much-appreciated ideas and have produced some remarkable outcomes, it is uncertain whether they will encourage the development of the emergent configurations and dynamics necessary for human-like general intelligence.

17.4.2 Emergent Architectures

The emergent architecture help to sustain cognitive agent's autonomy as humans do with the brain. Therefore, the cognitive agent is able to make interactions with the environment. They are intended to mimic neural networks or other characteristics of the human brain. Similar to traditional AI, emergent cognitive architectures have proven to be robust in identifying patterns in high-dimensional facts, associative memory configuration, reinforcement learning (Fodar and Pylyshyn, 1988). Nevertheless, perhaps due to the shallow analogy to brain circuits, these architectures have not yet led to the high-level emergence of any brain-like tasks.

17.4.3 Hybrid Architecture

From the work done by researchers, there are equal merits and demerits to symbolic and emergent methods. So, researchers fully dedicated themselves to, and focused on, developing cognitive architecture on integrative, hybrid architectures. So, this links the strengths of both designs, focusing on functioning as described above. Research scientists suggest groupings in various ways based on top-down and bottom-up approaches. Examples of popular and latest hybrid cognitive architectures are ACT-R, CLARION, and LIDA.

17.5 Cognitive Computing Agent Systems

The architecture of the cognitive computing agent system is the main focus of this chapter as a whole and is shown in Figure 17.5. It signifies the essential idea about cognitive computing agent-based human-like decision-making processes.

Compared to traditional intelligent and cognitive agent systems, the cognitive computing agent framework is a hybrid cognitive agent plus cognitive computing technology centered model (Chandiok and Chaturvedi, 2018). The agent-centric prototype gives importance to human perception, learning, memory, knowledge, and decision-making

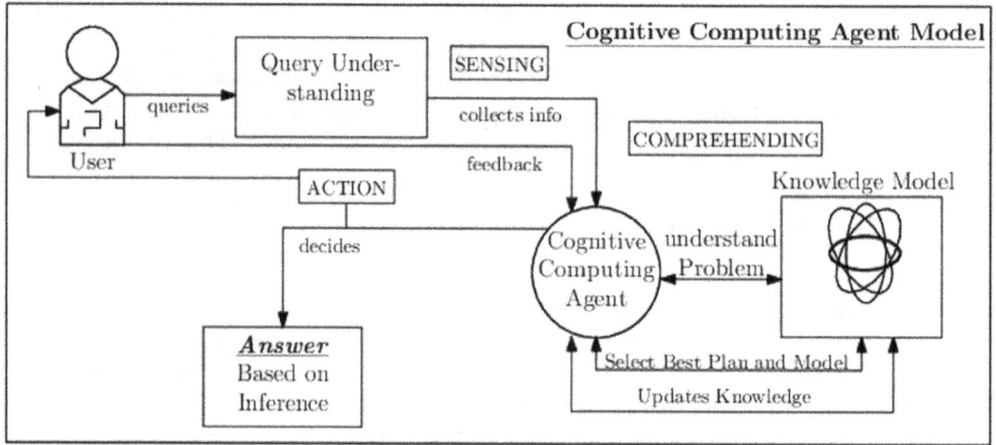

FIGURE 17.5
Cognitive computing agent model.

ability as a planning viewpoint for designing and developing an intelligent system. On the other hand, the technology-centered design emphasizes the values of technologies regarding system functionality. The cognitive computing agent framework includes the cognitive agent ability of situation understanding, building knowledge models, and associated cognitive values. The technology ability provides effective information collection for knowledge-building and Human-Computer Interaction (HCI) with emotional, cognitive values.

The cognitive computing technologies incorporate deep learning (DL), machine learning (ML), and natural language processing (NLP), knowledge reasoning, speech recognition, machine and computer vision, as well as all hardware and software for HCI, developed inside one system (Chandiok and Chaturvedi, 2015). Based on a computing science approach, the technologies provide new cognitive computing standards, programming models, architecture, processor designs, and reliable and fast interfaces. The application developed using the above hybrid model of cognitive computing agent is concerned with decision-making in an uncertain situation. Therefore, the hybrid agent and technology-centered design in the cognitive computing agent framework is illustrated by assisting the user in making improved decisions and enabling cognitive decisions in the real-world process (Chandiok and Chaturvedi, 2017). The architecture performs certain tasks in cognitive decision support to reach a decision and is as follows.

1. **Describing Experience and Query for the Decision Situation:** A cognitive computing agent-based system employs natural language process (NLP) based techniques that offer the user a natural language based interactive interface via which s/he can define knowledge or ask his/her query or decision situation in the form of human language. The NLP base techniques decrease the technical complication of the modern intelligent system interface and help the user to more easily develop and enrich his/her SA and mental models for decision-making.

2. **Creating Query and Developing Experience based Knowledge:** A cognitive computing agent creates experiences according to user inputs and develops an experience-based semantic knowledge model for decision making.

3. **Retrieving Relevant Knowledge:** Effective decision-making requires the system to attain adequate circumstance information as well as extract relevant domain

knowledge. The cognitive computing agent framework employs semantic knowledge extraction technology according to the user situation description or query.

4. **Inference the Query:** A cognitive computing agent develops a reward-based knowledge inference method according to current situation information in an intuitive way. This technique applies the human mental model of comprehending knowledge. So, this enables the agent to perceive and understand the decision situation more quickly.

Let us consider a case study of implementing cognitive computing agent in clinical decision-making in the following steps.

Step 1. Describing Experience and Query as the part of Decision Situation (**DS**):
The user can specify a situation for knowledge-building and input his/her query with natural language. An appropriate way to denote the query through natural language is with plain text. Therefore, a decision situation can be described and queried as natural language questions and sentences.

Definition 17.1. User Decision Situation Query (Q) and Experience (E) inputs as shown in Equations (17.1–17.3).

$$Q := (Q_1, Q_2 \dots Q_n), \tag{17.1}$$

$$E := (E_1, E_2 \dots E_n), \tag{17.2}$$

$$DS = \{Q_i, E_i\} \tag{17.3}$$

where, $n \geq 0$, Q_i is a natural language query and E_i is a natural language experience about a certain situation.

A decision situation (**DS**) denotes an n-tuple comprising of n inputs. Each input of (E_i) and (Q_i) Is termed a (**DS**) *comprising* experience and query input *which* is in the form of natural language. The following example is a clinician's (**DS**) consisting of two inputs:

Example 17.1. A Clinician's Query and Experience Decisions are as follow:

1. $Q_1 =$ *What is the treatment for the flu?*
2. $E_1 =$ *(Ashish was having symptoms like cold, cough, and mild fever in August 2018. He was suffering from flu. However, AntiViral has taken control of flu.)*

Step 2. Creating Query and Developing Experience based Knowledge:
Logically when humans are interacting with each other, they have a habit of highlighting particular matters in the domain of interest and expressing significant information about these entities in a situation from a specific opinion. Consider the following example of Decision situation (**DS**) sentence:

Example 17.2. A Decision situation having Experience sentence E_i

$E_a=$"*Ashish was having symptoms like cold, cough, and mild fever in August 2018.*"
$E_b=$"*He was suffering from flu.*"
$E_c=$ *However, AntiViral has taken control of the flu.*
$E_1=\{E_a, E_b, E_c\}$

In this experience sentence, *Ashish* represents a patient name. From *experience* (E_1), we can deduce the subsequent information the expert is trying to convey. In the perspective of the *patient*, the expert is talking about *Ashish* which is a particular patient name. In the context of patient *symptom*, the expert is talking about, cough, cold, and mild fever as a disease.

In the context of disease date, the expert is speaking about August *2018*, which is a disease time. The experience describes the basic info units of Situation Decision as knowledge *triples*.

To create the knowledge inside the knowledge base, the cognitive computing agent creates a triple. The triple comprises of (Entity, Association, Entity) or (Entity, Association, Phrase).

Definition 17.2. A Decision Situation Experience as Triple (*T*) represented as shown in Equation (17.4):

$$T := (e, a, e) \mid (e, a, p) \tag{17.4}$$

where, e, a, and p denote the *entity*, *association*, and *phrase* correspondingly.

A Decision Situation triple has three components: entity (e), association (a), and phrase (p). The third part of the triple represents an entity (single word) or a phrase (multiple words), which is the actual knowledge info the DS triple indicates. For the experience sentence in Example 2 denoted as (E_a, E_b, E_c), *Ashish, (cough, cold and mild fever), August 2018, flu, and AntiViral* are respectively the third part of the five different DS Experience Knowledge triples.

(Patient, Patient Name, Ashish)

(Patient, Patient Symptom, cold cough and mild fever)

(Patient, Time Report, August 2018)

(Patient, Patient Disease, Flu)

(Patient, Patient Treatment, AntiViral)

If the user input is a query (Q_i), then it is necessary to understand the question type, subject, predicate, and the answer type.

Q_1 = *What is the treatment of the flu?*

Let us consider (Q_1) as the query input given by the user. The cognitive computing agent develops the understanding of the query as follows:

Question Type = What (Fact)

Subject = Flu

Predicate = treatment

Answer Type = treatment

A Decision Situation experience and the query may consist of more than one sentence, each of which may have the same entity as the first term of the triple. Individual entities are either identical or dissimilar varying from individual experience. We refer to the overall Experience of a decision situation as a complete unit of knowledge.

Let $\mathbf{E} = (E_1, E_2, ..., E_n)$ be a DS. The Knowledge experience **(K(E))** of the decision situation is defined as follows.

Definition 17.3. Knowledge Experience (K(E)) shown by Equation (17.5):

$$K(E) := \{K(E_i) \mid E_i \in \textbf{DS}\} \qquad (17.5)$$

where $\textbf{E}_i$ is a collection of experience about decision situation DS.

$$\textbf{E}_i := \{\textbf{E}_a, \textbf{E}_b, \textbf{E}_c, ...\} \qquad (17.6)$$

Each experience in (E_i) represents the understanding of the present situation.

The clinician (E_i) signifies his/her understanding of the present decision situation. In the cognitive computing agent knowledge-based experience model, the experience comprised of several experience statements, each of which relates to a local entity triple. The local entity triple of experience denotes the collective context in which the experienced triple carries definite facts. Therefore, a local entity triple is a sub-experience or sub-information of the decision situation.

Step 3. Retrieving Situation Knowledge

The purpose of knowledge retrieval is to acquire appropriate experience which can be used to assist present decision-making. Therefore, it is sensible to form the knowledge retrieval based on decision situation triples stored in the knowledge base. The following description shows the method of knowledge need extraction.

Definition 17.4. Knowledge $(KR(Q))$ shown in Equation (17.7):

$$KR(Q) := \{Q(E_i) \mid E_i \in \textbf{DS}\}, \qquad (17.7)$$

where Q is a query, Ei is an experience of Decision Situation (DS), and $Q(E_i)$ is the knowledge experience retrieved during the process of searching for best experiences based on certain criteria.

The knowledge denotes the clinician's experience which is warehoused in the experience base of the cognitive computing agent. The concepts about clinical practices and the experiences create the domain knowledge base So, the response for clinical query is retrieved from the knowledge base created by storing experiences from clinicians.

Example 17.3. Knowledge retrieval according to query (Q)

$Q_1 =$ *What is the treatment of the flu?*

The cognitive computing agent applies natural language processing and understands the meaning of the question.

Question Type = What (Fact)
Subject = Flu
Predicate = treatment
Answer Type = treatment

The cognitive computing agent retrieves the knowledge according to the situation decision based on the query input.

$E_a =$ *"Ashish was having symptoms like cold, cough, and mild fever in August 2018."*
$E_b =$ *"He was suffering from flu."*
$E_c =$ *However, AntiViral has taken control of the flu.*
$E_1 = \{E_a, E_b, E_c\}$

The cognitive computing agent extracts experience information from the selected experience base

Selected Experience = E_c = *However, AntiViral has taken control of the flu.*
Triple T = (*Patient, Patient Treatment, AntiViral*)

The cognitive computing agent creates the answer according to the knowledge base.
Answer for Decision Situation = *AntiViral*

Step 4. The inference of Decision Query

The clinician's decision requirement imitates his/her belief in the situation information input and inference inside the cognitive computing agent. Situation information is stored in the data warehouse (Chandiok and Chaturvedi, 2016). Thus, information inference is one of the important features of a cognitive computing system. We define inference decision (I(K(E)) as follows:

Definition 17.4. Inference for Decision Query *I(K(E))* shown in Eq. (17.8):

$$I(K(E)) := I\{Q_1, Q_2, \ldots, Q_n\} => D, n \geq 0, \tag{17.8}$$

Where, Q_i is a query and the inference *I(Q)* outcomes in a set of decisions (D) based on the rankings. Rendering to this definition, the generation of inference is equal to the creation of the relevant decision.

17.6 Summary

This chapter presents intelligent agent, cognitive agent, and cognitive computing agent architectures for building intelligent applications. The cognitive computing agent framework performs best in developing decisions for future intelligent applications. The cognitive computing agent framework outlines two elementary features of a cognition-based decision development: cognitive agent and cognitive computing framework. According to this framework, the model proposes four features. The four important features of the model are knowledge creation, query understanding, knowledge retrieval, and inference.

In this sense, the cognitive computing agent decision method is determined by human-like cognition for achieving the best decision. Although the cognitive computing agent model is autonomous and its application does not depend on a specific field, the discussion of this model is grounded in the clinical domain for the sake of practical understanding. We believe the cognitive computing agent model can extend into other domains, such as finance, business, and education. This chapter embodies the theoretical as well as the practical portion of this cognition decision support for developing intelligent applications. Surely, cognitive computing agents are the future for building intelligent applications.

References

Anderson, J.R., Bothell, D., Byrne, M.D., Douglass, S., Lebiere, C., and Qin, Y. 2004. An integrated theory of the mind. *Psychological Review*, 111(4), p. 1036.

Chandiok, A. and Chaturvedi, D.K. 2018. CIT: Integrated cognitive computing and cognitive agent technologies based cognitive architecture for human-like functionality in artificial systems. *Biologically Inspired Cognitive Architectures*, 26, pp. 55–79.

Chandiok, A. and Chaturvedi, D.K. 2017. Biologically inspired techniques for cognitive decision-making. In *Proceedings of the International Conference on Data Engineering and Communication Technology* (pp. 353–361). Springer, Singapore.

Chandiok, A. and Chaturvedi, D.K. 2017. Cognitive decision making for object recognition by humanoid system. In *Proceedings of International Conference on Communication and Networks* (pp. 533–541). Springer, Singapore.

Chandiok, A. and Chaturvedi, D.K. 2016, March. Cognitive decision support system for medical diagnosis. In *2016 International Conference on Computational Techniques in Information and Communication Technologies (ICCTICT)* (pp. 337–342). IEEE.

Chandiok, A. and Chaturvedi, D.K. 2015, December. Machine learning techniques for cognitive decision making. In *2015 IEEE Workshop on Computational Intelligence: Theories, Applications and Future Directions (WCI)* (pp. 1–6). IEEE.

Chong, H.Q., Tan, A.H., and Ng, G.-W. 2009. Integrated cognitive architectures: A survey. *Artificial Intelligence Review*, 28(2), pp. 103–130.

De Carvalho, L.L., Pereira, D.J., and Coelho, S.A. 2016. Origins and evolution of enactive cognitive science: Toward an enactive cognitive architecture. *Biologically Inspired Cognitive Architectures*, 16, pp. 169–178.

Duch, W., Oentaryo, R.J., and Pasquier, M. 2008. Cognitive architectures: Where do we go from here? In Wang, P., Goertzel, B., and Franklin, S. (Eds.), *Artificial General Intelligence 2008: Proceedings of the First AGI Conference*, Vol. 171 of Frontiers in Artificial Intelligence and Applications, pp. 122–136. IOS Press.

Fodor, J.A. and Pylyshyn, Z.W. 1988. Connectionism and cognitive architecture: A critical analysis. *Cognition*, 28(1), pp. 3–71.

Helie, S. and Sun, R. 2014. Autonomous learning in psychologically-oriented cognitive architectures: A survey. *New Ideas in Psychology*, 34(1), pp. 37–55.

Kajdocsi, L. and Pozna, C.R. 2014, September. Review of the most successfully used cognitive architectures in robotics and a proposal for a new model of knowledge acquisition. In *2014 IEEE 12th International Symposium on Intelligent Systems and Informatics (SISY)* (pp. 239–244). IEEE.

Kakkasageri, M., Hridya, C., Vibha, N., Manvi, S., and Basarkod, P. 2012. A survey on cognitive agent architectures. *IETE Journal of Education*, 53(1), pp. 21–37.

Langley, P., Choi, D., and Shapiro, D. 2004. A cognitive architecture for physical agents. *Computational Learning Laboratory, CSLI*, Stanford University, CA, Tech. Rep.

Langley, P., Laird, J.E., and Rogers, S. 2009. Cognitive architectures: Research issues and challenges. *Cognitive Systems Research*, 10(2), 141–160.

Lucentini, D.F. and Gudwin, R.R. 2015. A comparison among cognitive architectures: A theoretical analysis. *Procedia Computer Science*, 71, pp. 56–61.

Maes, P. 1991. The agent network architecture (ANA). *Acm Sigart Bulletin*, 2(4), pp. 115–120.

Samsonovich, A.V. 2011, October. Comparative analysis of implemented cognitive architectures. In *BICA* (pp. 469–479).

Stollberg, M. and Rhomberg, F. 2006. Survey on goal-driven architectures. Technical Report DERI-TR-2006-06-04, DERI.

Sun, R. 2001. Artificial intelligence: Connectionist and symbolic approaches. *International Encyclopedia of the Social and Behavioral Sciences*, In: N.J. Smelser and P.B. Baltes (eds.),

International Encyclopedia of the Social and Behavioral Sciences. Elsevier, Oxford, UK, pp. 783–789.

Thagard, P. 2012. Cognitive architectures. *The Cambridge Handbook of Cognitive Science*, In K. Frankish & W. Ramsay (Eds.), The Cambridge handbook of cognitive science, Cambridge University Press, Cambrdige, UK, pp. 50–70.

Vernon, D., Metta, G., and Sandini, G. 2007. A survey of artificial cognitive systems: Implications for the autonomous development of mental capabilities in computational agents. *IEEE Transactions on Evolutionary Computation*, 11(2), pp. 151–180.

Wooldridge, M. and Jennings, N.R. 1995. Intelligent agents: Theory and practice. *The Knowledge Engineering Review*, 10(2), pp. 115–152.

Wray, R., Chong, R., Phillips, J., Rogers, S., and Walsh, B. 1994. *A Survey of Cognitive and Agent Architectures*. The University of Michigan.

18

Controllability of Quasilinear Stochastic Fractional Dynamical Systems in Hilbert Spaces

R. Mabel Lizzy and K. Balachandran

CONTENTS

18.1 Introduction...275
18.2 Preliminaries...276
18.3 Controllability Result for Stochastic Fractional Dynamical Systems 278
18.4 Stochastic Dynamical Systems 282
18.5 Controllability Result...284
Acknowledgments...289
References..289

18.1 Introduction

Quasilinear partial differential equations such as symmetric hyperbolic systems of the first order, wave equations, the Korteweg–de Vries equation, Navier–Stokes and Euler equations, equations for compressible fluids, magnetohydrodynamic equations, coupled Maxwell and Dirac equations, etc. occur in mathematical physics. The issue of controllability of such quasilinear equations has been well established for deterministic equations. See for instance [1, 2, 3] and references therein. When a physical state is modeled, external random noise can be a serious disturbance and affects controllability of the model. At the same time fractional derivatives are effectively used to model complex phenomenon by incorporating non-local relations of space and time [4, 5]. This motivates the study of qualitative property, namely controllability of the stochastic fractional quasilinear system of the form

$$^{C}D^{\alpha}x(t) = A(t, x(t))x(t) + Bu(t) + f(t, x(t)) + \sigma(t, x(t))\frac{dW(t)}{dt}, \ t \in [0, T], \tag{18.1}$$

$$x(0) = x_0,$$

where $0 \leq \alpha < 1$. For finite-dimensional systems, the notion of controllability of stochastic differential equations (SDEs) has been studied by many authors [6, 7]. The controllability of stochastic fractional dynamical systems in infinite spaces with $A(t,x(t)) = A$ and

$A(t,x(t)) = A(t)$ being a bounded linear operator is studied in [8, 9–11]. In particular when $\alpha = 1$ the system (18.1) takes the form

$$dx(t) + A(t, x(t))x(t)dt = f(t, x(t))dt + Bu(t)dt + \Phi dW(t), \quad t \in [0, T],$$

$$x(0) = x_0$$

(18.2)

and the controllability results of Equation (18.2) in some separable Hilbert space X with

$A(t,x(t)) = A$ in finite and infinte dimensional settings are studied in [6] and [7] respectively. Here A generates a C_0-semigroup and $W(t)$ is a Wiener process with finite trace. In this chapter, we discuss the controllability of the above system with the linear operator A depending on t and $x(t,\omega) \in X$, generating a C_0-semigroup (hyperbolic case) and perturbed with cylindrical Wiener process.

Kato [12, 13] presented a unified treatment of the Cauchy problem (local in time) for various quasilinear equations that appear in mathematical physics, based on the theory of abstract evolution equations. The stochastic counterpart of the same is considered in the Hilbert space setting [14] and in reflexive UMD Banach space [15], wherein the existence of local-time mild solutions is studied. There are many papers that have discussed the controllability of specific quasilinear partial differential equations [16]. In this chapter we obtain the controllability results for stochastic fractional system (18.1) by assuming that the operator $A(t,b)$ is a bounded linear operator for every $(t,b) \in J \times Ba$, where Ba is a ball in X. An equivalent integral equation to Equation (18.1) is obtained and Banach's contraction principle is used to obtain the desired controllability result. For the particular case of $\alpha = 1$ we study the controllability results by assuming that $A(t,b)$ generates a C_0 semigroup for every $(t,b) \in J \times Ba$. An equivalent integral equation to the quasilinear system is obtained by showing the existence of random evolution system as in [15] and applying Banach's contraction principle to arrive at the controllability result.

18.2 Preliminaries

Let $(\Omega, \mathscr{F}, (\mathscr{F}_t)_{t\geq 0}, P)$ be given probability space. Let X, H, U be separable Hilbert spaces. For any Banach spaces Z the norm of Z is denoted by $\|\cdot\|_Z$. The space of all bounded linear operators from X to Y is denoted by $\mathbb{L}(X, Y)$ and $\mathbb{L}(X, X)$ is denoted by $\mathbb{L}(X)$.

Consider the quasilinear stochastic fractional system of the form

$$^C D^\alpha x(t) = A(t, x(t))x(t) + Bu(t) + f(t, x(t)) + \sigma(t, x(t))\frac{dW(t)}{dt},$$

$$x(0) = x_0,$$

(18.3)

where $0 < \alpha \leq 1$, the state variable $x(t,\omega)$ is random and takes values in a separable Hilbert space X, and the control function $u(t,\omega)$ is from U_{ad}, a Hilbert space of admissible control functions and takes values in a separable Hilbert space U. $A(t,x)$ is a bounded linear operator on X for every $(t,x) \in J \times X$, $B: U \to X$ is a bounded linear operator, $f: J \times X \to X$ is a continuous function, $W(t)$ is a H-valued Wiener process, and $\sigma(t,x)$ is a Hilbert–Schmidtz operator from H to X for every $(t,x) \in J \times X$. Taking I^α on both the sides we get,

$$x(t) = x_0 + \frac{1}{\Gamma(\alpha)} \int_0^t (t-s)^{\alpha-1} A(s, x(s)) x(s) ds$$

$$+ \frac{1}{\Gamma(\alpha)} \int_0^t (t-s)^{\alpha-1} Bu(s) ds \qquad (18.4)$$

$$+ \frac{1}{\Gamma(\alpha)} \int_0^t (t-s)^{\alpha-1} \sigma(s, x(s)) dW(s).$$

The above integral equation is equivalent to the system given in Equation (18.3). Let $Ba := \{x \in X : \|x - x_0\|_X^2 \le r\}$. Since $A(t,b)$ is a bounded linear operator there exists a constant $C_1 > 0$ such that $\|A(t,b)\|_{\mathbb{L}(X,X)}^2 \le C_1$, for every $(t, x) \in J \times Ba$. Assume the following hypotheses:

(H1) For every $(t, b) \in J \times Ba$ there exists a constant $L_1 > 0$ such that the bounded linear operator $A(t,b)$ satisfies

$$\|A(t, b_1) - A(t, b_2)\|_{\mathbb{L}(X,X)}^2 \le L_1 \|b_1 - b_2\|_X^2$$

for every $b_1, b_2 \in Ba$.

(H2) The functions $f: J \times X \to X$ and $\sigma: J \times X \to \mathbb{L}_2^0$, where $\mathbb{L}_0^2$ denotes the space of Hilbert–Schmidt operators (see [17]) are continuous and there exist constants $L_2, L_3 > 0$ such that

$$\|f(t, b_1) - f(t, b_2)\|_X^2 \le L_2 \|b_1 - b_2\|_X^2$$

$$\|\sigma(t, b_1) - \sigma(t, b_2)\|_{\mathbb{L}_2^0(H,X)}^2 \le L_3 \|b_1 - b_2\|_X^2$$

for all $b_1, b_2 \in Ba$.

Define the sequence of stopping times τ_N by,

$$\tau_N := \inf_{t \ge 0} \{t : C_1 \vee L_1 \vee L_2 \vee L_3 \ge N\} \qquad (18.5)$$

and assume $\tau_N > 0$. Since $x(\cdot)$ is random, the constants C_1, L_1, L_2 in the assumptions depend on N, where N is from stopping time. We use the notation $\tilde{T} = T \wedge \tau_N$ and I to denote the time interval $[0, \tilde{T}] \subset J$.

Let us consider the following space settings. Denote,

- $\mathbb{L}_2^{\mathscr{F}}(\Omega, C(I, X))$ is the space of $\{\mathscr{F}_t\}_{t \ge 0}$-adapted processes having continuous paths a.s. and endowed with the norm,

$$\|\phi\|_{\mathbb{L}_2^{\mathscr{F}}(\Omega, C(I, X))}^2 = E\left(\sup_{t \in I} \|\phi(t)\|_X^2 \right),$$

where E denotes expectation with respect to P.

- $U_{ad} := \mathbb{L}_2^{\mathscr{F}}(I, U)$, which is a Hilbert space of all functions in $\mathbb{L}_2(I \times \Omega, U)$ that are $\{\mathscr{F}_t\}_{t \geq 0}$-adapted and endowed with the norm,

$$\|u\|_{U_{ad}}^2 = E\left(\int_0^{\tilde{T}} \|u(t)\|_U^2 \, dt\right).$$

- $\mathbb{L}_2(\Omega, \mathscr{F}_0, X)$, which is the Hilbert space of all $\mathscr{F}_0$-measurable square integrable random variables with values in X.
- $\mathbb{L}_2(\Omega, \mathscr{F}_T, X)$ is a closed subspace of $\mathbb{L}_2(\Omega, X)$ consisting of all $\mathscr{F}_T$-measurable square integrable random variables with values in X.

Similar to the conventional controllability concept, the controllability of the stochastic fractional dynamical system is defined as follows:

The set of all states attainable from x_0 in time $t > 0$ is given by the set

$$\mathscr{R}_t(x_0) = \{x(t; x_0, u) : u \in U_{ad}\}.$$

Definition 18.1. [7] The stochastic fractional dynamical system is said to be completely controllable on the interval I if for every $x_1 \in \mathbb{L}_2(\Omega, \mathscr{F}_T, X)$, there exists a control $u \in U_{ad}$ such that the solution $x(t)$ satisfies $x(T) = x_1$.

In other words,

$$\mathscr{R}_{\tilde{T}}(x_0) = \mathbb{L}_2(\Omega, \mathscr{F}_{\tilde{T}}, X).$$

18.3 Controllability Result for Stochastic Fractional Dynamical Systems

In this section we obtain the controllability result for the stochastic fractional dynamical system given in Equation (18.3). Let us further assume the following:

(H3) The operator $B: U \to X$ is a bounded linear operator satisfying $\|B\|_{\mathbb{L}(U,X)}^2 \leq K_1$.

(H4) The linear operator $\mathscr{W}_{\tilde{T}}: U_{ad} \to X$,

$$\mathscr{W}_{\tilde{T}} u = \frac{1}{\Gamma(\alpha)} \int_0^{\tilde{T}} (\tilde{T} - s)^{\alpha - 1} B u(s) ds$$

induces an invertible operator $\mathscr{W}_{\tilde{T}}$ defined on $U_{ad} / \ker(\mathscr{W}_{\tilde{T}})$ and there exists a positive constant K such that $\left\|\mathscr{W}_{\tilde{T}}^{-1}\right\|^2 \leq K$.

(H5) There exists $r > 0$ such that

$$\frac{4\tilde{T}^{2\alpha - 1}}{(\Gamma(\alpha))^2 (2\alpha - 1)}\left(1 + 5K_1 K \frac{T^{2\alpha + 1}}{(2\alpha - 1)}\right)$$

$$\left((6N\tilde{T} + 4N)(r + \| x_0 \|^2) + 2\tilde{T}\sup_{t\in I} \| f(t,0) \|^2 + 2\sup_{t\in I} \| \sigma(t,0) \|^2 \right) \le r$$

(H6) Let

$$\rho_1 = \frac{5\tilde{T}^{2\alpha-1}}{(\Gamma(\alpha))^2(2\alpha-1)} \left(\frac{\tilde{T}^{2\alpha}KK_1}{(\Gamma(\alpha))^2} + 1 \right) (L_1\tilde{T}r + N\tilde{T} + L_2 + L_3),$$

be such that $0 < \rho_1 < 1$.

Theorem 18.1. *If the hypotheses (H1)–(H6) are satisfied then the stochastic fractional dynamical system is completely controllable.*

Proof. Let x_1 be an arbitrary random variable in $\mathbb{L}_2(\Omega, \mathscr{F}_{\tilde{T}}, X)$. Define the operator Λ on $\mathbb{L}_2^{\mathcal{F}}(\Omega, C(I, X))$ by

$$\Lambda x(t) = x_0 + \frac{1}{\Gamma(\alpha)} \int_0^t (t-s)^{\alpha-1} A(s, x(s)) x(s) \mathrm{d}s + \frac{1}{\Gamma(\alpha)} \int_0^t (t-s)^{\alpha-1} Bu(s) \mathrm{d}s$$

$$+ \frac{1}{\Gamma(\alpha)} \int_0^t (t-s)^{\alpha-1} f(s, x(s)) \mathrm{d}s + \frac{1}{\Gamma(\alpha)} \int_0^t (t-s)^{\alpha-1} \sigma(s, x(s)) \mathrm{d}W(s).$$

The invertibility of $\mathscr{W}_{\tilde{T}}$ in (H4) allows us to choose the control process u as

$$u(t) = \mathscr{W}_{\tilde{T}}^{-1}\left(x_1 - x_0 - \frac{1}{\Gamma(\alpha)} \int_0^{\tilde{T}} (\tilde{T}-s)^{\alpha-1} A(s, x(s)) x(s) \mathrm{d}s \right.$$

$$\left. - \frac{1}{\Gamma(\alpha)} \int_0^{\tilde{T}} (\tilde{T}-s)^{\alpha-1} f(s, x(s)) \mathrm{d}s - \frac{1}{\Gamma(\alpha)} \int_0^{\tilde{T}} (\tilde{T}-s)^{\alpha-1} \sigma(s, x(s)) \mathrm{d}W(s) \right).$$

The control u steers the nonlinear system from the initial state x_0 to x_1 at time $\tilde{T}$, provided we can obtain a fixed point of the nonlinear operator Λ. First we show that Λ maps functions in $\mathbb{L}_2^{\mathcal{F}}(\Omega, C(I, X))$ having values in Ba into itself. Consider

$$\| \Lambda x(t) - x_0 \|^2 \le 4 \left(\frac{1}{(\Gamma(\alpha))^2} \left\| \int_0^t (t-s)^{\alpha-1} A(s, x(s)) x(s) \mathrm{d}s \right\|^2 \right.$$

$$+ \frac{1}{(\Gamma(\alpha))^2} \left\| \int_0^t (t-s)^{\alpha-1} Bu(s) \mathrm{d}s \right\|^2$$

$$+ \frac{1}{(\Gamma(\alpha))^2} \left\| \int_0^t (t-s)^{\alpha-1} f(s, x(s)) \mathrm{d}s \right\|^2$$

$$+ \frac{1}{(\Gamma(\alpha))^2} \left\| \int_0^t (t-s)^{\alpha-1} \sigma(s, x(s)) \mathrm{d}W(s) \right\|^2 \bigg)$$

$$\leq \frac{4}{(\Gamma(\alpha))^2} \left(\left[\sup_{t \in I} \|A(t, x(t))\|^2_{\mathbb{L}(X,X)} \sup_{t \in I} \|x(t)\|^2_X + \sup_{t \in I} \|f(t, x(t))\|^2 \right. \right.$$

$$\left. + T^{-1} \int_0^t \|B\|^2_{\mathbb{L}(U,X)} \|u(s)\|^2_U \, \mathrm{d}s \right] \frac{\tilde{T}^{2\alpha}}{(2\alpha - 1)}$$

$$\left. + \left\| \int_0^t (t-s)^{\alpha-1} \sigma(s, x(s)) \mathrm{d}W(s) \right\|^2 \right).$$

Therefore using Itô isometry and hypotheses (H1), (H2), and (H3) we have

$$\mathbf{E} \sup_{t \in I} \|\Lambda x(t) - x_0\|^2 \leq \frac{4}{(\Gamma(\alpha))^2} \left(\left[\mathbf{E} C_1 \sup_{t \in I} \|x(t)\|^2_X + 2 \mathbf{E} L_2 \sup_{t \in I} \|x(t)\|^2 \right. \right.$$

$$+ 2 \sup_{t \in I} \|f(t, 0)\|^2 + T^{-1} K_1 \mathbf{E} \int_0^{\tilde{T}} \|u(s)\|^2_U \, \mathrm{d}s \bigg] \frac{\tilde{T}^{2\alpha}}{(2\alpha - 1)}$$

$$\left. + \frac{\tilde{T}^{2\alpha-1}}{(2\alpha - 1)} \left[2 \mathbf{E} L_3 \sup_{t \in I} \|x(t)\|^2_X + 2 \sup_{t \in I} \|\sigma(t, 0)\|^2 \right] \right).$$

Using the definition of stopping time, Equation (18.5), and the fact that $\|x(t) - x_0\|^2_X \leq r$ we obtain

$$\mathbf{E} \sup_{t \in I} \|\Lambda x(t) - x_0\|^2 \leq \frac{4}{(\Gamma(\alpha))^2} \left(\frac{\tilde{T}^{2\alpha}}{(2\alpha - 1)} \left[6N\left(r + \|x_0\|^2\right) + 2 \sup_{t \in I} \|f(t, 0)\|^2 \right. \right.$$

$$\left. + K_1 \mathbf{E} \int_0^{\tilde{T}} \|u(s)\|^2_U \, \mathrm{d}s \right] + \frac{\tilde{T}^{2\alpha-1}}{(2\alpha - 1)} \left[4N(r + \|x_0\|^2) + 2 \sup_{t \in I} \|\sigma(t, 0)\|^2 \right] \right) \left(5\tilde{T}K + 1 \right). \qquad (18.6)$$

Also we have

$$\mathbf{E} \int_0^{\tilde{T}} \|u(t)\|^2_U \, \mathrm{d}t \leq 5\tilde{T}K \left(\|x_1\|^2_X + \|x_0\|^2_X + \left[(6N\tilde{T} + 4N)\left(r + \|x_0\|^2\right) \right. \right.$$

$$\left. + 2\tilde{T} \sup_{t \in I} \|f(t, 0)\|^2 + 2 \sup_{t \in I} \|\sigma(t, 0)\|^2 \right] \frac{\tilde{T}^{2\alpha-1}}{(2\alpha - 1)} \right). \qquad (18.7)$$

From the above inequalities, Equations (18.6) and (18.7), and (H5) it follows that

$$\mathbf{E}\sup_{t\in J}\|\Lambda x(t)-x_0\|^2 \le \frac{4\tilde{T}^{2\alpha-1}}{(\Gamma(\alpha))^2(2\alpha-1)}\left(1+5K_1K\frac{T^{2\alpha+1}}{(2\alpha-1)}\right)$$

$$\left((6N\tilde{T}+4N)(r+\|x_0\|^2)+2\tilde{T}\sup_{t\in I}\|f(t,0)\|^2+2\sup_{t\in I}\|\sigma(t,0)\|^2\right)\le r.$$

Thus Λ maps functions in $\mathbb{L}_2^{\mathscr{F}}(\Omega,C(I,X))$ having values in Ba into itself. Now for $x_1,x_2\in\mathbb{L}_2^{\mathscr{F}}(\Omega,C(I,X))$ having values in Ba, we have

$$\mathbf{E}\sup_{t\in I}\|\Lambda x_1(t)-\Lambda x_2(t)\|_X^2$$

$$=\mathbf{E}\sup_{t\in I}\left\|\frac{1}{\Gamma(\alpha)}\int_0^t(t-s)^{\alpha-1}\left[A(s,x_1(s))x_1(s)-A(s,x_2(s))x_1(s)\right]ds\right.$$

$$+\frac{1}{\Gamma(\alpha)}\int_0^t(t-s)^{\alpha-1}\left[A(s,x_2(s))x_1(s)-A(s,x_2(s))x_2(s)\right]ds$$

$$\frac{1}{(\Gamma(\alpha))^2}\int_0^t(t-s)^{\alpha-1}B\mathscr{W}_{\tilde{T}}^{-1}$$

$$\left[\int_0^T(T-s)^{\alpha-1}\left[A(s,x_1(s))x_2(s)-A(s,x_1(s))x_1(s)\right]ds\right.$$

$$+\int_0^T(T-s)^{\alpha-1}\left[f(s,x_2(s))-f(s,x_1(s))\right]ds$$

$$+\int_0^T(T-s)^{\alpha-1}\left[\sigma(s,x_2(s))-\sigma(s,x_1(s))\right]dW(s)\bigg]ds$$

$$+\frac{1}{\Gamma(\alpha)}\int_0^t(t-s)^{\alpha-1}\left[f(s,x_1(s))-f(s,x_2(s))\right]ds$$

$$\left.+\frac{1}{\Gamma(\alpha)}\int_0^t(t-s)^{\alpha-1}\left[\sigma(s,x_1(s))-\sigma(s,x_2(s))\right]dW(s)\right\|^2$$

$$\le\frac{5\tilde{T}^{2\alpha-1}\left(L_1\tilde{T}\|x_1\|_X^2+N\tilde{T}+L_2+L_3\right)}{(\Gamma(\alpha))^2(2\alpha-1)}\left(\frac{\tilde{T}^{2\alpha}KK_1}{(\Gamma(\alpha))^2}+1\right)\mathbf{E}\sup_{t\in I}\|x_1(t)-x_2(t)\|_X^2$$

$$\le\rho_1\mathbf{E}\sup_{t\in I}\|x_1(t)-x_2(t)\|_X^2.$$

By (H6) we conclude that Λ is a contraction mapping and hence there exists a unique fixed point $x \in \mathbb{L}_2^{\mathscr{F}}(\Omega, C(I, X))$ having value in Ba for Λ. Therefore the system in Equation (18.3) is completely controllable.

18.4 Stochastic Dynamical Systems

Now we consider the special case of stochastic quasilinear evolution system for $\alpha = 1$ with additive cylindrical Wiener noise of the form

$$dx(t) + A\left(t, x(t)\right)x(t)dt = f(t, x(t))dt + Bu(t)dt + \Phi dW(t), \ t \in J = [0, T],$$

$$x(0) = x_0,$$

(18.8)

where $A(t,x)$ is the infinitesimal generator of a C_0-semigroup in X and Φ is a Hilbert–Schmidt operator from H to X. Let Y be a separable Hilbert space densely and continuously embedded in X, i.e., there is a constant $c > 0$ such that $\|\cdot\|_X \le c\|\cdot\|_Y$. We recall some definitions and known facts from Pazy [18].

Definition 18.2. Let A be a linear operator in X and let Y be a subspace of X. The operator $\tilde{A}$ defined by $D(\tilde{A}) = \{x \in D(A) \cap Y : Ax \in Y\}$ and $\tilde{A}x = Ax$ for $x \in D(\tilde{A})$ is called the part of A in Y.

Definition 18.3. Let Ba be a subset of X and for every $t \in J$ and $b \in$ Ba, let $A(t,b)$ be an infinitesimal generator of a C_0-semigroup $S_{t,b}(s)$, $s \in J$ on X. The family of operators $A = \{A(t,b)\}_{(t,b) \in J \times \text{Ba}}$ is said to be stable if there are $M \ge 1$ and β such that

$$\left\| \prod_{j=1}^{k} \left(A(t_j, b_j) + \lambda \right)^{-1} \right\|_{\mathbb{L}(X)} \le M(\lambda - \beta)^{-k}, \quad \lambda > \beta,$$

(18.9)

for every finite sequence $0 \le t_1 \le \cdots \le t_k \le T$, $b_j \in$ Ba; $1 \le j \le k$. The pair (M, β) are called stability index for $\{A(t)\}$. In (18.9), the operator product is time-ordered, i.e., if $t_i > t_j$, then the operator $A(t_i)$ will be on the left of the operator $A(t_j)$.

Definition 18.4. Let $S_{t,b}(s)$, $s \in J$, be the C_0-semigroup generated by $A(t,b), (t,b) \in J \times \mathbb{B}$. A subspace Y of X is called $A(t,b)$-admissible if Y is invariant subspace of $S_{t,b}(s)$ and the restriction of $S_{t,b}(s)$ to Y is a C_0-semigroup in Y. Here we make the following assumptions.

(H7) For every $(t,b) \in J \times$ Ba, $A(t,b)$ is the infinitesimal generator of a C_0-semigroup $S_{t,b}(s)$, $s \in J$ on X. The family $\{A(t,b)\}$, $(t,b) \in J \times$ Ba is stable with stability constants (M, β).

(H8) Y is $A(t,b)$-admissible for $(t,b) \in J \times$ Ba and the family $\{\tilde{A}(t,b)\}$, $(t,b) \in J \times$ Ba of parts $\tilde{A}(t,b)$ of $A(t,b)$ in Y, is stable in Y.

(H9) For $(t,b) \in J \times$ Ba, $D(A(t,b)) \supset Y$, $A(t,b)$ is a bounded linear operator from Y to X with $C_2 > 0$ such that $\|A(t,b)\|_{\mathbb{L}(Y,X)}^2 \le C_2$ and $t \to A(t,b)$ is continuous in the $\mathbb{L}(Y, X)$ norm for every $b \in \mathbb{B}$. There exists a constant $L_3 > 0$ such that

$$\left\|A(t,b_1) - A(t,b_2)\right\|^2_{\mathbb{L}(Y,X)} \le L_3 \left\|b_1 - b_2\right\|^2_X$$

holds for every $t \in J$ and $b_1, b_2 \in$ Ba.

(H10) The nonlinear function $f: J \times X \to X$ satisfies (H2) with Lipschitz constant L_2 and there exists constants $C_3 > 0$ such that and for all $b \in$ Ba, $\left\|f(t,b)\right\|^2_Y \le C_3$.

(H11) The operator $\Phi \in \mathbb{L}^0_2(H, Y)$.

From (H11) we obtain $\left\|\int_0^{\tilde{T}} \Phi dW(\theta)\right\|^2_Y$ is bounded a.s. by using the Markov inequality and we denote it by M_1. Let us define the sequence of stopping times τ_N by,

$$\tau_N := \inf_{t \ge 0}\left\{t: M \vee \beta \vee C_2 \vee C_3 \vee L_2 \vee L_3 \vee M_1 \vee C \ge N\right\}, \tag{18.10}$$

where the constant C appears later in Lemma 18.1 and we assume $\tau_N > 0$. Since $x(\cdot,\cdot)$ is random, the constants $M, \beta, C_2, C_3, L_2, L_3, M_1$ in the assumptions depend not only on the parameters of $\mathbb{B}$ but also on N, where N is from stopping time. We use the notation $\tilde{T} = T \wedge \tau_N$ and I to denote the time interval $[0, \tilde{T}] \subset J$.

We now provide the definition of random evolution system and provide its existence.

Definition 18.5. A two parameter family of bounded linear operators $U(t,s)$, $0 \le s \le t \le \tilde{T}$, on X is called an evolution system if the following two conditions are satisfied:

(i) $U(s,s) = I$ (identity operator), $U(t,r)U(r,s) = U(t,s)$ for $0 \le s \le r \le t \le \tilde{T}$.

(ii) $(t,s) \to U(t,s)$ is strongly continuous for $0 \le s \le t \le \tilde{T}$.

Theorem 18.2. *Let τ_N be the stopping time defined in (18.10). Let* Ba $\subset X$ *and let* $\{A(t,b)\}$, $(t,b) \in I \times$ Ba *be a family of operators satisfying* (H7)–(H9). *If $x \in \mathbb{L}^x_2(\Omega, C(I, X))$ then there is a unique random evolution system* $\{U(t,s;x)\} := \{U(t,s,(x(r,\omega))_{s \le r \le t})\}$ *defined on the triangle* $\Delta := 0 \le s \le t \le \tilde{T}$ *with the following properties satisfied pointwise on Ω.*

(P1) $\sup_{t,s \in \Delta} \left\|U(t,s;x)\right\|_{\mathbb{L}(X,X)} \le Me^{\beta \tilde{T}}$, where M and β are stability constants.

(P2) $\dfrac{\partial}{\partial t} U(t,s;x)y = A(s, x(s))U(t,s;x)y$ for $y \in Y$, $0 \le s \le t \le \tilde{T}$.

(P3) $\dfrac{\partial}{\partial s} U(t,s;x)y = -U(t,s;x)A(s,x(s))y$ for $y \in Y$, $0 \le s \le t \le \tilde{T}$.

Proof of the above theorem is similar to that of Theorem 2.13 in [15] and Theorem 5.3.1 in [18] and hence omitted.

The random evolution operator constructed is only $\mathscr{F}_t$-measurable and generally not $\mathscr{F}_s$-measurable. Hence the stochastic integral $\int_0^t U(t,s;x)\Phi dW(s)$ is not defined in Itô sense.

To rectify this issue, we apply integration by parts (as in [19]) and obtain terms which do not involve the stochastic integration of non-adapted integrands as seen below.

$$\int_0^t U(t,s;x)\Phi dW(s) = U(t,0;x)\int_0^t \Phi dW(\theta) + \int_0^t U(t,s;x)A(s,x(s))\left(\int_s^t \Phi dW(\theta)\right)ds. \tag{18.11}$$

An $\mathscr{F}_t$-measurable and continuous function $x(\cdot)$ is called a mild solution to Equation (18.8) if $x(0) = x_0$ and the following integral equation

$$x(t) = U(t,0;x)x_0 + \int_0^t U(t,s;x)Bu(s)ds + \int_0^t U(t,s;x)f(s,x(s))ds$$

$$+ U(t,0)\int_0^t \Phi dW(\theta) + \int_0^t U(t,s)A(s,x(s))\left(\int_s^t \Phi dW(\theta)\right)ds$$

(18.12)

is satisfied.

Lemma 18.1. *Let τ_N be the stopping time defined in (18.10). Let $Ba \subset X$ and let $\{A(t,b)\}$, $(t,b) \in I \times Ba$ be a family of operators satisfying (H7)–(H9). There exists a constant C such that for every $x_1, x_2 \in \mathbb{L}_2^{\mathscr{F}}(\Omega, C(I,X))$ having values in Ba and every $y \in Y$ we have pointwise on Ω,*

$$\left\|U_{x_1}(t,s)y - U_{x_2}(t,s)y\right\|_X^2 \leq C\tilde{T}\|y\|_Y^2 \int_s^t \|x_1(r) - x_2(r)\|_X^2 \, dr,$$

(18.13)

where C depends only on the stability constants of $\{A(t,b)\}$ and $\{\tilde{A}(t,b)\}$.

18.5 Controllability Result

To study the controllability problem we assume the following conditions.

(H12)   The operator $B: U \to X$ is a bounded linear operator satisfying $\|B\|_{L(U,Y)}^2 \leq K_2$.

(H13)   Let

$$10NT^2\|x_0\|_Y^2(1 + 5TN^2e^{2NT}K_2K) \leq 1/2,$$

$$25TN^2e^{2NT}K_2\mathbf{K}(E\|x_1\|_X^2 + a) + 5a \leq r/2,$$

where

$$a = 2\mathbf{E}\|U(t,0;x_0)x_0 - x_0\|_X^2 + N^2e^{2NT}\|\Phi\|_{\mathbb{L}_2^0(H,Y)}^2 + T^2N^3e^{2NT}\left[1 + \|\Phi\|_{\mathbb{L}_2^0(H,Y)}^2\right]$$

(H14)   Let

$$\rho_2 = 5b(1 + 2T^4NK_2K),$$

where $b = NT^2\|x_0\|_Y^2 + T^4N^2 + 2T^2N^3e^{2NT} + NT^3M_1 + T^5N^2M_1 + 2T^2N^3e^{2NT}M_1$ be such that $0 < \rho_2 < 1$.

Theorem 18.3. If the hypotheses (H7)–(H14) are satisfied and $\|x_0\|_Y^2 < \infty$ a.s. then Equation (18.8) is completely controllable on I.

Proof. Let $x_1 \in \mathbb{L}_2(\Omega, \mathscr{F}_{\tilde{T}}, X)$ be an arbitrary point. Define the operator Λ on the functions in $\mathbb{L}_2^{\mathscr{F}}(\Omega, C(I, X))$ having values in Ba by

$$
\Lambda x(t) = U(t,0;x)x_0 + \int_0^t U(t,s;x)Bu(s)ds + \int_0^t U(t,s;x)f(s,x(s))ds
$$

$$
+ U(t,0;x)\int_0^t \Phi dW(\theta) + \int_0^t U(t,s;x)A(s,x(s))\left(\int_s^t \Phi dW(\theta)\right)ds, \tag{18.14}
$$

where the control variable $u(\cdot)$ is defined using hypothesis (H8) as

$$
u(t) = \mathscr{W}_{\tilde{T}}^{-1}\left[x_1 - U(\tilde{T},0;x)x_0 - \int_0^{\tilde{T}} U(\tilde{T},s;x)f(s,x(s))ds - U(\tilde{T},0;x)\int_0^{\tilde{T}} \Phi dW(\theta) \right.
$$

$$
\left. - \int_0^{\tilde{T}} U(\tilde{T},s;x)A(s,x(s))\left(\int_s^{\tilde{T}} \Phi dW(\theta)\right)ds \right]. \tag{18.15}
$$

It is clear that $\Lambda x(t)$ is $\mathscr{F}_t$-adapted and we now prove that Λ maps functions in $\mathbb{L}_2^{\mathscr{F}}(\Omega, C(I, X))$ having values in Ba into itself.

$$
\|\Lambda x(t) - x_0\|_X^2 \le 5\left(\|U(t,0;x)x_0 - x_0\|_X^2 + \left\|\int_0^t U(t,s;x)Bu(s)ds\right\|_X^2 \right.
$$

$$
+ \left\|\int_0^t U(t,s;x)f(s,x(s))ds\right\|_X^2 + \left\|U(t,0;x)\int_0^t \Phi dW(\theta)\right\|_X^2
$$

$$
+ \left\|\int_0^t U(t,s;x)A(s,x(s))\left(\int_s^t \Phi dW(\theta)\right)ds\right\|_X^2 \right)
$$

$$
\le 5\left(2\|U(t,0;x)x_0 - U(t,0,x_0)x_0\|_X^2 + 2\|U(t,0;x_0)x_0 - x_0\|_X^2 \right.
$$

$$
+ \|U(t,0;x)\|_{\mathbb{L}(X,X)}^2 \left\|\int_0^{\tilde{T}} \Phi dW(\theta)\right\|_Y^2 + \sup_{0 \le s \le \tilde{T}} \|U(t,s,x)\|_{\mathbb{L}(X,X)}^2 \|B\|_{\mathbb{L}(U,X)}^2
$$

$$
\sup_{t \in I} \int_0^t \|u(s)\|_U^2 ds + \tilde{T}^2 \sup_{0 \le s \le \tilde{T}} \|U(t,s,x)\|_{\mathbb{L}(X,X)}^2 \left[\sup_{t \in I} \|f(t,x(t))\|_X^2 \right.
$$

$$
\left. \left. + \sup_{t \in I} \|A(t,x(t))\|_{\mathbb{L}(Y,X)}^2 \left\|\int_0^{\tilde{T}} \Phi dW(\theta)\right\|_Y^2 \right] \right).
$$

Using Lemma 18.1 in the first term and the bound assumptions in (H9), (H10), (H12), and (P1) in the other terms we get

$$
\leq 5\left(2C\tilde{T}\|x_0\|_Y^2 \int_0^{\tilde{T}} \|x(s)-x_0\|_X^2\, ds + 2\|U(t,0;x_0)x_0 - x_0\|_X^2 \right.
$$

$$
+ M^2 e^{2\beta\tilde{T}} \left\| \int_0^{\tilde{T}} \Phi dW(\theta) \right\|_Y^2 + M^2 e^{2\beta\tilde{T}} K_2 \sup_{t\in I} \int_0^t \|u(s)\|_U^2\, ds
$$

$$
\left. + \tilde{T}^2 M^2 e^{2\beta\tilde{T}} \left[C_3 + C_2 \left\| \int_0^{\tilde{T}} \Phi dW(\theta) \right\|_Y^2 \right] \right).
$$

Using the stopping time definition in Equation (18.10) we see that the random variables C, M, β, C_2, C_3 are bounded by N. Further using the fact that $\tilde{T} \leq T$ and the assumption that $\|x_0\|_Y < \infty$ a.s. we have

$$
\mathbf{E}\sup_{t\in I}\|\Lambda x(t)-x_0\|_X^2 \leq 5\left(2NT^2\|x_0\|_Y^2 \, \mathbf{E}\sup_{t\in I}\|x(t)-x_0\|_X^2 \right.
$$

$$
+ 2\mathbf{E}\|U(t,0;x_0)x_0 - x_0\|_X^2
$$

$$
+ N^2 e^{2NT}\mathbf{E}\left\| \int_0^{\tilde{T}} \Phi dW(\theta) \right\|_Y^2 + N^2 e^{2NT} K_2 \mathbf{E}\int_0^{\tilde{T}} \|u(s)\|_U^2\, ds
$$

$$
\left. + T^2 N^3 e^{2NT}\left[1 + \mathbf{E}\left\| \int_0^{\tilde{T}} \Phi dW(\theta) \right\|_Y^2 \right] \right).
$$

Using Itô's isometry we have

$$
\begin{aligned}
\mathbf{E}\sup_{t\in I}\|\Lambda x(t)-x_0\|_X^2 \leq 5\bigg(&2NT^2\|x_0\|_Y^2 \, \mathbf{E}\sup_{t\in I}\|x(t)-x_0\|_X^2 \\
&+ 2\mathbf{E}\|U(t,0;x_0)x_0 - x_0\|_X^2 \\
&+ N^2 e^{2NT}\|\Phi\|_{\mathbb{L}_2^0(H,Y)}^2 + TN^2 e^{2NT} K_2 \mathbf{E}\int_0^{\tilde{T}} \|u(s)\|_U^2\, ds \\
&+ T^2 N^3 e^{2NT}\left[1 + \|\Phi\|_{\mathbb{L}_2^0(H,Y)}^2 \right]\bigg) \\
\leq 5\bigg(&2NT^2\|x_0\|_Y^2 \, \mathbf{E}\sup_{t\in I}\|x(t)-x_0\|_X^2 \\
&+ TN^2 e^{2NT} K_2 \mathbf{E}\int_0^{\tilde{T}} \|u(s)\|_U^2\, ds + a\bigg).
\end{aligned}
$$

$$(18.16)$$

Also we have estimated similarly

$$\|u(t)\|_{U_{ad}}^2 \le 5\left\|\mathscr{W}_{\tilde{T}}^{-1}\right\|^2 \mathbf{E}\bigg(\|x_1\|_X^2 + \left\|U(\tilde{T},0;x)x_0\right\|_X^2$$

$$+\left\|\int_0^{\tilde{T}} U(\tilde{T},s;x)\,f(s,x(s))\mathrm{d}s\right\|_X^2 + \left\|U(\tilde{T},0;x)\int_0^{\tilde{T}}\varPhi\mathrm{d}W(\theta)\right\|_X^2$$

$$+\left\|\int_0^{\tilde{T}} U(\tilde{T},s;x)\,A(s,x(s))\left(\int_s^{\tilde{T}}\varPhi\mathrm{d}W(\theta)\right)\mathrm{d}s\right\|_X^2\bigg)$$

(18.17)

$$\le 5K\bigg(\mathbf{E}\|x_1\|_X^2 + 2NT^2\|x_0\|_Y^2\,\mathbf{E}\sup_{t\in I}\|x(t)-x_0\|_X^2 + a\bigg)$$

From Equations (8.16) and (8.17) we have

$$\mathbf{E}\sup_{t\in I}\|\varLambda x(t)-x_0\|_X^2 \le 10NT^2\|x_0\|_Y^2\left(1+5TN^2e^{2NT}K_2K\right)\mathbf{E}\sup_{t\in I}\|x(t)-x_0\|_X^2$$

$$+ 25TN^2e^{2NT}K_2K\left(\mathbf{E}\|x_1\|_X^2+a\right)+5a.$$

From assumption (H13) we get $\sup_{t\in I}\|\varLambda x(t)-x_0\|_X^2 \le r$. Thus $\varLambda$ maps functions in $\mathbb{L}_2^{\mathscr{F}}(\varOmega,C(I,X))$ having value in Ba into itself. We now show that $\varLambda$ is a strict contraction. For $x_1,x_2\in\mathbb{L}_2^{\mathscr{F}}(\varOmega,C(I,X))$ and having values in Ba we have,

$$\|\varLambda x_1(t)-\varLambda x_2(t)\|_X^2 \le 5\bigg(\left\|[U(t,0;x_1)-U(t,0;x_2)]x_0\right\|_X^2$$

$$+\left\|\int_0^t [U(t,0;x_1)-U(t,0;x_2)]Bu(s)\mathrm{d}s\right\|_X^2$$

$$+2\left\|\int_0^t [U(t,s;x_1)-U(t,s;x_2)]f(s,x_1(s))\mathrm{d}s\right\|_X^2$$

$$+2\left\|\int_0^t U(t,s;x_2)\big[f(s,x_1(s))-f(s,x_2(s))\big]\mathrm{d}s\right\|_X^2$$

$$+\left\|[U(t,0;x_1)-U(t,0;x_2)]\int_0^t\varPhi\mathrm{d}W(\theta)\right\|_X^2$$

$$+2\left\|\int_0^t [U(t,s;x_1)-U(t,s;x_2)]A(s,x_1)\left(\int_s^t\varPhi\mathrm{d}W(\theta)\right)\mathrm{d}s\right\|_X^2$$

$$+2\left\|\int_0^t U(t,s;x_2)\big[A(s,x_1)-A(s,x_2(s))\big]\left(\int_s^t\varPhi\mathrm{d}W(\theta)\right)\mathrm{d}s\right\|_X^2\bigg).$$

Using Lemma 18.1 we have

$$
\leq 5\Bigg(C\tilde{T}\|x_0\|_Y^2 \int_s^t \|x_1(r)-x_2(r)\|_X^2\, dr
$$

$$
+ 4\tilde{T}^2 C\|B\|_{\mathbb{L}(U,Y)}^2 \int_0^t\int_s^t \|x_1(r)-x_2(r)\|^2\, \|u(s)\|_U^2\, drds
$$

$$
+ 2\tilde{T}^2 C \sup_{t\in I}\|f(t,x(t))\|_Y^2 \int_0^{\tilde{T}}\int_s^t \|x_1(r)-x_2(r)\|_X^2\, drds
$$

$$
+ 2\tilde{T}^2 L_2 \sup_{s,t\in\Delta}\|U(t,s,x_2)\|_{\mathbb{L}(X,X)}^2 \sup_{t\in I}\|x_1(t)-x_2(t)\|_X^2
$$

$$
+ C\tilde{T}\int_s^t \|x_1(r)-x_2(r)\|_X^2\, dr \left\|\int_0^t \Phi dW(\theta)\right\|_Y^2
$$

$$
+ 2\tilde{T}^2 C \sup_{t\in I}\|A(t,x(t))\|_{\mathbb{L}(Y,X)}^2 \sup_{t,s\in\Delta}\left\|\int_s^t \Phi dW(\theta)\right\|_Y^2 \int_0^{\tilde{T}}\int_s^t \|x_1(r)-x_2(r)\|_X^2\, drds
$$

$$
+ 2\tilde{T}^2 L_3 \sup_{s,t\in\Delta}\|U(t,s,x_2)\|_{\mathbb{L}(X,X)}^2 \sup_{t,s\in\Delta}\left\|\int_s^t \Phi dW(\theta)\right\|_Y^2 \sup_{t\in I}\|x_1(t)-x_2(t)\|_X^2 \Bigg).
$$

Using the assumptions (H9), (H10), (H11), (H12) and the property (P1) we have

$$
\mathbf{E}\sup_{t\in I}\|\Lambda x_1(t)-\Lambda x_2(t)\|_X^2 \leq 5\mathbf{E}\Bigg(C\tilde{T}^2\|x_0\|_Y^2 + 4\tilde{T}^2 CK_2\int_0^{\tilde{T}}\|u(s)\|_U^2\, ds + \tilde{T}^4 CC_3
$$

$$
+ 2\tilde{T}^2 L_2 M^2 e^{2\beta\tilde{T}} + C\tilde{T}^2\left\|\int_0^{\tilde{T}}\Phi dW(\theta)\right\|_Y^2 + \tilde{T}^4 CC_2\left\|\int_0^{\tilde{T}}\Phi dW(\theta)\right\|_Y^2
$$

$$
+ 2\tilde{T}^2 L_3 M^2 e^{2\beta\tilde{T}}\left\|\int_0^{\tilde{T}}\Phi dW(\theta)\right\|_Y^2 \Bigg)\mathbf{E}\sup_{t\in I}\|x_1(t)-x_2(t)\|_X^2
$$

$$
\leq 5b\Big(1+2T^4 NK_2 K\Big)\mathbf{E}\sup_{t\in I}\|x_1(t)-x_2(t)\|_X^2
$$

$$
\leq \rho_2\, \mathbf{E}\sup_{t\in I}\|x_1(t)-x_2(t)\|_X^2.
$$

Hence by the contraction mapping principle, Λ has a unique fixed point x in $\mathbb{L}_2^{\mathscr{F}}(\Omega, C(I,X))$ having value in Ba which is the solution of the quasilinear stochastic differential system given by Equation (18.8). Clearly, $\Lambda x(\tilde{T}) = x_1$, which means that the control u steers the nonlinear system from the initial state x_0 to x_1 at the time $\tilde{T}$. Hence the system is completely controllable in I.

Acknowledgments

The work of the first author was supported by the University Grants Commission under grant number: MANF-2015-17-TAM-50645 from the Government of India. The second author is grateful to the University Grants Commission for providing the UGC-BSR Faculty Fellowship to carry out this work.

References

1. Balachandran, K., Dauer, J.P.: Controllability of nonlinear systems in Banach spaces: A survey. *Journal of Optimization Theory and Applications* **115**(1), 7–28 (2002).
2. Balachandran, K., Park, J.Y., Park, S.H.: Controllability of nonlocal impulsive quasilinear integrodifferential systems in Banach spaces. *Reports on Mathematical Physics* **65**(2), 247–257 (2010).
3. Balachandran, K., Balasubramainiam, P., Dauer, J.P.: Controllability of quasi-linear delay systems in Banach spaces. *Optimal Control Applications and Methods* **16**, 283–290 (2010).
4. Kilbas, A., Srivastava, H.M., Trujillo, J.J.: *Theory and Application of Fractional Differential Equations*. Elsevier, New York (2006).
5. Podlubny, I.: *Fractional Differential Equations*. Academic Press, New York (1999).
6. Balachandran, K., Karthikeyan, S.: Controllability of nonlinear Itô stochastic integrodifferential systems. *Journal of the Franklin Institute* **345**(4), 382–391 (2008).
7. Mahmudov, N.I.: Controllability of linear stochastic systems in Hilbert spaces. *Journal of Mathematical Analysis and Applications* **259**(1), 64–82 (2001).
8. Balachandran, K., Matar, M., Trujillo, J.J.: Note on controllability of linear fractional dynamical systems. *Journal of Control and Decision* 3(4), 267–279 (2016).
9. Mabel Lizzy, R.: Controllability of nonlinear stochastic fractional integrodifferential systems in Hilbert spaces. *Lecture Notes in Electrical Engineering* **407**, 345–356 (2016).
10. Mabel Lizzy, R., Balachandran, K., Suvinthra, M.: Controllability of nonlinear stochastic fractional systems with distributed delays in control. *Journal of Control and Decision* 4(3), 153–167 (2017).
11. Mabel Lizzy, R., Balachandran, K., Trujillo, J.J.: Controllability of nonlinear stochastic fractional neutral systems with multiple time varying delays in control. *Chaos, Solitons and Fractals* **102**, 162–167 (2017).
12. Kato, T.: Quasilinear equations of evolution, with applications to partial differential equations. *Spectral Theory Differential Equations* **448**, 25–70 (1975).
13. Kato, T.: Linear evolution equations of hyperbolic type. *Journal of Faculty of Science: University of Tokyo Section I* 17, 241–258 (1970).
14. Fernando, B.P.W., Sritharan, S.S.: Stochastic quasilinear partial differential equations of evolution. *Infinite Dimensional Analysis Quantum Probability and Related Topics* **18**, 1550021 (13 pages) (2015).
15. Mohan, M.T., Sritharan, S.S.: Stochastic quasilinear evolution equations in UMD Banach spaces. *Mathematische Nachrichten* **290**(13), 1–20 (2017).

16. Kim, J.U.: Approximate controllability of a stochastic wave equation. *Applied Mathematics and Optimization* **49**(1), 81–98 (2004).
17. Da Prato, G., Zabczyk, J.: *Stochastic Equations in Infinite Dimensions.* Cambridge University Press, Cambridge (1990).
18. Pazy, A.: *Semigroups of Linear Operators and Applications to Partial Differential Equations.* Springer-Verlag, New York (1983).
19. Pronk, M., Veraar, M.: A new approach to stochastic evolution equations with adapted drift. *Journal of Differential Equations* **256**(11), 3634–3683 (2014).

19

The Advection-Dispersion Equation for Various Seepage Velocity Patterns in a Heterogeneous Medium

Amit Kumar Pandey and Mritunjay Kumar Singh

CONTENTS

19.1 Introduction 291
19.2 The Advection-Dispersion Equation with Linear-Isotherm 293
    19.2.1 Numerical Values, Results, and Discussion 299
    19.2.2 Sinusoidal Velocity Pattern 299
19.3 The Concentration Distribution for the Sinusoidal Velocity Pattern 299
19.4 The Concentration Distribution for Exponentially Decreasing Velocity Pattern 304
19.5 The Concentration Distribution for Exponentially Increasing Velocity Pattern 305
19.6 Conclusions 307
References 307

19.1 Introduction

As far as the domain of the information regarding solute transport in the soil medium is concerned, a vast form of the literature has been observed over the last four to five decades. The downstream movements of heavy metals, pesticides, agricultural waste, waste materials as a result of continuous industrial growth, and the reduced tolerance of the earth due to construction that requires hollowing out the earth's surface, are some of the major factors that affect the groundwater environment. Solute transport relies not only on complex geo-hydrological structures of soil aquifers, but also on the anomalous behavior of the solute particles due to which the process of solute spreading takes place. Enormous efforts have been made to develop the deterministic/stochastic mathematical model to observe the temporal-spatial distribution of contaminants in natural geological formations. The heterogeneous nature of the transport medium causes spreading of a solute in terms of the spatially or temporally dependent dispersion and seepage velocity. Variations in seepage velocity are due to varying hydraulic conductivity. A relatively lower positional change in porosity is observed than in hydraulic conductivity; in the solute transport process therefore, uniform porosity is frequently considered for the sake of numerical calculations.

As the initial framework on transport models, the advection–dispersion/diffusion equation based on the fundamental concept of mass conservation has been proven to be of great importance to provide deterministic/stochastic outcomes of the solute transport models under different initial and boundary conditions. A sequence of continuous attempts regarding the solute transport model has been reported in the literature, as Ebach and White (1958) concentrated their work on the longitudinal dispersion for a periodically varying input and plotted a curve of modified Peclet numbers as a function of Reynolds number. Ogata and Banks (1961) have obtained non-symmetric concentration distribution curves and claimed that the curves show a symmetric nature for a vanished value of dispersion in a region away from the source of concentration. The porous medium was assumed homogeneous and isotropic, and the effect of adsorption was not considered in the model. Bruce and Street (1967) discussed both longitudinal and lateral dispersion effect in the semi-infinite non-adsorbing porous media in a steady unidirectional fluid flow for a fixed input concentration. Marino (1974) discussed the contaminants distribution in the porous media, considering temporal variation of the contaminant concentration. The solution obtained by Marino was valid for constant seepage velocity and dispersion, although in natural geological formations; flow parameters are usually non-uniform and spatially or temporally varying. Crank (1975) presented the mathematical background behind the process of diffusion in the transport mechanism and made use of various advection–diffusion models with their analytical and numerical solutions. Genuchten and Alves (1982) presented the analytical solutions of a set of one-dimensional convective–dispersive transport equation with linear equilibrium adsorption, first-order decay, and zero-order production in some cases. Kumar (1983) presented an illuminating discussion on the solute transport against dispersion and introduced a new time variable to solve the unsteady flow problem. Yates (1990) presented the analytical solution for one-dimensional transport in a heterogeneous porous medium assuming spatially dependent dispersion; however, solute dispersion is also influenced by temporal change up-to some extent. Gelher et al. (1992) presented a critical review of the data on field-scale dispersion in aquifer formations and concluded that longitudinal dispersivities with diminished values are applicable to field studies. Basha and El-Habel (1993) obtained the analytical solution of a one-dimensional advection–dispersion equation for time-dependent dispersion and uniform seepage velocity, employing Green's function method, but seepage velocity may vary spatially or temporally in the heterogeneous medium. Fry et al. (1993)presented the analytical solution of the solute transport equation with rate-limited desorption and decay applying an eigen function integral function method. The contaminant flux at the inlet boundary was assumed to be zero. Logan and Zlotnik (1995) discussed the effect of convection diffusion on the solute transport for periodic boundary conditions in the presence of adsorption and decay and studied their interaction with the media in-homogeneity. Logan (1996) considered the effect of scale-dependent dispersion on the solute transport for periodic boundary conditions and obtained the solution in terms of hypergeometric functions; it was concluded that wave amplitude in the homogeneous medium was lower than that in the heterogeneous medium but not in general. Huang et al. (1996) obtained the analytical solution of the ADE with asymptotically saturated dispersion. Aral and Liao (1996) have discussed the effect of constant, linear, asymptotic, exponentially time dependent dispersion on the two dimensional solute transport. The obtained solutions have restricted applications or applicable to specific boundary conditions. Dispersivity may be increasing non-linearly with distance in natural geological formations. Zoppou and Knight (1997) presented the analytical solution for the advection and advection–diffusion equation with spatially changing variables. In the considered model the diffusion coefficient was assumed to be directly proportional to the square of the seepage velocity. Neelz (2006) discussed the limitations of the analytical solution for the spatially changing flow parameters and compared the analytical solution with the obtained numerical solution. Although the spatially varying deterministic dispersion and seepage velocity reflect the influence of the media heterogeneity on the solute transport, the stochastic behavior of the solute advancement is not efficiently captured by them. Singh et al. (2006) developed a physically based fractional advection–dispersion model for overland solute transport taking into account quadratic variation of dispersion with time; the semi-Lagrangian approach was employed to obtain the numerical solution. Singh et al. (2009) obtained the analytical solution of the ADE with time-dependent sinusoidal, exponentially decreasing velocity pattern in the homogeneous medium. Kumar et al. (2009) presented the analytic solution for the spatially or temporally dependent solute dispersion of pulse type input concentration in a one-dimensional finite medium; the dispersion was considered directly proportional to the square of the seepage velocity. Kumar et al. (2010) made use of the analytical solution to 1D-ADE for continuous point sources of uniform and increasing nature; the spatial and temporal dependence of solute dispersion was assumed. Guerrero and Skaggs (2010) presented an analytic solution for the one-dimensional advection–dispersion equation with spatially dependent coefficients in a finite heterogeneous medium, using a generalized integral transform technique. Gao et al. (2010) obtained the solution of a one-dimensional advection–diffusion equation for the dispersivity increasing linearly with distance. Singh et al. (2011) considered the ADE with longitudinal dispersion and time-varying seepage velocity, and presence and absence of solute source was assumed in the considered model. Singh et al. (2012) discussed their work on one-dimensional pollutant's advective-diffusive transport from a varying pulse-type point source through a medium of linear heterogeneity. Van Genuchten et al. (2013) made available one-dimensional and multi-dimensional solutions to the advection–dispersion equation in the presence and absence of the terms accounting for first-order decay and zero-order production. Singh and Das (2015) presented the analytical solution to a one-dimensional advection–dispersion equation with zero-order production term in the presence of linear isotherm in a heterogeneous medium. Pandey et al. (2018) obtained solution to advection dispersion equation for the heterogeneous medium employing Duhamel's principle. The functional combination of source term and zero order production term with spatial variable was considered to depict initial solute contamination in the transport medium. The spatial-temporal variation of the dispersion and seepage velocity was assumed. Due to the high computational cost of numerical solutions of the ADE for a long run time, the focus has been shifted to analytical solutions of the ADE. The present chapter carries out the analytical solution of a 1D ADE in the presence of a zero-order production term, in a heterogeneous medium for various seepage velocity patterns.

The main aim of this chapter is (1) to obtain the analytical solution of the ADE for exponentially increasing/decreasing, sinusoidal seepage velocity pattern in a heterogeneous medium for the Dirichlet, the Cauchy-type boundary conditions; (2) to analyze the effect of media heterogeneity on the solute transport, in terms of spatially or temporally dependent flow parameters; (3)to discuss the influence of the in-homogeneity parameter(b*) on the solute transport in the gravel, sand, clay medium; (4) to explore the effect of the non-dimensional numbers(Peclet number, Courant number) on the contaminant transport; (5) to present the practical application of the considered model.

19.2 The Advection-Dispersion Equation with Linear-Isotherm

Due to the heterogeneous nature of a porous medium, the advection–dispersion equation for solute transport in the medium is modeled by considering the spatial-temporal variability of seepage velocity and solute dispersion, with almost uniform porosity due to its negligible spatial change in comparison with dispersion. The ADE with varying dispersion and seepage velocity, in the presence of zero-order production term assumed to depict continuous supply of contaminants to the flow, is considered as follows:

$$\frac{\partial c}{\partial t} + \frac{1-n}{n}\frac{\partial F}{\partial t} = \frac{\partial}{\partial x}(D\frac{\partial c}{\partial x} - uc) + \gamma \tag{19.1}$$

Where, D [L^2T^{-1}] is the longitudinal dispersion coefficient, C [ML^{-3}] is the volume averaged dispersing solute concentration (liquid phase). u [LT^{-1}] is the unsteady seepage velocity, F [ML^{-3}] is the solute concentration in the solid phase, γ [$ML^{-3}\,T^{-1}$] is the zero order production rate coefficient for solute production in the liquid phase, n is the porosity of the geological formation considered constant throughout, t and x are temporal and spatial variables respectively.

For the sake of ease in the numerical calculation, a linear isotherm $F = K_d c$ is taken into account, where F is the solid phase solute concentration, c is the liquid phase solute concentration, $K_d(\frac{mg}{g})(\frac{dm^3}{g})^n$ is the isotherm constant related to the adsorption capacity of the medium, $b^*$the is non linearity factor corresponding to the in-homogeneity of the medium.

Equation (19.1) is considered with following initial and boundary conditions.

$$c(x,0) = c_i f(\alpha x) + \frac{\gamma x g(\alpha x)}{u} \tag{19.2}$$

$$c(0,t) = c_i e^{-\delta t}, t > 0 \tag{19.3}$$

$$\frac{\partial c}{\partial x} = 0; x \to \infty \tag{19.4}$$

Where, $\alpha[L^{-1}]$ is the in-homogeneity parameter, $\delta[t^{-1}]$ is the decay constant.

Using the relation $F = K_d c$ in equation (19.1), we obtain,

$$\left(1 + \frac{1-n}{n}K_d\right)\frac{\partial c}{\partial t} = \frac{\partial}{\partial x}(D\frac{\partial c}{\partial x} - uc) + \gamma \tag{19.5}$$

Further equation (19.5) becomes

$$r\frac{\partial c}{\partial t^*} = \frac{\partial}{\partial x}\left[D_0 f(pt)(a + b^*x)^2\frac{\partial c}{\partial x} - u_0(a + b^*x)c\right] + \gamma^* \tag{19.6}$$

Where

$$t^* = \int_0^t f(pt)dt \tag{19.7}$$

$$\frac{\gamma}{f(pt)} = \gamma^* \tag{19.8}$$

$$r = 1 + \frac{1-n}{n}K_d \tag{19.9}$$

t^*, γ^*, $r$are new temporal variable, new production constant, and the retardation factor of the medium respectively.

Considering a linear variation in the seepage velocity with distance and time and assuming the dispersion coefficient directly proportional to the square of the seepage velocity (Kumar et al., 2009), the two are written with further modifications as

$$u = u_0 f(pt)(a + b^* x) \tag{19.10}$$

Where, the velocity increases from $f(pt)au_0$, at $x=0$ to a value $f(pt)(a + e)u_0$ at $x=L$, such that $b^* = \dfrac{e}{L}$ is the in-homogeneity parameter of the medium; for the sake of validity of Darcy's law (i.e. variation in the velocity should not be very large) in the medium, it is assumed that as $x \to \infty$, $b^*x \to M$, where $b^* = \dfrac{M}{x}$. Using Equation (19.10), the coefficient of dispersion is written as follows:

$$D = D_0 f^2(pt)(a + b^* x)^2 \tag{19.11}$$

Using the transformation

$$y = \log(a + b^* x)^2 \tag{19.12}$$

Equation (19.6) becomes

$$r \frac{\partial c}{\partial t^*} = \left[4\left(b^*\right)^2 D_0 f(pt) \frac{\partial^2 c}{\partial y^2} - 2u_0 g(pt) \frac{\partial c}{\partial y} - b^* u_0 c + \gamma^* \right] \tag{19.13}$$

$$\text{Where, } g(pt) = \left(1 - \frac{D_0 f(pt)\left(b^*\right)}{u_0} \right) \tag{19.14}$$

Now, employing the transformation

$$z = \frac{1}{2b^*} \int \frac{g(pt)}{f(pt)} dy \tag{19.15}$$

Equation (19.13) becomes

$$r \frac{f(pt)}{g^2(pt)} \frac{\partial c}{\partial t^*} = \left[D_0 \frac{\partial^2 c}{\partial z^2} - u_0 \frac{\partial c}{\partial z} - b^* u_0 c \frac{f(pt)}{(g(pt))^2} + \gamma^* \frac{f(pt)}{(g(pt))^2} \right] \tag{19.16}$$

Further; changing the temporal variable as

$$t^{**} = \int_0^t \frac{g^2(pt)}{f(pt)} dt^* \tag{19.17}$$

Equation (19.16) reduces to

$$r \frac{\partial c}{\partial t^{**}} = \left[D_0 \frac{\partial^2 c}{\partial z^2} - u_0 \frac{\partial c}{\partial z} - b^* u_0 c \frac{f(pt)}{g(pt)^2} + \gamma^* \frac{f(pt)}{g(pt)^2} \right] \tag{19.18}$$

Moreover, the initial and boundary conditions (19.2) to (19.4) transform to

$$c(z,0) = c_i + \left[\frac{1 - a(1 - (1 + \frac{D_0}{u_0} b^*)b^* z)}{b^*} \right] \frac{\gamma^*}{u_0} \tag{19.19}$$

$$c(0,t^{**}) = c_i + c_0(1 - \delta t^{**}), t^{**} > 0 \tag{19.20}$$

$$\frac{\partial c}{\partial z} = 0, z \to \infty \tag{19.21}$$

For the sake of simplicity, a transformation $k(z,t^{**})$ is used as follows:

$$c(z,t^{**}) = k(z,t^{**}) \exp[\frac{u_0}{2D_0} z - \frac{1}{r} \{ \frac{u_0^2}{4D_0} + \frac{f(pt)u_0}{g^2(pt)} b^* \} t^{**}] + \frac{\gamma^*}{b^* u_0} \tag{19.22}$$

Therefore, Equation (19.18) transforms to

$$r \frac{\partial k}{\partial t^{**}} = D_0 \frac{\partial^2 k}{\partial z^2} + \frac{\gamma^* f(pt)u_0}{g^2(pt)} (1 - b^*)k \tag{19.23}$$

$$r \frac{\partial k}{\partial t^{**}} = D_0 \frac{\partial^2 k}{\partial z^2} + \phi_1 k \tag{19.24}$$

Where

$$\frac{\gamma^* f(pt)u_0}{g^2(pt)} (1 - b^*) = \phi_1 \tag{19.25}$$

Equations (19.19) to (19.21), due to (19.22), are transformed as follows

$$k(z,0) = c_i e^{-\frac{u_0}{2D_0} z} + \left[\frac{1 - a(1 - (1 + \frac{D_0}{u_0} b^*)b^* z)}{b^*} \right] \frac{\gamma^*}{u_0} e^{-\frac{u_0}{2D_0} z} - \frac{\gamma^*}{b^* u_0} \tag{19.26}$$

$$k(0,t^{**}) = \left[c_i - \frac{\gamma^*}{b^* u_0} + c_0(1 - \delta t^{**}) \right] \exp(\phi t^{**}), t^{**} > 0 \tag{19.27}$$

$$\frac{\partial k}{\partial z} = -\frac{ku_0}{2D_0}, z \to \infty \tag{19.28}$$

Using the Laplace transformation technique on Equation (19.24) with Equations (19.26) to (19.28) and transforming back to the initial concentration function $c(x,t)$, the analytical solution to Equation (19.1) is written as follows:

$$c(z,t^{**}) = \left[(c_i + c_0 - \frac{\gamma^*}{b^* u_0}) A(z,t^{**}) - \delta c_0 B(z,t^{**}) \right] \exp\left(\frac{u_0}{2D_0} z - \phi t^{**} \right)$$

$$+ \left(-c_i + \frac{a\gamma^*}{b^* u_0} \right) C(z,t^{**}) \exp\left(\frac{u_0}{2D_0} z - \phi t^{**} \right)$$

$$+ \frac{a\gamma^*}{r} \left(1 + \frac{D_0 b^*}{u_0} \right) E(z,t^{**}) \exp\left(\frac{u_0}{2D_0} z - \phi t^{**} \right) \tag{19.29}$$

$$+ \left(c_i - a\frac{\gamma^*}{b^* u_0} + a\frac{\gamma^*}{u_0} \left(1 + \frac{D_0 b^*}{u_0} \right) z - \frac{a\gamma^*}{r} \left(1 + \frac{D_0 b^*}{u_0} \right) t^{**} \right) \exp\left(\frac{u_0^2}{4rD_0} - \frac{\phi_1}{r} \right) t^{**} \exp\left(-\phi t^{**} \right)$$

$$+ \frac{\gamma^*}{b^* u_0}$$

$$A(z,t^{**}) = \frac{1}{2} \exp\left(\left(\phi_2 - \frac{\phi_1}{r} \right) t^{**} - \sqrt{\frac{r\phi_2}{D_0}} z \right) erfc\left(\frac{1}{2} \sqrt{\frac{r}{D_0 t^{**}}} z - \sqrt{\phi_2 t^{**}} \right)$$

$$+ \frac{1}{2} \exp\left(\left(\phi_2 - \frac{\phi_1}{r} \right) t^{**} + \sqrt{\frac{r\phi_2}{D_0}} z \right) erfc\left(\frac{1}{2} \sqrt{\frac{r}{D_0 t^{**}}} z + \sqrt{\phi_2 t^{**}} \right) \tag{19.30}$$

$$B(z,t^{**}) = \frac{1}{2} \left(t^{**} + \sqrt{\frac{r}{D_0 \phi_2}} z \right) \exp\left(\left(\phi_2 - \frac{\phi_1}{r} \right) t^{**} + \sqrt{\frac{r\phi_2}{D_0}} z \right)$$

$$erfc\left(\frac{1}{2} \sqrt{\frac{r}{D_0 t^{**}}} z + \sqrt{\phi_2 t^{**}} \right) + \frac{1}{2} \left(t^{**} - \sqrt{\frac{r}{D_0 \phi_2}} z \right) \tag{19.31}$$

$$\exp\left(\left(\phi_2 - \frac{\phi_1}{r} \right) t^{**} - \sqrt{\frac{r\phi_2}{D_0}} z \right) erfc\left(\frac{1}{2} \sqrt{\frac{r}{D_0 t^{**}}} z - \sqrt{\phi_2 t^{**}} \right)$$

$$C(z,t^{**}) = \frac{1}{2} \exp\left(\left(-\frac{\phi_1}{r} + \frac{u_0^2}{4rD_0} \right) t^{**} - \frac{u_0}{2D_0} z \right)$$

$$erfc\left(\frac{1}{2} \sqrt{\frac{r}{D_0 t^{**}}} z - \frac{u_0}{2} \sqrt{\frac{t^{**}}{D_0 r}} \right) + \frac{1}{2} \exp\left(\left(-\frac{\phi_1}{r} + \frac{u_0^2}{4rD_0} \right) t^{**} + \frac{u_0}{2D_0} z \right) \tag{19.32}$$

$$erfc\left(\frac{1}{2} \sqrt{\frac{r}{D_0 t^{**}}} z + \frac{u_0}{2} \sqrt{\frac{t^{**}}{D_0 r}} \right)$$

$$E(z,t^{**}) = \begin{bmatrix} \frac{1}{2} \left(t^{**} + \frac{rz}{u_0} \right) \exp\left(\frac{u_0^2}{4rD_0} t^{**} + \frac{u_0}{2D_0} z - \frac{\phi_1}{r} t^{**} \right) erfc\left(\frac{1}{2} \sqrt{\frac{r}{D_0 t^{**}}} z + \frac{u_0}{2\sqrt{rD_0}} \sqrt{t^{**}} \right) + \\ \frac{1}{2} \left(t^{**} - \frac{rz}{u_0} \right) \exp\left(\frac{u_0^2}{4rD_0} t^{**} - \frac{u_0}{2D_0} z - \frac{\phi_1}{r} t^{**} \right) erfc\left(\frac{1}{2} \sqrt{\frac{r}{D_0 t^{**}}} z - \frac{u_0}{2\sqrt{rD_0}} \sqrt{t^{**}} \right) \end{bmatrix} \tag{19.33}$$

$$\text{Where, } \phi = \frac{1}{r}\{\frac{u_0^2}{4D_0} + \frac{f(pt)u_0}{g^2(pt)}b^*\} \tag{19.34}$$

$$\phi_2 = \phi + \frac{\phi_1}{r} \tag{19.35}$$

Equation (19.1) is also solved for the Cauchy-type boundary conditions. Due to increasing human interference on the earth's surface/subsurface, a continuous growth in the contamination level is noticed; this situation is appropriately modeled by the third type or the Cauchy-type boundary condition. At origin, the mathematical form of the mixed-type boundary condition is:

$$-D\frac{\partial c}{\partial x} + uc = u\left(c_i + c_0 e^{-\delta t}\right); x = 0, t > 0 \tag{19.36}$$

Solving Equation (19.24) by the Laplace transformation technique, the solution is obtained as

$$c(z,t^{**}) = \frac{-2D_0}{qu_0}\left[\begin{array}{l}(c_i + c_0 - \frac{\gamma^*}{b^*u_0})A(z,t^{**}) \\ \\ -\delta c_0 B(z,t^{**})\end{array}\right]$$

$$\exp\left(\frac{u_0}{2D_0}z - \phi t^{**}\right) - \frac{2D_0}{qu_0}\left(\frac{qa\gamma^*b^*}{r}\left(1 + \frac{D_0 b^*}{u_0}\right) + \left(-c_i + \frac{a\gamma^*}{b^*u_0}\right)\right)$$

$$E(z,t^{**})\exp\left(\frac{u_0}{2D_0}z - \phi t^{**}\right)$$

$$-\frac{4D_0\sqrt{rD_0}}{qu_0^2}\left(\left(c_i + c_0 - \frac{\gamma^*}{b^*u_0}\right)A_1(z,t^{**}) - \delta c_0 B_1(z,t^{**})\right)\exp\left(\frac{u_0}{2D_0}z - \phi t^{**}\right)$$

$$+\left(-\frac{a\gamma^*}{r}\left(1 + \frac{D_0 b^*}{u_0}\right)t^{**}\right)\exp\left(\frac{u_0^2}{4rD_0} - \frac{\phi_1}{r}\right)t^{**}\exp\left(-\frac{u_0}{2D_0}z\right)\exp\left(\frac{u_0}{2D_0}z - \phi t^{**}\right)$$

$$+\left(c_i - a\frac{\gamma^*}{b^*u_0} + a\frac{\gamma^*}{u_0}\left(1 + \frac{D_0 b^*}{u_0}\right)z\right)\exp\left(\frac{u_0^2}{4rD_0} - \frac{\phi_1}{r}\right)t^{**}\exp\left(-\frac{u_0}{2D_0}z\right)\exp\left(\frac{u_0}{2D_0}z - \phi t^{**}\right) \tag{19.37}$$

$$+\frac{\gamma^*}{b^*u_0}$$

$$B(z,t^{**}) = \frac{1}{2}\left(t^{**} + \sqrt{\frac{r}{D_0\phi_2}}z\right)\exp\left(\left(\phi_2 - \frac{\phi_1}{r}\right)t^{**} + \sqrt{\frac{r\phi_2}{D_0}}z\right)$$

$$erfc\left(\frac{1}{2}\sqrt{\frac{r}{D_0 t^{**}}}z + \sqrt{\phi_2 t^{**}}\right) + \frac{1}{2}\left(t^{**} - \sqrt{\frac{r}{D_0\phi_2}}z\right) \tag{19.38}$$

$$\exp\left(\left(\phi_2 - \frac{\phi_1}{r}\right)t^{**} - \sqrt{\frac{r\phi_2}{D_0}}z\right)erfc\left(\frac{1}{2}\sqrt{\frac{r}{D_0 t^{**}}}z - \sqrt{\phi_2 t^{**}}\right)$$

Where

$$q = \frac{D_0}{u_0}\left(1 + \frac{D_0 b^*}{u_0}\right)\left(pt^{**} - 1\right) \tag{19.39}$$

19.2.1 Numerical Values, Results, and Discussion

Due to the irregular soil structure, an aquifer medium shows spatial-temporal variation of the flow parameters. A 1D- ADE has been considered with varying dispersion and seepage velocity. The ADE is solved by the Laplace transform technique. The in-homogeneity parameter (b*) is taken into account corresponding to the spatial heterogeneity of the aquifer system. The obtained analytical solution has been discussed for $c_i = 0.01; c_0 = 1.0; D_0 = 0.1(km^2 / year); u_0 = 0.7(km / year)$ $\delta = 0.02; \gamma = 0.0005(1 / km), K_d = 2.5; n = 0.32; n = 0.37; n = 0.55 \quad p = 0.01(year)^{-1}$(Singh and Das, 2015) for various seepage velocity patterns.

19.2.2 Sinusoidal Velocity Pattern

In the tidal region of flow, a sinusoidal form of the non-dimensional seepage velocity can be of great importance; therefore the velocity pattern is assumed to be as follows:

$$f = A\sin(pt) + L \tag{19.40}$$

Where, p [s^{-1}] is the flow resistance coefficient.

19.3 The Concentration Distribution for the Sinusoidal Velocity Pattern

Figures 19.1 to 19.3 represents the solute concentration in the gravel medium, the sand medium, and the clay medium, for the period of five, six, and seven years, respectively. It is observed that the concentration magnitude at each point of the gravel formation is lower than that in the sand medium and clayey medium. Due to the low porosity but active permeability of the gravel medium and sand medium, solute transport is less resistive in the mediums; on the other hand, the clayey formation with diminished permeability allows solute to move in a resistive way and therefore an accumulation of the contaminant particles is noticed there. Since solute moves in the micro pore through the process of mixing and in macrospores, it moves via advection, hence the dispersion pathway is longer than the advection pathway; as a result, a long-tailed concentration curve is remarked. It is also noticed that in the gravel, sand, and clay formation at each point the concentration increases with increasing temporal variables. The concentration decreases more rapidly in the gravel medium than in the sand medium (Figures 19.1 and 19.2) or in the clay medium (Figure 19.3). Therefore the solute is more sensitive to the clay formation. Further, the concentration curve in the gravel, sand, clay concentration curve shows positive skewed nature.

Figure 19.4 depicts the concentration distribution for different values of the Peclet number. A Peclet number can relate the effectiveness of mass transport by advection to the effectiveness of mass transport by either dispersion or diffusion.

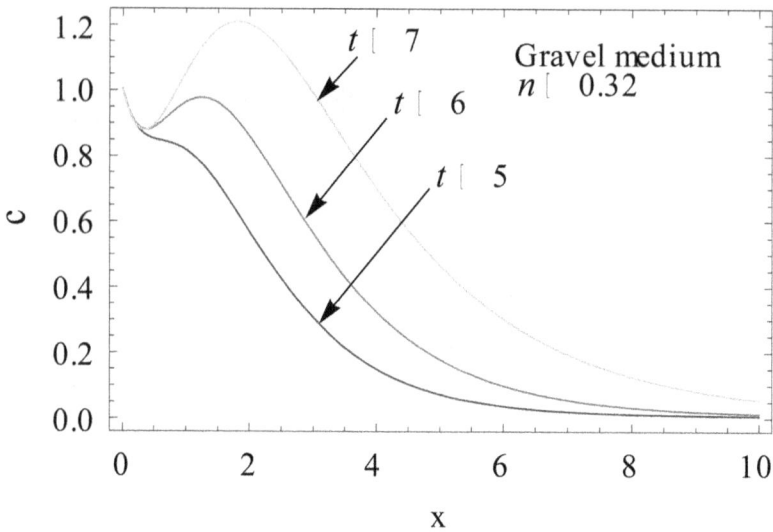

FIGURE 19.1
The concentration profile for the sinusoidal velocity pattern in the gravel medium at t=5, 6, 7 years.

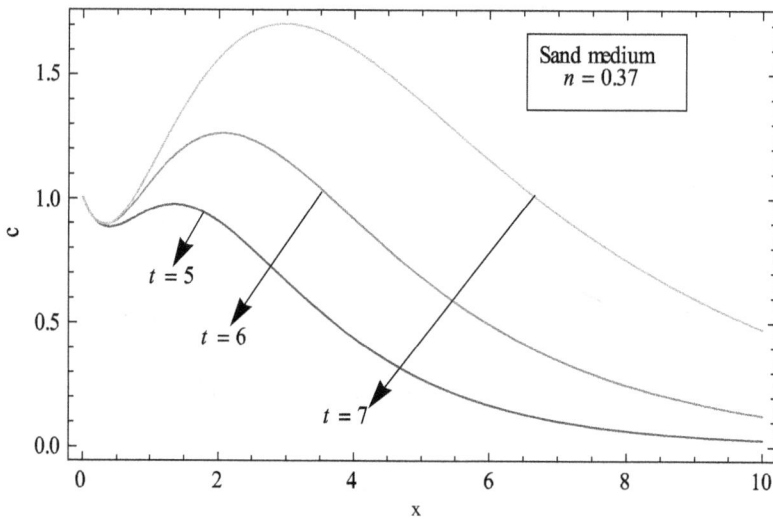

FIGURE 19.2
The concentration profile for the sinusoidal velocity pattern in the sand medium at t=5, 6, 7 years.

There are several definitions of Peclet number; depending on various factors governing the flow, $Pe = u_0 x/nD_0$, suits the given scenario well. It is observed that with increasing value of the porosity, the Peclet number decreases and seepage velocity is directly proportional to the Peclet number. Therefore, the factor that may affect the efficiency of the Peclet number is the velocity with which solute advances and the distance it travels.

The ballooned value of the Peclet number represents the domination of advection over dispersion, while the decreased numerical value of the number implies that the dispersion overcomes the effect of advection in solute transport process. The solute depicts more sensitivity to the clay medium (Aquitard, a zone within the earth that restricts the flow of ground water from one aquifer to another) than to the gravel medium or sand medium;

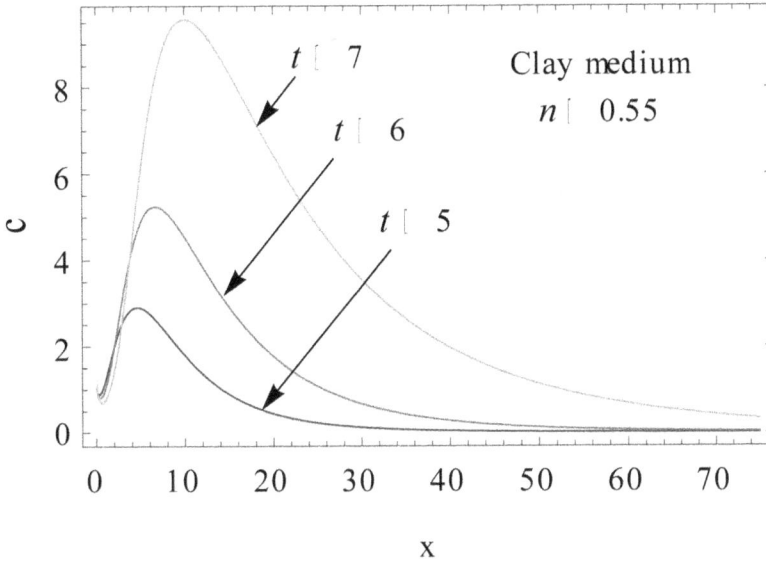

FIGURE 19.3
The concentration profile for the sinusoidal velocity pattern in the clay medium at t=5, 6, 7 years.

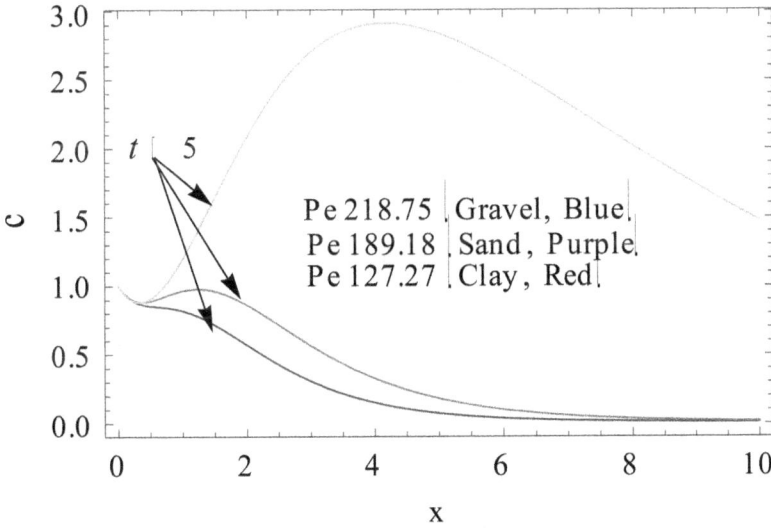

FIGURE 19.4
The concentration distribution at t=5 years for the gravel, sand, and the clay medium with corresponding values of the Peclet number.

although choice of the dispersion or diffusion coefficient does not have a significant effect on the ability of a Peclet number to determine the importance of advection in the low permeable environment (Fetter, 1999).

Figure 19.5 shows the concentration distribution at t=4 years for varying value of the Courant number in the gravel, sand, and clay medium. It is clear from the expression $(C_r = u_0 t/nx)$ that the average porosity of the medium affects the Courant number to a great extent. An increment in the solute concentration for decreased value of the Courant number at each place in the domains is noticed. The increment in the concentration in the

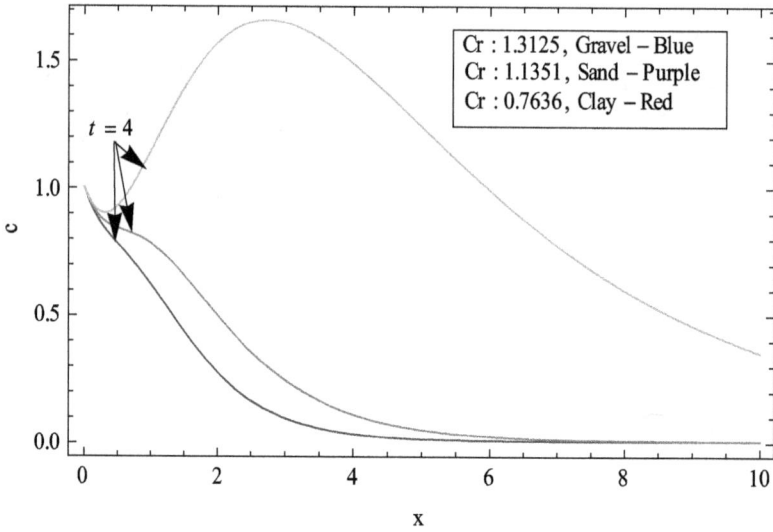

FIGURE 19.5
The concentration distribution at t=4 years for the gravel, sand, and the clay medium with corresponding values of the Courant number.

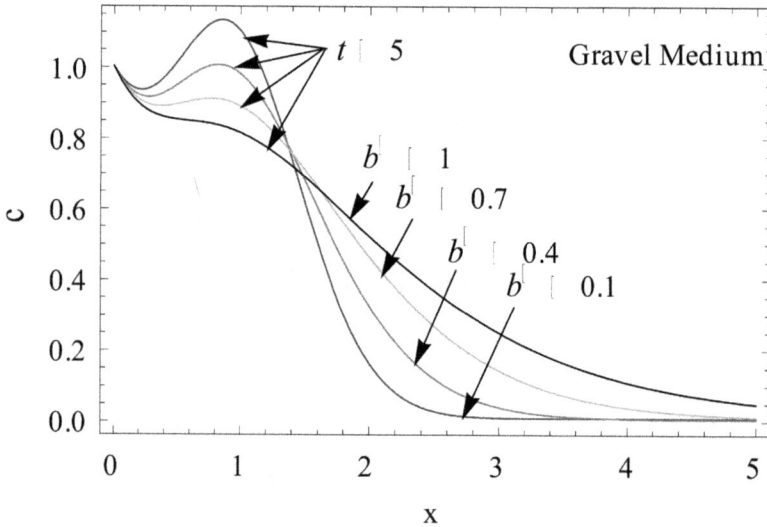

FIGURE 19.6
The concentration distribution in the gravel medium at t=5 years for different values of the in-homogeneity parameter, for the sinusoidal velocity pattern.

clay medium is relatively more than that in the gravel or sand medium. The solute after a significant distance from origin of contamination attains a saturated level that further reaches to almost vanished value.

Figure 19.6 depicts the solute concentration distribution with distance for t=5 years in the gravel medium, for different values of the in-homogeneity parameter (b*).It is noticed that the concentration increases with decreasing value of b* in each of the mediums (gravel, sand, clay) (Figure 19.6, Figure 19.7, Figure 19.8). The peak concentration magnitude is higher in the clayey formation, for diminished values of b*.This is due to the intrinsic

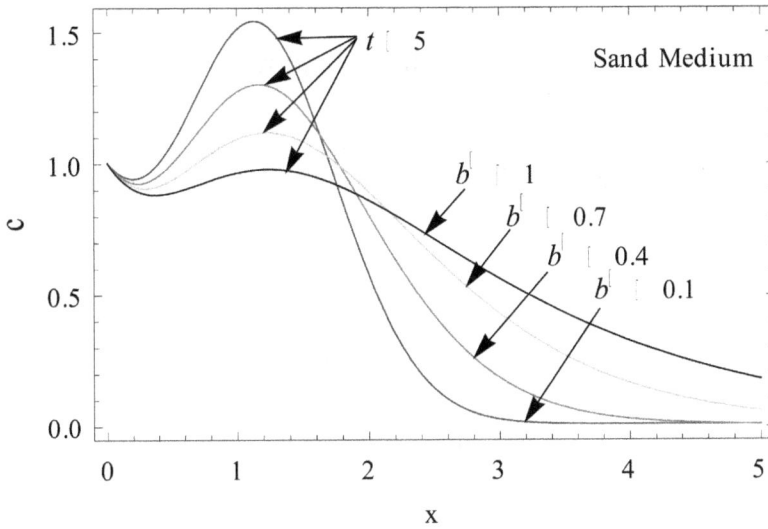

FIGURE 19.7
The concentration distribution for the sand medium at t=5 years with different values of the in-homogeneity parameter, for sinusoidal velocity pattern.

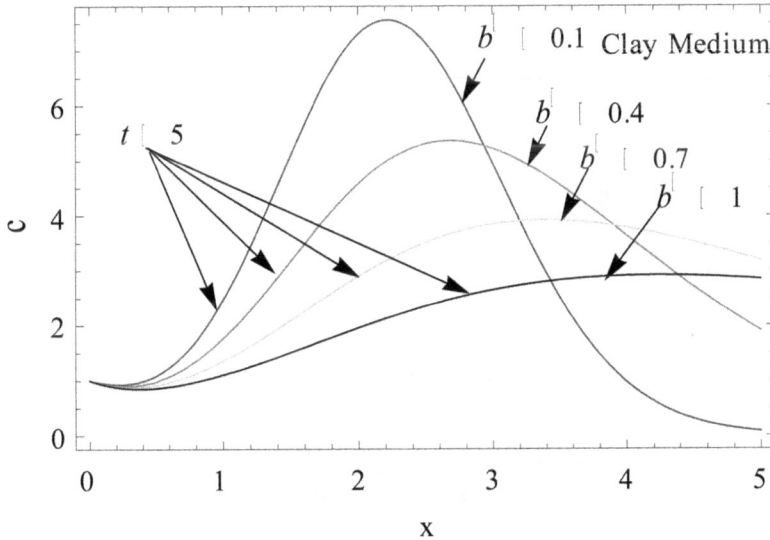

FIGURE 19.8
The concentration distribution in the clay medium at t=5 years for different values of the in-homogeneity parameter and for sinusoidal velocity pattern.

heterogeneity and vanished permeability of the clayey soil formation. The decreased value of the in-homogeneity parameter reflects resistivity in the contaminant transport and is due to it; the solute attenuates at a very slow rate in the formations (gravel, sand, clay).

An excess number of macrospores in the clay structures reflects the domination of the seepage velocity and hence enhanced advection level is noticed. Temporal change also affects the heterogeneity of the aquifer to some extent and thus the effect of time variability on the transport of the solute flow is remarked (Figure 19.7, Figure 19.8).

19.4 The Concentration Distribution for Exponentially Decreasing Velocity Pattern

Figures 19.9 to 19.11 show the solute concentration distribution for the gravel, sand, and clay mediums respectively, at $t = 4$ years, and it is observed that the concentration in the

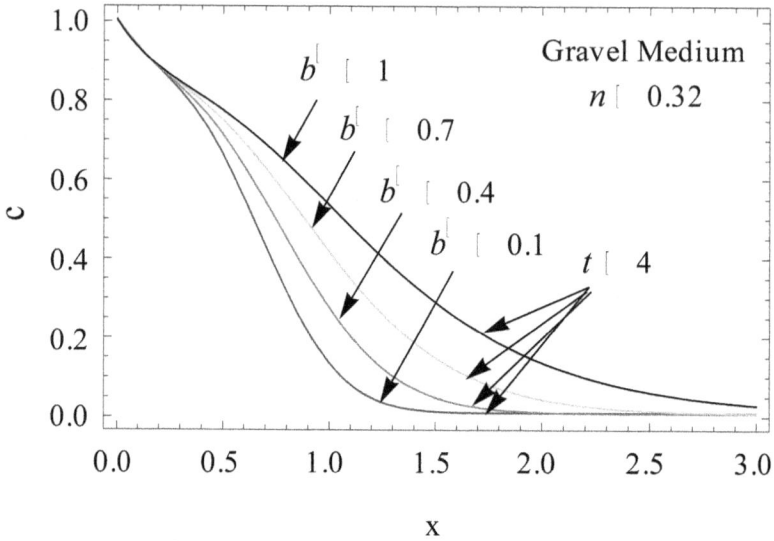

FIGURE 19.9
The concentration profile for exponentially decreasing velocity pattern and various values of the in-homogeneity parameter in the gravel medium.

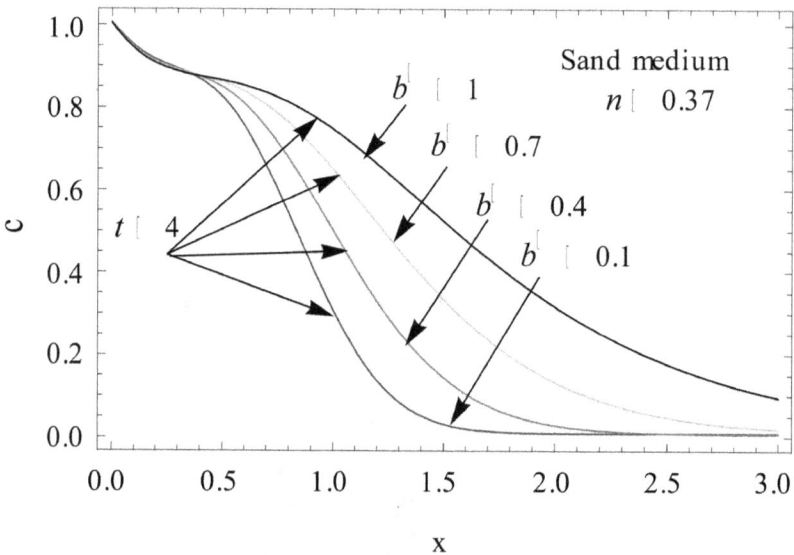

FIGURE 19.10
The concentration distribution for the sand medium at $t = 4$ years and for different values of the in-homogeneity parameter and for the exponentially decreasing velocity pattern.

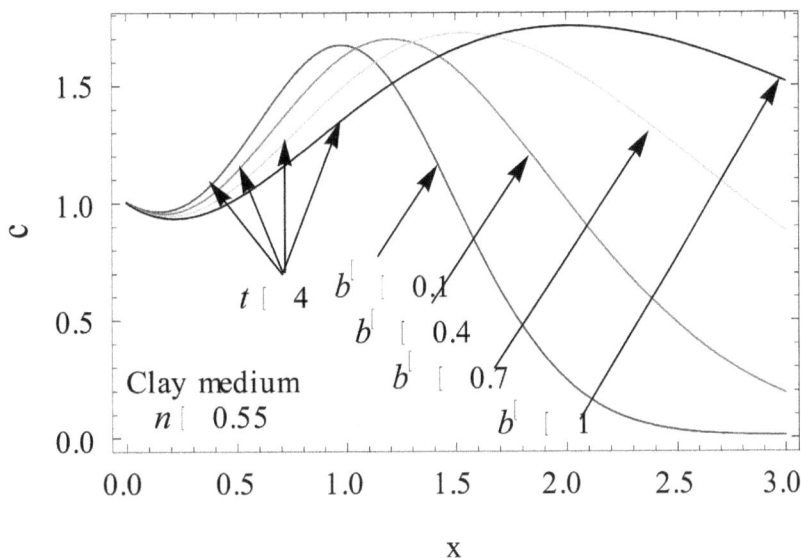

FIGURE 19.11
The concentration distribution at t=4 years for the clay medium, for different values of the in-homogeneity parameter and for the exponentially decreasing velocity pattern.

clay medium at each place in the domain is higher than that that in the sand or clayey medium. It is also observed that at the beginning the concentration decreases with increasing in-homogeneity parameter(b*) and after the attainment of concentration peak, the solute density increases with increasing b*that further reduces to a diminished value.

19.5 The Concentration Distribution for Exponentially Increasing Velocity Pattern

Figures 19.12 to 19.14 depict the concentration distribution in the gravel, sand, and clay medium. It is observed that the solute concentration shows more sensitivity to the decreased value of the in-homogeneity parameter. It is noticed that at each position the concentration density is more in the clay medium with increasing temporal variable. It is also remarked that solute sensitivity increases with the decreasing in-homogeneity parameter (b*).

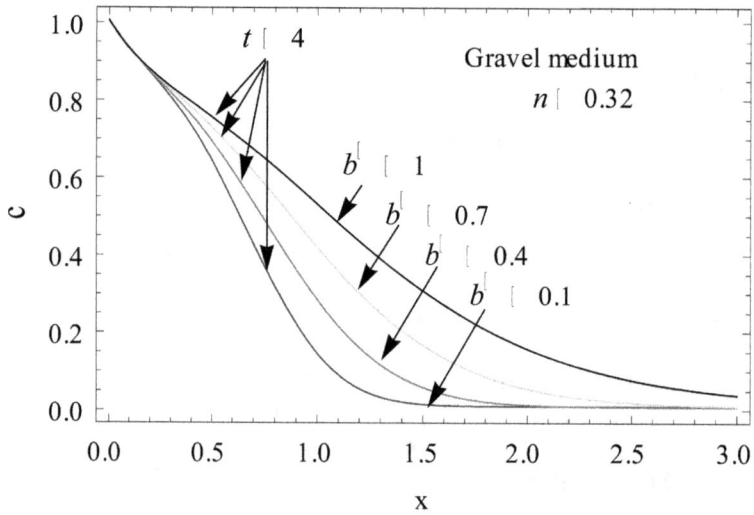

FIGURE 19.12
The concentration distribution for the Gravel medium at t=4 years for different values of the in-homogeneity parameter and for the exponentially increasing velocity pattern.

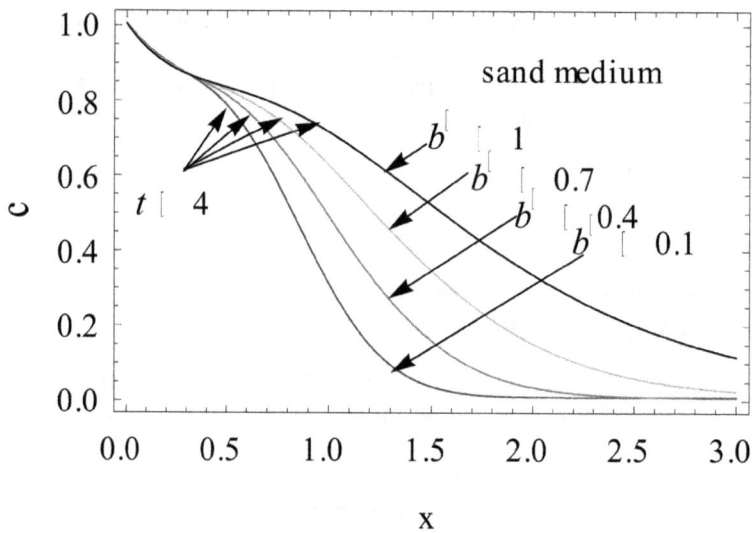

FIGURE 19.13
The concentration distribution for the sand medium for the exponentially increasing velocity pattern at=4 years and for different values of the in-homogeneity parameter.

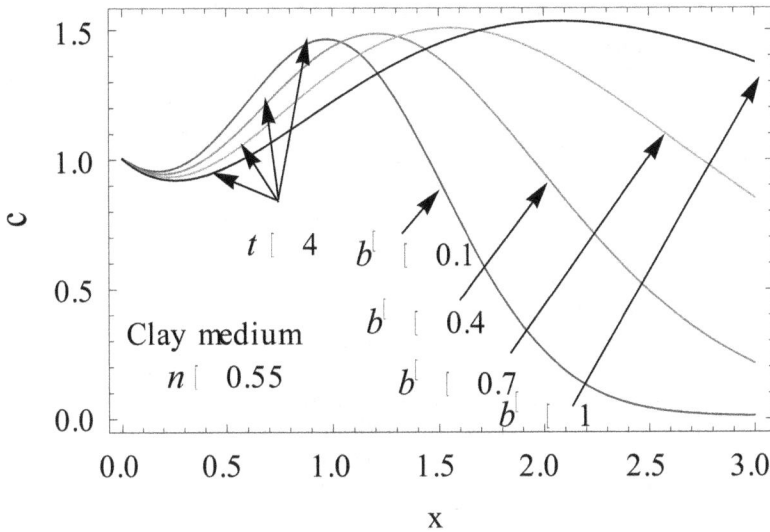

FIGURE 19.14
The concentration distribution for the clay medium at t=4 years for different values of the in-homogeneity parameter and for the exponentially inc.

19.6 Conclusions

This chapter makes some important observations, they are as follows:

The solute density in the gravel, sand, and clay formations increases with increasing temporal variables; moreover the increment in concentration in the clay medium is relatively more than that in the gravel or sand medium. The concentration decreases more rapidly in the gravel medium than in sand or clay medium. The solute is more sensitive to the clayey formation. Further, the concentration curve in the gravel, sand, clay concentration curve shows positive skewed nature for the sinusoidal velocity pattern of solute particles. The solute under transport shows the highest sensitivity to the clayey formations and least sensitivity to the gravel formation. The velocities with which a solute advances and the distance it travels affect the Peclet number. The solute concentration increases with a decreased value of the Courant number at each place in the domain. The contaminant concentration increases with the decreasing value of the in-homogeneity parameter (b^*) in the gravel, sand, and clay mediums. The peak concentration magnitude is higher in the clayey formation, for diminished values of b^*. At the beginning the concentration decreases with the increasing in-homogeneity parameter (b^*) and after the attainment of the concentration peak, the solute density increases with increasing b^*, for decreased seepage velocity.

References

Aral, M. M., Liao, B., 1996. Analytical solutions for two-dimensional transport equation with time-dependent dispersion coefficients. *Journal of Hydraulic Engineering*1:20-32.

Basha, H. A., El-Habel, F. S., 1993. Analytical solution of one dimensional time-dependent transport equation.*Water Resources Research*29:3209–3214.

Bruce, J. C., Street, R. L., 1967. Studies of free surface flow and two dimensional dispersion in porous media. Report No. 63, Civil Engineering Department, Stanford University, Stanford, CA.

Crank, J., 1975. *The Mathematics of Diffusion*. London: Oxford University Press.

Ebach, E. H., White, R., 1958. Mixing of fluids flowing through beds of packed solids. *Transactions of the American Institute of Chemical Engineers*4:161–169.

Fetter, C. W., 1999. *Contaminant Hydrogeology. Edition, 2, Illustrated*. Prentice Hall, Original from, the University of Michigan, New Jersey. Digitized, December 4, 2007.

Fry, A., Istok, J. D., Guenther, R. B., 1993. Analytical solutions to the solute transport equation with rate-limited desorption and decay. *Water Resources Research*29:3201–3208.

Gelhar, L. W., Claire, W., Kenneth, R. R, 1992. Critical review of data on field-scale dispersion n Aquifers. *Water Resources Research*28:1713–1723.

Guerrero, J. S. P., Skaggs, T. H., 2010. Analytical solution for one-dimensional advection—dispersion transport equation with space-dependent coefficients. *Journal of Hydrology*390:57–65.

Gao, G., Zhan, H., Feng, S., Fu, Boje., Ma, Y., Huang, G., 2010. A new mobile-immobile model for reactive solute transport with scale-dependent dispersion. *Water Resources research* 46:1–16.

Huang, K., Van Genuchten, M. Th., Zhang, R., 1996. Exact solutions for one dimensional transport with asymptotic scale dependent dispersion. *Applied Mathematical Modeling*20:298–308.

Kumar, A., Jaiswal, D. K., Kumar, N., 2009. Analytic solutions of one-dimensional advection-diffusion equation with variable coefficients in a finite domain. *Journal of Earth System Science*118:539–549.

Kumar, A., Jaiswal, D. K., Kumar, N., 2010. Analytical solutions to one-dimensional advection-diffusion equation with variable coefficients in semi-infinite media. *Journal of Hydrology*380:330–337.

Kumar, N., 1983. Unsteady flow against dispersion in finite porous media. *Journal of Hydrology*63:345–358.

Logan, J. D., 1996. Solute transport in porous media with scale-dependent dispersion and periodic boundary conditions. *Journal of Hydrology*184:261–276.

Logan, J. D., Zlotnik, V., 1995. The convection diffusion equation with periodic boundary conditions. *Applied Mathematics Letter*8:55–61.

Marino, M. A., 1974. Distribution of contaminants in porous media flow. *Water Resources Research*10:1013–1018.

Neelz, S., 2006. Limitations of an analytical solution for advection-diffusion equation with variable coefficients. *Communications in Numerical Methods in Engineering*22:387–396.

Ogata, A., Banks, R. B., 1961. A solution of the differential equation of longitudinal dispersion in porous media. *US Geological Survey Professional Papers*, United States Government Printing Office, Washington 25 D.C. 411A.

Pandey, A.K., Kumar, R., Singh, M.K., 2018. Solution to Advection-Dispersion Equation for the Heterogeneous Medium Using Duhamel's Principle. Applications of Fluid Dynamics. Lecture Notes in Mechanical Engineering. Springer, Singapore: 559–571.

Sing, M. K., Das, P., 2015. Scale dependent solute dispersion with linear isotherm in heterogeneous medium. *Journal of Hydrology*520:289–299.

Singh, M. K., Mahto, N. K., Singh, P., 2011. Longitudinal dispersion with constant source concentration along unsteady groundwater flow in finite aquifer: Analytical solution with pulse type boundary condition. *Natural Science*3:186–192.

Singh, M. K., Singh, V. P., Singh, P., Shukla, D., 2009. Analytical solution for conservative solute transport in one dimensional homogeneous porous formations with time dependent velocity. *Journal of Engineering Mechanics*135(9), 1015–1021.

Singh, P., Yadav, S. K., Kumar, N., 2012. One-dimensional pollutant's advective-diffusive transport from a varying pulse -type point source through a medium of linear-heterogeneity. *Journal of Hydraulic Engineering* 17:1047–1052.

Singh, V. P., Deng, Z.-Q., Lima, J.L.M.P.de., Lima, M.I.P.de., 2006. A fractional dispersion model for over land solute transport. *Water Resources Research*42:W034161-14.

Van Genuchten, M. Th., Alvesm, W. J., 1982. Analytical solutions of the one-dimensional convective-dispersive solute transport equation. *U.S. Department of Agriculture, Technical Bulletin No.* 1661, 151p.

Van Genuchten, M. Th., Leij, F. J., Skaggs, T. H., et al., 2013. Exact analytical solutions for contaminant transport in rivers 1. The equilibrium advection–dispersion equation. *Journal of Hydrology and Hydromechanics*61:146–160.

Yates, S. R., 1990. An analytical solution for one-dimensional transport in heterogeneous porous media. *Water Resources Research*26:2331–2338.

Zoppou, C., Knight, J. H., 1997. Analytical solution for advection and advection-diffusion equations with spatially variable coefficients. *Journal of Hydraulic Engineering – ASCE*123:144–148.

20

On the Application of Genetic Algorithm and Support Vector Machine for Classification of MRI Images for Brain Tumors

D. Gupta and M. Ahmad

CONTENTS

20.1 Introduction.. 311
20.2 Literature Review.. 312
20.3 Proposed Methodology... 314
    20.3.1 MRI and Its Preprocessing 316
    20.3.2 Filtering 316
    20.3.3 Skull Masking... 316
    20.3.4 Morphological Operation 317
    20.3.5 Dilation.. 319
    20.3.6 Erosion... 319
    20.3.7 Feature Extraction.. 319
    20.3.8 Gray-Level Co-Occurrence Matrix (GLCM) Features 320
    20.3.9 Gray-Level Run-Length Matrix (GLRLM)....................................... 321
    20.3.10 Short-Run Emphasis 321
    20.3.11 Long-Runs Emphasis... 321
    20.3.12 Gray-Level Non-Uniformity.. 321
    20.3.13 Run-Length Non-Uniformity.. 322
20.4 The Kernel Trick.. 322
20.5 Training... 322
20.6 Testing 322
20.7 Results and Discussion 323
References.. 323

20.1 Introduction

In the medical domain, many critical human diseases are making an appearance. CT scans, MRI images, and ECGs are well-known tools used for the detection of such human diseases [1]. The utilization of MRI is very important in view of its major feature of soft tissue discretization, high contrast and high spatial resolution , and the fact that no harmful ionizing radiation is used for imaging [2]. Our work deals with the classification of tumor and non-tumor MRI images of the brain. A brain tumor is one of the major reasons for death. It is caused due to abnormal growth of cells in the brain. To spot the presence of a tumor, the radiologist visually examines MRI pictures and interprets them. If a large volume of MRI images are analyzed at the same time, it can lead to a wrong diagnosis because the analysis is done by visual detection and the sensitivity of the human eye decreases if the same task is performed continuously for a long time. So, an economical automated system is needed for analysis and classification of medical pictures. Our work considers both normal and abnormal MRI images to check the validation of the proposed technique. A brain tumor is a collection, or mass, of abnormal cells in the brain. The skull, which encloses the brain, is very rigid. Any growth inside the skull, which has restricted space, can cause problems. A brain tumor can be cancerous (malignant) or non-cancerous (benign) [3]. When a benign or a malignant tumor grows, it can cause the pressure inside the skull to increase. This can cause brain damage, and it can be life-threatening [4]. A brain tumor can be categorized as primary or secondary. A primary brain tumor originates in the brain. Many primary brain tumors are benign. A secondary brain tumor, also known as a metastatic brain tumor, occurs when cancer cells spread to the brain from another organ, such as the lung or breast. Types of brain tumor are shown in Figure 20.1, according to the World Health Organization [5].

20.2 Literature Review

Bhanumurthy and Koteswararao have developed a Graphical User Interface (GUI) with an automated artificial intelligence system that uses a neuro-fuzzy classifier to detect and segment tumors in MR images of the brain. Padmakant, Dhage, and Phegade illustrate the watershed segmentation technique to segregate the abnormal tissue from the adjacent normal tissue to make an actual identification of allied and non-allied portions that help doctors to establish the relevant area precisely. This technique can accurately localize activated abnormal tissue such as tumors. Lakshmi et al. introduced an efficient denoising technique using curvelet transform and mathematical morphology to enhance the quality of the MR image of the brain. After the removal of noise, skull stripping is done for accurate diagnosis and analysis of the brain tumor. Deepthi Murthy and Sadashivappa

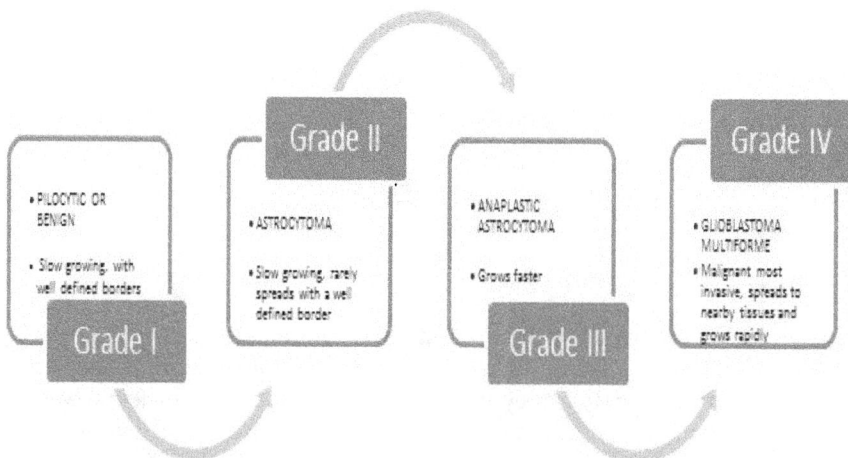

FIGURE 20.1
Types of brain tumor.

have presented an efficient method by using Sobel filter for image preprocessing, histogram equalization for image enhancement, thresholding, and morphological operations for image segmentation. Tumor regions, especially, are intensified, and noises are reduced by a Gaussian equation-based algorithm. Hybrid filters are used in a cascaded method with spatial domain filters to eliminate time-domain noises [6]. To enhance the efficiency of brain tumor detection, Charutha and Jayashree have suggested an automated approach of image segmentation, modified texture-based segmentation unified with cellular automata edge detection. Swarup Roy and Dhruba Bhattacharyya have a Java-based tool for a density-based clustering technique (DenSeg) to segment gray and white matter with tumor from MR images. This method also explores a preprocessing method for extracting cortex from the skull. Dharshini and Hemanandhini have applied Self Organizing Map (SOM) for classifying voxels, and Learning Vector Quantization (LVQ) with high-diversity contour deformation data like tumor figuration. Discrete Wavelet Transform (DWT) coefficients have been used for feature extraction and have identified tissue types White Matter (WM), Grey Matter (GM), Cerebrospinal Fluid (CSF), and sometimes pathological tissues [7,8,9]. A fuzzy image enhancement is actuated on T1 weighted magnetic resonance (MR), and fluid attenuated inversion recovery (FLAIR) images for preferable segmentation [10]. An overlap performance of 84% to 93% was achieved with an emphasis on robustness with respect to distinct and heterogeneous lesion types.

To achieve the higher accuracy, a new data-mining technique, based on a combination of support vector machine (SVM) and Fuzzy C-Means, is implemented for brain tumor classification. Heena Hooda et al have compared three image segmentation techniques—K-Means Clustering, Fuzzy C-Means Clustering, and Region Growing—for detection of brain tumors and claim that Fuzzy C-Means is the best technique for image segmentation out of these techniques. For better accuracy and to shorten the processing time by up to 80% compared to the available global threshold segmentation technique, Khamy et al. have used a rectangular window to calculate the number of clusters and have proposed a segmentation method implementing FCM and a conformed threshold. Discrete Wavelet Transform (DWT)-based, soft thresholding technique is applied for denoising brain MR, and image segmentation is done through a Grammatical Swarm-based clustering algorithm to distinguish the tumor region from the non-diseased objects in the image[11,12]. Yehualashet Megersa and Getachew Alemu have suggested a hybrid method based on Fuzzy C-Means and Hopfield Neural Network for image segmentation [13].

A hybrid segmentation that merges threshold segmentation, watershed segmentation, edge detection, and morphological operators gets better results for classification of brain MR images [14]. To reduce the data dimensionally, (PCA) principle component analysis is used [15]. Martinez and KAK showed it performed well when the feature dimension were higher than the training sets. Linear Discriminant Analysis (LDA) has been applied for decrees the features for training and testing through support vector machine. Therefore, the time of calculations and complexity will decrease by this technique [16]. For various applications like identification of hand-written digits, face recognition, speaker, text classification, and for medical picture classifiers like SVM, K-nearest neighbor (KNN), Artificial Neural Network (ANN), Hidden Mathematician Model (HMM), Probabilistic Neural Network (PNN), etc. are used. Each of these classification schemes has individual characteristics and related advantages and disadvantages. The huge restriction with KNN is that it considers all features for distantly intensive computation, particularly for increasing the size of the training set. The presence of noisy or unrelated features also degraded the accuracy of K-nearest neighbor classification, for higher number of attributes PNN performance degraded as compared to multilayer perceptron networks to classify new cases and needs additional memory to store the model [17]. ANN performs classification faster than the other classification strategies with high-dimensional features and contradictory data. But expensive computing due to high consumption of CPU and usage of physical memory is the main drawback of ANN [18,19]. Bayesian approach is mostly implemented in a different class of domain and applications because of its easiness and low calculation cost. But this method is less accurate than SVM [20,21].

20.3 Proposed Methodology

The methodology is applied to distinguish MR images of the human brain as non-tumorous, as shown in Figure 20.3 and tumorous, as shown in Figure 20.4. The proposed methodology involves training and testing of preprocessed images and for preprocessing of images we have performed some steps like filtering, skull masking, feature extraction, and feature reduction before training, and testing. A flow chart of this technique is presented

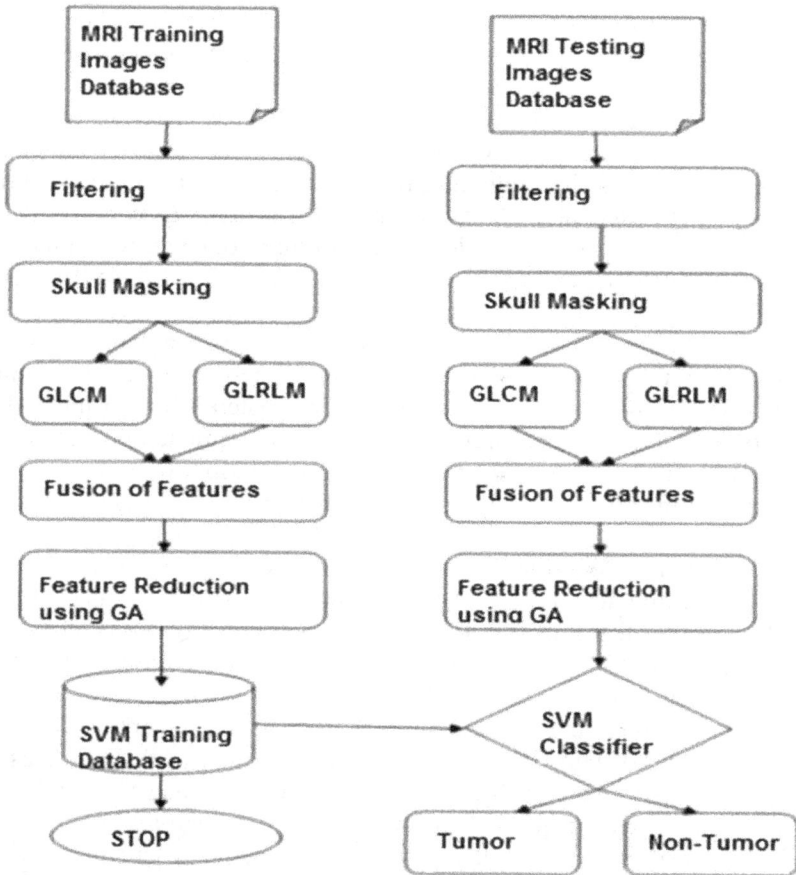

FIGURE 20.2
Flow chart of proposed methodology.

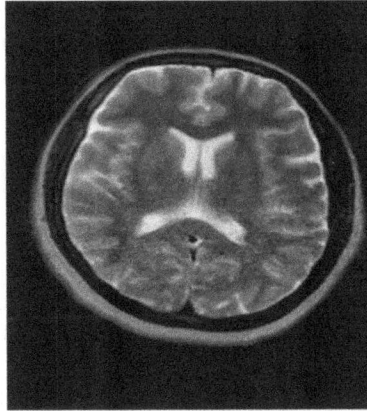

FIGURE 20.3
Non-tumor brain image.

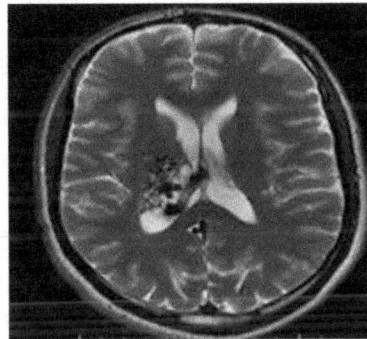

FIGURE 20.4
Tumor brain image.

in Figure 20.2, and Figure 20.5 shows gray scale conversion of images. First, the inputs of MRI pictures are filtered and then skull masking is finished; then the output pictures are used to calculate the GLCM and GLRLM feature and then fusion of both the features is done. After the feature extraction and its fusion, the genetic algorithmic program is used to decrease the size of the feature in each sample. The chosen feature from the genetic

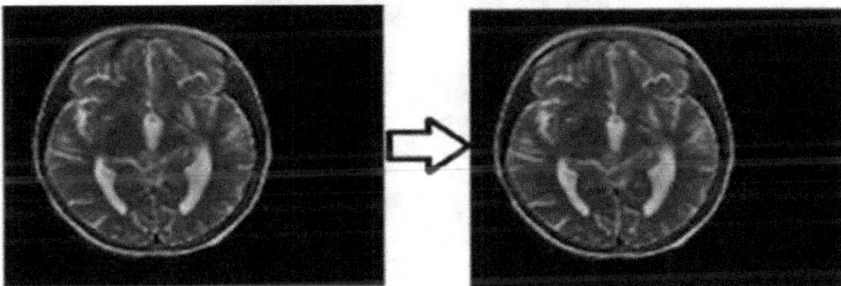

FIGURE 20.5
Gray-scale image conversion.

algorithm (GA) is then given to the SVM classifier for training; then new MRI pictures are set as input, to test the performance of the trained classifier. Its performance is better than other alternative approaches of classification [22,23]. Its disadvantages are high computing costs and physical memory usage.

20.3.1 MRI and Its Preprocessing

MRI is a safe and useful technique for medical diagnosis by reason of its ability to distinguish between various organs and tissues. Also, more biologically variable parameters, variable image contrast, transversal relaxation time (T1), proton density (PD), longitudinal relaxation time (T2), are often achieved by using various pulse sequences and varying the image parameters. Three pictures PD, T1, and T2 are produced and their signal strength used to specify tissue characteristics. Any one of the following can be used to take pictures: (Neck to head), coronal (front to back), and sagittal (ear to ear) [24]. Brain MRI pictures may be color pictures. The color picture is initially converted into grey scale or an intensity image. The gray scale image is the array of groups of pixel intensity values that specifies the intensity strength. The range of the intensity is (0,1) for single and double cases. For unsigned integer 8-bits, the values change from [0,255]. For unsigned integer 16-bits, the values also change from [0, 65535]. For signed integer 16-bits, the values change from [−32768, 32767]. Strength or brightness of a picture as two-dimensional continuous function F (x, y) wherever (x, y) denotes the spatial coordinates [25].

20.3.2 Filtering

Filtering is the procedure of eliminating noise from MRI pictures. In general, medical pictures contain several noises at the time of acquisition. In this work, a median filter is applied to separate noise from the MRI pictures (Figure 20.6).

20.3.3 Skull Masking

An exclusion of non-brain tissue like scalp, skull, fat, eyes, neck, etc. from MRI brain pictures is known as skull masking. It is used for enhancing the speed and accuracy, in medical applications, to diagnose and predict. Two elementary morphological operations

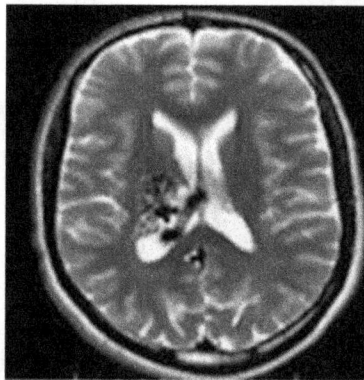

FIGURE 20.6
Filter image.

erosion and dilation are applied for skull masking. As shown in Figure 20.7 the erosion is applied to the removal of pixels on the boundaries of an object. Also, we use a median filter and the boundaries are added by dilation. The sum of pixels included or removed from the picture depends on the size and the shape of the configuring part included in processing the picture, and to fill holes within the brain, the region filling (Figure 20.8) is used. The task of image processing is extremely important for image enrichment that state precisely more subjective decision about photographs. For image enrichment power, law Transformation is used, as presented in Figure 20.9. The transformation of a picture m into image n using R is image enrichment. p and q denote the values of pixels in image m and n, the relation of constituent values p and q is expressed by

$$q = R(p) \tag{20.1}$$

Where, in this R denote transform that map a pixel value p into a pixel value q.

20.3.4 Morphological Operation

Mathematical morphology is a theory used especially in image analysis, whose aim is the study of entities in accordance with shape, size, and relationships with their neighbors, texture, and gray scale. This theory proposes several transformations which can be

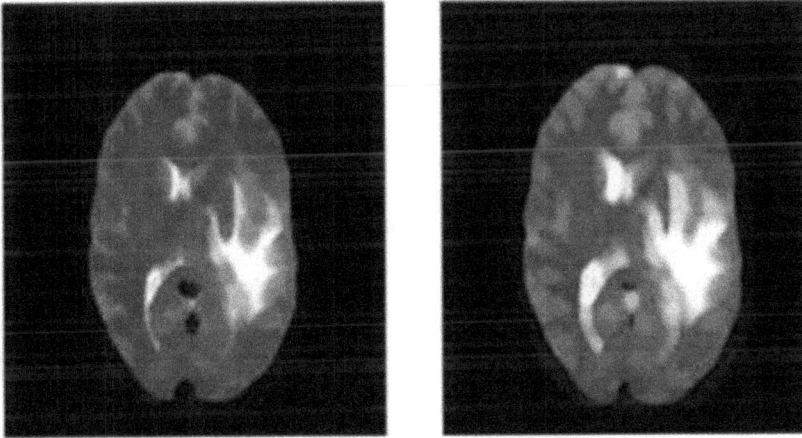

FIGURE 20.7
Erosion and dilation of brain tumor image.

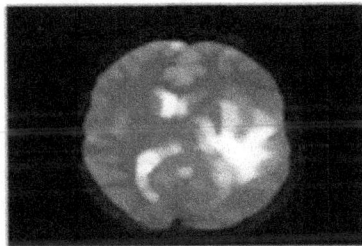

FIGURE 20.8
Region filling of tumor image.

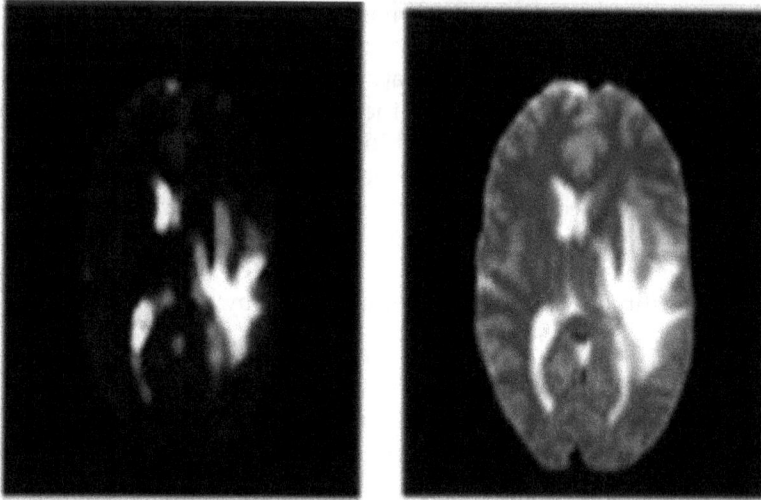

FIGURE 20.9
Power law Transformation and skull-masked image.

functional at several levels of image processing, filtering, segmentation [26], and texture analysis, etc. [3]. In our work, we will use this theory as a post-segmentation operation to clean segmentation results and fill in some cavities. Morphologic operations are especially suited to binary image processing, which is the case in our study. Within the common morphological operations, we cite: dilation and erosion. In fact, dilation increases the size of small regions by adding pixels to object boundaries, while erosion reduces the size of regions by removing pixels on object boundaries. A primordial part of these morphological operations is the structuring element (SE). It consists of a matrix of 0s and 1s. The center pixel is called the origin. It represents the pixel of interest of the structuring element. All pixels containing 1s in the neighborhood of SE. The idea behind using such a geometrical set is that the value of each pixel in the input image is matched to SE and, depending on the transformation (dilation or erosion), this pixel remains as it is or changes and becomes part of the background or object pixels, and Figure 2.10 shows a diamond structuring element.

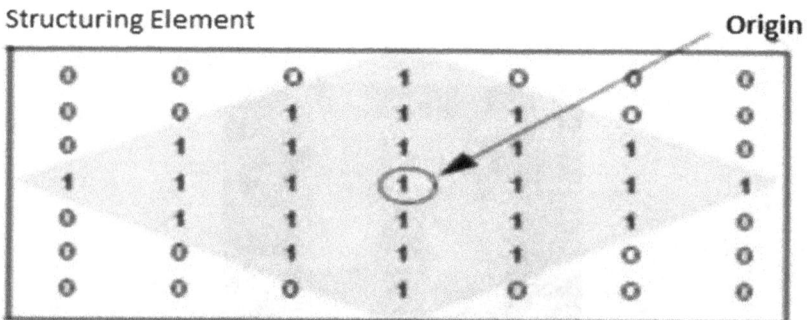

FIGURE 20.10
Illustration of a diamond structuring element.

20.3.5 Dilation

There are two inputs: the binary image to be dilated composed of both background and object regions, and the structuring element. Note that the result of dilation depends only on the structuring element. The mathematical definition of dilation is the following: Suppose that X is the set of points corresponding to the input image and K is the set of points of SE [27]. Let Y be the translation of K in position x. Then dilation of X by K is the set of points such that the intersection between Y and X is not empty. More precisely, for each background pixel of the input image, we superimpose it with the origin of the structuring element. Two cases can be identified: If at least one pixel in SE coincides with an object pixel then this pixel is set to be within object pixels. If all pixels of SE are background, so the input pixel remains background; as Figure 20.11 demonstrates, the effect of dilation using a 3×3 square structuring element such as background is represented by 0 and objects are represented by 1.

20.3.6 Erosion

Erosion is the opposite of dilation. Let X and K represent the input image and SE, respectively. The erosion of X by K is the subset of X. More precisely, for each object pixel, we superimpose it on the structuring element. Two cases can be presented: If every pixel of SE is an object pixel then the input pixel remains as it is (i.e. an object pixel). If any of the SE pixels are background, then the input pixel is set to background. Figure 20.12 demonstrates the effect of erosion using a 3×3 square structuring elements. Note that dilation and erosion operations change the object's shape by adding or removing boundaries pixels. Figure 20.13 shows this process before and after dilation and erosion using a circle SE.

20.3.7 Feature Extraction

Statistical texture analysis was used to excerpt the features from the given image and computed based on statistical distribution of pixel intensity at a specified position relative to alternative pixels within the matrix. We utilize the first-order statistics, second-order statistics, or high-order statistics based on the sum of pixels or dots in each combination.

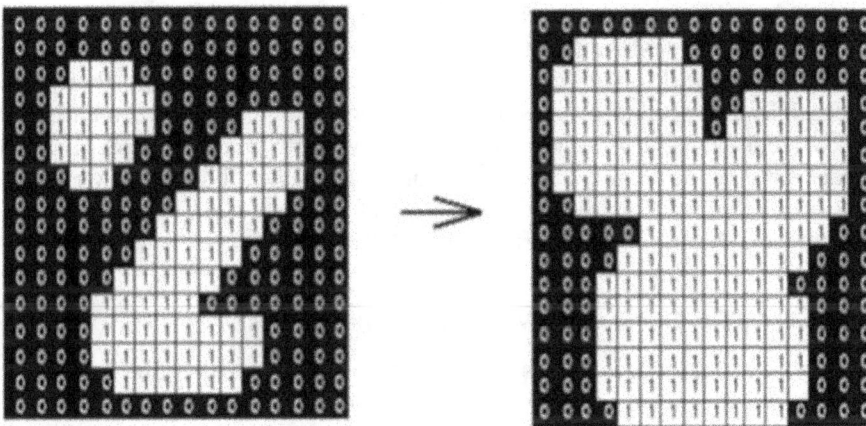

FIGURE 20.11
Illustration of the effect of dilation using a square SE.

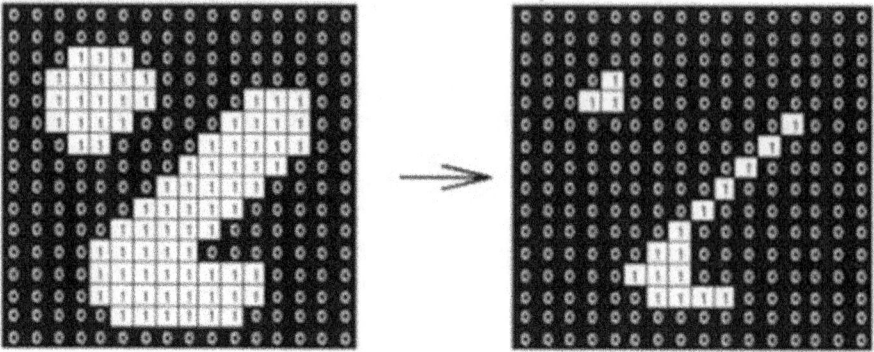

FIGURE 20.12
Illustration of the effect of erosion using a square SE.

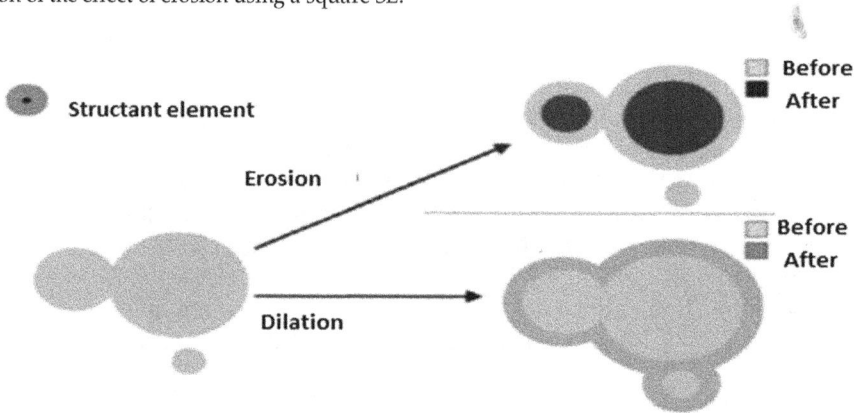

FIGURE 20.13
Illustration of dilation and erosion operations.

Statistical calculation is used to calculate the primary-order statistics' texture measurements from the original image values, like variance, without consideration of the relationship with the neighboring pixel, but the relationship between the two groups is needed for the second- order measurements; on the other hand, it was not possible to implement the third- and higher-order texture (relationship among three or a lot of pixels) due to interpretation difficulties and longer time-consuming calculation, therefore their theoretical use was possible. The second-order statistics analyze image as a texture based on gray-level co-occurrence matrix (GLCM). Gray-level run-length matrix (GLRLM) is used for higher-order statistics to analyze the photographs. In this explicit stage, we examined a set of 33 features that were applied to the Region of Interest (ROI), 6 features from FO, 20 features from GLCM, and 7 features from GLRLM [28].

20.3.8 Gray-Level Co-Occurrence Matrix (GLCM) Features

The gray-level co-occurrence matrix is a statistical method, also called the gray-level spatial dependence matrix because it considers the spatial relationship of pixels. Spatial relationship is specified between the pixel and the adjacent pixel. GLCM features extracted in this chapter are: under dis-similarity Energy, contrast, cluster prominence, Autocorrelation,

Correlation, cluster shade, entropy, maximum probability, sum Average, add entropy, difference entropy, system of measurement of correlation, Homogeneity, add of squares, difference variance, and inverse distinction normalized.

20.3.9 Gray-Level Run-Length Matrix (GLRLM)

For texture-based analysis, the texture feature can be extracted by gray-level run-length matrix (GLRLM). Texture is a design of grey strength in a unique direction from the reference pixels. Run length is the number ranges of the adjacent pixels having identical grey intensity in a specific direction. It is a two-dimensional matrix in which every element $p(i, j/\theta)$ is the number of elements with the intensity j in the direction θ. Figure 20.14(a) shows a matrix of size 4×4 pixel image with 4 gray levels. Figure 20.14(b) is the representation matrix GLRL (gray-level run-length) in the direction of 0^0 [p $(i, j/\theta = 0^o)$]. In addition to the 0° direction, GLRL matrix can similarly create in the other directions, i.e. 45°, 90°, or 135°. Texture features can be obtained from the GLRL matrix on the basis of run percentage (RP), low gray-level run emphasis (LGRE), run-length non-uniformity (RLN), short-run emphasis (SRE), long-run emphasis (LRE), high gray-level run emphasis (HGRE), and gray-level non-uniformity (GLN). GLRLM features description is given as follows.

20.3.10 Short-Run Emphasis

This function is a ratio of each run length and the length of the run squared. This ratio stands for short runs. The total number of runs in the picture is of use as a normalizing factor.

$$F_1 = \frac{\sum_{i=1}^{N_o}\sum_{j=1}^{N_r} \frac{p(i,j)}{j^2}}{\sum_{i=1}^{N_o}\sum_{j=1}^{N_r} p(i,j)} \tag{20.2}$$

20.3.11 Long-Runs Emphasis

This following function F_2 stands for long-run emphasis

$$F_2 = \frac{\sum_{i=1}^{N_o}\sum_{j=1}^{N_r} j^2 p(i,j)}{\sum_{i=1}^{N_o}\sum_{j=1}^{N_r} p(i,j)} \tag{20.3}$$

20.3.12 Gray-Level Non-Uniformity

To scale the gray-level non-uniformity of the picture, the sum of squares of the number of run lengths for each gray level is divided by the normalizing factor

$$F_3 = \frac{\sum_{i=1}^{N_o}\left(\sum_{j=1}^{N_r} p(i,j)\right)^2}{\sum_{i=1}^{N_o}\sum_{j=1}^{N_r} p(i,j)} \tag{20.4}$$

20.3.13 Run-Length Non-Uniformity

The following function F_4 is used to measure the non-uniformity of the run lengths, so that this function will have least-value equally distributed runs throughout the lengths. Maximum value of function will be attained at large run counts [19].

$$F_4 = \frac{\sum_{j=1}^{N_r} \left(\sum_{i=1}^{N_o} p(i,j) \right)^2}{\sum_{i=1}^{N_o} \sum_{j=1}^{N_r} p(i,j)} \tag{20.5}$$

20.4 The Kernel Trick

Sometimes linear classification is not a possibility, not even with a fair amount of slack. This issue can be handled with something called a kernel trick [15,29]. The kernel trick basically takes all the points and maps them into a higher dimensional space. To do so, a kernel K is defined such that points x and x' have a kernel value $K(\mathbf{x}; \mathbf{x}')$ which is equal to an inner product of Φ) and $\Phi\mathbf{x}'$). That is

$$K(x,x') = \left(\Phi(x), \Phi(x') \right) \tag{20.6}$$

In practice, this means that instead of using $\mathbf{x}$ in Equation (20.6) we will use Φ). In this work, we have applied Linear, Quadratic, and Radial Basis Function (RBF) kernel, sometimes also known as a Gaussian kernel. It can be thought of simply as a transformation of $\mathbf{x}$ into an infinite dimensional space, allowing the linear classification which is the basis of SVM.

20.5 Training

The training and the testing are the two important parts of the classification process. For the step of training, the set of data which is already known are fed to the algorithm that is used for classification. For this experiment 96 MRI pictures have been used for the training of the classifier.

20.6 Testing

In the testing step the dataset other than the training is get to the classifier. The classification is done after the training part; the efficiency of the training is responsible for error rates and the success rate of the classification. For the testing set, 24 unknown MRI images are taken, out of which how many are correctly detected and how many are incorrectly detected is analyzed.

TABLE 20.1

SVM Classifier Results

S. No.	Kernel Function	Specificity	Sensitivity	Accuracy
1.	Linear	95.12%	30.25%	90%
2.	RBF	98.56%	50%	95%
3.	Quadratic	96.24%	32.43%	92%

20.7 Results and Discussion

In the biomedical field the research on MRI pictures is so extensive and much work is taking place on automated detection of tumors in the brain. Here, in this chapter, we have proposed a technique with the fusion of two feature extraction algorithms, GLCM and GLRLM, and SVM is used as the classifier. We have considered three kernel functions, i.e. Linear, RBF, and Quadratic. RBF kernel has produced the maximum accuracy as displayed in Table 20.1. There were 24 MRI images considered for the testing to be identified as tumor or non-tumor images using SVM classifier, and the training is done using a set of 96 images. There are certain preprocessing steps involved, which are performed on the MRI images before the feature extraction and training.

References

1. F. F. Chamasemani and Y. P. Singh, "Multi-class Support Vector Machine (SVM) Classifiers - An Application in Hypothyroid Detection and Classification," in *International Conference on Bio-Inspired Computing: Theories and Applications*, 2011, pp. 351–356.
2. N. B. Bahadure, A. K. Ray, and H. P. Thethi, "Image Analysis for MRI Based Brain Tumor Detection and Feature Extraction Using Biologically Inspired BWT and SVM," *Int. J. Biomed. Imaging*, vol. 1, no. 1, pp. 1–12, 2017.
3. K. Machhale, H. B. Nandpuru, V. Kapur and L. Kosta, "MRI Brain Cancer Classification Using Hybrid Classifier (SVM-KNN)," in International Conference on Industrial Instrumentation and Control, 2015, pp. 60–65.
4. X. Wang and J. Tian, "A gene selection method for cancer classification," *Mach. Learn.*, vol. 46, no. 1, pp. 389–422, 2002.
5. A. Gooya, G. Biros, and C. Davatzikos, "Automatic Classification and Segmentation of Brain Tumor in {CT} Images Using Optimal Dominant Gray Level Run Length Texture Features," *Int. J. Adv. Comp. Sci. Appl.*, vol. 2, no. 10, pp. 53–59, 2011.
6. O. Chapelle, P. Haffner, and V. N. Vapnik, "Support Vector Machines for Histogram based Image Classification," *IEEE Trans. Neural Networks*, vol. 10, no. 5, pp. 1055–1064, 1999.
7. P. B. Tamsekar and V. S. Gomase, "Machine Intelligence Approach for Optimization of Cranial Tumor Image," *Int. J. Mach. Intell.*, vol. 1, no. 2, pp. 50–54, 2009.
8. Y. hui Liu, M. Muftah, T. Das, L. Bai, K. Robson, and D. Auer, "Classification of MR Tumor Images Based on Gabor Wavelet Analysis," *J. Med. Biol. Eng.*, vol. 32, no. 1, pp. 22–28, 2012.
9. S. Tong and E. Chang, "Support Vector Machine Active Learning for Image Retrieval," in *Proceedings of the international conference on Multimedia*, 2001, pp. 107–118.
10. G. Chen and W. Xie, "Multiwavelet support vector machines," in *Proceedings of Image and Vision Computing*, 2005, pp. 2–7.

11. A. Guillermo, GarcíaJosu Maiora and M. D. B. Tapia, "Textural Classification of Abdominal Aortic Aneurysm after Endovascular Repair: Preliminary Results," *Comput. Anal. Images Patterns. Lect. Notes Comput. Sci.*, vol. 6854, pp. 537–544, 2011.

12. H. H. Loh, J. G. Leu, and R. C. Luo, "The Analysis of Natural Textures Using Run Length Features," *IEEE Trans. Ind. Electron.*, vol. 35, no. 2, pp. 323–328, 1988.

13. J. Sjolund, A. E. Jarlideni, M. Andersson, H. Knutsson, and H. Nordstrom, "Skull Segmentation in MRI by a Support Vector Machine Combining Local and Global Features," *Proc. - Int. Conf. Pattern Recognit.*, 2014, pp. 3274–3279.

14. Y. Liu and Y. F. Zheng, "FS_SFS: A Novel Feature Selection Method for Support Vector Machines," *Pattern Recognit.*, vol. 39, no. 7, pp. 1333–1345, 2006.

15. Y. Zhang and L. Wu, "An MR Brain Images Classifier via Principal Component Analysis and Kernel Support Vector Machine," *Prog. Electromagn. Res.*, vol. 130, no. 2, pp. 369–388, 2012.

16. S. A. Arezki, H. B. Djamila, and B. C. Bouziane, "Aquazone: A Spatial Decision Support System for Aquatic Zone Management," *Int. J. Inf. Technol. Comput. Sci.*, vol. 7, no. 4, pp. 1–13, 2015.

17. E. Sivasankar, H. Sridhar, V. Balakrishnan, K. Ashwin, and R. S. Rajesh, "Comparison of Dimensionality Reduction Techniques Using a Backpropagation Neural Network Based Classifier," *Int. J. Inf. Acquis.*, vol. 8, no. 2, pp. 161–169, 2011.

18. Y. Fuqing, *Failure Diagnostics Using Support Vector Machine*, Lulea University of Technology, pp. 51–57, 2011.

19. Mary M.Galloway, "Texture Analysis Using Gray Level Run Lengths," *Comput. Graph. Image Process.*, vol. 4, no. 2, pp. 172–179, 1975.

20. W. Lanlan and W. Youxian, "Application of Support Vector Machine for Identifying Single Corn/Weed Seedling in Fields using Shape Parameters," in International Conference on Information Science and Engineering, 2010, pp. 1–4.

21. D. K. Srivastava and L. Bhambhu, "Data Classification Using Support Vector Machine," *J. Theor. Appl. Inf. Technol.*, vol. 12, pp. 1–7, 2009.

22. D. Boswell, "Introduction to Support Vector Machines," in *IEEE Conference on Computer Vision and Pattern Recognition*, 2002, pp. 1–15.

23. U. Acharya, "An Integrated Index for The Identification of Diabetic Retinopathy Stages Using Texture Parameters," *J. Med. Syst.*, vol. 1, no. 1, pp. 1–10, 2011.

24. H. B. Nandpuru, "MRI Brain Cancer Classification Using Support Vector Machine," in *IEEE Students' Conference on Electrical, Electronics and Computer Science*, 2014, pp. 25–30.

25. N. Abdullah, U. K. Ngah, and S. A. Aziz, "Image Classification of Brain MRI Using Support Vector Machine," in *IEEE International Conference on Imaging Systems and Techniques*, 2011, pp. 242–247.

26. H. Selvaraj, "Support Vector Machine Based Automatic Classification of Human Brain Using MR Image Features," *Int. J. Comput. Intell. Appl.*, vol. 2, no. 3, pp. 16–21, 2006.

27. M. Sasikala and N. Kumaravel, "Wavelet Based Automatic Segmentation of Brain Tumors Using Optimal Texture Features," in *International Conference on Biomedical Engineering*, 2008, pp. 637–640.

28. S. L. A. Mohanty, M. Senapati and S. Beberta, "Texture Based Features for Classification of Mammograms Using Decision Tree," *Neural Comput. Appl.*, vol. 23, no. (3–4), pp. 1011–1017, 2013.

29. B. Scholkopf, I. Guyon, and J. Weston, "Statistical learning and kernel methods in bioinformatics," *Artif. Intell. Heuristic Methods Bioinforma.*, vol. 1, no. 2, pp. 1–21, 2003.

21

Proximinality and Remotality in Abstract Spaces

Sangeeta and T. D. Narang

CONTENTS

21.1 Introduction...325
21.2 Notations and Definitions 326
21.3 Proximinality and Remotality in Linear Metric Spaces.............................329
21.4 Unique Remotality in Convex Spaces..331
References...334

21.1 Introduction

Approximation theory consists of the theory of nearest points (best approximation) and the theory of farthest points (worst approximation). The theory started with the concept of best approximation, which is concerned with the problem of describing the elements of a space X that may be approximated by the elements of a subset M of X. Given a non-empty subset M of a metric space (X, d) and $x \in X$, an element $y \in M$ satisfying $d(x, y) = d(x, M) \equiv \inf\{d(x, m) : m \in M\}$ is called a nearest point or a best approximation to x in M. The set of all such nearest points in M is denoted by $P_M(x)$. The set M is called proximinal (Chebyshev) if $P_M(x) \neq \phi$ (a singleton) for each $x \in X$. It is well known that a closed convex subset of a Hilbert space is Chebyshev (see e.g. [Deutsch 2001]. p. 302). Whether every Chebyshev subset of a Hilbert space is convex is still an open problem in Approximation Theory. Various partial answers to this problem are known in the literature (see e.g. [Deutsch 2001], [Narang 1977], and [Singer 1970]). As in the theory of nearest points, a basic problem in the theory of farthest points is:

Given a bounded subset K of a metric space (X, d) and $p \in X$, determine a point in K farthest from p i.e. to find $k_o \in K$ such that $d(p, k_o) = \sup \{d(p, k) : k \in K\}$. Such a farthest point may or may not exist. If such an k_o exists for each $p \in X$, then K is said to be remotal and if it is also unique for each $p \in X$ then K is said to be uniquely remotal.

One of the most interesting and hitherto unsolved problems in the theory of farthest points, known as the "farthest point problem" (f. p. p.), is: If every point of a normed linear space X admits a unique farthest point in a given bounded subset K, then must K be a singleton? There are some partial affirmative answers to this problem and there are many special cases in which the answer is negative (see e.g. [Baronti and Papini 2001], [Klee 1961], [Narang 1990], [Narang 1991] and references cited therein). The problem is not solved in general even in Hilbert spaces. Most natural examples of uniquely remotal sets in normed linear spaces are the singletons.

The importance of this question grew more when [Klee 1961] proved: Singletonness of uniquely remotal sets in Hilbert spaces is equivalent to convexity of Chebyshev sets.

Most of the literature available in approximation theory is in spaces which are normed linear spaces. The absolute homogeneity of norm function, convexity of balls, and existence of non-trivial dual spaces of normed linear spaces have helped a lot in developing a fairly large theory of approximation in normed linear spaces. On the other hand, such properties are generally not available in more general spaces. Moreover, in these spaces the spheres are not generally convex and there are linear metric spaces whose dual spaces consist of zero element alone.

So, the consideration of problems of approximation in abstract spaces like linear metric spaces, convex metric spaces, and metric spaces is quite challenging. It is interesting in itself to see which of the results about normed linear spaces survive in the somewhat more general settings. Moreover, often one has to consider approximation problems in spaces which are not normed linear spaces. With this aim, in this chapter we consider nearest points and farthest points in abstract spaces which are more general than normed linear spaces. We obtain results for subsets of these spaces to be proximinal, Chebyshev, remotal, and uniquely remotal, and also discuss singletonness of uniquely remotal sets. The results proved in this chapter extend results proved in [Mousavi et al. 2017], [Sababheh et al. 2010], and [Sain et al. 2017] for normed linear spaces to other abstract spaces.

21.2 Notations and Definitions

In this section we give some notations and recall a few definitions to be used in the sequel.

Definition 21.1 Let (X, d) be a metric space and $x, y, z \in X$. We say that z is *between* x and y if $d(x,z) + d(z,y) = d(x,y)$. For any two points $x, y \in X$, the set

$$\{z \in X : d(x,z) + d(z,y) = d(x,y)\}$$

is called a *metric segment* and is denoted by $[x, y]$. The set $[x, y, -[= \{z \in X : d(x,y) + d(y,z) = d(x,z)\}$ denotes a half ray starting from x and passing through y. Also $(x,y]$ denotes $[x,y]/\{x\}$.

Definition 21.2 A metric space (X, d) is called *externally convex* [Khalil 1988] if for all distinct points $x, y \in X$ such that $d(x, y) = \lambda$ and $r > \lambda$ there exists a unique $z \in X$ such that $d(x,y) + d(y,z) = d(x,z) = r$.

> **Example 21.1** (Khalil 1988) Consider the space (X,d) consisting of points on the lines $y = 1$ and $y = 2$ in the cartesian plane with $x \geq 0$. Define a metric d as
>
> $$d((x_1, y_1), (x_2, y_2)) = \begin{cases} |x_1| - |x_2|, & \text{if } y_1 = y_2, \\ 1 + |x_1| + |x_2|, & \text{if } y_1 \neq y_2. \end{cases} \qquad (21.1)$$
>
> Then (X, d) is externally convex. X is not a normed linear space as it is not a linear space.

Let (X, d) be a metric space, $x \in X$ and $r \in \mathbb{R}^+$, we set

$$B[x,r] = \{z \in X; d(x,z) \le r\}$$

$$B^c[x,r] = \{z \in X; d(x,z) \ge r\}$$

For a subset K of a metric space (X, d), K' denotes the set of all limit points of K and $\overline{K}$ denotes closure of K.

Definition 21.3 A subset V of a metric space (X, d) is said to be *proximinal* (see [Phelps 1957]) if for each $x \in X$ there exists a point $v_o \in V$ which is nearest to x, i.e.

$$d_x = d(x, v_o) = d(x, V) \equiv \inf\{d(x,v) : v \in V\}. \tag{21.2}$$

Every element $v_o \in V$ satisfying Equation (21.2) is called a *best approximation* or a *nearest point* or a *closest point* to x in V.

We denote by $P_V(x)$, the set of all best approximants to x in V, i.e.

$$P_V(x) = \{v_o \in V : d(x, v_o) = d(x, V)\}.$$

The set V is said to be *proximinal* or an *existence set* if $P_V(x) \ne \phi$ for each $x \in X$. The set V is said to be *antiproximinal* if $P_V(x) = \phi$ for each $x \in X / V$.

In case $P_V(x)$ is exactly a singleton (at most a singleton) for each $x \in X$, we have the following:

Definition 21.4 A subset V of a metric space (X, d) is said to be *Chebyshev* or *uniquely proximinal (respectively, semi-Chebyshev)* if $P_V(x)$ is exactly a singleton (at most a singleton) for each $x \in X$ i.e. for each $x \in X$ there exists exactly one (respectively, at most one) point v_o in V such that $d(x, v_o) = d(x, V)$.

Definition 21.5 The set-valued mapping $P_V : X \to 2^V \equiv$ the set of all subsets of V, which associates with each $x \in X$, the set $P_V(x)$ is called the *metric projection* or the *best approximation map* or *the nearest point map*.

From the definition of P_V, it is clear that P_V exists if and only if V is proximinal and that P_V is single-valued if and only if V is Chebyshev.

Definition 21.6 Let (X, d) be a metric space and K a non-empty bounded subset of X. The set K is said to be *remotal (uniquely remotal)* if for each $x \in X$ there exists at least one (exactly one) $k \in K$ such that $d(x,k) = \sup\{d(x,y) : y \in K\} \equiv \delta_x \equiv \delta(x, K)$. Such a point k is called a *farthest point* of x in K. The set-valued map $F_K : X \to 2^K \equiv$ the collection of all subsets of K, defined by

$$F_K(x) = \{k \in K : d(x,k) = \delta(x, K)\}$$

is called the *farthest point map (f.p.m.)*. The set K is said to be *antiremotal* if $F_K(x) = \phi$ for each $x \in X$. The farthest distance of x from K may or may not be attained by an element of K. If this distance is attained, then collection of all such points of K is denoted by $F(x, K)$, i.e.

$$F(x,K) = \{k \in K : d(x,k) = \delta(x,K)\}.$$

The collection of all such points of K for which farthest distance of some x from K is attained is denoted by *FarK*, i.e

$$FarK = \{k \in K : k \in F(x,K) \text{ for some } x \in X\}.$$

Definition 21.7 A *center c* of a subset M of a metric space (X, d) is an element $c \in X$ such that

$$\delta(c,M) = \inf\{\delta(x,M) : x \in X\}.$$

A space X is said to *admit centers* if every non-empty bounded subset of X has a center.

Centers of sets have played a major role in the study of uniquely remotal sets (see [Astaneh 1983], [Astaneh 1986] and [Baronti 1993]). Whether a set has a center or not, is another question. However, in inner product spaces, any closed bounded set does have a center(see [Astaneh 1983]).

Definition 21.8 A subset V of a metric space (X, d) is said to be *M-compact* if for every $x \in X$ and every sequence $< v_n >$ in V with

$$\lim d(x,v_n) = \delta(x,V) \tag{21.3}$$

there exists a subsequence $< v_{n_i} >$ converging to an element of V. Any sequence satisfying Equation (21.3) is called a *maximizing sequence* for x in V.

Clearly every compact set is M-compact.

Definition 21.9 Let (X, d) be a metric space. A mapping $W : X \times X \times [0,1] \to X$ is said to be a *convex structure* on X if for all $x, y \in X$ and $\lambda \in [0,1]$

$$d(u,W(x,y,\lambda)) \le \lambda d(u,x) + (1-\lambda)d(u,y) \tag{21.4}$$

holds for all $u \in X$. A metric space (X, d) together with a convex structure is called a *convex metric space* [Takahashi 1970] and is denoted by (X, d, W).

The following properties (see [Takahashi 1970]) are direct consequences of Equation (21.4):

$$W(x,y,1) = x, \quad W(x,y,0) = y, \quad d(W(x,y,\lambda),y) = \lambda d(x,y),$$

$$d(W(x,y,\lambda),x) = (1-\lambda)d(x,y), d(x,y) = d(x,W(x,y,\lambda)) + d(W(x,y,\lambda),y).$$

Every normed linear space is a convex metric space, however there are plenty of convex metric spaces which are not normed linear spaces (see [Takahashi 1970]).

Definition 21.10 A metric space (X, d) is called a *linear metric space* if (i) X is a linear space (ii) addition and scalar multiplication in X are continuous, and (iii) d is translation invariant i.e. $d(x+z,y+z) = d(x,y)$ for all $x,y,z \in X$. A complete locally convex linear metric space is called a *Fréchet space*.

Example 21.2 Let X be the collection of all sequences of scalars made into a vector space with the usual vector space operations. Define $d : X \times X \to [0,1]$ by $d(x,y) = \sum_n \frac{1}{2^n} \min$

$\{1, |x_n - y_n|\}$, where $x = <x_n>$, $y = <y_n>$. Then d is an invariant metric and (X, d) is a Fréchet space but X is not normable.

Definition 21.11 A linear metric space (X, d) is said to be *convex* if for all $x, y \in X$, $\lambda \in [0,1]$ we have $d(u, \lambda x + (1 - \lambda)y) \le \lambda d(u, x) + (1 - \lambda)d(u, y)$ for all $u \in X$.

Every normed linear space is convex linear metric space with $W(x, y, \lambda) = \lambda x + (1 - \lambda)y$, and $d(x, y) = ||x - y||$.

21.3 Proximinality and Remotality in Linear Metric Spaces

Theorem 21.1 *Let (X, d) be a linear metric space and A a closed subset of X.*

1. The set A is proximinal if and only if for every $x \in X$, $x \in H_{d_x}$, where

$$H_{d_x} = A + B[0, d_x].$$

2. The set A is antiproximinal if and only if for every $x \in X$, $x \notin H_{d_x}$.

3. The proximinal set A is Chebyshev if and only if for every $x \in X$, $x \in H_{d_x}^{\oplus}$ (where $\oplus$ means that the sum decomposition of each element $x \in X$ is unique).

Proof. (a) Suppose A is proximinal and $x \in X / A$. If we take $d_x = d(x, A)$, then it is clear that $d_x > 0$ and there exists an $a \in A$ such that $d(x, a) = d_x$. It follows that $x = a + (x - a) \in H_{d_x}$. Also, if $x \in A$, then $d_x = 0$ and $B[0, d_x] = \{0\}$, therefore $x = x + 0 \in H_{d_x}$. Conversely, suppose for every $x \in X$, $x \in H_{d_x}$. It follows that for some $a \in A$, $x = a + x - a \in A + B[0, d_x]$ and so $d(x, a) \le d_x \le d(x, a)$ i.e. $d(x, a) = d_x$. Therefore $a \in P_A(x)$ and A is proximinal.

(b) Suppose A is antiproximinal and $x \in X / A$. then $P_A(x) = \phi$. If $x \in H_{d_x}$, then there exists an $a \in A$ such that $x = a + (x - a) \in A + B[0, d_x]$ and $d(x, a) = d(x - a, 0) \le d_x \le d(x, a)$. Therefore $a \in P_A(x)$, a contradiction. It follows that $x \notin H_{d_x}$. Conversely, suppose for every $x \in X$, $x \notin H_{d_x}$. To show A is antiproximinal. Suppose it is not. Then for some $x \in X / A$ there exists an $a \in P_A(x)$. But then $x = a + (x - a) \in H_{d_x}$, which is a contradiction. It follows that $P_A(x) = \phi$ for every $x \in X / A$ and hence A is antiproximinal.

(c) Suppose that the set A is Chebyshev. Since A is proximinal, for each $x \in X / A$, $x \in H_{d_x}$. Suppose there exist two representations of x i.e. $x = a_1 + u_1 = a_2 + u_2$ where $a_1, a_2 \in A$ and $u_1, u_2 \in B[0, d_x]$. Since $d(x, a_i) \le d_x \le d(x, a_i)$ for $i = 1, 2$, i.e $a_1, a_2 \in P_A(x)$. Since A is Chebyshev, it follows that $a_1 = a_2$ and $u_1 = u_2$. Therefore $x \in H_{d_x}^{\oplus}$.

Conversely, if for some $x \in X$ there exists $a_1, a_2 \in P_A(x)$, then $u_i = x - a_i \in B[0, d_x]$ and $x = a_1 + u_1 = a_2 + u_2$. Since $x \in H_{d_x}^{\oplus}$, we have $a_1 = a_2$. Hence A is Chebyshev.

Theorem 21.2 *Let (X, d) be a linear metric space and A a non-empty bounded subset of X.*

1. The set A is remotal if and only if for every $x \in X$, $x \in K_{\delta_x}$, where $K_{\delta_x} = A + B^c[0, \delta_x]$.

2. The set A is antiremotal if and only if for every $x \in X$, $x \notin K_{\delta_x}$.

3. The remotal set A is uniquely remotal if and only if for every $x \in X$, $x \in K_{\delta_x}^{\oplus}$ (where $\oplus$ means that sum decomposition of each element $x \in X$ is unique).

Proof. (a) Suppose that the set A is remotal, then for arbitrary $x \in X$, there exist $a \in A$ such that $d(x,a) = \delta_x$. It follows that $x = a + (x-a) \in K_{\delta_x}$. Conversely, suppose for every $x \in X$, $x \in K_{\delta_x}$. It follows that for some $a \in A$, $d(x,a) \geq \delta_x \geq d(x,a)$ i.e. $d(x,a) = \delta_x$. Therefore $a \in F_A(x)$ and A is remotal.

(b) Suppose A is antiremotal, then for every $x \in X$, $F_A(x) = \phi$. Suppose for some $x \in X$, $x \in K_{\delta_x}$, then there exists an $a \in A$ such that $x - a \in B^c[0, \delta_x]$. But then $a \in F_A(x)$, a contradiction. It follows that for every $x \in X$, $x \notin K_{\delta_x}$.

Conversely, suppose for every $x \in X$, $x \notin K_{\delta_x}$. Suppose for some $x \in X$ there exists an $a \in F_A(x)$. Then $x = a + (x-a) \in K_{\delta_x}$, a contradiction. It follows that $F_A(x) = \phi$ for every $x \in X$ i.e. A is antiremotal.

(c) Since A is remotal and $x \in X$, then $x \in K_{\delta_x}$. If there exist two representations for x i.e. $x = a_1 + u_1 = a_2 + u_2$, where $a_1, a_2 \in A$ and $u_1, u_2 \in B^c[0, d_x]$. But then $a_1, a_2 \in F_A(x)$. Since A is uniquely remotal, it follows that $a_1 = a_2$ and $u_1 = u_2$. Therefore $x \in K_{\delta_x}^{\oplus}$.

Suppose A is remotal and for some $x \in X$ there exist $a_1, a_2 \in F_A(x)$. Then $u_i = x - a_i \in B^c[0, d_x]$ and $x = a_1 + u_1 = a_2 + u_2$. Since $x \in K_{\delta_x}^{\oplus}$ we have $a_1 = a_2$. Therefore A is uniquely remotal.

The following well-known result can be obtained from the above theorems.

Theorem 21.3 *Every non empty compact set in a linear metric space (X, d) is proximinal and remotal.*

Proof. Suppose A is a non-empty compact set in X and $x \in X$. For $n \in \mathbb{N}$, consider the set $H_n(x) = \left\{ a \in A : d(x,a) \leq d_x + \dfrac{1}{n} \right\}$. From the definition of d_x, $H_n(x) \neq \phi$, $H_{n+1}(x) \subseteq H_n(x)$ and $H_n(x) \subseteq A$ is compact. So, there exists an $a \in A$ such that for all $n \in \mathbb{N}, n \geq 1$, $a \in H_n(x)$ i.e. for all $n \geq 1$, $d(x,a) \leq d_x + \dfrac{1}{n}$. Therefore $d(x,a) \leq d_x$ and $x \in H_{d_x}$. It follows that A is proximinal.

For $n \in \mathbb{N}$, consider the set $K_n(x) = \left\{ a \in A : d(x,a) \geq \delta_x - \dfrac{1}{n} \right\}$. From the definition of δ_x, $K_n(x) \neq \phi$, $K_{n+1}(x) \subseteq K_n(x)$ and $K_n(x) \subseteq A$ is compact. So there exist $a \in A$ such that for all $n \geq 1$, $a \in K_n(x)$ i.e. for all $n \geq 1$, $d(x,a) \geq \delta_x - \dfrac{1}{n}$. Therefore $d(x,a) \geq \delta_x$ and $x \in K_{\delta_x}$. It follows that A is remotal.

Theorem 21.4 *Let (X, d) be a linear metric space and A a closed remotal(proximinal) subset of X, $0 \in A$ and $x \in X$. If the set $K_{\delta_x}(H_{d_x})$ is compact, then $F_A(x)(P_A(x))$ is compact.*

Proof. We know that the set $K_{\delta_x} \neq \phi(H_{d_x} \neq \phi)$. Suppose that the set $K_{\delta_x}(H_{d_x})$ is compact. If $\{g_n\} \subseteq F_A(x)$ ($\{g_n\} \subseteq P_A(x)$) then $d(x, g_n) = d(x - g_n, 0) = \delta_x$ ($d(x, g_n) = d(x - g_n, 0) = d_x$).

Since $0 \in A$, we have $\{x - g_n\} \in K_{\delta_x}$ ($\{x - g_n\} \in H_{d_x}$) has a convergent subsequence $\{x - g_{n_k}\}$. Therefore there exists a $z_0 \in K_{\delta_x}$ ($z_0 \in H_{d_x}$) such that $x - g_{n_k} \to z_0$, $g_{n_k} \to x - z_0$ as $k \to \infty$. Since $g_{n_k} \subseteq A$ and A is closed, we have $g_0 = x - z_0 \in A$ and $d(x, g_{n_k}) = \delta_x$ ($d(x, g_{n_k}) = d_x$). It follows that $d(x, g_0) = \delta_x$ ($d(x, g_0) = d_x$) and $g_0 \in F_A(x)$ ($g_0 \in P_A(x)$). Therefore the set $F_A(x)$ ($P_A(x)$) is compact.

Note 21.1 For normed linear spaces these results were proved in Mousavi et al. [2017].

21.4 Unique Remotality in Convex Spaces

Theorem 21.5 *Let (X, d) be a convex linear metric space and A a closed bounded remotal subset of X admitting a center c. If A is not a singleton, then for any extracted function F of $F(x, A)$ there exists a $\rho > 0$ such that $d(F(x), F(c)) > \rho$ for all x in a neighborhood of c on $(c, F(c)]$.*

Proof. Let F be any extracted function from $F(., A)$ and $x \in (c, F(c)]$ then

$$d(x, F(x)) = d(tc + (1-t)F(c), F(x))$$

$$\leq td(c, F(x)) + (1-t)d(F(c), F(x))$$

$$\leq td(c, F(c)) + (1-t)d(F(c), F(x))$$

$$\leq td(x, F(x)) + (1-t)d(F(c), F(x)).$$

which implies

$$d(x, F(x)) - td((x, F(x)) \leq (1-t)d(F(c), F(x))$$

$$\text{i.e } (1-t)d(x, F(x)) \leq (1-t)d(F(c), F(x))$$

$$\text{i.e } d(x, F(x)) \leq d(F(c), F(x))$$

Therefore

$$d(F(x), F(c)) \geq d(x, F(x)) > d(c, F(c)) \equiv r.$$

Thus ρ can be chosen so that $\rho \leq r$, the Chebyshev radius of A.

Theorem 21.6 *Let A be a non-singleton closed and bounded subset of a convex linear metric space (X, d) admitting a center c and let ρ be as in Theorem 21.1. If the distance $\delta(c, A / B(F(c), \rho))$ is attained, then A can not be uniquely remotal*

Proof. Suppose on contrary that A is uniquely remotal and assume $c = 0$. Let $a \in (0, F(0))$ be such that $d(F(x), F(0)) > \rho$ for every $x \in (0, a]$. Let $x_n \in (0, a]$ be such that $x_n \to 0$. By isolation result $F(x_n) \in A / B(F(c), \rho)$. Consider

$$\delta(0, A) = \lim \delta(x_n, A)$$

$$= \lim \delta(x_n, A / B(F(0), \rho))$$

$$= \delta(0, A / B(F(0), \rho))$$

Since $\delta(0, A / B(F(0), \rho))$ is attained, there exists $a_o \in A / B(F(0), \rho))$ such that $\delta(0, A) = d(0, a_o)$. Since $a_o \in A / B(F(0), \rho))$, $a_o \neq F(0)$, contradicting the assumption that A is uniquely remotal.

Corollary 21.1 If a compact subset A with center c is uniquely remotal in a convex linear metric space (X, d), then A must be a singleton.

Proof. Assume that A is not a singleton and ρ is as above. Since A is compact, $A \,/\, B(F(c), \rho)$ is compact and hence remotal i.e. $\delta(c, A \,/\, B(F(c), \rho))$ is attained. Consequently, A is not uniquely remotal by Theorem 21.2, contradicting our assumption.

Note 21.2 For normed linear spaces these results were proved in Sababheh et al. [2010].
   The following propositions will be used in the sequel.

Proposition 21.1 (Sangeeta and Narang 2014) Let K be a non-empty uniquely remotal subset of an externally convex metric space (X, d), then $cl(K) \equiv \overline{K}$ is also uniquely remotal.

Proposition 21.2 Let (X, d) be a metric space. If $K \subseteq X$ is uniquely remotal and M-compact, then $FarK = \{k \in K : k$ is farthest point for some $x \in X\}$ is M-compact.
   Proof. Let $\{k_n\}$ be a maximizing sequence in $FarK$. So there exists $x \in X$ such that $\lim d(x, k_n) = \delta(x, FarK) = \delta(x, K)$. Then $\{k_n\}$ is a maximizing sequence for x in K. Since K is M-compact, $\{k_n\}$ has a convergent subsequence $\{k_{n_l}\}$. As K is uniquely remotal, $\{k_{n_l}\} \to k_x \in K$. Now $\lim d(x, k_{n_l}) = d(x, k_x) = \delta(x, K) = \delta(x, FarK)$ and $k_x \in FarK$, so $\{k_n\}$ has a convergent subsequence in $FarK$. Therefore $FarK$ is M-compact.

Proposition 21.3 Let (X, d) be a metric space. If $K \subseteq X$ is uniquely remotal and M-compact, then $\overline{K}$ is uniquely remotal.
   Proof. As $\delta(x, K) = \delta(x, \overline{K})$, for every $x \in X$ and K is uniquely remotal, we obtain that $\overline{K}$ is remotal(see [Sangeeta and Narang 2014]). Suppose there exists $x_o \in X$ such that $F(x_o, \overline{K})$ is not a singleton. Let $k_1, k_2 \in \overline{K}$ be such that $k_1, k_2 \in F(x_o, \overline{K})$. Since $k_1, k_2 \in \overline{K}$, there exist sequences $\{x_n\} \subseteq K$ and $\{y_n\} \subseteq K$ such that $\{x_n\} \to k_1$ and $\{y_n\} \to k_2$. Clearly $\{x_n\}$ and $\{y_n\}$ are maximizing sequences of $x_o \in K$ and K is M-compact so $\{x_n\}$ and $\{y_n\}$ have convergent subsequences in K which implies $k_1, k_2 \in K$ and this contradicts the fact that K is uniquely remotal. Hence $\overline{K}$ is uniquely remotal.

Note 21.3 For normed linear spaces these results were proved in Sain et al.[2017].

Theorem 21.7 *Let (X, d) be an externally convex Frechet space. If $K \subseteq X$ is such that K is uniquely remotal and $\overline{K}$ is a totally bounded subset of X, then K is a singleton.*
   Proof. By Proposition 4.1, $\overline{K}$ ia uniquely remotal. Since $\overline{K}$ is a closed subset of the complete space X, $\overline{K}$ is complete. Now $\overline{K}$ is a totally bounded complete subset of X, $\overline{K}$ is compact and hence M-compact. Since a totally bounded, M-compact uniquely remotal subset of a Frechet space is a singleton [Narang 1990], $\overline{K}$ is a singleton and so K is a singleton.
   In normed linear spaces the closure of M-compact set K is M-compact but converse is not true. Next we show that the converse holds if the space is an externally convex metric space and the set K is uniquely remotal.

Theorem 21.8 *Let (X, d) be an externally convex metric space. If $K \subseteq X$ is uniquely remotal and $\overline{K}$ is M-compact, then K is also M-compact.*
   Proof. By Proposition 4.1. $\overline{K}$ is also uniquely remotal. Suppose K is not M-compact. Then there exists a maximizing sequence $\{x_n\} \subseteq K$ such that $\{x_n\}$ has no convergent subsequence. Since $\{x_n\}$ is a maximizing sequence in K, there exists $x_o \in X$ such that $\lim d(x_n, x_o) = \delta(x_o, K)$ and $\delta(x_o, K) = \delta(x_o, \overline{K})$. So $\{x_n\}$ is a maximizing sequence for x_o in $\overline{K}$. Since $\overline{K}$ is M-compact,

$\{x_n\}$ has a convergent subsequence $\{x_{n_k}\}$ in $\overline{K}$. Suppose $\{x_{n_k}\} \to w_o,\ w_o \in \overline{K}/K$. Then $d(x_o, w_o) = \lim d(x_o, x_{n_k}) = d(x_o, \overline{K})$. Since K is uniquely remotal, there exists $k_{x_o} \in K$ such that $d(x_o, k_{x_o}) = \delta(x_o, K)$, so we have $d(x_o, k_{x_o}) = \delta(x_o, K) = \delta(x_o, \overline{K}) = d(x_o, w_o)$ which contradicts the fact that $\overline{K}$ is uniquely remotal. Hence K is M-compact.

Theorem 21.9 *Let (X, d, W) be a convex metric space. If $K \subseteq X$ is a uniquely remotal and M-compact set and $FarK \not\subseteq K'$ then K is a singleton.*

Proof. As $FarK \not\subseteq K'$, there exists $k_o \in FarK / K'$. As K is uniquely remotal, there exists $x \in X$ such that $d(x, k_o) = \sup\{d(x, k) : k \in K\}$. Consider the set $C = \{0 \le t \le 1, k_o$ is farthest point for $y' = W(x, k_o, t)$ such that $d(y', x) = td(x, k_o)$ and $d(y', k_o) = (1-t)d(x, k_o)\}$. Clearly $C \ne \phi$ as $1 \in C$. We claim that $\inf C > 0$, otherwise k_o is farthest point of itself which implies that K is a singleton. Suppose $\inf C = t_o > 0$. Consider $x_o \in W(x, k_o, t_o)$ such that $d(x_o, x) = t_o d(x, k_o))$. We claim that there exist $\varepsilon_o > 0$ such that $d(x_o, k) < d(x_o, k_o) - \varepsilon_o$ for every $k \in K / \{k_o\}$. If not, then there exists $\{k_n\} \subseteq K / \{k_o\}$ such that $d(x_o, k_n) \to d(x_o, k_o) = \delta(x_o, K)$. So $\{k_n\}$ is maximizing sequence for x_o. Since K is M-compact, there exists $\{k_{n_a}\} \subseteq \{k_n\}$ such that $k_{n_a} \to k'$ Since k_o is not a limit point of $K, k' \ne k_o$. This contradicts the fact that K is uniquely remotal. Now, take $y'' = W(k_o, x_o, t), 0 < t < 1$. For $k \in K / \{k_o\}$, consider

$$d(y'', k) = d(W(x_o, k_o, t), k)$$

$$\le (1-t)d(k_o, k) + td(x_o, k)$$

$$< (1-t)diamK + t(d(x_o, k_o) - \varepsilon_o)$$

$$= td(x_o, k_o) - \varepsilon_o t + (1-t)diamK$$

$$= td(x_o, k_o) - [\varepsilon_o t - (1-t)diamK]$$

$$= d(W(k_o, x_o, t), k_o) - [\varepsilon_o t - (1-t)diamK].$$

Now consider the mapping $f : R \to R$ defined as $f(t) = \varepsilon_o t - (1-t)diamK$. Clearly f is continuous and $f(1) = \varepsilon_o > 0$ and so there exists a $\delta > 0$ such that $f(t) > 0$ for all $t \in (1-\delta, 1+\delta)$. Therefore there exists at least one $t' \in (1-\delta, 1)$ such that $f(t') > 0$ i.e. $\varepsilon_o t' - (1-t')diamK > 0$, so $d(W(k_o, x_o, t'), k) < d(W(k_o, x_o, t'), k_o)$ for all $k \in K / \{k_o\}$. Therefore, k_o is a farthest point for $W(k_o, x_o, t')$ in K which contradicts the fact that $\inf C = t_o > 0$. So $\inf C = 0$. i.e. k_o is farthest point of itself. Hence K is a singleton.

Theorem 21.10 *Let (X, d, W) be a convex metric space. If $K \subseteq X$ is uniquely remotal and M-compact, then K' is either empty or M-compact and uniquely remotal.*

Proof. If $FarK \not\subseteq K'$ then K is a singleton by Theorem 21.5. Suppose $FarK \subseteq K'$. We claim that $\delta(x, K) = \delta(x, K')$. First we show that $\delta(x, K) \ge \delta(x, K')$ for all $x \in X$. For $x \in X$, there exists $\{k'_n\} \subseteq K'$ such that $d(x, k'_n) \to \delta(x, K')$ as $n \to \infty$. As $k'_n \in K'$, for each $n \in \mathbb{N}$ there exists $k_n \in K$ such that $d(k_n, k'_n) < \dfrac{1}{n}$. Now $d(x, k'_n) \le d(x, k_n) + d(k_n, k'_n)$, implies $d(x, k_n) \ge d(x, k'_n) - d(k_n, k'_n) > d(x, k'_n) - \dfrac{1}{n}$. So $\lim d(x, k_n) \ge \delta(x, K')$ and therefore $\delta(x, K) \ge \lim d(x, k_n) \ge \delta(x, K')$ for all $x \in X$.

We next show that $\delta(x, K') \ge \delta(x, K)$ for all $x \in X$. Since K is uniquely remotal, there exists $k_x \in K$ such that $d(x, k_x) = \delta(x, K)$. Then $k_x \in K'$, since $FarK \subseteq K'$. So $\delta(x, K') \ge d(x, k_x) = \delta(x, K)$. Hence $\delta(x, K') = \delta(x, K)$.

We next prove that K' is M-compact. Let $\{x_n\} \subseteq K'$ be a maximizing sequence in K'. So there exists $x \in X$ such that $\lim d(x, x_n) = \delta(x, K') = \delta(x, K)$. Therefore for each $x_n \in K'$ there exists a $k_n \in K$ such that $d(x_n, k_n) < \dfrac{1}{n}$ for all $n \in \mathbb{N}$. Now $d(x, k_n) \leq d(x, x_n) + d(x_n, k_n) < d(x, x_n) + \dfrac{1}{n}$ implies $\lim d(x, k_n) \leq \lim d(x, x_n)$. Again $d(x, x_n) \leq d(x, k_n) + d(k_n, x_n) < d(x, K_n) + \dfrac{1}{n}$, so $\lim d(x, x_n) \leq \lim d(x, k_n)$. Therefore

$$\lim d(x, k_n) = \lim d(x, x_n) = \delta(x, K') = \delta(x, K).$$

Thus k_n is a maximizing sequence in K for $x \in X$. As K is M-compact, $\{k_n\}$ has a convergent subsequence $\{k_{n_l}\}$ converging to k_o(say) in K. Since $d(x_n, k_n) < \dfrac{1}{n}$ for all $n \in \mathbb{N}$, the subsequence $\{x_{n_l}\}$ of sequence $\{x_n\}$ converges to $k_o \in K'$ as K' is closed. Thus every maximizing sequence in K' has a convergent subsequence. Hence K' is M-compact.

Now we shall prove that K' is uniquely remotal. As $\delta(x, K') = \delta(x, K)$ and K is uniquely remotal, K' is remotal. Suppose K' is not uniquely remotal. Then there exists $x_o \in X$ such that $F(x_o, K')$ is not a singleton. Let $k_1, k_2 \in K'$, then there exist $\{a_n\} \in K$ and $\{b_n\} \in K$ such that $\{a_n\} \to k_1$ and $\{b_n\} \to k_2$ as $n \to \infty$. Then $\lim d(x_o, a_n) = d(x_o, k_1) = d(x_o, k_2) = \lim d(x_o, b_n) = \delta(x, K') = \delta(x, K)$ and so $\{a_n\}$ and $\{b_n\}$ are maximizing sequences for x_o in K. Since K is M-compact $\{a_n\}$ and $\{b_n\}$ have convergent subsequences in K which implies $k_1, k_2 \in K$. This contradicts the fact that K is uniquely remotal. Hence K' is uniquely remotal.

Note 21.4 For normed linear spaces these results were proved in Sain et al. [2017].

References

Astaneh, A. A. 1983. On uniquely remotal subsets of Hilbert spaces, *Indian J. Pure Appl. Math.*14: 1311–1317.

Astaneh, A. A. 1986. On singletoness of uniquely remotal sets, *Indian J. Pure Appl. Math.*17: 1137–1139.

Baronti, Marco 1993. A note on remotal sets in Banach spaces, *Publ. Inst. Math.*53: 95–98.

Baronti, Marco and Papini, P. L. 2001. Remotal sets revisited, *Taiwan. J. Math.*5(2): 367–373.

Deutsch, Frank 2001. *Best Approximation in Inner Product Spaces*. Springer-Verlag. New York.

Khalil, Roshdi 1988. Best approximation in metric spaces, *Proc. Amer. Math. Soc.*103(2): 579–586.

Klee, V. 1961. Convexity of Chebyshev sets, *Math. Ann.*142(3): 292–304.

Mousavi, S. M., Abad, S., Mazaheri, H. and Dehghan, M. A. 2017. Proximinality and remotality in normed linear spaces, *J. Mahani Math. Res. Cent.*6: 73–80.

Narang, T. D. 1977. Convexity of Chebyshev sets, *Nieuw Arch. Wisk.*25: 377–404.

Narang, T. D. 1990. On singletonness of uniquely remotal sets, *Period. Math. Hung.*21(1): 17–19.

Narang, T. D. 1991. Uniquely remotal sets are singletons, *Nieuw Arch. Wisk.*9: 1–12.

Phelps, R. R. 1957. Convex sets and nearest points I, *Proc. Amer. Math. Soc.*8(4): 790–797.

Sababheh, M., Yousef, A. and Khalil, R. 2010. Uniquely remotal sets in Bananch spaces, *J. Comp. Anal.*13: 1233–1239.

Sain, Debamalya, Paul, Kallol and Ray, Anubhab 2017. Farthest point problem and M-compact sets, *J. Nonlinear Convex Anal.*18: 451–457.

Sangeeta and Narang, T. D. 2014. On the farthest points in convex metric spaces and linear metric spaces, *Publ. Inst. Math.*95(109): 229–238.

Singer, Ivan 1970. *Best Approximation in Normed Linear Spaces by Elements of Linear Subspaces*. Springer-Verlag. New York.

Takahashi, W. 1970. A convexity in metric space and non-expansive mappings I. *Kodai Math. Sem. Rep.*22: 142–149.

22

Conformable Fractional Laguerre and Chebyshev Differential Equations with Corresponding Fractional Polynomials

Ajay Dixit and Amit Ujlayan

CONTENTS

22.1 Introduction 335
22.2 Conformable Fractional Laguerre Equation and Its Solution 336
22.3 Conformable Fractional Chebyshev Equation and Its Solution................340
    22.3.1 Another Form of Conformable Fractional Chebyshev Differential
        Equation 341
    22.3.2 The Rodrigues' Formula for Chebyshev Fractional Polynomial 342
    22.3.3 Generating Function for $U_n(x^\beta)$ 343
22.4 Conclusion 343
Acknowledgment...343
References..343

22.1 Introduction

The exact solution of every differential solution dose not exists though it is of integer order, and then we have to turn toward the numerical solution or series solution. The same is the case with fractional order differential equation, likewise differential equations; series solutions also exist for the fractional ordered differential equations. A series solution for a fractional differential equation presented in terms of Mittag–Leffler function [1,2,3,4] to begin with, but when the concept of conformable fractional derivative [5] took place, it made a revolution in the field of fractional calculus by producing so many fundamental properties [6,7]. Now researchers are taking a keen interest in developing the theory as this definition has resolved the problem of computation. Fractional power series [8,9,10] is one of the applications of conformable fractional derivative. Here we study the fractional power series solution of Laguerre and Chebyshev conformable differential equations using Katugampola's derivative [11] which is the generalized form of conformable derivative.

Existence and uniqueness theorem of linear conformable fractional differential equations around a point has been given in [12,9] and the existence of fractional power series solutions about a singular point of sequential conformable differential equation have been presented in [13].

Definition 22.1. For $h : [0, \infty) \rightarrow R$, and $x > 0$ Conformable fractional derivative of order β is given as

$$D^\beta h(x) = \lim_{\epsilon \to 0} \frac{h(x + \epsilon x^{1-\beta}) - h(x)}{\epsilon} \quad \forall x > 0, \beta \in (0, 1]$$

Definition 22.2. For $h : [0, \infty) \rightarrow R$, and $x > 0$ Fractional derivative of order β is given as:

$$D^\beta h(x) = \lim_{\epsilon \to 0} \frac{h(xe^{\epsilon x^{-\beta}}) - h(x)}{\epsilon} \quad \forall x > 0, \beta \in (0, 1]$$

The above two definitions of fractional derivative are given by R. Khalil and U. N. Katugampola respectively and both possess the following property:

If $h(x)$ is differentiable then β derivative is $D^\beta h(x) = \dfrac{d^\beta h(x)}{dx^\beta} = x^{1-\beta} \dfrac{dh(x)}{dx}$.

Definition 22.3. (Fractional Integral) Let $0 \le a \le x$, and h be a function defined on $(a, x]$, then β–fractional integral, $\beta \in (0, 1]$ is defined by

$$I_a^\beta(x) = \int_a^x \frac{h(t)}{t^{1-\beta}} dt$$

provided integral exists.

Definition 22.4. (Fractional power series) A power series is called a fractional power series about origin, if it can be written in the form of $\displaystyle\sum_{r=0}^{\infty} c_r x^{\beta r}$ for all $\beta \in (0, 1]$.

22.2 Conformable Fractional Laguerre Equation and Its Solution

Consider the following sequential conformable Laguerre equation

$$x^{2\beta} D^\beta D^\beta y + \beta(1 - x^\beta) + \beta^2 \lambda y = 0$$

where $0 < \beta \le 1$ and λ is a positive integer.

It is obvious that at $\beta = 1$; this equation becomes a classical Laguerre differential equation.

Now we are interested to find a series solution about β singular point $x = 0$ so consider the Frobenious method for solving fractional differential equation as

$$y = \sum_{r=0}^{\infty} c_r x^{(m+r)\beta} = x^{m\beta} \sum_{r=0}^{\infty} c_r x^{r\beta}$$

So that $D^\beta y = \sum_{r=0}^{\infty} \beta(m+r)c_r x^{(m+r-1)\beta}$ and $D^\beta D^\beta y = \sum_{r=0}^{\infty} \beta^2(m+r)(m+r-1)c_r x^{(m+r-2)\beta}$ using the definition of conformable derivative.

Substituting the above values of the fractional derivative in our equation, we get

$$\sum_{r=0}^{\infty} \beta^2(m+r)(m+r-1)c_r x^{(m+r-1)\beta} + \sum_{r=0}^{\infty} \beta^2(m+r)c_r x^{(m+r-1)\beta}$$

$$-\sum_{r=0}^{\infty} \beta^2(m+r)c_r x^{(m+r)\beta} + \lambda \sum_{r=0}^{\infty} \beta^2 c_r x^{(m+r)\beta} = 0$$

$$\Rightarrow \sum_{r=0}^{\infty} (m+r)(m+r-1)c_r x^{(m+r-1)\beta} + \sum_{r=0}^{\infty} (m+r)c_r x^{(m+r-1)\beta}$$

$$-\sum_{r=1}^{\infty} (m+r-1)c_{r-1} x^{(m+r-1)\beta} + \lambda \sum_{r=1}^{\infty} c_{r-1} x^{(m+r-1)\beta} = 0$$

To find the indicial equation we equate to zero the coefficient of the smallest power of x i.e. $x^{(m-1)\beta}$ and get

$$m(m-1)c_0 + mc_0 = 0 \Rightarrow c_0 m^2 = 0 \Rightarrow m = 0,0$$

now put the coefficient of $x^{(m+r-1)\beta}$ equal to zero to obtain the recurrence relation

$$(k(k-1)+k)c_r - (k-1-\lambda)c_{r-1} = 0$$

$$\Rightarrow c_r = \frac{(k-1-\lambda)}{k^2}c_{r-1} = \frac{(m+r-1-\lambda)}{(m+r)^2}c_{r-1}$$

where $k = m+r$.

Since the roots of indicial equation are multiple, two independent solutions will be $(y)_{m=0}$

and $\left(\frac{\partial y}{\partial m}\right)_{m=0}$. But for $\left(\frac{\partial y}{\partial m}\right)_{m=0}$ the solution will be unbounded due to log x.

So we consider only $\left(\frac{\partial y}{\partial m}\right)_{m=0}$

For $m=0$, $c_r = \frac{r-1-\lambda}{r^2}c_{r-1}$

putting $r = 1,2,3...$ we get

$$c_1 = \frac{-\lambda}{1^2}c_0 = (-1)\lambda c_0, \; c_2 = \frac{1-\lambda}{2^2}c_1 = \frac{(-1)^2}{(2!)^2}\lambda(\lambda-1)c_0,$$

$$c_3 = \frac{2-\lambda}{3^2}c_2 = (-1)^3 \frac{\lambda(\lambda-1)(\lambda-2)}{(3!)^2}c_0$$

And general term for $p \le \lambda$

$$c_p = (-1)^p \frac{\lambda(\lambda-1)(\lambda-2)...(\lambda-p+1)}{(p!)^2}$$

and obviously $c_{\lambda+1} = c_{\lambda+2} = c_{\lambda+3} = ...0$.

Hence we obtain

$$y = c_0 \left(1 - \frac{\lambda}{(1!)^2} x^\beta + \frac{\lambda(\lambda-1)}{(2!)^2} x^{2\beta} + ... + (-1)^p \frac{\lambda(\lambda-1)(\lambda-2)...(\lambda-p+1)}{(p!)^2} x^{p\beta} + ... \right)$$

$$\Rightarrow y = c_0 \sum_{p=0}^{\lambda} (-1)^p \frac{\lambda(\lambda-1)(\lambda-2)...(\lambda-p+1)}{(p!)^2} x^{p\beta} = c_0 \sum_{p=0}^{\lambda} (-1)^p \frac{\lambda!}{(\lambda-p)!(p!)^2} x^{p\beta}$$

Taking $c_0 = 1$ we get the corresponding Laguerre fractional polynomial of order λ denoted by

$$L_\lambda^\beta(x) = \sum_{p=0}^{\lambda} (-1)^p \frac{\lambda!}{(\lambda-p)!(p!)^2} x^{\beta p}$$

Theorem 22.1. Generating function for Laguerre fractional polynomial is

$$\frac{\exp\{-x^\beta t / (1-t)\}}{1-t} = \sum_{n=0}^{\infty} L_n(x^\beta) t^n$$

Proof. We have

$$\frac{1}{1-t} \sum_{r=0}^{\infty} \left(\frac{-x^\beta t}{1-t} \right)^r \frac{1}{r!} = \sum_{r=0}^{\infty} \frac{(-1)^r}{r!} x^{\beta r} t^r (1-t)^{-(r+1)}$$

$$= \sum_{r=0}^{\infty} \frac{(-1)^r}{r!} x^{\beta r} t^r \sum_{s=0}^{\infty} \frac{(r+s)}{r!s!} t^s$$

$$= \sum_{r=0}^{\infty} \sum_{s=0}^{\infty} \frac{(-1)^r}{(r!)^2} \frac{(r+s)}{s!} t^{r+s} x^{\beta r} = \sum_{r=0}^{\infty} L_\lambda(x^\beta) t^{r+s}$$

Let r be fixed using $s = \lambda - r$ such that $s \geq 0$ then the coefficient of t^λ is

$$\sum_{r=0}^{\lambda} (-1)^r \frac{\lambda!}{(r!)^2(\lambda-r)!} x^{\beta r}$$

Putting $\lambda = 0,1,2,3,...$ in the generating function we find the Laguerre's fractional polynomials

$$L_0(x^\beta) = 1$$

$$L_1(x^\beta) = 1 - x^\beta$$

$$L_2(x^\beta) = \frac{1}{2!}(2 - 4x^\beta + x^{2\beta})$$

$$L_3(x^\beta) = \frac{1}{3!}(6 - 18x^\beta + 9x^{2\beta} - x^{3\beta})$$

$$L_4(x^\beta) = \frac{1}{4!}(24 - 96x^\beta + 72x^{2\beta} - 16x^{3\beta} + x^{4\beta})$$

and so on.

Theorem 22.2. Orthogonal property of Laguerre's fractional polynomials.

$$\int_0^{\infty} e^{-x^{\beta}} x^{\beta-1} L_n(x^{\beta}) L_m(x^{\beta}) dx = \frac{1}{\beta} \delta_{mn} \begin{cases} 0, m \neq n \\ 1, m = n \end{cases}$$

Proof. Consider

$$\sum_{n=0}^{\infty} L_n(x^{\beta}) t^n = \frac{\exp\left(\dfrac{-x^{\beta}t}{1-t}\right)}{1-t} \text{ and } \sum_{m=0}^{\infty} L_m(x^{\beta}) s^m = \frac{\exp\left(\dfrac{-x^{\beta}s}{1-s}\right)}{1-s}$$

$$\sum_{n=0}^{\infty} \sum_{m=0}^{\infty} L_n(x^{\beta}) L_m(x^{\beta}) t^n s^m = \frac{\exp\left(\dfrac{-x^{\beta}t}{1-t} + \dfrac{-x^{\beta}s}{1-s}\right)}{(1-t)(1-s)}$$

Multiplying by $x^{\beta-1} e^{-x^{\beta}}$ and integrating

$$\sum_{n=0}^{\infty} \sum_{m=0}^{\infty} \int_0^{\infty} e^{-x^{\beta}} x^{\beta-1} L_n(x^{\beta}) L_m(x^{\beta}) t^n s^m dx = \frac{1}{(1-t)(1-s)} \int_0^{\infty} \exp\left(-x^{\beta} + \frac{-x^{\beta}t}{1-t} + \frac{-x^{\beta}s}{1-s}\right) dx$$

$$= \frac{1}{\beta(1-t)(1-s)} \left[\frac{e^{-x^{\beta}\{1+t/(1-t)+s/(1-s)\}}}{\{1+t/(1-t)+s/(1-s)\}} \right]_{\infty}^{0}$$

$$= \frac{1}{\beta(1-st)}$$

$$= \frac{1}{\beta} \sum_{n=0}^{\infty} t^n s^n$$

Now equating the coefficients of $t^n s^n$ both sides we get

$$\int_0^{\infty} e^{-x^{\beta}} x^{\beta-1} L_n(x^{\beta}) L_m(x^{\beta}) dx = 0, m \neq n$$

$$\int_0^{\infty} e^{-x^{\beta}} x^{\beta-1} L_n(x^{\beta}) L_m(x^{\beta}) dx = \frac{1}{\beta}, m = n$$

Theorem 22.3. $L_n(x^{\beta})$ has the following recurrence relations:

(a) $(n+1)L_{n+1}(x^{\beta}) = (2n+1-x^{\beta})L_n(x^{\beta}) - nL_{n-1}(x^{\beta})$

(b) $L'_{n+1}(x^{\beta}) = L'_n(x^{\beta}) - L_n(x^{\beta})$

(c) $x^{\beta} L'_n(x^{\beta}) = nL_n(x^{\beta}) - nL_{n-1}(x^{\beta})$

Proof.   (a) We have

$$\sum_{n=0}^{\infty} L_n(x^\beta) t^n = \frac{1}{1-t} e^{-\frac{x^\beta t}{1-t}}$$

Differentiating w.r.t 't'

$$\sum_{n=0}^{\infty} L_n(x^\beta) n t^{n-1} = \frac{1}{(1-t)^2} e^{-\frac{x^\beta t}{1-t}} - \frac{x^\beta}{(1-t)^3} e^{-\frac{x^\beta t}{1-t}}$$

$$= \frac{1}{1-t} \sum_{n=0}^{\infty} L_n(x^\beta) t^n - \frac{x^\beta}{(1-t)^2} \sum_{n=0}^{\infty} L_n(x^\beta) t^n$$

$$(1-2t+t^2)\sum_{n=0}^{\infty} L_n(x^\beta) n t^{n-1} = (1-t)\sum_{n=0}^{\infty} L_n(x^\beta) t^n - x^\beta \sum_{n=0}^{\infty} L_n(x^\beta) t^n$$

Equating the coefficients of t^n both sides

$$(n+1)L_{n+1}(x^\beta) = (2n+1-x^\beta)L_n(x^\beta) - nL_{n-1}(x^\beta)$$

(b) Operating β derivative on both sides in the definition of generating function

$$\sum_{n=0}^{\infty} L_n'(x^\beta) t^n = \frac{-t}{(1-t)^2} e^{-\frac{x^\beta t}{1-t}} \text{ which implies}$$

$$(1-t)\sum_{n=0}^{\infty} L_n'(x^\beta) t^n = -t \sum_{n=0}^{\infty} L_n(x^\beta) t^n$$

Equating the coefficients of t^{n+1}

$$L_{n+1}'(x^\beta) = L_n'(x^\beta) - L_n(x^\beta) \text{ and}$$

(c) $x^\beta L_n'(x^\beta) = nL_n(x^\beta) - nL_{n-1}(x^\beta)$ follows from (a) and (b).

22.3 Conformable Fractional Chebyshev Equation and Its Solution

$$(1-x^{2\beta})D^{2\beta}y - \beta x^\beta D^\beta y + \beta^2 p^2 y = 0$$

where $x>0$ and p is a real constant.

We are interested to find the series solution about ordinary point $x=0$ so consider

$$y = \sum_{r=0}^{\infty} c_r x^{r\beta}$$

So that $D^\beta y = \sum_{r=0}^{\infty} \beta r c_r x^{(r-1)\beta}$ and $D^{2\beta} y = \sum_{r=0}^{\infty} \beta r (r-1) c_r x^{(r-2)\beta}$ using the definition of conformable derivative.

$$\Rightarrow \sum_{r=2}^{\infty} \beta^2 r(r-1) c_r x^{(r-2)\beta} - \sum_{r=2}^{\infty} \beta^2 r(r-1) c_r x^{r\beta} - \sum_{r=1}^{\infty} \beta^2 r c_r x^{r\beta} + p^2 \sum_{r=0}^{\infty} \beta^2 c_r x^{r\beta} = 0$$

$$\Rightarrow \sum_{r=2}^{\infty} r(r-1) c_r x^{(r-2)\beta} - \sum_{r=2}^{\infty} r(r-1) c_r x^{r\beta} - \sum_{r=1}^{\infty} r c_r x^{r\beta} + p^2 \sum_{r=0}^{\infty} c_r x^{r\beta} = 0$$

$$\Rightarrow 2c_2 + p^2 c_0 + \left(6c_3 - c_1 + p^2 c_0\right) x^\beta + \sum_{r=0}^{\infty} \left[(r+2)(r+1) c_{n+2} - \left\{ r(r-1) + r - p^2 \right\} c_n \right] x^{\beta r} = 0$$

putting the constant and coefficient of $x^{\beta r}$ equal to zero, we get

$$2c_2 + p^2 c_0 = 0 \Rightarrow c_2 = \frac{-p^2}{2} , \dots c_{r+2} = \frac{r^2 - p^2}{(r+1)(r+2)} c_r$$

and putting $r = 1, 2, 3 \dots$ we get

$$c_3 = \frac{1 - p^2}{3!} c_1, c_4 = \frac{(2^2 - p^2)(-p^2)}{4!} c_1, c_5 = \frac{(3^2 - p^2)(1 - p^2)}{5!} c_1, c_6 = \frac{(4^2 - p^2)(2^2 - p^2)(-p^2)}{6!} c_0$$

in general

$$c_{2r} = \frac{[(2r-2)^2 - p^2][(2r-4)^2 - p^2]\dots[2^2 - p^2](-p^2)}{(2r)!} c_0$$

$$c_{2r+1} = \frac{[(2r-1)^2 - p^2][(2r-3)^2 - p^2]\dots(3^2 - p^2)(1^2 - p^2)}{(2r+1)!} c_1$$

Hence the complete solution is given by

$$y = A \left(1 + \sum_{r=1}^{\infty} \frac{[(2r-2)^2 - p^2][(2n-4)^2 - p^2]\dots[2^2 - p^2](-p^2)}{(2n)!} x^{2\beta r} \right)$$

$$+ B \left(x + \sum_{r=1}^{\infty} \frac{[(2r-1)^2 - p^2][(2n-3)^2 - p^2]\dots(3^2 - p^2)(1^2 - p^2)}{(2n+1)!} x^{(2r+1)\beta} \right)$$

where A, B are constants.

22.3.1 Another Form of Conformable Fractional Chebyshev Differential Equation

$$(1 - x^{2\beta}) D^\beta D^\beta y - \beta x^\beta D^\beta y + \beta^2 p^2 y = 0$$

Using the transformation $x^\beta = \cos t$; $D^\beta y = \dfrac{d^\beta y}{dx^\beta} = \dfrac{-\beta}{\sqrt{1-x^{2\beta}}} \dfrac{d^\beta t}{dx^\beta}$ and

$$D^\beta D^\beta y = \frac{d^\beta}{dx^\beta}\left(\frac{-\beta}{\sqrt{1-x^{2\beta}}}\frac{d^\beta t}{dx^\beta}\right) = \frac{-\beta^2 x^\beta}{(1-x^{2\beta})^{3/2}}\frac{d^\beta y}{dt^\beta} - \frac{\beta}{\sqrt{1-x^{2\beta}}}\frac{d^\beta}{dt^\beta}\frac{d^\beta y}{dt^\beta}$$

putting all the values of the derivative we get

$$\frac{d^\beta}{dt^\beta}\frac{d^\beta y}{dt^\beta} + p^2 y = 0$$

having solution $y = c_1 \cos\left(\dfrac{p}{\beta}\cos^{-1} x^\beta\right) + c_2 \sin\left(\dfrac{p}{\beta}\cos^{-1} x^\beta\right)$ or equivalently

$$y(x^\beta) = c_1 U_p(x^\beta) + c_2 V_p(x^\beta)$$

where U_p and V_p are defined as Chebyshev fractional polynomials of the first and second kinds respectively in x^β

for $0 < x^\beta < 1$, $U_p(x^\beta) \pm iV_p(x^\beta) = (\cos t \pm i\sin t)^p = (x^\beta \pm i\sqrt{1-x^{2\beta}})^p$

So that $U_p(x^\beta) = \dfrac{1}{2}\left[\left((x^\beta + i\sqrt{1-x^{2\beta}})^p\right) + (x^\beta - i\sqrt{1-x^{2\beta}})^p\right]$ and

$V_p(x^\beta) = \dfrac{1}{2i}\left[\left((x^\beta + i\sqrt{1-x^{2\beta}})^p\right) - i(x^\beta - i\sqrt{1-x^{2\beta}})^p\right]$

22.3.2 The Rodrigues' Formula for Chebyshev Fractional Polynomial

$$U_p(x^\beta) = \frac{(-2)^p p!}{(2p)!\,\beta^p}\sqrt{1-x^{2\beta}}\,\frac{d^\beta}{dx^\beta}(1-x^{2\beta})^{p-1/2}$$

The Chebyshev fractional polynomials of the first kind can be obtained by putting $p = 0, 1, 2.3...$ as

$$U_0(x^\beta) = 1$$

$$U_1(x^\beta) = x^\beta$$

$$U_2(x^\beta) = 2x^{2\beta} - 1$$

$$U_3(x^\beta) = 4x^{3\beta} - 3x^\beta$$

$$U_4(x^\beta) = 8x^{4\beta} - 8x^{2\beta} + 1$$

and so on.

22.3.3 Generating Function for $U_n(x^\beta)$

The Chebyshev fractional polynomials of the first kind can be developed by the following:

$$\frac{1-tx^\beta}{1-2tx^\beta+t^2} = \sum_{n=0}^{\infty} U_n(x^\beta)t^n$$

22.4 Conclusion

We have discussed the fractional power series solution as well as corresponding functions of conformable fractional Laguerre and Chebyshev differential equations. The work reflects that at $\beta = 1$ all the solutions and functions behave like classical functions. In addition, we present some properties of the special fractional functions. The findings of the whole work indicate that the results of the fractional case conform with the results of the ordinary case.

Acknowledgment

The authors are grateful to the reviewers for their valuable and constructive suggestions in the direction of improving the manuscript. Also, we would like to express our gratitude to the contributors for their cooperation, directly or indirectly, when this work was being carried out.

References

1. Podlubny, Igor. *Fractional Differential Equations: An Introduction to Fractional Derivatives, Fractional Differential Equations, to Methods of Their Solution and Some of Their Applications.* Vol. 198.
2. Kilbas, A., Anatolii Aleksandrovich, Hari Mohan Srivastava, and Juan J. Trujillo. *Theory and Applications of Fractional Differential Equations.* Vol. 204, 2006.
3. Debnath, Lokenath. "A brief historical introduction to fractional calculus." *International Journal of Mathematical Education in Science and Technology* 35, no. 4 (2004): 487–501. doi:10.1080/0020739 0410001686571
4. Tomovski, Živorad, Rudolf Hilfer, and H. M. Srivastava. "Fractional and operational calculus with generalized fractional derivative operators and Mittag–Leffler type functions." *Integral Transforms and Special Functions* 21, no. 11 (2010): 797–814. doi:10.1080/10652461003675737
5. Khalil, Roshdi, Mohammed Al Horani, Abdelrahman Yousef, and Mohammad Sababheh. "A new definition of fractional derivative." *Journal of Computational and Applied Mathematics* 264 (2014): 65–70. doi:10.1016/j.cam.2014.01.002
6. Atangana, Abdon, Dumitru Baleanu, and Ahmed Alsaedi. "New properties of conformable derivative." *Open Mathematics* 13, no. 1 (2015).

7. Abdeljawad, Thabet. "On conformable fractional calculus." *Journal of Computational and Applied Mathematics* 279 (2015): 57–66. doi:10.1016/j.cam.2014.10.016.

8. Hammad, Abu M., and Khalil R. "Legendre fractional differential equation and Legender fractional polynomials." *International Journal of Applied Mathematics Research* 3, no. 3 (2014): 214–219. doi:10.14419/ijamr.v3i3.2747

9. Ünal, Emrah, Ahmet Gökdoğan, and Ercan Çelik. "Solutions of sequential conformable fractional differential equations around an ordinary point and conformable fractional Hermite differential equation." *arXiv preprint arXiv:1503.05407* (2015).

10. Gökdoğan, Ahmet, Emrah Ünal, and Ercan Çelik. "Conformable fractional Bessel equation and Bessel functions." *arXiv preprint arXiv:1506.07382* (2015).

11. Katugampola, Udita N. "A new fractional derivative with classical properties." *arXiv preprint arXiv:1410.6535* (2014).

12. Gökdoğan, Ahmet, Emrah Ünal, and Ercan Çelik. "Existence and uniqueness theorems for sequential linear conformable fractional differential equations." *arXiv preprint arXiv:1504.02016* (2015).

13. Ünal, Emrah, Ahmet Gökdoğan, and Ercan Çelik. "Solutions around a regular {\alpha} singular point of a sequential conformable fractional differential equation." *arXiv preprint arXiv:1505.06245* (2015).

23

Computational Linguistics: Inverse Problems and Emerging Mathematical Concepts

Sana A. Ansari and Masood Alam

CONTENTS

23.1 Introduction...345
23.2 What is Linguistics? 346
    23.2.1 Core Areas of Linguistics 346
    23.2.2 Main Branches of Linguistics..347
23.3 What Is Computational Linguistics?...348
23.4 History of Computational Linguistics..348
    23.4.1 1940s and 1950s 348
    23.4.2 1960s...349
    23.4.3 1970s and 1980s...349
    23.4.4 1990s to the present...349
23.5 Applications of Computational Linguistics..350
    23.5.1 Machine Translation..350
    23.5.2 Information Retrieval and Clustering Applications.......................350
    23.5.3 Chatbots and Companionable Dialogue Agents............................351
    23.5.4 Virtual Worlds, Games, and Interactive Fiction 351
    23.5.5 Natural Language User Interfaces 351
    23.5.6 Knowledge Extraction...352
    23.5.7 Sentiment Analysis..352
    23.5.8 Language-Enabled Robots...353
23.6 Mathematics and Computational Linguistics 353
23.7 Scope of Computational Linguistics 354
References...354

23.1 Introduction

Computational linguistics is the scientific discipline concerned with understanding written and spoken language from a computational perspective, and building devices that usefully process and produce language in various settings. It is basically a branch of linguistics which deals with applications of techniques of computer science to the analysis and synthesis of language and speech. It can also be understood as a conglomerate of mathematics and linguistics with computer science. As far as it is known, language is a mirror of mind and a computational understanding of language provides insight into thinking and intelligence. Language being our most natural and versatile means of communication, linguistically competent computers greatly facilitate our interaction with machines and software of all sorts.

To have a better understanding of computational linguistics, knowing what linguistics is, is a prerequisite. A brief introduction to linguistics follows.

23.2 What is Linguistics?

Language is one of the fundamental aspects of human behavior and a major component of our lives. In written form it serves as a long-term record of knowledge from one generation to the next. In spoken form it serves as our primary means of coordinating our day-to-day interaction with others. Linguistics is the scientific study of language. Anderson calls linguistics a science because the discipline of linguistics consists of a set of true facts that can be proven objectively; it uses scientific methods in its various dealings; and it uses empirical evidences to develop theories of language (Anderson 2018). The study of linguistics involves an analysis of language form, language meaning, and language in context. Therefore, the approach of linguists is of a descriptive, not a prescriptive kind (Anderson 2018). Their work is to understand why human language is the way it is. They do not prescribe rules to be followed but describe language in all aspects.

23.2.1 Core Areas of Linguistics

The core areas of linguistics are phonetics, phonology, morphology, syntax, semantics, and pragmatics (Crystal 2008).

- **Phonetics:** It is the subfield of linguistics that deals with the physical aspect of sounds. It involves the study of characteristics of human sound production and provides methods for their description, classification, and transcription.
- **Phonology:** It deals with the sound systems of specific languages focusing on the study of the patterns of distinctive sounds found in a language and generalizing the nature of sound systems in the languages of the world.

Phonetics and phonology are the study of language at the level of sounds.

- **Morphology:** It is the study of language at the level of words dealing with the formation of words from smaller units called morphemes.
- **Syntax:** Syntax is the study of language at the level of sentences. It is the subfield of linguistics that studies the rules governing the ways in which words combine to form sentences.
- **Semantics:** It is the subfield of linguistics that deals with the study of meaning in the language.
- **Pragmatics:** Pragmatics is the study of language meaning and use in context. It is the study of language from the point of view of the speakers, the choices they make, the problems they come across while using language in social situations and the effect their use of language has on others (Figure 23.1).

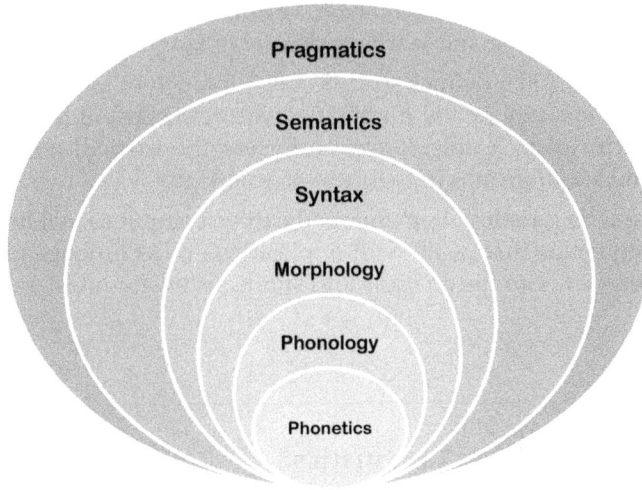

FIGURE 23.1
Core areas of linguistics.

23.2.2 Main Branches of Linguistics

There are a number of branches of linguistics. Computational linguistics is one of the main branches. Others include sociolinguistics, psycholinguistics, historical linguistics, applied linguistics, stylistics, anthropological linguistics, philosophical linguistics, forensic linguistics, neurolinguistics, etc. (Crystal 2008). Some of the branches of linguistics are:

- **Sociolinguistics:** It is the branch of linguistics that studies the relationship between language and society.
- **Anthropological linguistics:** It is the branch of linguistics that uses theories and methods of anthropology to study the role of language in relation to human cultural patterns and beliefs (Figure 23.2).
- **Psycholinguistics:** It is the branch of linguistics that studies the relation between linguistic behavior and the psychological processes that underlie these behaviors.
- **Applied Linguistics:** It is the branch of linguistics that deals with the application of linguistic theories, methods, and findings to the elucidation of language

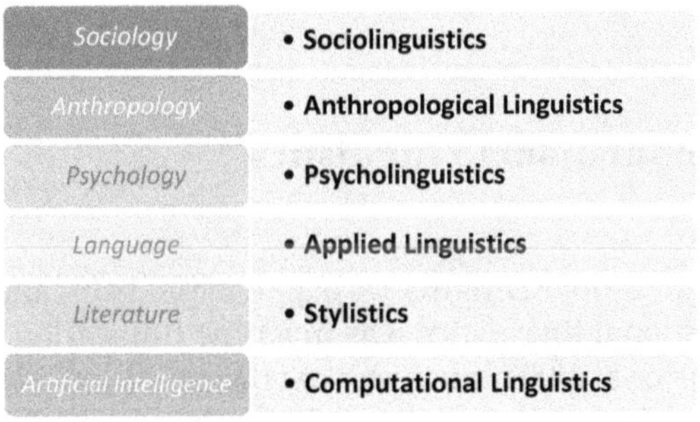

FIGURE 23.2
Some of the branches of linguistics.

problems that arise in other related fields. The most well-developed area of applied linguistics is the teaching and learning of foreign languages.

- **Stylistics:** Crystal (2008) defines stylistics as "a branch of linguistics which studies the features of situationally distinctive uses (varieties) of language, and tries to establish principles capable of accounting for the particular choices made by individual and social groups in their use of language."
- **Computational linguistics:** As mentioned earlier, computational linguistics is the branch of linguistics that deals with applications of techniques of mathematics and computer science to the analysis and synthesis of language and speech.

23.3 What Is Computational Linguistics?

Computational linguistics is the simulation of language by the use of computers (Finch 2000). As the name itself suggests, it comprises of two main components—computer science and linguistics. It involves a formal, rigorous, computationally based investigation that intercedes with linguistics. The term computational linguistics, according to Roland Hausser, refers to that subarea of non-numerical computer science that deals with language production and language understanding (Hausser 2001). Thus, the primary concern of computational linguistics is computational modeling of linguistic processes to support the development of software applications that do useful things with language.

Natural languages such as English, Hindi, Spanish, French, etc. are supple and depend upon apparently obvious situations of utterances. Favorable circumstances for such languages also include common background knowledge and content of earlier conversations, among other things. On the other hand, programming languages are inflexible and entirely depend upon the operations of the machine. Moreover, the expressions of the programming languages differ from those of everyday languages and also the use of the programming languages requires great precision. Computational linguists not only bridge this gap but also work to build devices that make use of language in innovative modes.

Computational linguistics is a highly interdisciplinary field. It takes inputs from different other fields besides computer science and linguistics. Cognitive science, philosophy, psychology and mathematics are some of the areas from which insights are taken in computational linguistics.

23.4 History of Computational Linguistics

23.4.1 1940s and 1950s

Study in the field of computational linguistics began in the 1940s and 1950s. The first project undertaken in computational linguistics was machine translation (MT) (Bates 1995, Jones 2001, Manaris 1998, etc.). Jones (2001) suggests that the obsession with MT was prompted by Weaver's (1949) memorandum on translation, although a couple MT projects were already in the works at that time. One of the early works was the efforts of a group of scholars who were experienced in decryption. They aimed to apply their knowledge base to translation

(Manaris 1998). One of the notable events of the times was the IBM-Georgetown demonstration in 1954. It was an experiment in automatic translation from Russian to English (Jones 2001). Although the 1940s and 1950s were still the era of extremely primitive processing, with no higher-level programming languages and restricted access to machines and storage, the researchers persisted, with special focus on syntax.

23.4.2 1960s

During the 1960s, the focus of the researchers into computational linguistics turned from machine translation to question answering (Bates 1995). A little flavor of artificial intelligence (AI) also seeped in as companies looked for a product that could deal with some of the more basic customer queries without having to pay the salaries of an actual human being (Jones 2001, Wahlster 1989). Perhaps the most important event of the 1960s was the 1966 report released by the Automatic Language Processing Advisory Council (ALPAC) sponsored by the US National Academy of Sciences (Manaris 1998). The report was intended to critically survey the natural language processing field at the time to see if it was actually making valid progress and was worth government funding being spent on it. However, things took a turn in 1966 when a report on the survey done in the field of natural language processing by the Automatic Language Processing Advisory Council (ALPAC) condemned the field of machine translation, which eventually resulted in the termination of funding for all MT projects in the US (Manaris 1998). MT gradually died out in favor of other more profitable-looking topics of research.

23.4.3 1970s and 1980s

The search that began in the 1960s for a system that could replace a basic customer service representative continued in the 1970s (Wahlster 1989). Furthermore, additional tools were developed to allow for more advanced stochastic processes, including Hidden Markov Models and the Viterbi algorithm (Manaris 1998).

The 1980s brought a major change that included a switch from complete understanding programs to partial understanding programs (Bates 1995). Earlier the complete understanding programs used to colligate the meaning of every single part of the input. On the other hand, the partial understanding programs just sought to understand the overall gist of the input and ignored non-essential elements. Another major element of the 1980s was the return of machine translation. Not only this but the work done in the field of machine translation in the difficult circumstances of the 1940s and 1950s gained appreciation from researchers. It was also expanded in order to provide more practical and easily accessible tools. However, researches in most of the natural language systems failed on broader discourse levels. Consequently, steps were taken to deal with those shortcomings.

23.4.4 1990s to the present

The 1990s saw a considerable expansion in the field of computational linguistics. The return to MT in the 1980s witnessed more popularity in the 1990s. Along with the stochastic techniques, new evaluation techniques were also developed in order to judge the viability of each product or program (Manaris 1998). However, work in the field of natural language processing was not that smooth. Nonetheless work continued to overcome the loopholes and, to a great extent, things smoothened out in the field of natural language processing. On the other hand, development of learning algorithms took a jump from lexicon and syntax to semantics and pragmatics. Other popular topics in the 1990s were spoken language systems combining speech recognition with language understanding, language generation, text processing, and interactive NL interfaces.

After the 1990s the 2000s focused on the need for more user-friendly systems that did not require learning complicated programming (Frappaola 2000, Bose 2004).

23.5 Applications of Computational Linguistics

Applications of computational linguistics techniques are varied. From those depending on linguistic structure and meaning, such as document retrieval and clustering, to those that use comprehension and communication, such as dialogue agents, there are a number of applications of computational linguistics. The following section briefly elucidates some of the applications.

23.5.1 Machine Translation

Machine translation (MT) is the subfield of computational linguistics that uses specific software to translate speech or text from one language to another. SYSTRAN, one of the oldest MT systems was developed as a rule-based system in the 1960s, and had been in vogue in US and European government agencies until recently (Arnold, Balkan, Meijer, Humphreys, and Sadler 2001). In the beginning of the twenty-first century, it was hybridized with statistical MT techniques. Google Translate currently uses a phrase-based machine translation technique. Microsoft's Bing Translator together with statistical MT employs dependency structure analysis. Other compendious translation systems include Asia Online and WorldLingo. Many other translation systems exist for small groups of languages such as OpenLogos, IdiomaX, and GramTrans.

Although translations done through MT techniques fail to remain error-free but they suffice in situations where only a restricted understanding is required, such as international web browsing where the readers are required to grasp just a general drift of the source contents. Also, MT applications on hand-held devices, meant to aid international travelers with enquiries such as asking for directions or help, interacting with transportation personnel, or making purchases or reservations, often serve their purpose successfully. For purposes where high-quality translations are required, automatic methods can be used as an aid to human translators, but it can be cumbersome at times.

23.5.2 Information Retrieval and Clustering Applications

Information retrieval, a central theme of information science, covers retrieval of both structured data as well as unstructured text documents (e.g., Salton 1989). Retrieval of both structured and unstructured data require *content-directed* methods. Retrieving employee records by the unique name/number of employees is an example of retrieving information from structured data. Text documents might also be retrieved using some unique label. Also, information can be retrieved with its relevance to a certain query or topic header. The idea behind it is that the documents should contain the terms of the query.

Document clustering is used when a large number of data needs to be organized for easy access. For example, in collections of patent descriptions, medical histories, or captioned images. Clustering also indirectly aids various NLP applications. Clustering is widely used in other areas, such as biological and medical research and epidemiology, market research and grouping and recommendation of shopping items, educational research, social network analysis, geological analysis, and many others.

Document retrieval and clustering are the preliminary steps in information extraction or text mining. Information extracting or text mining involves extracting useful knowledge from documents of named entities (category, roles in relation to other entities, location, dates, etc.) or of particular types of events, or inferring correlations between relational terms (e.g., that purchasing of one type of product correlates with purchase of another), etc.

23.5.3 Chatbots and Companionable Dialogue Agents

Chatbots are used for entertainment purposes or to engage the interest of visitors of certain websites. Current chatbots are the descendants of Weizenbaum's ELIZA. They are equipped with large readymade scripts that enable them to answer simple inquiries about the concerned topic. They are also designed to exchange greetings and pleasantries. They can also be a part of interactive gaming sites or can be used to recommend websites or products to users.

Companionable dialogue agents use chatbot techniques in engaging in authored input patterns and corresponding outputs. The higher goal, however, is to transcend these techniques and to create agents with personality traits and capable of showing emotion and empathy. They are required to have semantic and episodic memory, learning about the user over the long term and providing services to the user. Those services might include advice in some areas of life, health and fitness, schedule maintenance, reminders, question answering, game playing, and internet services (Wilks 2010).

Other uses include systems aimed at clinically well-founded health counseling (Bickmore et al. 2011). Such systems have proved to simulate human counselors to a large extent but since they consist of scripted system utterances paired with multiple-choice lists of responses offered to the client, they cannot be efficient all the time.

23.5.4 Virtual Worlds, Games, and Interactive Fiction

Text-based adventure (quest) games generally feature textual descriptions of the setting and challenges confronting the player, and allow for simple command-line input from the player to select available actions. Description of the settings is often accompanied by pictures and can be elaborate, as in adventure fiction. However, the input options available to the player can be restricted to simple utterances which can be anticipated by the programmers of the game and be manually prepared responses. Nonetheless text-based games are no longer in vogue. Likewise, interactive fiction development softwares that enable online dialogue are also elaborately scripted programs, allowing only for inputs that can be heuristically mapped to one of various preprogrammed responses. Many games in this genre also make use of chatbot-like input–output response patterns. At present, much advanced research is being done in this area.

23.5.5 Natural Language User Interfaces

This topic includes a variety of natural language applications, ranging from text-based systems dependent on minimal understanding to systems with significant comprehension and inference capabilities in text- or speech-based interactions. Some of the traditional and current applications are discussed.

- **Text-based question answering:** Text-based QA involves storing ready-made answers in the text corpora being accessed by the QA system for a certain type of question being asked. Questions need to be factual in nature and straightforward and not ones requiring inference. Text-based QA begins with question classification (e.g., yes–no questions, who-questions, what-questions, when-questions, etc.), followed by information retrieval for the identified type of question, followed by narrowing of the search to paragraphs and finally sentences that may contain the answer to the question (Maybury 2004).

- **Inferential (knowledge-based) question answering:** Inferential QA systems rely on the ability to confirm entailment relations between candidate answers and questions. They make use of simple sorts of semantic relations among the terms involved. Other things that can be done through these systems are sorting or categorizing data sets from databases, computing averages, creating statistical charts, etc. However, these systems rely on quite limited and specialized inference methods as opposed to general reasoning based on symbolic knowledge.

- **Voice-based web services and assistants:** Voice-based services on mobile devices can serve as organizers (for grocery lists, meeting schedules, reminders, contact lists, etc.), to in-car "infotainment" (routing, traffic conditions, hazard warnings, iTunes selection, finding nearby restaurants and other venues, etc.). A number of other uses include dialing contacts, financial transactions, reservations and placing orders, Wikipedia access, help-desk services, health advising, and general question answering. Some of these services (such as dialing and iTunes selection) fall into the category of hands-free controls, and such controls are becoming increasingly important in transport (including driverless or pilotless vehicles), logistics, and manufacturing.

The key technology in these services is speech recognition. One of the examples of advanced voice-based assistants is iPhone's Siri. Siri adds personality and improved dialogue handling and service integration, which give users the feeling of interacting with a lively synthetic character rather than an app.

23.5.6 Knowledge Extraction

Extracting knowledge or producing summaries from unstructured text are very important applications. One commonly used method in knowledge extraction is the use of extraction patterns. The patterns are designed to match the kinds of conventional linguistic patterns typically used by authors to express the concerned information. For example, text corpora or news wires might be mined for information about companies, by keying in on known company names and terms such as "Corp.," ".com," "headquartered at," etc.

23.5.7 Sentiment Analysis

Sentiment analysis is understood as the detection of positive or negative attitudes on the part of authors of articles or blogs toward commercial products, films, organizations, persons, ideologies, etc. This has become a very active area of applied computational linguistics.

Its importance largely lies with the marketing and ranking of products, analysis of social networks, classification of personality types or disorders based on writing samples, and other areas. The techniques that are used in such analyses are based on sentiment lexicons that classify the affective polarity of vocabulary items. Also, supervised machine learning applied to texts is used from which word and phrasal features are extracted that have been labeled as expressing positive or negative attitudes towards some theme.

For example, frequent occurrences of phrases such as *wonderful match* or *great match* or *terrible match* in blogs concerning some football match might suggest that *wonderful*, *great*, and *terrible* belong to a contrast spectrum ranging from a very positive to a very negative polarity. However, sentiment analysis based on lexical and phrasal features has obvious limitations, such as obliviousness to sarcasm and irony, and lack of understanding of entailments. The paper (Zhao, Qin, & Liu 2010) surveys the state of art of sentiment analysis. There are three main areas of sentiment analysis, namely entraction, classification, and extraction. The paper also presents evaluation and corpus for sentiment analysis.

23.5.8 Language-Enabled Robots

Robots today are equipped with web services, question-answering abilities, chatbot techniques (for fallback and entertainment), tutoring functions, and so on. They are equipped with the hardware and software needed for basic visual perception, speech recognition, exploratory and goal-directed navigation (in the case of mobile robots), and object manipulation. However, the keen public interest in intelligent robots and their enormous economic potential (for household help, eldercare, medicine, education, entertainment, agriculture, industry, search and rescue, military missions, space exploration, and so on) surely has the potential to continue to energize the drive toward greater robotic intelligence and linguistic competence.

The challenges in real-time interaction involve speech recognition errors, non-fluent and complex multi-clause utterances, perspective-dependent utterances, and deixis. Other problems include recognition of speech in the presence of noise, better and broader linguistic coverage, parsing, and dialogue handling, mental modeling of the interlocutor and other humans in the environment, and greater general knowledge about the world and the ability to use it for inference and planning.

23.6 Mathematics and Computational Linguistics

Mathematical tools like inverse problems, wavelets and fractals, fractional derivatives are discussed in Meyer (1993), Daubechies (1992), and Siddiqi (2018). A beautiful account of application of mathematical tools in speech recognition is given in Suksiri (2016). These artificial neural network tools such as wavelet theory, neural network, are nicely elaborated in this context. Against this background, classical wavelet theory is being replaced by wavelet constructed by Walsh function; see Farkov, Manch, Siddiqi (2019). It is well known that many physical laws are modeled by classical derivatives and differential equations. In recent years the effect of replacing classical derivatives with fractional derivatives has been studied in a large number of research problems; see, for example, Siddiqi, Alam, and Tripathi (2019). Somehow or other such studies in computational linguistics have escaped the attention of researchers of this field.

23.7 Scope of Computational Linguistics

The scope of computational linguistics includes systems such as automatic fact retrieval systems, question-answering systems, the production of computer generated abstracts, indexes and catalogues, and the analysis of the information content of literature by statistical, syntactic, and logical methods.

Although it is a challenging field, skilled computational linguists are in demand and are highly paid. Computational linguists need a good understanding of both programming and linguistics. They are hired to develop computer systems that deal with human language, to build systems that can perform tasks such as speech recognition, speech synthesis, machine translation, grammar checking, text mining, etc.

People with a degree in computational linguistics can work in research groups in universities, governmental research labs, and large enterprises. The demand for computational linguists is quite high in the public sector. With the ever-increasing availability of unstructured data, governments are searching for ways to better process and understand this information and, therefore, are looking for skilled computational linguists to hire.

References

Anderson, C. (2018). *Essentials of Linguistics*. Open Educational Resource, McMaster University, Hamilton, ON.

Arnold, D., Balkan, L., Meijer, S., Humphreys, R. L., & Sadler, L. (2001). *Machine Translation: An Introductory Guide*. London: Blackwell.

Bates, Madeleine. (1995). Models of natural language understanding. *Proceedings of the National Academy of Science of the UnitedStates of America*, 92(22), 9977–9982.

Bickmore, T. W., Schulman, D., & Sidner, C. L. (2011). A reusable framework for health counseling dialogue systems based on a behavioral medicine ontology. *Journal of Biomedical Informatics*, 44(2), 183–197.

Bose, Ranjit. (2004). Natural language processing: Current state and future directions. *International Journal of the Computer, the Internet and Management*, 12(1), 1–11.

Crystal, David. (2008). *A Dictionary of Linguistics and Phonetics* (6th edition). Oxford, MA: Blackwell Publishing.

Daubechies, I. (1992). *Ten Lectures on Wavelets*. Philadelphia: SIAM.

Farkov, Y. A., Manchanda, P., & Siddiqi, A. H. (2019). *Construction of Wavelets Through Walsh Function*. Singapore: Springer Nature.

Finch, Geoffrey. (2000). Linguistics - The main branches. In: *Linguistic Terms and Concepts*. New York: Palgrave. 10.1007/978-1-349-27748-3.

Frappaolo, Carl. (2000). Now it's personal. *Intelligent Enterprise*, 3(17).

Hausser, R. (2001). *Foundations of Computational Linguistics* (2nd edition). New York: Springer.

Jones, Karen Sparck. (2001). Natural language processing: A historical review. Current issues in computational linguistics. In: *Honour of Don Walker*, Dordrecht, Germany: Springer.

Manaris, Bill. (1998). Natural language processing: A human-computer interaction perspective. *Advances in Computers*, 47, 1–66.

Maybury, M. T. (ed.) (2004). *New Directions in Question Answering*. Cambridge, MA: AAAI and MIT Press.

Meyer, Y. (1993). *Wavelets Algorithms and Applications SIAM*, Philadelphia, PA.

Salton, G. (1989). *Automatic Text Processing: The Transformation, Analysis, and Retrieval of Information by Computer*. Boston: Addison-Wesley Longman.

Siddiqi, A. H. (2018). *Functional Analysis with Applications*. Springer Nature.

Siddiqi, A. H., Alam, M., & Tripathi, P. (2019). Inverse problems for conformable differential equations. *Indian Journal of Industrial and Applied Mathematics*, 2.

Suksiri, B. Voice recognition using distributed artificial neural network for multiresolution wavelet transform decomposition. Master's Thesis. Kochi University of Technology, https://pdfs semantic scholar.org/d7f1.

Wahlster, Wolfgang. (1989). Natural language systems: Some research trends. *Logic and Linguistics: Research Directions in Cognitive Science*, (2), 171–183.

Weaver, W. (1949). Translation. In: Repr. *Machine Translation of Languages: Fourteen Essays*, ed. W. N. Locke and A. D. Booth, 15–23. Cambridge, MA: Technology Press of M.I.T., 1955.

Wilks, D. S. (2010). Use of stochastic weather generators for precipitation downscaling. Interdisciplinary Reviews. *Climate Change*, 1(6), 898–907.

Zhao, Yan, Qin, Bin, & Live, Tin. (2010). Sentiment analysis. *Journal of Software*, 21(8), 1834–1848.

24

Legendre Wavelets Method for Solution of First-Kind Volterra Problems

Pooja, J. Kumar, and P. Manchanda

CONTENTS

24.1 Introduction...357
24.2 Legendre Wavelet 358
    24.2.1 Approximation of Function...359
    24.2.2 Operational Matrix of Integration..360
24.3 Numerical Technique..360
24.4 Error Analysis 363
24.5 Numerical Examples 365
24.6 Conclusion 369
Acknowledgment...369
References..369

24.1 Introduction

Volterra integral equations of the first kind arise naturally in various problems. Many problems in mathematical physics, engineering, and integral geometry are often reduced to first-kind Volterra integral equations. First-kind Volterra integral equations appear in problems like determination of density distribution of a sphere by transmitting X-rays, and distribution of plutonium in a small sphere [14]. The inverse heat-conduction problem can also be converted into a first-kind Volterra integral equation [4]. Inverse problems like evaluating particle size distributions in aerosols using time-resolved laser-induced incandescence data are also represented as Volterra integral equations of the first kind.

A linear Volterra integral equation of the first kind is of the form

$$g(x) = \int_0^x K(x,t)f(t)dt, \ x \in [0,1), \tag{24.1}$$

where $f(x) \in L^2[0,1)$. A nonlinear Volterra integral equation of the first kind is given as

$$g(x) = \int_0^x K(x,t)Y(f(t))dt, \ x \in [0,1), \tag{24.2}$$

where $Y(f(x))$ is a nonlinear function in $f(x)$. The problem is to determine function $f(x)$ using known values of function $g(x)$ and the kernel $K(x, t)$, which is an inverse problem. It is an ill-posed problem because the solution $f(x)$ does not depend continuously on the data $g(x)$. Therefore, regularization techniques are necessary to find a stable solution to the problem. The regularization method converts an ill-posed problem into a well-posed problem and the approximate solution to the original problem can be found by solving this well-posed problem. A survey of regularization methods for first kind Volterra equations is given by Lamm [4].

Volterra problems are casual in nature, which means that the solution at any point depends only on the past and present data values. Classical regularization methods like Tikhonov, bounded variation regularization methods, etc destroy its casual behaviour as these methods use adjoint operator which is non-casual in nature. So these methods convert a Volterra problem into a non-Volterra problem, i.e., evaluation of a solution at any point also requires the knowledge of future data values. In recent years, different local regularization methods have been developed which are effective in handling the Volterra problem as they preserve the causal structure of the Volterra problem. The classical methods require an initial guess of an unknown function to start the algorithm which strongly affects the performance of the method. However, no initial guess is required in the case of local regularization methods. Also, the computation cost of local regularization is lower than that of classical methods. In recent years, several local regularization techniques have been used for solving first-kind Volterra integral equations [5, 6, 7, 8].

Different types of orthogonal functions and wavelets have been used for numerical approximation of the ill-posed Volterra problem. Babolian et al. [1] have solved Volterra and integro-differential equations of the convolution type using Laplace transform and block pulse function. This problem has been investigated by Babolian et al. [2] using block pulse function. Maleknejad et al. [9] has used modified block pulse function to solve Volterra integral equation of the first kind. The Haar wavelet collocation method has been applied for solving first-kind Volterra integral equations by Maleknejad et al. [10] and Singh et al. [13]. Legendre wavelets have been successfully applied to solve systems of Volterra integro-differential equations [12], Volterra–Fredholm integro-differential equations [11], and first-kind Fredholm integral equations [3]. In this chapter, we have used Legendre wavelet to solve Volterra integral equations of the first kind.

The aim of this chapter is to present an efficient numerical method to find a stable solution of Volterra integral equations of the first kind. For this purpose, a combination of local regularization method and Legendre wavelets has been used. The method of local regularization helps in finding a stable solution by converting the problem into a well-posed problem and then after applying wavelets the equation is reduced to a system of algebraic equations which can then be solved to find the solution to the problem.

The outline of this chapter is as follows. In Section 24.2, we give an introduction to Legendre wavelets and an approximation of function using Legendre wavelets. Section 24.3 introduces a local regularization method and presents the numerical method to find an approximate solution to the given problem. The convergence and error analysis is considered in Section 24.4. In Section 24.5, some numerical examples are presented to verify the accuracy of the numerical method. A conclusion is given in the last section.

24.2 Legendre Wavelet

In this section, we present a brief introduction to Legendre wavelets. Let $P_m(x)$ be Legendre polynomials of order $m \geq 0$, then the Legendre wavelets $\psi_{n,m}(x)$ on interval $[0,1)$ are defined by

$$\psi_{n,m}(x) = \begin{cases} \sqrt{(m+\frac{1}{2})}2^{k/2}P_m(2^k x - \hat{n}), & \text{for } \dfrac{\hat{n}-1}{2^k} \le x < \dfrac{\hat{n}+1}{2^k}, \\ 0, & \text{otherwise} \end{cases} \quad (24.3)$$

where $\hat{n} = 2n-1, n = 1,2,\dots,2^{k-1}; k \ge 1$. The polynomials $P_0(x)$ and $P_1(x)$ are given as

$$P_0(x) = 1$$

$$P_1(x) = x$$

and all higher degree Legendre polynomials are obtained using recursive formula

$$P_{m+1}(x) = \left(\frac{2m+1}{m+1}\right)xP_m(x) - \left(\frac{m}{m+1}\right)P_{m-1}(x), \text{ for } m = 1,2,3,\dots.$$

Legendre wavelets have compact support and form a complete orthonormal basis for $L^2[0,1)$[3, 12].

24.2.1 Approximation of Function

Any square integrable function $f(x)$ defined over $[0,1)$ can be written as the sum of Legendre wavelets as

$$f(x) = \sum_{n=1}^{\infty}\sum_{m=0}^{\infty} c_{n,m}\psi_{n,m}(x), \quad (24.4)$$

where $\psi_{n,m}(x)$ are Legendre wavelets and $c_{n,m}$ are Legendre wavelet coefficients given by

$$c_{n,m} = \langle f(x), \psi_{n,m}(x) \rangle = \int_0^1 f(x)\psi_{n,m}(x)dx. \quad (24.5)$$

After truncating the series at $m = 0,1,2,\dots,M-1$ and $n = 1,2,\dots,2^{k-1}$, Equation (24.4) can be written as the sum of series of $2^{k-1}M$ terms.

$$f(x) \simeq \sum_{n=1}^{2^{k-1}}\sum_{m=0}^{M-1} c_{n,m}\psi_{n,m}(x), \quad (24.6)$$

Writing it in matrix notation,

$$f(x) = C\psi^T(x),$$

where C and $\psi(x)$ are $1 \times 2^{k-1}M$ matrices and are given by

$$C = [c_{1,0}, c_{1,1}, \dots, c_{1,M-1}, c_{2,0}, \dots, c_{2,M-1}, \dots, c_{2^{k-1},0} \dots c_{2^{k-1},M-1}], \quad (24.7)$$

$$\psi(x) = [\psi_{1,0}(x), \psi_{1,1}(x), \dots, \psi_{1,M-1}(x), \psi_{2,0}(x), \dots, \psi_{2,M-1}(x), \dots$$
$$\dots, \psi_{2^{k-1},0}(x), \dots, \psi_{2^{k-1},M-1}(x)]. \quad (24.8)$$

24.2.2 Operational Matrix of Integration

Let $\psi(x) = \{\psi_{1,0}, \ldots, \psi_{1,m}, \psi_{2,0}, \ldots, \psi_{2,m}, \ldots, \psi_{n,0} \ldots \psi_{n,m}\}$ be the Legendre wavelet. Then integration of $\psi(x)$ over interval $[0, x]$ is given by

$$\int_0^x \psi(x)dx = P\psi(x),$$

where P is square matrix of order $2^{k-1}M$ and is given by

$$P = \frac{1}{2^k} \begin{bmatrix} L & F & F & \cdots & F \\ 0 & L & F & \cdots & F \\ 0 & 0 & L & \cdots & F \\ \vdots & \vdots & \vdots & \ddots & \vdots \\ 0 & 0 & \cdots & 0 & L \end{bmatrix}.$$

The matrix P is known as operational matrix of integration. The matrix L and F are of order $M \times M$ and are given as

$$L = \begin{bmatrix} 1 & \frac{1}{\sqrt{3}} & 0 & \cdots & 0 & 0 & 0 \\ -\frac{\sqrt{3}}{3} & 0 & \frac{\sqrt{3}}{3\sqrt{5}} & 0 & \cdots & 0 & 0 \\ 0 & -\frac{\sqrt{5}}{5\sqrt{3}} & 0 & \cdots & 0 & 0 & 0 \\ \vdots & \vdots & \vdots & \vdots & \vdots & \vdots & \vdots \\ 0 & 0 & 0 & \cdots & -\frac{\sqrt{2M-3}}{(2M-3)\sqrt{2M-5}} & 0 & \frac{\sqrt{2M-3}}{(2M-3)\sqrt{2M-1}} \\ 0 & 0 & 0 & \cdots & 0 & -\frac{\sqrt{2M-1}}{(2M-1)\sqrt{2M-3}} & 0 \end{bmatrix}$$

$$F = \begin{bmatrix} 2 & 0 & \cdots & 0 \\ 0 & 0 & \cdots & 0 \\ \vdots & \vdots & \vdots & \vdots \\ 0 & 0 & \cdots & 0 \end{bmatrix}.$$

24.3 Numerical Technique

In the Volterra problem the solution $f(x)$, $x \in [0,1)$ on interval $[0, x]$ depends only on the values of $g(x)$ on this interval but for the non-Volterra problem the values on interval $(x, 1]$ are also used i.e., it also uses future data values. The classical regularization method converts the Volterra problem into a non-Volterra problem. To overcome the difficulty arising in classical regularization techniques, local regularization methods are used. We have used a local regularization method to find the approximate solution to the ill-posed Volterra problem. A brief introduction to local regularization methods is given in this section.

To apply a local regularization let us first extend the domain of x from $[0,1)$ to $[0,1+r)$, where $r > 0$ is a small fixed constant. Again assume that Equation (24.1) holds on extended domain, then the solution $f(x)$ satisfies

$$\int_0^{x+p} K(x+p,t)f(t)dt = g(x+p), \ x \in [0,1), \ p \in [0,r).$$

Splitting the integration at x,

$$\int_0^x K(x+p,t)f(t)dt + \int_x^{x+p} K(x+p,t)f(t)dt = g(x+p) \quad (24.9)$$

$$\int_0^x K(x+p,t)f(t)dt + \int_0^p K(x+p,x+t)f(t+x)dt = g(x+p) \quad (24.10)$$

Integrating w.r.t to 'p' and interchanging order of integration in the first integral, we get

$$\int_0^x \int_0^r K(x+p,t)dpf(t)dt + \int_0^r \int_0^p K(x+p,x+t)f(t+x)dtdp = \int_0^r g(x+p)dp, \quad (24.11)$$

The idea of local regularization is to assume that function $f(x)$ is constant over an extended local interval $[x, x+r)$, where r is the length of local interval and is known as a regularization parameter. Replacing $f(x+t)$ by $f(x)$ in second integral, Equation (24.11) becomes

$$\int_0^x \int_0^r K(x+p,t)dpf(t)dt + f(x)\int_0^r \int_0^p K(x+p,x+t)dtdp = \int_0^r g(x+p)dp, \quad (24.12)$$

Equation (24.12) represents a regularized form of Equation (24.1) which is a second kind integral equation. Like other regularization methods, it also converts first-kind integral equations into second-kind integral equations but it also retains the Volterra properties of the original problem. For more details about this regularization method and its convergence, see [4] and references therein.

Now we will apply Legendre wavelets to Equation (24.12). Expanding $f(x) \in L^2[0,1)$ as sum of series of Legendre wavelets, we get

$$f(x) = \sum_{n=1}^{2^{k-1}} \sum_{m=0}^{M-1} c_{n,m}\psi_{n,m}(x). \quad (24.13)$$

Putting this values of $f(x)$ in Equation (24.12), we get

$$\int_0^x \int_0^r K(x+p,t)dp \sum_{n=1}^{2^{k-1}} \sum_{m=0}^{M-1} c_{n,m}\psi_{n,m}(t)dt + \quad (24.14)$$

$$\sum_{n=1}^{2^{k-1}} \sum_{m=0}^{M-1} c_{n,m}\psi_{n,m}(x)\int_0^r \int_0^p K(x+p,x+t)dtdp = \int_0^r g(x+p)dp,$$

$$\sum_{n=1}^{2^{k-1}} \sum_{m=0}^{M-1} c_{n,m}\left(\int_0^x L(x,t)\psi_{n,m}(t)dt + \psi_{n,m}(x)\alpha\right) = G(x), \quad (24.15)$$

where $L(x,t) = \int_0^r K(x+p,t)dp$, $G(x) = \int_0^r g(x+p)dp$ and $\alpha = \int_0^r \int_0^p K(x+p,x+t)dtdp$.

$$\sum_{n=1}^{2^{k-1}} \sum_{m=0}^{M-1} c_{n,m} \left(H_{n,m}(x) + \psi_{n,m}(x)\alpha \right) = G(x), \tag{24.16}$$

where $H_{n,m}(x) = \int_0^x L(x,t)\psi_{n,m}(t)dt$. Collocation points are defined as

$$x_i = \frac{2i-1}{2p}, \quad i = 1, 2, \ldots, p,$$

where $p = 2^{k-1}M$. Satisfying Equation (24.15) at collocation points x_i, yields a linear system of equations. In matrix notation, we write

$$C(H + \Psi\alpha) = G,$$

where $G = [G(x_0), G(x_1), G(x_2), \ldots, G(x_p)]^T$,

$$\Psi = \begin{bmatrix} \psi_{1,0}(x_1) & \psi_{1,0}(x_2) & \cdot & \cdot & \cdot & \cdot & \cdot & \psi_{1,0}(x_p) \\ \psi_{1,1}(x_1) & \psi_{1,1}(x_2) & & \cdot & \cdot & \cdot & \cdot & \psi_{1,1}(x_p) \\ \cdot & \cdot & \cdot & \cdot & \cdot & \cdot & \cdot & \\ \psi_{1,M-1}(x_1) & \psi_{1,M-1}(x_2) & \cdot & \cdot & \cdot & \cdot & \cdot & \psi_{1,M-1}(x_p) \\ \psi_{2,0}(x_1) & \psi_{2,0}(x_2) & \cdot & \cdot & \cdot & \cdot & \cdot & \psi_{2,0}(x_p) \\ \cdot & \cdot & \cdot & \cdot & \cdot & \cdot & \cdot & \cdot \\ \cdot & \cdot & \cdot & \cdot & \cdot & \cdot & \cdot & \cdot \\ \psi_{2^{k-1},M-1}(x_1) & \psi_{2^{k-1},M-1}(x_2) & \cdot & \cdot & \cdot & \cdot & \cdot & \psi_{2^{k-1},M-1}(x_p) \end{bmatrix}$$

and

$$H = \begin{bmatrix} H_{1,0}(x_1) & H_{1,0}(x_2) & \cdot & \cdot & \cdot & \cdot & \cdot & H_{1,0}(x_p) \\ H_{1,1}(x_1) & H_{1,1}(x_2) & & \cdot & \cdot & \cdot & \cdot & H_{1,1}(x_p) \\ \cdot & \cdot & \cdot & \cdot & \cdot & \cdot & \cdot & \\ H_{1,M-1}(x_1) & H_{1,M-1}(x_2) & \cdot & \cdot & \cdot & \cdot & \cdot & H_{1,M-1}(x_p) \\ H_{2,0}(x_1) & H_{2,0}(x_2) & \cdot & \cdot & \cdot & \cdot & \cdot & H_{2,0}(x_p) \\ \cdot & \cdot & \cdot & \cdot & \cdot & \cdot & \cdot & \cdot \\ \cdot & \cdot & \cdot & \cdot & \cdot & \cdot & \cdot & \cdot \\ H_{2^{k-1},M-1}(x_1) & H_{2^{k-1},M-1}(x_2) & \cdot & \cdot & \cdot & \cdot & \cdot & H_{2^{k-1},M-1}(x_p) \end{bmatrix}.$$

$$C = G(H + \Psi\alpha)^{-1}. \tag{24.17}$$

The wavelet coefficients $C = \{c_{n,m}\}$ are calculated by using Equation (24.17). Then using the values of c_{nm}, the function $f(x)$ can be obtained by using Equation (24.13).

Nonlinear Volterra integral equations can also be handled using the proposed method. First, we will convert a nonlinear equation into a linear Volterra integral equation by substituting

$$Y(f(x)) = v(x).$$

Then after finding solution to this linear equation using the method explained above, the value of $f(x)$ can be obtained by

$$f(x) = Y^{-1}(v(x)).$$

To check the accuracy of the proposed method, we use root mean square error (RMS) and maximum absolute error (MAE) given by

$$\text{RMS} = \left(\frac{1}{N} \sum_{i=1}^{N} |f(x_i) - f_{ex}(x_i)|^2 \right)^{1/2},$$

$$\text{MAE} = \max_{1 \le i \le N} |f(x_i) - f_{ex}(x_i)|,$$

where $f(x_i)$ and $f_{ex}(x_i)$ are numerical and exact solution and N is the number of points.

Now the question is what value of r should be choosen to find accurate results. We start with $r = 0.1$ and decrease the value of α until the numerical solution converges to the exact solution of integral equation. We have choosen that value of r for which the root mean square (RMS) attained is minimum. For practical problems, we do not know the exact solution and so finding the optimal value of α is difficult. In that case, the value of the regularization parameter is chosen depending upon the a priori information about the solution.

24.4 Error Analysis

We present the following convergence theorem for Legendre wavelet basis.

Theorem 24.1 [11] Let $f(x)$ be a function defined on $[0,1)$ with bounded second derivative, say $|f''(x)| \le M$, and $\sum_{n=1}^{\infty} \sum_{m=0}^{\infty} c_{n,m} \psi_{n,m}(x)$ be its infinite Legendre wavelets expansion, then

$$|c_{n,m}| \le \frac{\sqrt{12}M}{(2n)^{5/2}(2M-3)^2}.$$

This means the Legendre wavelets series converges uniformly to $f(x)$ and

$$f(x) = \sum_{n=1}^{\infty} \sum_{m=0}^{\infty} c_{n,m} \psi_{n,m}(x).$$

Theorem 24.2 [11] Suppose $f(x)$ be a continuous function defined on $[0,1)$, with a second derivative $f''(x)$ bounded by M, then we have the following accuracy estimation

$$\| f(x) - f_{n,m}(x) \| = \| e_n \|_2 \leq \left(\frac{3M^2}{2} \sum_{n=1}^{\infty} \sum_{m=M}^{\infty} \frac{1}{n^5(2m-3)^4} \right.$$

$$\left. + \frac{3M^2}{2} \sum_{n=2^{k-1}+1}^{\infty} \sum_{m=0}^{M-1} \frac{1}{n^5(2m-3)^4} \right)^{\frac{1}{2}}.$$

(24.18)

We have applied Legendre wavelets to Equation (24.12) which is a second-kind Volterra integral equation. Writing Equation (24.12) in simplified form we get

$$\int_0^x K(x,t)f(t)dt + \alpha f(x) = g(x)$$

(24.19)

Let $f_{n,m}(x)$ be the approximate value of function $f(x)$ obtained by truncating the Legendre wavelet for some suitable value of M and k. If we use the approximate value of $f(t)$ in Equation (24.19), the function $g(x)$ will also differ from its actual value (let us denote it by $g_{n,m}(x)$), we have

$$\int_0^x K(x,t)f_{n,m}(t)dt + f_{n,m}(x)\alpha = g_{n,m}(x)$$

(24.20)

Subtracting Equation (24.20) from Equation (24.19), we get

$$g_{n,m}(x) - g(x) = \int_0^x K(x,t)(f_{n,m}(t) - f(t))dt + (f_{n,m}(x) - f(x))\alpha$$

which is again a second-kind integral equation. Taking L^2 norm, we have

$$\| g(x) - g_{n,m}(x) \| = \| \int_0^x K(x,t)(f_{n,m}(t) - f(t))dt + (f_{n,m}(x) - f(x))\alpha \|$$

$$\leq \| \int_0^x K(x,t)(f_{n,m}(t) - f(t))dt \| + \| (f_{n,m}(x) - f(x))\alpha \|$$

$$\leq \int_0^x \| K(x,t)(f_{n,m}(t) - f(t)) \| dt + \| (f_{n,m}(x) - f(x))\alpha \|$$

$$\leq \int_0^x \| K(x,t) \| \| (f_{n,m}(t) - f(t)) \| dt + \| (f_{n,m}(x) - f(x)) \| | \alpha |$$

$$= \| (f_{n,m}(x) - f(x)) \| (\int_0^x \| K(x,t) \| dt + | \alpha |)$$

Assuming $K(x, t)$ to be bounded by U on $[0,1)$, and using Equation (24.18), we have

$$\| g(x) - g_{n,m}(x) \| \leq (| \alpha | + U) \left(\frac{3M^2}{2} \sum_{n=1}^{\infty} \sum_{m=M}^{\infty} \frac{1}{n^5(2m-3)^4} \right.$$

$$\left. + \frac{3M^2}{2} \sum_{n=2^{k-1}+1}^{\infty} \sum_{m=0}^{M-1} \frac{1}{n^5(2m-3)^4} \right)^{\frac{1}{2}}$$

(24.21)

Equation (24.21) gives the upper bound for the error. We can see that as the value of m and n become large, the function $g_{n,\,m}(x)$ converges to function $g(x)$.

24.5 Numerical Examples

We will present the numerical solution to some problems of Volterra integral equations of the first kind. The numerical results presented in this chapter are obtained using $k=3$, $M=3$, i.e., 12 collocation points. All the computations are performed with help of Matlab.

Example 5.1 Solve the following integral equation

$$\int_0^x \cos(x-t)f(t)dt = x\sin x,\ 0 \leq x < 1.$$

The exact solution is $f(x) = 2\sin x$.

The numerical solution converges to the exact solution for $r = 10^{-3}$. The RMS and MAE obtained are 9.90872e-05 and 1.57312e-04 respectively. Table 24.1 shows the numerical results for Example 1. The error obtained is less than the method discussed in [2] using 64 and 128 collocation points. Maleknejad [10] used Haar wavelets with 128 elements and MAE obtained is 0.01, which is very large as compared to our method where we have used 12 collocation points. The results obtained by the proposed method are better as compared to the methods presented in [2] and [10].

Example 5.2 Solve the following first kind Volterra integral equation:

$$\int_0^x e^{x+t}f(t)dt = xe^x,\ 0 \leq x < 1.$$

The exact solution is $f(x) = e^{-x}$.

TABLE 24.1

Numerical Results for Example 1

x	Exact Solution $f_{ex}(x)$	Approximate Solution $f(x)$	Error $|f_{ex}(x) - f(x)|$
0	0	-0.00007965	7.96550793e-05
0.1	0.19966683	0.19963722	2.96138005e-05
0.2	0.39733866	0.3974936	1.54965758e-04
0.3	0.59104041	0.5908831	1.57311985e-04
0.4	0.77883668	0.77886882	3.21418644e-05
0.5	0.95885107	0.95880658	4.44903973e-05
0.6	1.12928494	1.12925794	2.70043350e-05
0.7	1.28843537	1.28856879	1.33420386e-04
0.8	1.43471218	1.43457363	1.38551225e-04
0.9	1.56665381	1.56669203	3.82122340e-05

The exact and approximate solutions are compared in Table 24.2. For $r=10^{-7}$, the RMS value obtained is 4.25832179e-05. In [2], 32 and 64 collocation points are used to solve the equation and the error obtained is of the order 10^{-4}, which is larger than the error obtained by our method which uses only 12 collocation points. Also, the MAE obtained is 6.66114205e-04 whereas the MAE in [10] using 128 elements is 0.005. So, we can say that the present method works better than the methods in [2] and [10].

Example 5.3 Consider the following integral equation:

$$\int_0^x \cos(x-t)f''(t)dt = 2\sin x,\ 0\le x<1,$$

with initial conditions $y(0)=0,\ y'(0)=0$.
The exact solution to this equation is given by $f(x)=x^2$.
The RMS and MAE for $r=10^{-3}$ are given as

$$RMS = 1.09713545e-18 \text{ and } MAE = 3.46944695e-18.$$

Table 24.3 shows the numerical results using the proposed method which are in good agreement with the exact solution. In [13] this problem is solved using different collocation points and MAE obtained are of the order 10^{-16} whereas our method gives a maximum absolute error of the order 10^{-18}. We can see that our method gives more accuracy than the Haar wavelet method given in [13].

Example 5.4 Consider the following integral equation:

$$\int_0^x \cos(x-t)f''(t)dt = 6(1-\cos x),\ 0\le x<1,$$

with initial conditions $y(0)=0,\ y'(0)=0$ and exact solution is given by $f(x)=x^3$.
For $r=10^{-5}$, the RMS value is 4.15754221e-04. The results obtained are shown in Table 24.4 which shows that the method gives good results. Singh et al. [13] have solved this example using Haar wavelet with 16 collocation points and the MAE obtained is 2.1e-03 whereas the MAE in our method using 16 collocation points is 7.81250052e-04.

TABLE 24.2

Numerical Results for Example 2

x	Exact Solution $f_{ex}(x)$	Approximate Solution $f(x)$	Error $\|f_{ex}(x)-f(x)\|$
0	1	0.99994741	5.25849051e-05
0.1	0.90483742	0.90482634	1.10749066e-05
0.2	0.81873075	0.81879610	6.53475231e-05
0.3	0.74081822	0.74075160	6.66114204e-05
0.4	0.67032004	0.67033623	1.61860716e-05
0.5	0.60653065	0.60654908	1.84196344e-05
0.6	0.54881163	0.54879850	1.31330883e-05
0.7	0.49658530	0.49663733	5.20355253e-05
0.8	0.44932896	0.44927604	5.29170127e-05
0.9	0.40656965	0.40658567	1.60138249e-05

TABLE 24.3

Numerical Results for Example 3

x	Exact Solution $f_{ex}(x)$	Approximate Solution $f(x)$	Error $\lvert f_{ex}(x) - f(x) \rvert$
0	0.0	0.0	3.46944695e-18
0.1	0.01	0.01	0.0
0.2	0.04	0.04	0.0
0.3	0.09	0.09	0.0
0.4	0.16	0.16	0.0
0.5	0.25	0.25	0.0
0.6	0.36	0.36	0.0
0.7	0.49	0.49	0.0
0.8	0.64	0.64	0.0
0.9	0.81	0.81	0.0

TABLE 24.4

Numerical Results for Example 4

x	Exact Solution $f_{ex}(x)$	Approximate Solution $f(x)$	Error $\lvert f_{ex}(x) - f(x) \rvert$
0	0	0.00078125	7.81250000e-04
0.1	0.001	0.00078125	2.18749991e-04
0.2	0.008	0.00828125	2.81250018e-04
0.3	0.027	0.02671875	2.81249968e-04
0.4	0.064	0.06421875	2.18750042e-04
0.5	0.125	0.12578125	7.81250052e-04
0.6	0.216	0.21578125	2.18749938e-04
0.7	0.343	0.34328125	2.81250074e-04
0.8	0.512	0.51171875	2.81249910e-04
0.9	0.729	0.72921875	2.18750103e-04

Example 5.5 Solve the following integral equation:

$$\int_0^x \exp(x-t) f(t)dt = x \ \ 0 \le x < 1,$$

with the exact solution $f(x) = 1 - x$.

The minimum value of RMS is achieved for $r = 10^{-6}$ and its value is 3.16342271e-12. The MAE obtained in [13] using Haar wavelet with 16 collocation points is 4.5e-04 whereas in our method the MAE obtained is 6.89326373e-12 using 12 collocation points. The accuracy achieved by the present method is much higher than the method presented in [13]. The numerical results are presented in Table 24.5.

Example 5.6 Solve the following integral equation:

$$\int_0^x e^{(x-t)} \ln(f(t))dt = e^x - x - 1, \ 0 \le x < 1.$$

TABLE 24.5

Numerical Results for Example 5

x	Exact Solution $f_{ex}(x)$	Approximate Solution $f(x)$	Error $\| f_{ex}(x) - f(x) \|$
0	1.0	0.99999999	6.09234884e-12
0.1	0.9	0.90000000	9.84123893e-13
0.2	0.8	0.79999999	4.49462772e-12
0.3	0.7	0.70000000	1.80295778e-12
0.4	0.6	0.59999999	6.74704736e-13
0.5	0.5	0.49999999	6.89326373e-12
0.6	0.4	0.40000000	1.08926201e-12
0.7	0.3	0.29999999	1.66601177e-12
0.8	0.2	0.20000000	2.01687000e-12
0.9	0.1	0.09999999	7.95899457e-13

The exact solution is $f(x) = e^x$.

Numerical results obtained by applying proposed method for $r = 10^{-5}$ are shown in Table 24.6. The RMS and MAE values are given as

$$RMS = 4.66296850e - 10 \text{ and } MAE = 1.14091358e - 09.$$

We can see that the results obtained are very accurate. The MAE in [13] is of the order 10^{-7} using 1024 points whereas our method gives accuracy of the order 10^{-9} using 12 collocation points. As the error is very low as compared to [13], we can say that our method is more accurate.

Example 5.7 Solve the following integral equation:

$$\int_0^x (\sin(x-t)+1)\cos(f(t))dt = \frac{x\sin x}{2} + \sin x, \ 0 \le x < 1,$$

whose exact solution is $f(x) = x$.

TABLE 24.6

Numerical Results for Example 6

x	Exact Solution $f_{ex}(x)$	Approximate Solution $f(x)$	Error $\| f_{ex}(x) - f(x) \|$
0	1.0	1.00000000	6.09190697e-10
0.1	1.10517091	1.10517091	1.08664632e-10
0.2	1.22140275	1.22140275	1.76841208e-10
0.3	1.34985880	1.34985880	2.42655007e-10
0.4	1.49182469	1.49182469	1.00434327e-10
0.5	1.64872127	1.64872127	1.14091358e-09
0.6	1.82211880	1.82211880	1.98843386e-10
0.7	2.01375270	2.01375270	3.34571037e-10
0.8	2.22554092	2.22554092	4.48094006e-10
0.9	2.45960311	2.45960311	1.92915461e-10

TABLE 24.7

Numerical Results for Example 7

x	Exact Solution $f_{ex}(x)$	Approximate Solution $f(x)$	Error $\mid f_{ex}(x) - f(x) \mid$
0	0	0.00078202	7.82023733e-04
0.1	0.1	0.10002678	2.67860935e-05
0.2	0.2	0.20005707	5.70792487e-05
0.3	0.3	0.29999642	3.57536325e-06
0.4	0.4	0.39998663	1.33609094e-05
0.5	0.5	0.49974519	2.54809881e-04
0.6	0.6	0.60001056	1.05615279e-05
0.7	0.7	0.70003245	3.24508884e-05
0.8	0.8	0.79997837	2.16264481e-05
0.9	0.9	0.89998556	1.44378332e-05

Comparison of the exact and the numerical solutions is given in Table 24.7. RMS and MAE for $r = 0.005$ are 2.61246394e-04 and 7.82023733e-04 respectively. However, if we calculate error at collocation points then MAE = 1.14e-04 which is lower than the method discussed in [13] where MAE obtained is 3.1e-004.

24.6 Conclusion

An approximate solution to Volterra integral equations of the first kind is obtained by combining a local regularization method with Legendre wavelet. The numerical solution to some of the Volterra problems is approximated using the proposed method and the results obtained are in good agreement with exact solutions. Also, the numerical results obtained by the presented method are compared with existing methods which shows that our method is better as compared to other methods. Only a small number of basis functions are required to achieve satisfactory results. Also the concept of regularization is used, which leads to stable results. The described method is very simple and reliable. All the calculations are done using $M = 3$, $k = 3$ and very good results are obtained. However, the values of k and M can be increased to get more accuracy.

Acknowledgment

The authors are grateful to the University Grants Commission, India, for providing financial assistance for the preparation of the manuscript.

References

1. E. Babolian, A. S. Shamloo, Numerical solution of Volterra integral and integro-differential equations of convolution type by using operational matrices of piecewise constant orthogonal functions, *J. Comp. Appl. Math.*, **214** (2008), pp. 495–508.

2. E. Babolian, Z. Masouri, Direct method to solve Volterra integral equation of the first kind using operational matrix with block-pulse functions, *J. Comp. Appl. Math.*, **220** (1–2) (2008), pp. 51–57.

3. J. Kumar, P. Manchanda, Pooja, Numerical solution of Fredholm integral equations of the first kind using Legendre wavelet collocation method, *Int. J. Pure Appl. Math.*, **117** (1) (2017), pp. 33–43.

4. P. K. Lamm, A survey of regularization methods for first-kind Volterra equations, in *Surveys on Solution Methods for Inverse Problems*, D. Colton, H. W. Engl, A. Louis, J. R. McLaughlin and W. Rundell, eds., Springer, Vienna, 2000.

5. P. K. Lamm, Variable-smoothing local regularization methods for first-kind integral equations, *Inverse Probl.*, **19** (1) (2003), pp. 195–216.

6. P. K. Lamm, Full convergence of sequential local regularization methods for Volterra inverse problems, *Inverse Probl.*, **21** (3) (2005), pp. 785–803.

7. P. K. Lamm, T. L. Scofield, Local regularization methods for the stabilization of linear ill-posed equations of Volterra type, *Numer. Funct. Anal. Optim.*, **22** (7–8) (2008), pp. 913–940.

8. P. K. Lamm, Z. Dai, On local regularization methods for linear Volterra equations and nonlinear equations of Hammerstein type, *Inverse Probl.*, **21** (2005), pp. 1773–1790.

9. K. Maleknejad, B. Rahimi, Modification of block pulse functions and their application to solve numerically Volterra integral equation of the first kind, *Commun. Nonlinear Sci. Numer. Simul.*, **6** (6) (2011), pp. 2469–2477.

10. K. Maleknejad, R. Mollapourasl, M. Alizadeh, Numerical solution of Volterra type integral equation of the first kind with wavelet basis, *Appl. Math. Comp.*, **61** (2011), pp. 2821–2828.

11. F. Mohammadi, A computational wavelet method for numerical solution of stochastic Volterra-Fredholm integral equations, *Wavelets and Linear Algebr.*, **3** (2016), pp. 13–25.

12. P. K. Sahu, S. S. Ray, Legendre wavelets operational method for the numerical solutions of nonlinear Volterra integro-differential equations system, *Appl. Math. Comp.*, **256** (2015), pp. 715–723.

13. I. Singh, S. Kumar, Haar wavelet method for some nonlinear Volterra integral equations of the first kind, *J. Comp. Appl. Math.*, **292** (2016), pp. 541–552.

14. G. M. Wing, J. D. Zahrt, *A Primer on Integral Equations of the First Kind: The Problem of Deconvolution and Unfolding*, Society for Industrial and Applied Mathematics, Philadelphia, 1991.

25

Computational Linguistic Analysis of Retail E-Commerce

Rashmi Bhardwaj and Aashima Bangia

CONTENTS

25.1 Introduction...371
25.2 Methodology 372
    25.2.1 Text Data Handling 372
    25.2.2 Word Cloud..373
    25.2.3 Designing Text Matrix 374
    25.2.4 Calculation of Reduction Score (R)..374
25.3 Mathematical Case Study of the E-Commerce Market Reviews 375
    25.3.1 Algorithm...375
    25.3.2 Flowchart 376
25.4 Results and Discussion 377
Acknowledgment...378
References...378

25.1 Introduction

The computational perspective of scientific and engineering discipline in relation to understanding all forms of language and constructing artifacts that useably process and produce language is computational linguistics. As we all know, language is a way to express our thinking, therefore, a computational aspect also provides insight into thinking and intelligence. As language is the most obvious means of communication, linguistic-sensitive processors need to integrate human interactions with machines and related procedures. This would, in a true sense, meet requirements, as it would make the vast textual and other resources of the internet easily accessible.

The conjectural objectives of computational linguistics are about devising syntactic and semantic structures for portraying languages using techniques aiding computationally manageable implementations of grammatical, semantic analyses; the unearthing of handling practices and learning moralities which expose the structural and arithmetical properties of linguistics; and the advancement of cognitive neuron-scientific simulating models of various possible ways in which dispensing and learning might occur in the brain.

The hands-on objectives of this are broad and varied. Some of the most obvious are about the competent retrieval of text on any favorable matter; machine translation (MT); question-answering (QA), that is, alternating from simple realistic questions to others demanding implication with descriptive responses; summaries of text; scrutiny of either written or spoken material for various sentiments or psychosomatic traits; dialogues for undertaking particular tasks; and eventually, the conception of computational classifications with human-like proficiency in acquiring language, and in gaining knowledge from manuscripts.

Computational linguistics is very important, as human knowledge is expressed in languages and sentiments. Sentiments are the views or opinions expressed about a person, situation, place, or thing. Social sentimentality is all about categorizing the triumphs and conflicts of the human race and indicates an assurance to human contentment within the laws of nature. Human thoughts have found expression in various ideas all through the centuries and have continually put a lot of stress on the fundamental value of life. However, the manifestation of human emotion is primarily narrowed to the territory of notions.

Simulation of language helps to decide the emotional tendency hidden in the series of words that is devoted to understanding the outlooks, opinions, and emotions expressed. The circumstantial quarrying of text detects and extracts subjective evidence in the the source and helps the corporates to catch the social nerve of their brand, product, or service while observing online banter. The most obvious text cataloging tool that analyses an arriving message and informs about the underlying sentiment is the constructive, destructive, or neutral study of sentiments. One can input a sentence of choice and measure the hidden sentimentality.

The capability of algorithms of simulating text has become more efficient with innovations in deep learning. The advanced artificial intelligence techniques have proven to be an effective tool for carrying out study. It is essential to sort inbound consumer conversation about a brand with the help of vital facets of goods and services that consumers consider and the underlying meanings and reactions related to specific aspects.

Anderson and Bower [1] discussed in detail associative human memory. Anderson [2-4] gave a detailed explanation of phonology and morphous morphology. Angluin [5] surveyed for selected bibliography under the study of computational learning theory. Antonelli [6] discussed the vigilant theory of defeasible significance for default reasoning via the view of general extension. Aronoff [7] described word formation in generative grammar. Austin [8] discussed the possible analysis that can be made with words. Barzilay *et al* [9] discussed conjecturing policies for sentence assembling in multi-document summarization. Bird and Ellison [10] discussed one-level phonology. Bloch and Trager [11] gave an outline for linguistic analysis. Chao [12] studied graphic and phonetic aspects of language with mathematical codes. Charniak [13] discussed the statistical inferences of language. Collins *et al* [14] explored parallel tag clouds and analyzed faceted text corpora. Cui *et al* [15] described the dynamic word cloud visualization of context-preserving text. Dumais [16] studied latent semantic analysis. Kornai [17] elaborately studied the mathematical aspects of language. Makkonen *et al* [18] discussed simple semantics in topic detection and tracking. White *et al* [19] discussed information extraction for multi-document summarization. Wu *et al* [20] designed semantic preserving clouds by stratum figurine. Xie *et al* [21] studied a layered mixture model for the discovery of patterns.

As per the literature survey carried out through various sources, no articles reported the word clouds or bag of clouds of sentimentality for the retail reviews recorded through ee-commerce site. None of the authors have studied the goods' reviews based on large text corpora yet.

25.2 Methodology

25.2.1 Text Data Handling

The knack of extracting acumens from shared information is being widely embraced by organizations around the world. However, exploration of media streams is usually restrained to fundamental sentimentality analysis and count based measures. In this chapter, we study sentiment across all the customer reviews by separating the text into word clouds.

Text data can be large and can contain lots of noise which can have undesirable effects on analysis. Text data can contain the following forms of noises:

(a) Variations in case i.e. lowercase and uppercase
(b) Variations in word forms i.e. verb forms, adjective forms etc.
(c) Words that add noise, stop words such as "the," "in," and "of" etc.
(d) Punctuation and special characters
(e) HTML and XML tags

The text data in such formed word cloud sis a visualization of word occurrence in a given document as a weighted list of words. The practice of visualizing the content on specific issues has recently gained popularity on various platforms.

25.2.2 Word Cloud

Display of data using font size and color (optional) to indicate statistical or arithmetical synthesis of information is known as word cloud. Analogous to tag cloud, it displays data as a weighted list of words in an innovative visual representation of text data. Classically, it depicts keyword metadata on different websites, and visualizing free forms of manuscript data. Different font sizes or colors highlight the importance of each tag in different orderss. This procedure is used to swiftly identify the most noticeable terms and for allocating the terms an alphabetic order to define their relative importance. In the case of website navigation, the expressions are hyperlinked to the ones associated with the tag. These word clouds illustrate word frequency analysis applied to some raw text data from various sources, and a preprocessed version of the same text data. The bag of words after all the cleaning is converted into a matrix. The matrices that encompass all occurrences of words are generated as the words are embodied in rows and the number of documents in columns.

The type size of a word existing in a word cloud is decided on the basis of its occurrence in the text under considerations. Word clouds of various classifications have frequencies of words that resemble the number of word admittances that are apportioned for classification. Font sizes, from one to some maximum font size, are specified for lesser frequencies, whereas a ruling should be made for greater frequencies. Under an undeviating regularization, the weight c_i of a describing word is mapped to a size scale of 1 through y_{max}, where c_{min} and c_{max} are defining the range of obtainable weights. This computation is carried out through the following formula:

$$F_i = \begin{cases} \left\lceil \dfrac{y_{max}(c_i - c_{min})}{c_{max} - c_{min}} \right\rceil & \text{for } c_i > c_{min}; \\ 1 & \text{else} \end{cases}$$

with
  F_i = display fontsize
  y_{max} = maximum fontsize
  c_i = count
  c_{min} = minimum count
  c_{max} = maximum count

The number of indexed items per descriptor is customarily circulated in accordance with power law for larger series of values, a logarithmic exemplification is found more suitable. Applications of words clouds include text parsing and straining out unaccommodating ones such as common words, numbers, and punctuation. Various programs used to generate word clouds are based on users' text data taken as input data.

25.2.3 Designing Text Matrix

Algorithm for design:

1. Represent each document as a vector D_i, with D_{iw}—count of word w in document i.
2. The word cloud, v_i is an arrangement consisting of:

$$v_i = \left(B_i,\ \{\psi_{iw}\},\ \{\vartheta_{iw}\},\ \{\alpha_{iw}\} \right)$$

   B_i—set of words that are to be displayed in cloud i; ψ_{iw}—position vector for each word; ϑ_{iw}—color representation of the words; α_{iw}—size of the words.
3. Define: $\vec{\psi}_{iw} = (\delta_{iw}, \gamma_{iw})$ as the position vector of w in the cloud v_i, for any word $w \in W_i$.
4. Let $\rho_i = \{\vec{\psi}_{iw} \mid w \in B_i\}$ denote the set of all word whereabouts in v_i.
5. The word clusters after sanitizing are transformed into a matrix. This matrix embeds all incidences of words that are generated,, with the words as rows and the document numbers as columns.

$$\text{Text Matrix}\,(T_m) = \begin{array}{ccccc} D_1 & D_2 & D_3 & D_4 & D_5 \end{array} \\ \begin{bmatrix} w_1 & w_2 & w_3 & w_4 & w_5 \\ w_1 & w_2 & w_3 & w_4 & w_5 \\ \vdots & \vdots & \vdots & \vdots & \vdots \\ \vdots & \vdots & \vdots & \vdots & \vdots \end{bmatrix}$$

   D_i: *ith document;*
   w_i: *ith word in the respective documents.*

6. The unlikeliness component, f_v among clouds can be defined as:

$$f_v(\sigma_i, \sigma_j) = \sum_{w \in B} (\alpha_{iw} - \alpha_{jw})^2 + \kappa \sum_{w \in B_i \cap B_j} (\delta_{iw} - \delta_{jw})^2 + (\gamma_{iw} - \gamma_{jw})^2,\ \text{where}\ \kappa \geq 0\ \kappa\text{-}$$ regulates weights assigned toward differences of sizes of the font and in locales, and other terms are as defined in the previous steps. Clearly, the first part of the sum includes each of the words in either cloud the second part contains only the words that are present in both. When any word is not present in any cloud, it is treated as size zero.

25.2.4 Calculation of Reduction Score (R)

Reduction score (R) is based on the subjectivity of the corpus set containing documents. It is dimensionless, quantifying the overall opinion of the corpus text under consideration. It determines the polarity of opinions in reviews to describe products or services. This can be applied to any form of recorded opinions, which can be blogs, reviews, and microblogs.

It can be calculated at document level or sentence level. In the first type, the whole transcript is evaluated to classify the opinion polarity, where the characteristics relating to the goods is highlighted primarily. Further, the document is diversified into five subsets of equal number of sentences and each is evaluated distinctively to discuss the polarization.

Steps for calculations are:

1. First determine the number of clean words (C) out of the total raw number of words (T) in the document set.
2. Then, find the ratio of number of clean words (C) and number of total raw words (T).
3. Given the ratio (r) of number of clean words (C) to number of total raw words (T), compute the reduction score, $R = 1-(C/T) = 1-r$.

Now, $0 < r < 1 \Rightarrow 1-0 > 1-r > 1-1 \Rightarrow 0 < R < 1$. When $0<r<0.5$, the polarity of opinions is toward the negative perspective and $0.5<r<1$, polarity moves toward the positive perspective. Thus, when $0< r<0.5 => 0.5<R<1$, the attitude toward the product or service is satisfied, whereas if $0.5 < r < 1 \Rightarrow 0 < R < 0.5$, which means that the attitude of reviews indicates dissatisfaction towards the product or service.

25.3 Mathematical Case Study of the E-Commerce Market Reviews

For this case study, the text data have been extracted from the enormous corpora of Amazon customer reviews for Amazon products: Amazon Fire TV stick, Amazon Fire HD 8 inch tablet, and Amazon Kindle Paper white [22]. Then, the dataset is clustered into five subsets of equal size. Each subset is transformed into an array of tokenized documents. Every such tokenized document is sensitized by removing the non-attributing words occurring in the document. This is followed by forming the word clusters. Now, word clusters are filtered by removing the infrequent words, etc. Generation of the word cloud is the final outcome of the process. The response can be easily seen through the words captured by the clouds or bag of words. If there is an affirmative vibe across all the reviews given by the users, then it can be decided in favor of buying product whose reviews have been taken into account. The two word bags containing raw and filtered data are compared, to visualize the analysis.

25.3.1 Algorithm

Steps involved in the generation of word cloud are as follows:

1. **Data extraction:**
   (i) Extract text data
   (ii) Prepare string data for tokenizing
   (iii) Create an array of tokenized documents
2. **Data cleaning:**
   (i) Remove stop words like "to," "the," "in," "and" etc.
   (ii) Remove unncessary long words or short words according to length of the words

3. **Forming the word cloud:**
  (i)   Develop the bag-of-words model with the tokenized words
  (ii)   Clean the bag by removing infrequent words and empty documents
  (iii) Visualize the word cloud generated,, after all,, the above-text filtering
  (iv)   Now, develop the bag-of words model with raw text data taken as input data
4. **Comparing the two word clouds**
  Compare this raw data word cloud with the clean tokenized data word cloud.

25.3.2 Flowchart

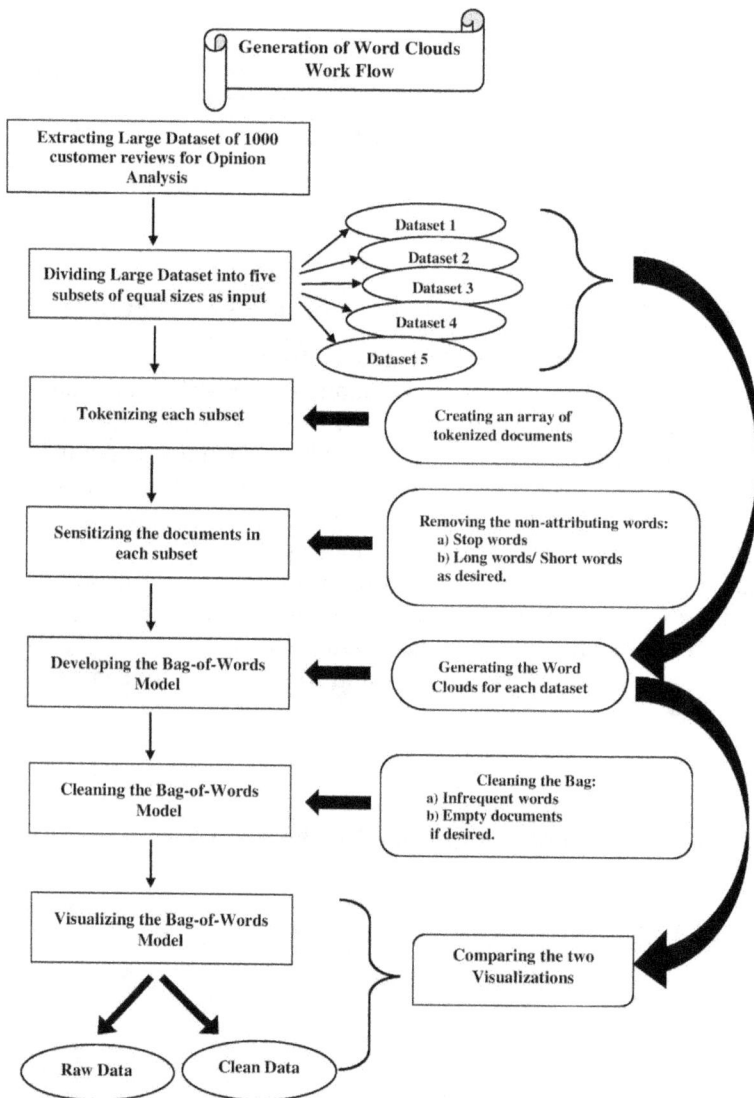

FIGURE 25.1
Workflow for generation of word clouds.

25.4 Results and Discussion

The exploration shows it is conceivable to accumulate, preprocess, simulate, and visualize metadata, as can be seen as a step-by-step procedure in Figure 25.1. It helps in analyzing customers' likes and dislikes about the brand or service. It is feasible to relate data excavation and computational linguistics scrutiny for e-commerce data to analyze consumer contributed reviews for goods. The e-commerce site reviews are the text data set for this study. The responses of various individuals about the products have been used as the input for the analysis and the outcomes have been studied for the overall opinion of the product. It captures the average sentiment of the users of the product and helps other prospective buyers in choosing the product best suited for themselves. Also, it improves market value and their consumer outlook, so the text made available is such that the outcomes are positively bent toward the product. Customer feedback from retail e-commerce is the source that consists of a treasure-trove of valuable business information. But, it is not enough to know what customers are talking about. One must also get their sentiments. Computational linguistics is one way to uncover those views. Table 25.1 shows the significance of the reduction score (R). The reduction score computed in this chapter proves to be helpful as when 0.5<R<1, the attitude toward the product or service is satisfied, whereas if 0< R<0.5, it means that the attitude of reviews indicates dissatisfaction towards the product or service. For Dataset 1, the value of R=0.7922, which is a fairly positive response. For Dataset 2, the value of R=0.8383 shows a highly positive response. For Dataset 3, the value of R=0.7805, which is a fairly positive response. For Dataset 4, the value of R=0.8041, which is a very positive response. For Dataset 5, the value of R=0.8122, which is a very positive response. As the value of R is very close to 1, it means that the customers have an opinion in favor of the product under scrutiny. This proves that the data provided by the retailers portray a more or less positive attitude of customers in competition with other service providers. In general, the discussed methodology is applicable to virtual media. The word cloud is impactful for the study as it is favorable to see and understand the attitude of people towards the products as claimed by the retailers. It highlights the most recurring words in the text data in a descending order of font size and word count.

Generally, retailers utilize their consumer notions produced from social media tracking and computations through adapting marketing plans, various merchandises, and innovations. An imperative perspective for upcoming work could be edifying virtual media

TABLE 25.1

Table Showing the Significance of Reduction Score (R)

Dataset	No. of Clean Words (C)	No. of Raw Words (T)	Ratio (r) = (C/T)	Reduction Score (R) = 1−(C/T)	Essence of R
Dataset 1	305	1468	0.2078	0.7922	As 0.5<R<1, it shows a fairly positive response
Dataset 2	178	1101	0.1617	0.8383	As 0.5<R<1, it shows a highly positive response
Dataset 3	369	1681	0.2195	0.7805	As 0.5<R<1, it shows a fairly positive response
Dataset 4	261	1332	0.1959	0.8041	As 0.5<R<1, it shows a very positive response
Dataset 5	243	1294	0.1878	0.8122	As 0.5<R<1, it shows a very positive response

tracking and controlling systems as thoughts are transforming with time. Moreover, the practice of unsupervised techniques in computational linguistics analysis and the estimation of quarrying for civilizing corporate competitiveness and consumer rapport should be increased. In addition, utilizing this technique predicting that the sales can be more consistent and precise. It is concluded that retailers only recorded affirmative reviews of their products as part of a marketing strategy to increase product sale in competition with other e-commerce retailers.

Acknowledgment

The authors are grateful to Guru Gobind Singh Indraprastha University for providing financial support and research facilities.

References

1. Anderson, J.R. and Bower, G.H. 1973. *Human Associative Memory*. Winston, Washington, DC.
2. Anderson, S.R. 1982. Where is morphology? *Linguistic Inquiry*. 13:571–612.
3. Anderson, S.R. 1985. *Phonology in the Twentieth Century: Theories of Rules and Theories of Representations*. University of Chicago Press, Chicago.
4. Anderson, S.R. 1992. *A-Morphous Morphology*. Cambridge University Press, Cambridge.
5. Angluin, D. 1992. Computational learning theory: Survey and selected bibliography. *Proceedings of STOC*. 92:351–369.
6. Antonelli, G.A. 1999. A directly cautious theory of defeasible consequence fordefault logic via the notion of general extension. *Artificial Intelligence*. 109(1–2):71–109.
7. Aronoff, M. 1976. *Word Formation in Generative Grammar*. MIT Press, Cambridge, Massachusetts.
8. Austin, J.L. 1962. *How to Do Things with Words*. Clarendon Press, Oxford.
9. Barzilay, R., Elhadad, N., and McKeown, K. 2002. Inferring strategies for sentence ordering in multidocument summarization. *Journal of Artificial Intelligence Research*. 17:35–55.
10. Bird, S. and Ellison, T.M. 1994. One-level phonology: Autosegmental representations and rules as finite automata. *Computational Linguistics*. 20(1):55–90.
11. Bloch, B. and Trager, G.L. 1942. Outline of linguistic analysis. *Linguistic Society of America*. Waverly Press, Baltimore.
12. Chao, Y.R. 1961. Graphic and phonetic aspects of linguistic and mathematical symbols. In: *Structure of Language and Its Mathematical Aspects*, ed. R. Jakobson, 69–82. Linguistic Society of America, Washington, DC.
13. Charniak, E. 1993. *Statistical Language Learning*. MIT Press, Cambridge, MA.
14. Collins, C., Viegas, F.B., and Wattenberg, M. 2009. Parallel tag clouds to explore and analyze faceted text corpora. *IEEE Symposium on Visual Analytics Science and Technology*, United States, October 2009, 91–98.
15. Cui, W., Wu, Y., Liu, S., Wei, F., Zhou, M., and Qu, H. 2010. Context-preserving dynamic word cloud visualization. *IEEE Computer Graphics and Applications*. 30(6):42–53.
16. Dumais, S.T. 2004. Latent semantic analysis. *Annual Review of Information Science and Technology*. 38(1):188–230.
17. Kornai, A. 2008. *Mathematical Linguistics. Advanced Information and Knowledge Processing*. Springer-Verlag, London.

18. Makkonen, J., Myka, H.A., and Salmenkivi, M. 2004. Simple semanticsin topic detection and tracking. *Information Retrieval*. 7(3–4):347–368.
19. White, M., Korelsky, T., Cardie, C., Ng, V.D., and Wagstaff, K. 2001. Multidocument summarization via information extraction. *Proceedings of Human Language Technology Conference: (HLT-2001)*, United States, March 2001 1–7.
20. Wu, Y., Provan, T., Wei, F., Liu, S., and Ma, K.L. 2011. Semantic-preserving word clouds by seam carving. *Computer Graphics Forum*. 30(3):741–750.
21. Xie, L., Kennedy, L., Chang, S.F., et al. 2005. Layered dynamic mixture model for pattern discovery in asynchronous multi-modal streams. *IEEE International Conference on Acoustics, Speech, and Signal Processing (ICASSP)*, Massachusetts, December 2005, 1053–1056.
22. Sasikala Alaguvel. 2018. Accessed November 30, 2018. https://www.kaggle.com/sasikala11/amazon-customer-reviews.

26

Impact of Data Assimilation on Simulation of a Monsoonal Heavy Rainfall Event over India Using ARW Modeling System

Shilpi Kalra, Sushil Kumar, and A. Routray

CONTENTS

26.1 Introduction 381
26.2 Modeling System 382
    26.2.1 Advanced Research WRF Modeling System (ARW) 382
    26.2.2 3DVAR Analysis System 383
26.3 Synoptic Features Associated with the Heavy Rainfall Events 383
26.4 Numerical Experiments 384
26.5 Result and Discussions 384
    26.5.1 Impact of Assimilation on Model Initial Fields 384
        26.5.1.1 Verification of Wind, Temperature, and Specific Humidity
                Against GTS Observations 385
        26.5.1.2 Impact on Initial Vorticity Field 386
    26.5.2 Impact on Model Simulation 386
        26.5.2.1 Vertical Cross Section of Moisture Fluxes 386
        26.5.2.2 Intensity Prediction 387
        26.5.2.3 Model Simulated Rainfall 389
26.6 Conclusions 389
Acknowledgments 390
References 390

26.1 Introduction

The forecasting skill of numerical weather prediction (NWP) depends upon the accuracy of the initial conditions(ICs). If we know the present weather conditions, i.e. the IC of the model, the mathematical models of the earth's atmosphere system can be developed to predict the future weather conditions. These mathematical models use basic differential equations based on the physical principles of fluid dynamics.

In recent decades, the non-hydrostatic regional models have been developed and are capable of predicting the heavy rainfall events associated with extreme weather phenomenon over the Indian monsoon regime (Routray et al. 2005, 2010; Kalra et al. 2019). However, the forecasting skill of the regional models are dependent upon the initial and the lateral boundary conditions (IC and BCs). The ICs and BCs are provided from the analyses or forecasts of low-resolution global NWP models. These ICs have certain limitations such as coarse resolution and inadequate representation of localized mesoscale processes. Therefore, the assimilation approach is one of the most promising techniques to provide improved analyses by utilizing the high spatial and temporal observations from different platforms, which serve as ICs to the regional models (Daley 1991). The data assimilation is a mathematical technique to create the best estimate of the initial state of the atmosphere by inserting the observed information into the model background field.

India receives 80% to 90% of its annual rainfall during the southwest monsoon season (June–September). The major portion of the rainfall is received due to formation of monsoon depressions (MDs) moving across India. During the life span of the MDs, heavy to very rainfall occur along their paths, which create havoc across the region (Sikka and Rao, 2008; Chang et al., 2009). Therefore, the accurate prediction of the heavy to very heavy rainfall events associated with MDs is a demanding and challenging task for the Indian monsoon operational and research community. In recent years, the high-resolution mesoscale models with three/four dimensional variational (3/4DVAR) assimilation techniques have become more popular for such types of weather events (Routray et al. 2013, 2014,2016; Mohanty et al. 2013; Kumar et al. 2014; Dutta et al. 2019).

The main objective of this chapter is to improve the forecasting performance of the high-resolution mesoscale model Weather Research and Forecasting (WRF) with modified initial conditions by inserting the conventional and non-conventional observations through 3DVAR analysis technique. For this purpose, a heavy rainfall event (August 9–12, 2016) associated with the MD, formed over the Bay of Bengal (BoB), is considered. The WRF and 3DVAR modeling systems are briefly described in Section 26.2. The synoptic conditions associated with the heavy rainfall event are presented in Section 26.3, while Section 26.4 presents the numerical experiments performed in this study. Section 26.5 is presents the results and discussion and is followed by the overall conclusions drawn from this chapter in Section 26.6.

26.2 Modeling System

26.2.1 Advanced Research WRF Modeling System (ARW)

The study uses the advanced research core of WRF model (ARW version 3.4), which is derived from fully compressible, non-hydrostatic Eulerian equations, horizontal staggering from Arakawa C grid, and Runge–Kutta time integration scheme, etc. The ARW model is formulated using a terrain-following hydrostatic-pressure vertical co-ordinate (η) defined as

$$\eta = \frac{p_h - p_{ht}}{\mu} \text{ Where } \mu = p_{hs} - p_{ht}$$

p_h refers to the hydrostatic component of pressure, p_{hs} and p_{ht} are values along the surface and top boundaries respectively. As described in NCAR technical note (Skamarock et al. 2008), the governing equation of ARW model is expressed as:

$$\partial_t U = F_U - (\vec{\nabla}.\vec{V}u) + \partial_x p \partial_\eta \varphi - \partial_\eta p \partial_x \varphi \tag{26.1}$$

$$\partial_t V = F_V - (\vec{\nabla}.\vec{V}v) + \partial_y p \partial_\eta \varphi - \partial_\eta p \partial_y \varphi \tag{26.2}$$

$$\partial_t W = F_W - (\vec{\nabla}.\vec{V}w) + g(\partial_\eta p - \mu) \qquad (26.3)$$

$$\partial_t \Theta = F_\Theta - (\vec{\nabla}.\vec{V}\theta) \qquad (26.4)$$

$$\partial_t \mu + (\vec{\nabla}.\vec{V}) = 0 \qquad (26.5)$$

$$\partial_t \varphi + \mu^{-1}[(\vec{\nabla}.\vec{V}\theta) - gW] = 0 \qquad (26.6)$$

$$p = p_0 \left(\frac{R_d \theta}{p_0 \alpha} \right)^\lambda \qquad (26.7)$$

Along with the diagnostic relation $\partial_\eta \varphi = -\alpha\mu$

Where $\vec{V} = \mu v = (U,V,W)$ are the velocities in two horizontal and vertical directions, $\Theta = \mu\theta$, θ is the potential temperature, $\varphi = gz$ (the geopotential), p (pressure), and ρ (density). The detailed description of ARW dynamical core, model physics, and model discretization can be found in Skamarock et al. (2008).

26.2.2 3DVAR Analysis System

The standard formulation of 3DVAR analysis technique is derived from first principles by Lorenc using Bayesian probabilities and assuming Gaussian error distributions (Lorenc et al. 2000). The 3DVAR data assimilation system provides the analysis by minimizing the prescribed cost function (Ide et al. 1997) through iterative process $J(x)$.

$$J(x) = J^b + J^o = \frac{1}{2}\left(x - x^b\right)^T B^{-1}(x - x^b) + \frac{1}{2}(y^o - H(x))^T R^{-1}(y^o - H(x)) \qquad (26.8)$$

where J^b and J^o are cost function of background and observation respectively. B and R are background and observation error covariance matrix respectively. x^b is background model state and x is true model state. y^o is vector of observations, y vector of modeled observations, i.e. the outputs of the model ($y = Hx$); H is the observation operator. Further, details about the components and real time applications of the 3DVAR system have been reported in Barker et al. (2004) and Gu et al. (2005).

26.3 Synoptic Features Associated with the Heavy Rainfall Events

In this study, a heavy rainfall event (August 9–12 2016) that occurred along the east coast of India is considered. The case considered here is associated with MD leading to heavy rainfall occurrence.

A low pressure system (LPS) formed over the northwest BoB in the morning of August 9 2016 in association with the active monsoon conditions. Under the favorable environment factors, such as increase in low-level vorticity, lower-level convergence, upper-level divergence etc., it concentrated into a depression at 0900 UTC near 22.0°N, 88.5°E on August 9 2016. The depression intensified into a deep depression over south Bangladesh and neighborhood near 23.0°N and 89.4°E at 0300 UTC on August 10 2016. The satellite images (Figure 26.1(a) and (b)) show strong convective cloud bands over the regions

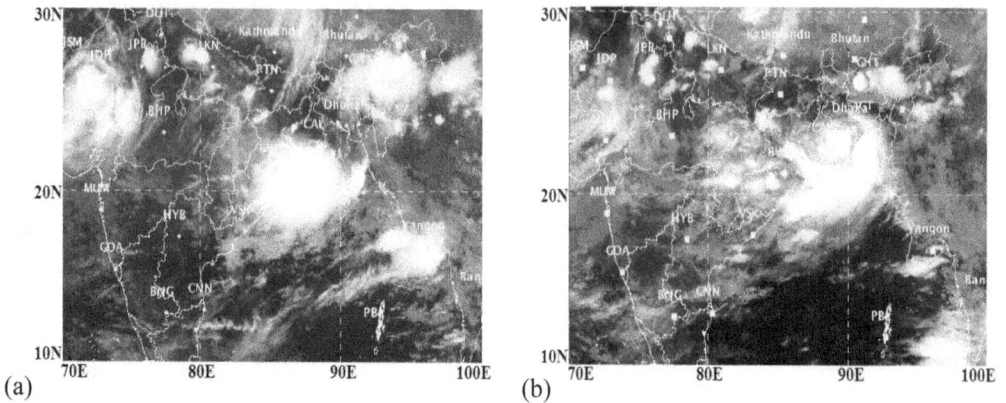

FIGURE 26.1
INSAT-3D satellite imageries valid at 0000 UTC for August (a) 9 and (b) 10 2016 respectively.

during August 9 and 10 2016. It then moved west-north-eastward as a deep depression until the morning of the August 11. It then weakened into a depression at 0300 UTC August 11 and further into a well-marked LPS on the morning of August 13 while continuing its westward movement up into south Bihar. The details about the heavy rainfall events can found in Regional Specialized Meteorological Centre (RSMC-2016; http://www.rsmcnewdelhi.imd.gov.in/) report.

26.4 Numerical Experiments

In this chapter, the ARW model is configured with a single domain at 9 km horizontal grid resolution and 42 sigma levels vertically. The details of physical parameterization schemes and model configuration are summarized in Table 26.1. In this study, two numerical experiments are performed to simulate two heavy rainfall events. The first experiment is named as a control (CNTL) experiment without data assimilation using global National Centers for Environmental Prediction (NCEP) FiNaL analysis (FNL, resolution $1° \times 1°$) with a 6-hourly interval as IC and BCs of the model. The second experiment is named as 3DV, where the model IC is modified by inserting Global Telecommunication System (GTS) observations through 3DVAR data assimilation system. The detail observations used in the data assimilation system are provided in Table 26.2. The observations used in the assimilation experiment are with a ± 3 h time window. The ARW model configuration and model physics are the same for both experiments.

26.5 Result and Discussions

26.5.1 Impact of Assimilation on Model Initial Fields

The impact of inserting observations through 3DVAR analysis system at model initial condition and model simulated meteorological fields is examined in this section. The

TABLE 26.1

WRF Model Configuration for Present Study

Dynamics	Non-Hydrostatic
Spatial coverage	(16°N-28°N, 78°E-94°E)
Number of grid points	(198×165)
Map projection	Mercator
Horizontal grid distribution	Arakawa C-grid
Integration time step	54 sec
Initial and boundary conditions	Three-dimensional real data (FNL:1°×1°)
Vertical coordinate	Terrain-following sigma coordinate
Time integration	3rd-order Runge–Kutta
Spatial differencing scheme	6th-order differencing
Surface layer parameterization	Thermal diffusion scheme
Planetary boundary layer	Yonsei University Scheme
Cumulus parameterization schemes	Betts–Miller–Janjic scheme
Microphysics	WSM class 6 scheme
Land surface option	Noah land surface option
Radiation scheme	• RRTM for long wave • Dudhia for short wave

TABLE 26.2

Observation Data Dets Used in 3DVAR Assimilation System

Serial No.	Name of Data Set	Description
1	SOUND	Upper-air observation from RS/RW
2	SONDE_SFC	Radiosonde at surface
3	SYNOP	Surface observations from land station
4	GEOAMV	Geo-stationary atmospheric motion vectors
5	PILOT	Wind profiles from optical theodolite
6	METAR	Aerodrome routine meteorological report
7	SHIPS	Surface observation from a sea station
8	BUOY	Drifting and moored buoy observations

3DVAR-based analyses are provided as model initial conditions to simulate the heavy rainfall event during August 9–12, 2016.

26.5.1.1 Verification of Wind, Temperature, and Specific Humidity Against GTS Observations

A statistical verification is carried out to assess the performance of 3DVAR during the analysis time 00 UTC August 9 2016. Root mean square (RMSE) of zonal (u) component of wind, meridional (v) component of wind, temperature (T), and specific humidity (q) of 3DV experiment is verified against the GTS observations used in the 3DVAR assimilation cycle. Figure 26.2 shows the averaged RMSE of difference between observations (O) and first guess (B) before assimilation i.e. (O–B) as well as the difference between observations (O) and analysis (A) i.e. (O–A) after assimilation of GTS observations for the meteorological variables. It is clearly seen that mean RMSE of (O–B) is higher than the RMSE of (O–A) for

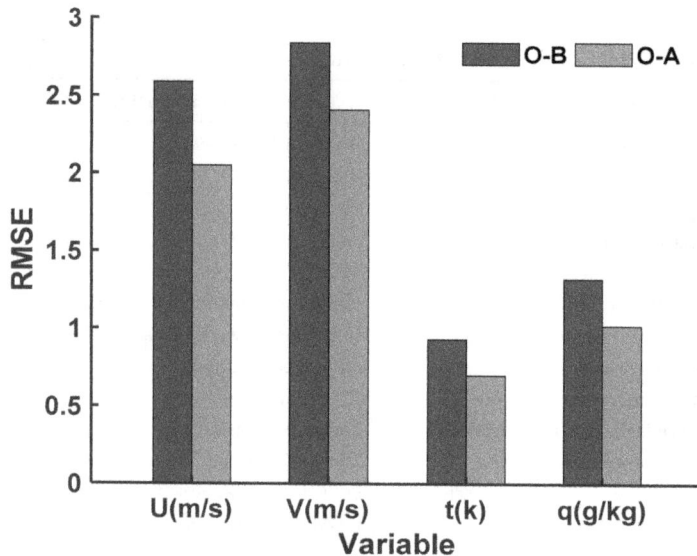

FIGURE 26.2
Averaged root mean square of O–A and O–B for wind (m/s), temperature (°K), and specific humidity (g/kg).

all the variables, which suggests abetter agreement of the analysis fields to the observations than first guess. This infers that the GTS observations are correctly assimilated and produced better analyses for ARW modeling system.

26.5.1.2 Impact on Initial Vorticity Field

Vorticity is an important parameter for weather and climate modeling. Due to the horizontal shear of the wind field, vorticity is developed around the system and provides circulation in the flow. The relative vorticity along with streamlines at different pressure levels (850 and 500 hPa) from the CNTL and 3DV analyses valid at 00 UTC August 9 2016 is depicted in Figure 26.3(a–b and c–d) respectively. From the figures, it is seen that both the analyses at 850 hPa are showing strong relative vorticity around the MD over head of BoB. However, the magnitude of the vorticity (10–20×10^{-5}) is higher in the 3DV analysis (Figure 26.3(b)) as compared to the analysis from CNTL (Figure 26.3(a)). Similar features are also observed: the relative vorticity is more around 10–15×10^{-5} in the 3DV analysis (Figure 26.3(d)) at 500 hPa pressure level as compared to CNTL (Figure 26.3(3c)). Overall, the stronger vorticity features with well-organized convergence zone around the MD are found in 3DV analysis as compared to CNTL analysis.

26.5.2 Impact on Model Simulation

26.5.2.1 Vertical Cross Section of Moisture Fluxes

Convection plays an important role in the development of meteorological systems. To diagnose the effect of assimilation of observations in the model simulation, the moisture convection of horizontal moisture flux and vertical moisture flux are analyzed during the life span of the MD (00 UTC 09 to 12 UTC August 11 2016). The time-pressure cross section of horizontal and vertical moisture fluxes from CNTL and 3DV simulations is depicted in Figure 26.4(a–b and c–d) respectively. During the depression stage (09 UTC 9 to

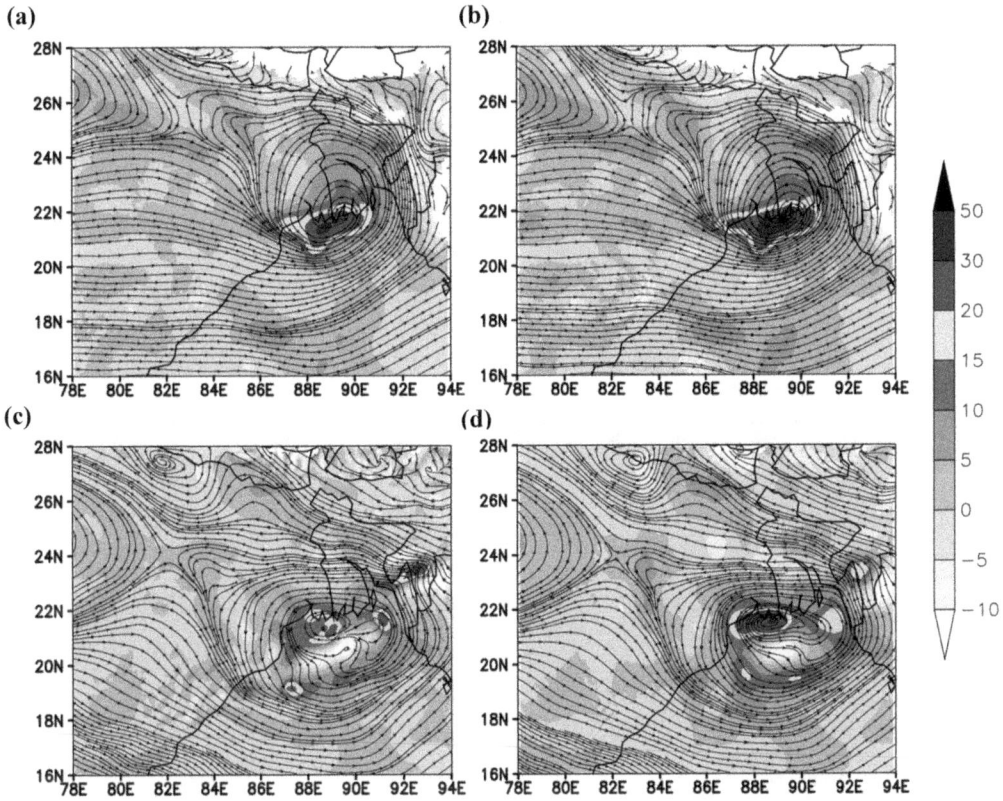

FIGURE 26.3
Vorticity (shaded; $\times 10^{-5}$) and streamlines at 850 hPa pressure level from (a) CNTL and (b) 3DV experiments valid at 00 UTC August 9 2016 (model initial time). (c–d) are the same as (a–b) but at 500 hPa pressure level.

00 UTC August 10 2016), the horizontal moisture flux obtained from the 3DV experiment (Figure 26.4(b)) shows higher positive fluxes (14×10^{-3} g kg^{-1}s^{-1}) extended up to 750 hPa as compared to the CNTL (Figure 26.4(a); 6×10^{-3} g kg^{-1}s^{-1}). It is also seen that the 3DV experiment simulated the maximum horizontal moisture flux (30–36×10^{-3} g kg^{-1}s^{-1}) extended upto 500 hPa during deep depression stage (03 UTC 10 to 03 UTC August 11 2016). The horizontal moisture flux is gradually decreased afterwards. The simulated horizontal moisture fluxes from the CNTL experiment also followed the same trend as was noticed in the 3DV experiment during different stages of the MD. However, the CNTL experiment is able to simulate the maximum horizontal moisture flux around 28–30×10^{-3} g kg^{-1}s^{-1} extended upto 550 hPa during the deep depression stage of the MD. At the same time, the vertical moisture fluxes extended from the lower to the upper atmospherein both the simulations (Figure 26.4(c) and(d)) during the deep depression stage of the MD. But the maximum value of vertical moisture flux (20×10^{-3} g kg^{-1}s^{-1})is simulated by the 3DV simulation (Figure 26.4(c)) as compared to the CNTL simulation (Figure 26.4(d); 7×10^{-3}g kg^{-1}s^{-1}).

26.5.2.2 Intensity Prediction

The time series of intensity in terms of 10-m maximum sustainable wind (MSW) and absolute error (AE) of MSW along with gain skill (%) of 3DV with respect to CNTL simulation is illustrated in Figure 26.5(a) and (b). The AE of MSW is calculated against the IMD

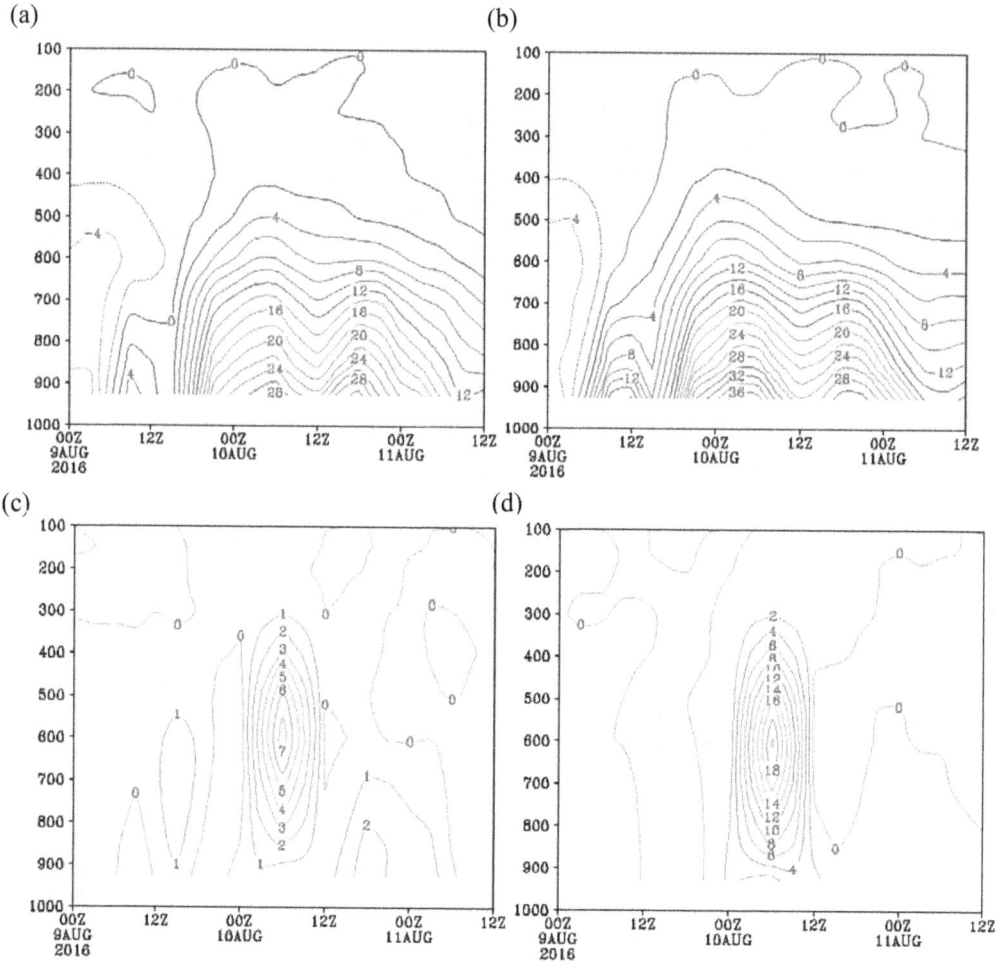

(a)

(b)

(c)

(d)

FIGURE 26.4

Time-pressure cross section of model simulated horizontal moisture flux ($\times 10^{-3}$ g.kg^{-1}s^{-1}) at the point (22.2°N and 88.6°E) of (a) CNTL and (b) 3DV. (c–d) are the same as (a–b) but for vertical moisture flux.

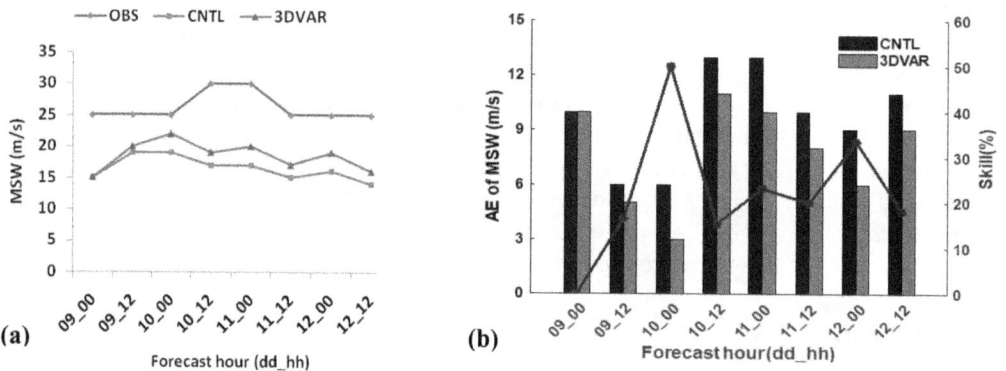

(a)

(b)

FIGURE 26.5

(a) Time series of maximum sustained wind 10 m and (b) Absolute error of maximum sustained wind 10 m.

observations. It is observed from Figure 26.5(a)that no simulations are able to capture the observed trend of MSW. However, the magnitude of the MSW is reasonably well simulated by the 3DV experiment throughout the forecast hours as compared to the CNTL simulation. From Figure 26.5(b), the AE (m/s) of MSW is significantly reduced in the 3DV simulation compared to the CNTL simulation throughout the forecast period. The mean gain skill of 3DV is about 22% with respect to the CNTL. Thus, 3DV simulation predicted the intensity better in comparison to the CNTL simulation.

26.5.2.3 Model Simulated Rainfall

A comparative study of location specific 24 hours accumulated observed from IMD and model-simulated rainfall (cm) is depicted in Figure 26.6(a) and (b) for Day 1 and Day 2 valid at 03 UTC of August 10 and 11 2016 respectively. It is clearly seen that the observed amount of rainfall over the maximum number of stations is well simulated in the 3DV experiment as compared to the CNTL for both the days. During the Day 1 forecast, the 3DV experiment overestimated the rainfall over the station at Salebhatta. Similarly, on Day 2 the CNTL experiment also overpredicted the rainfall over the station at Kalna.

26.6 Conclusions

The high-resolution state-of-the-art mesoscale model WRF and 3DVAR analysis systems with single domain (9 km horizontal resolution) are considered to simulate a heavy rainfall event (August 9–12 2016) associated with the MD during the summer monsoon season along the east coast of India. To this purpose, two numerical experiments, i.e CNTL without data assimilation where FNL analyses are provided as IC and BCs to the model and 3DV by assimilating observations from GTS through 3DVAR, are carried out to evaluate the impact of high-resolution analysis. The broad conclusions of this chapter are given below.

The performance of the 3DVAR analysis system is examined during the initial time through various statistical verifications, i.e innovation error of background (O–B) and analysis (O–A) for the meteorological variables, i.e u-wind, v-wind, temperature, and

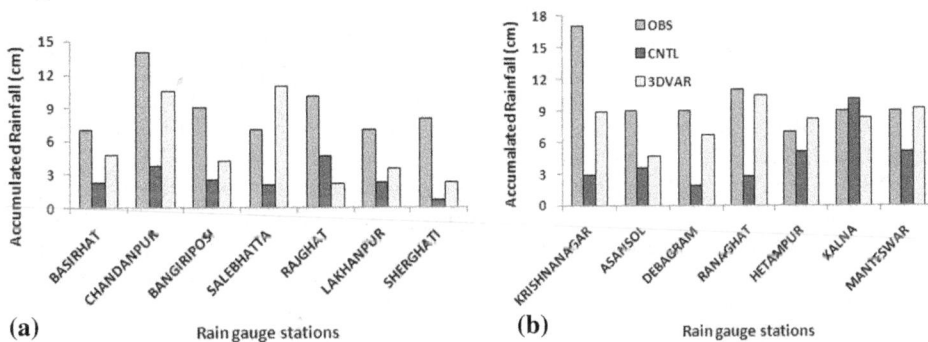

FIGURE 26.6
24 hrs observed and model simulated rainfall (cm) over different stations (a) Day 1 and (b) Day 2 valid at 03 UTC of August 10 and 11, 2016 respectively.

specific humidity. The RMSE of the background field departure of the variables is higher as compared to the RMSE of the departure of the analysis.

The relative vorticity around the MD is stronger in the 3DV analysis as compared to the CNTL analysis at different pressure levels. It is clearly suggested that the intensity of the MD is improved at model initial time after assimilation of observations through the 3DVAR analysis system.

The model-simulated horizontal and vertical moisture fluxes are reasonably well represented in both the simulations. However, the positive fluxes are stronger and extended up to the upper atmosphere in the 3DV simulation during the life span of the MD. It is also noticed in the 3DV simulation that horizontal and vertical moisture fluxes are gradually increased from depression stage to deep depression stage and decreased afterward (i.e. dissipation of MD). These features are not well represented in the CNTL simulation.

The magnitude of intensity and rainfall associated with MD are well simulated in the 3DV experiment as compared to the CNTL experiments.

This chapter clearly suggests the advantages of impact of the assimilation of observations through the 3DVAR analysis system for the simulation of a heavy rainfall event during the monsoon season. However, further improvement can be obtained by increasing the model resolution, assimilation of additional high-quality spatial and temporal observations like Doppler weather radar (DWR) radial velocity and reflectivity, radiance with more advance data assimilation techniques such as 4DVAR, Hybrid, Ensemble methods, etc.

Acknowledgments

The authors gratefully acknowledge the NCEP for providing FNL analysis datasets and GTS observations used in WRF and 3DVAR data assimilation system for this chapter. We express our sincere thanks to IMD for providing rain gauge stations data.

References

Barker, D.M., Huang, W., Guo, Y.R., Bourgeois, A.J., Xiao, Q.N. 2004. A three dimensional variational data assimilation system for use with MM5: Implementation and initial results. *Mon Weather Rev* 132: 897–914. doi:10.1175/1520-0493(2004)132%3C0897%3AATVDAS%3E2.0.CO%3B2

Chang, H.-I., Niyogi, D., Kumar, A., Kishtawal, C.M., Dudhia, J., Chen, F., Mohanty, U.C., Shepherd, M. 2009. Possible relation between land surface feedback and the post-landfall structure of monsoon depressions. *Geophys Res Lett* 36: L15826. doi:10.1029/2009GL037781

Daley, R. 1991. *Atmospheric Data Analysis*, Cambridge Atmospheric and Space Science Series, vol 12, issue 7. Cambridge University Press, New York, pp 763–764. doi:10.1002/joc.337012070

Dutta, Devajyoti, RoutrayA., Preveen Kumar D., George John P., Singh Vivek 2019. Simulation of a heavy rainfall event during southwest monsoon using high resolution NCUM modeling system: A case study. *Meteorol Atmos Phys*. doi:10.1007/s00703-018-0619-0

Gu, J., Xiao, Q., Kuo, Y.H., Barker, D.M., Xue, J., Ma, X. 2005. Assimilation and simulation of Typhoon Rusa (2002) using the WRF system. *Adv Atmos Sci* 22: 415–427. doi:10.1007/BF02918755

Ide, K., Courtier, P., Ghil, M., Lorenc, A.C. 1997. Unified notation for data assimilation: Operational sequential and variational. *J Meteor Soc Japan* 75: 181–189. doi:10.2151/jmsj1965.75.1B_181

Kalra, S., Kumar, S., Routray, A. 2019. Simulation of heavy rainfall event along east coast of India using WRF modeling system: Impact of 3DVAR data assimilation. *Model Earth Syst Environ* 5: 245–256. doi:10.1007/s40808-018-0531-0

Kumar, S., Routray, A., Chauhan, R., Panda, J. 2014. Impact of parameterization schemes and 3DVAR data assimilation for simulation of heavy rainfall events along west coast of India with WRF modeling system. *Int J Earth Atmos Sci* 01: 18–34.

Lorenc, A.C., Ballard, S.P., Bell, R.S., Ingleby, N.B., Andrews, P.L.F., Barker, D.M., Bray, J.R., Clayton, A.M., Dalby, T., Li, D., Payne, T.J., Saunders, F.W. 2000. The Met. Office global three-dimensional variational data assimilation scheme. *Quart J Roy Meteor Soc* 126: 2991–3012. doi:10.1002/qj.49712657002

Mohanty, U.C., Routray, A., Osuri, K.K., Prasad, S.K. 2013. A study on simulation of heavy rainfall events over Indian region with ARW-3DVAR modeling system. *Pure Appl Geophys* 169: 381–399.

Routray, A., Kar, S.C., Mali, P., Sowjanya, K. 2014. Simulation of monsoon depressions using WRF-VAR: Impact of different background error statistics and lateral boundary conditions. *Mon Weather Rev* 142: 3586–3613. doi:10.1175/MWR-D-13-00285.1

Routray, A., Mohanty, U.C., Das, A.K., Sam N.V. 2005. Study of heavy rainfall event over the west-coast of India using analysis nudging in MM5 during ARMEX-I. *Mausam* 56(1): 107–120.

Routray, A., Mohanty, U.C., Niyogi, D., Rizvi, S.R.H., Osuri, K.K. 2010. Simulation of heavy rainfall events over indian monsoon region using WRF-3DVAR data assimilation system. *Meteorol Atmos Phys* 106: 107–125. doi:10.1007/s00703-009-0054-3

Routray, A., Mohanty, U.C., Osuri, K.K., Prasad, S.K. 2013. Improvement of monsoon depressions forecast with assimilation of Indian DWR data using WRF-3DVAR analysis system. *Pure Appl Geophys* 170: 2329–2350.

Routray, A., Mohanty, U.C., Osuri, K.K., Kar, S.C., Niyogi, D. 2016. Impact of satellite radiance data on simulations of Bay of Bengal tropical cyclones using the WRF-3DVAR modeling system. *IEEE Trans Geosci Remote Sens* 54: 2285–2303. doi:10.1109/TGRS.2015.2498971

Sikka, D.R., Sanjeeva Rao, P. 2008. The use and performance of mesoscale models over the Indian region for two high impact events. *Nat Hazards* 44: 353–372. doi:10.1007/s11069-007-9129-y

Skamarock, W.C., Klemp, J.B., Dudhia, J., Gill, D.O., Barker, D.M., Duda, M., Huang, X.-Y., Wang, W., Powers, J.G. 2008. A description of the advanced research WRF Version 3. NCAR technical note. www.wrfmodel.org

Index

Ahmad, M., 233, 311
Alam, M., 345, 353
Al Eisa, H. N., 209
Algorithm
    genetic, 311, 315, 316
    iterative clipping, 225
    iteratively regularized Gauss–Newton, 86
Analysis
    error, 363
    Fourier, 1–3, 5, 6, 9–12
    Gabor, 13, 17
    sentiment, 352
    Spectral, 8
    wavelet, 213
Anandh, T., 149
Ansari, S. A., 345
Autism spectrum disorder, 209, 210

Balachandran, K., 275
Bangia, A., 371
Bhardwaj, R., 371
Bhatla, R., 243

Chandiok, A., 257
Chaturvedi, D. K., 257
Computational
    fluid dynamics, 149
    linguistics, 345–350, 353, 354, 371, 372, 376
Concentration distribution, 291, 299, 304, 305
Conformable
    fractional Chebyshev equation, 340
    fractional derivative, 103, 105, 106, 108–110,
        112, 336
    fractional heat equation, 103, 105, 106
    fractional Laguerre equation, 336
Crop yield forecast, 243, 244, 250

Dani, B., 243
Distribution theory, 1, 15
Dixit, A., 335
DNA rupture, 171, 179
Domain decomposition, 25, 32, 34

El-Rahman, S. A., 209
Electron Spin Resonance, 61, 64, 65
Elliptic curve group, 233, 235
Equation

advection-dispersion, 291, 293
elliptic partial differential, 78
partial differential, 25
radiative transport, 77, 82
Volterra integral, 357, 363, 365
Error
    mean bias, 245
    root mean square, 245, 250, 363

Feichtinger, H. G., 1
Free radical, 61–66, 70

Ganesan, S., 149
Gomrok, S., 61
Graphical user interface, 195, 196, 312
Guided image filter, 117, 125
Gupta, D., 311

Inequality
    Adams, 142, 143
    Moser–Trudinger, 135, 137, 139–142
    Sobolev, 135
Iteratively regularized Gauss–Newton, 86

Jacobi-type approach, 25, 34
Jahan, M. S., 61
Jamjoom, M., 209

Kalra, S., 381
Kazemi, M., 91
Khan, T., 77
Khokher, R., 195
Kumar, J., 357
Kumar, N., 225
Kumar, S., 103
Kumar, Sanjay, 171
Kumar, Sushil, 381
Kumar, V., 233

Legendre wavelet, 357–361, 363, 364
Leugering, G., 25
Lizzy, R. M., 275
Lozi, R., 41

Manchanda, P., 117, 357
Method
    finite element, 85

front tracking, 149, 158, 160
level set, 149, 158
Markov chain Monte Carlo, 88
with memory, 94, 97, 98
volume of fluid, 149, 156, 157
without-memory, 92
Model
Coupled Weighted PCR, 43
Panic–Control–Reflex, 42
ship hydrodynamics, 149, 150
Weighted Panic Control Reflex, 43
MRI images, 311, 312

Narang, T. D., 325
Nath, S., 171
Network modeling, 25, 26
Nonlinear kernel weights, 117, 126
Nonsubsampled
contourlet transform, 117, 119
directional filter bank, 117, 122
pyramid, 117, 120

Optimal control, 25, 26, 28, 34

Palmprint recognition, 195, 198, 199, 201
Pandey, A. K., 291
Pan-sharpening, 117, 119, 124
Phase symmetry, 195, 202
Pooja, 357
Prakash, A., 257
Problem
inverse, 77, 81, 86, 345
Neumann, 79, 80
statistical inverse, 77, 86
Provitolo, D., 41
Proximinality, 325, 329

Raman spectra, 225, 227, 228
Remotality, 325, 329, 331
Routray, A., 381

Sandeep, K., 135
Sangeeta, 325
Sharmin, A., 61
Siddiqi, A. H., 103, 225, 257
Similarity measures, 195, 204
Singh, M. K., 291
Singh, R. C., 103, 195
Singh, T., 117
Space
abstract, 84, 325
convex, 325, 331
linear metric, 325, 328–331
Support vector machine, 311, 313
Systems
artificial intelligent agent, 263, 265
cognitive agent, 257, 262, 265–267
stochastic dynamical, 275, 282

Tabassum, K., 209
Teja, B., 149
Time discretization, 25, 29, 164
Tomography
electrical impedance, 77, 78
diffuse optical 77, 78, 82
Torkashvand, V., 91
Total variation de-noising, 225, 228
Tric, E., 41
Tripathi, A., 243
Tripathi, P., 225
Tulshyan, R., 243

Ujlayan, A., 335
Upadhyaya, A., 171

Velocity pattern, 291, 299, 304, 305

Walters, B., 61
Word cloud, 371, 373, 374, 376

Zahra, N., 209

For Product Safety Concerns and Information please contact our EU
representative GPSR@taylorandfrancis.com
Taylor & Francis Verlag GmbH, Kaufingerstraße 24, 80331 München, Germany

www.ingramcontent.com/pod-product-compliance
Lightning Source LLC
Chambersburg PA
CBHW080652220326
41598CB00033B/5183